W9-DHV-865

# NMR APPLICATIONS IN BIOPOLYMERS

# BASIC LIFE SCIENCES

## Ernest H. Y. Chu, Series Editor

The University of Michigan Medical School
Ann Arbor, Michigan

## Alexander Hollaender, Founding Editor

*Recent volumes in the series:*

Volume 44  GENETIC MANIPULATION OF WOODY PLANTS
Edited by James W. Hanover and Daniel E. Keathley

Volume 45  ENVIRONMENTAL BIOTECHNOLOGY: Reducing Risks from Environmental
Chemicals through Biotechnology
Edited by Gilbert S. Omenn

Volume 46  BIOTECHNOLOGY AND THE HUMAN GENOME: Innovations and Impact
Edited by Avril D. Woodhead and Benjamin J. Barnhart

Volume 47  PLANT TRANSPOSABLE ELEMENTS
Edited by Oliver Nelson

Volume 48  HUMAN ACHONDROPLASIA: A Multidisciplinary Approach
Edited by Benedetto Nicoletti, Steven E. Kopits,
Elio Ascani, and Victor A. McKusick

Volume 49  OXYGEN RADICALS IN BIOLOGY AND MEDICINE
Edited by Michael G. Simic, Karen A. Taylor,
John F. Ward, and Clemens von Sonntag

Volume 50  CLINICAL ASPECTS OF NEUTRON CAPTURE THERAPY
Edited by Ralph G. Fairchild, Victor P. Bond,
and Avril D. Woodhead

Volume 51  SYNCHROTRON RADIATION IN STRUCTURAL BIOLOGY
Edited by Robert M. Sweet and Avril D. Woodhead

Volume 52  ANTIMUTAGENESIS AND ANTICARCINOGENESIS MECHANISMS II
Edited by Yukiaki Kuroda, Delbert M. Shankel,
and Michael D. Waters

Volume 53  DNA DAMAGE AND REPAIR IN HUMAN TISSUES
Edited by Betsy M. Sutherland and Avril D. Woodhead

Volume 54  NEUTRON BEAM DESIGN, DEVELOPMENT, AND PERFORMANCE
FOR NEUTRON CAPTURE THERAPY
Edited by Otto K. Harling, John A. Bernard, and Robert G. Zamenhof

Volume 55  *IN VIVO* BODY COMPOSITION STUDIES: Recent Advances
Edited by Seiichi Yasumura, Joan E. Harrison, Kenneth G. McNeill,
Avril D. Woodhead, and F. Avraham Dilmanian

Volume 56  NMR APPLICATIONS IN BIOPOLYMERS
Edited by John W. Finley, S. J. Schmidt, and A. S. Serianni

A Continuation Order Plan is available for this series. A continuation order will bring delivery of each new
volume immediately upon publication. Volumes are billed only upon actual shipment. For further information
please contact the publisher.

# NMR APPLICATIONS IN BIOPOLYMERS

## Edited by

### J. W. Finley
Nabisco Brands, Inc.
East Hanover, New Jersey

### S. J. Schmidt
University of Illinois
Urbana, Illinois

### and

### A. S. Serianni
University of Notre Dame
Notre Dame, Indiana

PLENUM PRESS • NEW YORK AND LONDON

Library of Congress Cataloging-in-Publication Data

---

American Chemical Society Symposium on Applications of NMR in
  Biopolymers (1988 : Los Angeles, Calif.)
   NMR applications in biopolymers / edited by J.W. Finley, S.J.
Schmidt, and A.S. Serianni.
      p.    cm. -- (Basic life sciences ; v. 56)
   "Proceedings based on an American Chemical Society Symposium on
Applications of NMR in Biopolymers, held September 25-28, 1988, in
Los Angeles, California, and an American Chemical Society Symposium
on Recent Developments in NMR Spectroscopy of Carbohydrates, held
June 5-8, 1988, in Toronto, Canada"--T.p. verso.
   Includes bibliographical references and index.
   ISBN 0-306-43719-8
   1. Nuclear magnetic resonance spectroscopy--Congresses.
2. Biopolymers--Structure--Congresses.    I. Finley, John W., 1942-
.  II. Schmidt, S. J.   III. Serianni, Anthony Stephen.   IV. American
Chemical Society.   V. American Chemical Society Symposium on Recent
Developments in NMR Spectroscopy of Carbohydrates (1988 : Toronto,
Ont.)  VI. Title.   VII. Series.
QP519.9.N83A44   1988
574.19'285--dc20                                           90-47280
                                                              CIP

---

QP
519
.9
.N83
A44
1988

Proceedings based on an American Chemical Society Symposium on Applications of NMR in
Biopolymers, held September 25–28, 1988, in Los Angeles, California, and an American
Chemical Society Symposium on Recent Developments in NMR Spectroscopy of Carbohydrates,
held June 5–8, 1988, in Toronto, Canada

ISBN 0-306-43719-8

© 1990 Plenum Press, New York
A Division of Plenum Publishing Corporation
233 Spring Street, New York, N.Y. 10013

All rights reserved

No part of this book may be reproduced, stored in a retrieval system, or transmitted
in any form or by any means, electronic, mechanical, photocopying, microfilming,
recording, or otherwise, without written permission from the Publisher

Printed in the United States of America

PREFACE

Elucidating the structures of biopolymers as they exist in nature has long been a goal of biochemists and biologists. Understanding how these substances interact with themselves, other solutes, and solvents can provide useful insights into many areas of biochemistry, agriculture, food science and medicine. Knowledge of the structure of a protein or complex carbohydrate in its native form provides guidelines for the chemical or genetic modifications often desired to optimize these compounds to specific needs and applications. For example, in the pharmaceutical industry, structure-function relationships involving biopolymers are studied routinely as a means to design new drugs and improve their efficacies.

The tools to conduct structure investigations of biopolymers at the molecular level are limited in number. Historically X-ray crystallography has been the most attractive method to conduct studies of this type. However, X-ray methods can only be applied to highly ordered, crystalline materials, thus obviating studies of solution dynamics that are often critical to attaining a global understanding of biopolymer behavior. In recent years, nuclear magnetic resonance (NMR) spectroscopy has evolved to become a powerful tool to probe the structures of biopolymers in solution and in the solid state. NMR provides a means to study the dynamics of polymers in solution, and to examine the effects of solute, solvent and other factors on polymer behavior. With the development of 2D and 3D forms of NMR spectroscopy, it is now possible to assess the solution conformations of small proteins, oligonucleotides and oligosaccharides.

This book grew from two recent symposia sponsored by the American Chemical Society: "Recent Developments in the NMR Spectroscopy of Carbohydrates" sponsored by the Division of Carbohydrate Chemistry at the Toronto meeting in June 1988, and "NMR Applications in Food Chemistry" sponsored by the Agricultural and Food Division at the Los Angeles meeting in September 1988. The complementarity in subject matter between these two events suggested that contributions be brought together under one cover. This volume provides the NMR spectroscopist, whether new to the field or experienced, with discussions of recent techniques and results pertaining to complex molecular systems such as proteins and complex carbohydrates. It also provides the practitioners of biochemistry, food chemistry and carbohydrate chemistry with valuable information on the behavior of various biopolymers in solution and in the solid state. Perhaps the most exciting feature is the wide range of NMR applications found in this volume, a testimony to the impact that the NMR method has made on many fields of scientific inquiry. It is our hope that this book will be intellectually rewarding to its readers, and that it will stimulate new, future applications of the NMR method.


J. Finley
A. Serianni
S. Schmidt

## ACKNOWLEDGMENTS

The editors gratefully acknowledge and thank all of the contributors who devoted their time to prepare the chapters of this book; without their significant effort, this book would not have been possible. Their efforts are particularly significant in light of the fact that the royalties from this book will be given to the Carbohydrate Chemistry, and Agricultural and Food, Divisions of the American Chemical Society to support future symposia.

The editors also thank Susie Anderson of EG&G Washington Analytical Services Center, Inc., for retyping the manuscripts in the correct format for publication and for her endless patience during the editorial process, and Fran Osborne of Nabisco Brands, Inc., for "reminding" the contributors to submit their contributions and for expediting the review process.

# CONTENTS

# APPLICATIONS OF NMR IN AGRICULTURE AND BIOCHEMISTRY

S. J. Schmidt(a), A. S. Serianni(b) and J. W. Finley(c)

(a) Division of Foods and Nutrition
    University of Illinois
    Urbana, IL
(b) Department of Chemistry and Biochemistry
    University of Notre Dame
    Notre Dame,  IN
(c) Nabisco Biscuit Company
    East Hanover, NJ

AN OVERVIEW

   The complex relationship between molecular structure and biological function is a central theme of most contemporary biochemical studies.  The specific interactions of biomacromolecules with themselves, solvents, substrates and other solutes determine their biological functions in living systems.  For decades organic chemists have explored structure-reactivity correlations in small organic molecules by studying the effects of thoughtful, systematic changes in molecular structure on chemical behavior.  Now, with the tools of modern molecular biology at their disposal, biochemists may systematically alter protein structure to assess the structure-function relationships in this important class of biopolymers.  Such studies promise to identify and quantify the molecular factors that confer specificity to protein-substrate binding, and the factors that are responsible for the rate enhancements of enzyme-catalyzed reactions.  This information is essential, for example, to the rational design of artificial enzymes and the development of specific enzyme inhibitors for use in the treatment of human metabolic disorders.

   With the ability to prepare proteins and other biopolymers of defined structure almost at will comes the need to establish their three-dimensional structures and their solution behaviors in a routine and timely fashion. Although other analytical methods are available to study biopolymer structure and behavior, in recent years nuclear magnetic resonance (NMR) spectroscopy has emerged as a very powerful and versatile tool to address these problems at the molecular level.  The NMR method is complemented frequently

*NMR Applications in Biopolymers*, Edited by
J. W. Finley *et al.*, Plenum Press, New York, 1990

by computer–assisted molecular modeling and studies of crystal structures. In contrast to the latter methods of X–ray and neutron diffraction, however, NMR allows the investigation of biopolymer structure in solution, the biologically–relevant medium. The solution conformation and dynamics of macromolecules, which are closely linked to their biological functions, can thus be examined as a function of solvent, pH, temperature, ligand (substrate) concentration, and ionic strength. This book describes a variety of studies that illustrate the diversity of problems that may be tackled by modern biological NMR spectroscopy, and the molecular detail that may be achieved when the full power of the technique is brought to bear on the problem at hand. These applications extend to fundamental biochemical investigations, to structure of complex carbohydrates, to magnetic resonance imaging of plants and animals, to investigative methods of food processing. As a background to the topics discussed in this book, the reader may wish to consult several recent reviews on the application of NMR to biopolymers (Table 1).

Over the last decade, NMR hardware, computer technology and experimental design have developed simultaneously to produce a modern generation of powerful commercial NMR spectrometers capable of probing complex biopolymer structures. Substantial improvements have been realized in the computerized processing and display of spectral data and the construction of high–field superconducting magnets. Most dramatic has been the development and propagation of special multi–pulse sequences, especially those that

Table 1.   Recent Review Articles on NMR of Biopolymers.

| | |
|---|---|
| Biopolymers | |
|     Natural Macromolecules, Reviews | 1–6 |
| Proteins | |
|     Structure | 7–11 |
|     Proteins in Solution | 12–19 |
| Nucleic Acids | 7–10 |
| Interactions | 20–21 |
| Hydration | 22–28 |
| Carbohydrates | 29–33 |
| Solid Macromolecules | 34–36 |
| Molecular Dynamics | 37 |
| Molecular Modeling | 38 |
| Food Chemistry | 39, 45, 57 |

permit the display of spectral correlations in two- and three-dimensions. These latter sequences have been essential to simplifying the complex NMR spectra encountered with biopolymers in order to facilitate signal assignments. Magnetic resonance imaging (MRI) has developed from a laboratory curiosity to a viable diagnostic tool in hospitals in less than one decade. The theory and applications of these hardware and software developments are discussed in detail in several comprehensive texts and reviews [40-55].

NMR has been central to establishing the solution structures of an increasing number of biologically important compounds and offers the opportunity to further enhance this knowledge in the future. The diverse applications of NMR spectroscopy to these problems are illustrated in this book, which is based on two ACS-sponsored symposia (Toronto, June 1988, Div. of Carbohydrate Chemistry; Los Angeles, September 1988, Div. of Agricultural and Food Chemistry). Linear and cyclic peptides have been used as model systems to develop experimental methods applicable to studies of larger biopolymers. The size and complexity of polypeptides that may be successfully probed by the method are growing rapidly (see James et al.). Several chapters of this book deal with the application of $^1$H and/or $^{13}$C 2D NMR to carbohydrates (Coxon, Lerner). Andersen et al. and James et al. discuss computerized approaches to the analysis of 2D NOE data to determine 3D structures of biomolecules in solution.

Other contributions focus on the use of high-field, multi-nuclear NMR to investigate biological metabolism. Vogel and Lundberg discuss the uses of $^{14}$N and $^{15}$N NMR to investigate the uptake and metabolism of nitrogen in the form of $NH_4^+$ and/or $NO_3^-$ in plants of agricultural importance. London and coworkers describe in vivo $^{31}$P NMR studies of glucosamine and galactosamine metabolism, and Becker and Ackerman use in vivo $^{13}$C NMR to monitor carbohydrate metabolism in whole animals.

Schmidt reviews the use of $^1$H, $^2$H and $^{17}$O NMR to investigate the behavior of water associated with various natural food components and model food systems. d'Avignon and coworkers have used $^1$H and $^2$H NMR relaxation methods to study water compartmentation and hydration of work free wheat flour doughs. Baianu and coworkers discuss the use of $^1$H, $^2$H and $^{17}$O nuclear relaxation rates to study the hydration of lysozyme in $H_2O$ and $^2H_2O$ with and without NaCl at various pH/p$^2$H. These authors also report the use of $^1$H, $^{17}$O and $^{23}$Na nuclei in examining muscle protein-water-NaCl interactions.

Several chapters focus on solid-state NMR studies of biologically important systems (Bryant et al., Cyr et al., Hiyama and Torchia, Jarrell and Smith), methods of water suppression (Sklenar), pulse-gradient NMR (Schmidt) and $^1$H and $^{17}$O NMR imaging (Schmidt), and the use of low-cost NMR spectrometers for process control in the food industry (Barker and Stronks, Pearson and Adams).

Because of its application to a wide variety of systems (from living to non-living and from liquids to solids) and because of its increasing usefulness in studying a wide array of problems (from investigating the molecular structure and conformation of a protein to imaging of the human body), NMR has great future potential in studies of biological systems. We have attempted in this work to bring together a number of these applications and present a state-of-the-art discussion of where NMR is today and to demonstrate the enormous potential of the technique for future investigations.

REFERENCES

1.  D. B. Davies, 1984, Natural Macromolecules, Nucl. Magn. Reson., 13:207-243.
2.  D. B. Davies, 1985, Nuclear Magnetic Resonance, Natural Macromolecules, Nucl. Magn. Reson., 14:211-249.
3.  D. B. Davies, 1986, Natural Macromolecules, Nucl. Magn. Reson., 15:191-215.
4.  D. B. Davies, 1987, Natural Macromolecules, Nucl. Magn. Reson., 16:191-222.
5.  G. E. Chapman, 1978, NMR of Natural Macromolecules, Nucl. Magn. Reson., 7:281-302.
6.  G. E. Chapman, 1979, NMR of Natural Macromolecules, Nucl. Magn. Reson., 8:242-65.
7.  J. DeVlieg, R. M. Scheek, W. F. Van Gunsteren, H. J. C. Berendsen, R. Kaptein, and J. Thomason, 1988, Combined Procedure of Distance Geometry and Restrained Molecular Dynamics Techniques for Protein Structure Determination from Nuclear Magnetic Resonance Data: Application to the DNA Binding Domain of LAC Repressor from Escherichia coli, Proteins: Struct., Funct., Genet., 3(4):209-18.
8.  K. Kanamori, and J. D. Roberts, 1983, Nitrogen-15 NMR Studies of Biological Systems, Acc. Chem. Res., 16(2):35-41.
9.  T. Endo, and T. Miyazawa, 1986, Application of NMR to Conformation Analysis of Biopolymers, Kagaku Sosetsu, 49:143-152.
10. T. A. Cross, J. A. Diverrdi, and S. J. Opella, 1982, Strategy for the Nitrogen NMR Analysis of Biopolymers, J. Am. Chem. Soc., 104(6):1759-61.
11. K. Wüthrich, 1986, NMR of Proteins and Nucleic Acids.
12. M. Ikura, 1988, Two Dimensional NMR of Proteins, Seibutsu Butsuri, 29(3):135-40.
13. Y. Kobayashi, 1988, Structure of Solution State of Proteins, NMR and Distance Geometry, Kessho Kaiseki Kenkyu Senta Dayori, 1., 9:5-16.
14. K. Wüthrich, 1987, A NMR View of Proteins in Solution, Springer Ser. Biophys., 1 (Struct., Dyn. Funct. Biomol.), 104-7.
15. S. W. Homans, A. L. DeVries, and S. B. Parker, 1985, Solution Structure of Antifreeze Glycopeptides, Determination of the Major Conformers of the Glycosidic Linkages, FEBS Lettr., 183(1): 133-7.
16. A. M. Gronenborn, 1987, Determination of Three-Dimensional Structures of Proteins in Solution by Nuclear Magnetic Resonance, Protein Eng., 1(4):275-288.
17. A. Eugster, 1988, The Structure of Noncrystalline Proteins. SLZ, Schweiz. Lab. Z., 45(1):14-18.
18. D. A. Torchia, and D. L. Vanderhart, 1979, High-Power Double-Resonance Studies of Fibrous Proteins, Proteoglucans and Model Membranes, Top. Carbon-13 NMR Spectrosc., 3:3225-60.

19. A. Graeslund, and R. Rigler, 1986, Biological Macromolecules in Solution - Structure and Dynamics, Kosmos (Stockholy), 63:103-14.
20. J. S. Cohen, 1973, Nuclear Magnetic Resonance Investigations of the Interactions of Biomolecules, Exp. Methods Biophys. Chem., 521-88, Ed. by Nicolau, C., Wiley, London.
21. M. A. Landau, 1976, Study of the Complexes of Drug Molecules with Biopolymers and Biomembranes by High-Resolution NMR, Khim.-Farm. Zh., 10(11):29-41.
22. G. M. Nikolaev, S. I. Aksenov, and V. S. Pshezhetskii, 1981, Hydration Model Studies of Biopolymers, Stud. Biophys., 85(1):1-2.
23. M. Aizawa, J. Mizuguchi, S. Suzuki, and S. Hayashi, 1972, Properties of Water in Macromolecular Gels, IV, Proton Magnetic Resonance of Water in Macromolecular Gels, Bull. Chem. Soc., Japan, 45(10):3031-4.
24. S. Takizawa, 1973, Applications of NMR to Biopolymers, Water and Hydration, I, in NMR No Seitai Kobunshi Eno Oyo, 64-74, Ed. by A. Wada, Kyoritsu Shuppansha, Tokyo.
25. H. Hayashi, 1973, Applications of NMR to Biopolymers, Water and Hydration, II, in NMR No Seitai Kobunshi Eno Oyo, 64-74, Ed. by A. Wada, Kyoritsu Shuppansha, Tokyo.
26. M. I. Burgar, 1982, Hydration Role of Water in Biological Systems as Determined by Oxygen-17 NMR, Stud. Biophys., 91(1):29-36.
27. S. J. Richardson, 1986, Molecular Mobility Characterization of Polymer and Solute Water States as Determined by Nuclear Magnetic Resonance, Rheology and Hydrodynamic Equilibrium, Diss. Abstr. Int. B., 1987, 47(7):2701.
28. V. Sklenar, 1987, Water Supression Using a Combination of Hard and Soft Pulses, J. Mag. Res., 75(2):352-7.
29. G. G. S. Dutton, 1981, Use of NMR Spectroscopy in the Study of the Molecular Structure of Biopolymers. Polym. Prep., Am. Chem. Soc., Div. Polym. Chem., 22(1):326.
30. M. K. McIntyre, and G. W. Small, 1987, Carbon-13 Nuclear Magnetic Resonance Spectrum Simulation Methodology for the Structure Elucidation of Carbohydrates, Anal. Chem., 59(14):1805-11.
31. B. Overdijk, E. P. Beem, Ge J. VanSteijn, L. A. W. Trippelvitz, J. J. W. Lisman, J. Paz Parente, P. Cardon, I. Leroy, and B. Fourtner, 1985, Biochem. J., 232(3):637-41.
32. P. Pfeffer, F. W. Parish, and J. Unruh, 1980, Deuterium Induced, Differential Isotope-Shift Carbon-13 NMR, Part 2, Effects of Carbohydrate-Structure Changes on Induced Shifts in Differential Isotope-Shift Carbon-13 NMR, Carbohydr. Res., 84(1):13-23.
33. P. Voss, 1979, Carbon-13 NMR Spectroscopy as a Method for the Elucidation of Structural Problems in Carbohydrate Chemistry, Starch/Staerke, 31(12):404-9.
34. M. Hatano, 1982, High Resolution NMR Approach to the Study of Solid Macromolecules, Kagaku (Kyoto), 37(9):703-6.
35. B. C. Gerstein, 1983, High-Resolution NMR Spectroscopy of Solids, Part II, Anal. Chem., 55(8):899A-900A.
36. S. J. Opella, J. G. Hexem, M. H. Frey, and T. A. Cross, Philos. Trans. R. Soc. London, Ser. A, 299(1452):665-83.
37. O. Jardetzky, 1981, NMR Studies of Macromolecular Dynamics, Acc. Chem. Res., 14(10):291-8.
38. B. Sheard, 1987, NMR and Molecular Modeling: A New Tool for Biotechnology, World. Biotech Rep., Vol. 1(2):117-22. Online Pugl., London.
39. I. Horman, 1984, NMR Spectroscopy in Analysis of Foods and Beverages, 205-263, Academic Press, New York.
40. A. Abragam, 1961, The Principles of Nuclear Magnetism, Clarendon Press, Oxford.
41. R. K. Harris, and B. E. Mann, (Eds.), 1978, NMR and the Periodic Table, Academic Press, New York.
42. L. J. Berliner, and J. Reuben, (Eds.), 1980, Biological Magnetic Resonance, Plenum Press, New York.
43. P. Laszlo, (Ed.), 1983, NMR of Newly Accessible Nuclei, Vols. 1 and 2, Academic Press, Inc., New York.

44. I. D. Campbell, R. A. Dwek, 1984, Biological Spectroscopy, Benjamin Cummings Publishing Company, Inc., Menlo Park, CA.

45. I. Horman, 1984, NMR Spectroscopy, in: "Analysis of Foods and Beverages; Modern Techniques," E. Charalambous, (Ed.), Academic Press, Inc., New York.

46. L. W. Jelinski, 1984, Modern NMR Spectroscopy, C & E News, November 5:26.

47. W. P. Rothwel, 1985, Nuclear Magnetic Resonance Imaging, Applied Optics, 24(23):3958.

48. F. D. Blum, 1986, Pulsed-gradient spin-echo nuclear magnetic resonance spectroscopy, Spectroscopy, 1(15):32.

49. W. Kemp, 1986, NMR in Chemistry; A Multinuclear Introduction, MacMillan Education Ltd., London.

50. G. A. Morris, 1986, Modern NMR Techniques For Structure Elucidation, Magnetic Resonance in Chemistry, 24:371.

51. P. G. Morris, 1986, Nuclear Magnetic Resonance Imaging in Medicine and Biology, Clarendon Press, Oxford.

52. Atta-Ur Rahman, 1986, Nuclear Magnetic Resonance, Springer-Verlag, New York.

53. A. E. Dermone, 1987, Modern NMR Techniques for Chemistry Research, Pergamon Press, New York.

54. R. R. Ernst, G. Bodenhausen, and A. Wokaun, 1987, Principles of Nuclear Magnetic Resonance in One and Two Dimensions, Clarendon Press, Oxford.

55. R. F. Bovey, L. Jelinski, and P. A. Mirau, 1988, Nuclear Magnetic Resonance Spectroscopy, 2nd Edition, Academic Press, Inc., New York.

56. J. Karger, H. Pfeifer, and W. Heink, 1988, Principles and application of self diffusion measurements by nuclear magnetic resonance, Advances in Magnetic Resonance, Vol. 12, Academic Press, Inc., New York.

57. F. E. Pfeifer, and W. V. Gerasimowicz, 1989, Nuclear Magnetic Resonance in Agriculture, CRC Press, Boca Raton, Florida.

# CARBOHYDRATE STEREOCHEMISTRY, AND NMR SPECTROSCOPY

Arthur S. Perlin

Department of Chemisty
McGill University
Montreal, Quebec, Canada H3A 2A7

## INTRODUCTION

This opening presentation is intended to draw attention to several aspects of carbohydrate stereochemistry that can profitably be examined by NMR spectroscopy which, as a consequence, has contributed abundantly to modern developments in the carbohydrate field. Some of the NMR applications are to be treated in greater depth by other participants in this symposium. Measurements of chemical shift, spin-spin coupling, relaxation parameters, and nuclear Overhauser enhancement are all utilized for the purpose. Among the kinds of stereochemical information accessible from NMR data are: configurational assignment, molecular conformation (including solvent influences), a comparison of solution and solid state conformations, orientation of substituent groups, and the geometry of interactions between a carbohydrate and other species. Examples of several of these are cited from our own work, and emphasis is given to certain observations that, although widely recognized for some time, are not well understood. Hopefully, this reminder will stimulate a further assessment, and a clarification, of those points.

### Chemical Shift Characteristics

A configurational inversion is associated [1-3] with extensive changes in chemical shift ($\Delta\delta$), as represented by the tautomerization of β-D-glucopyranose into its α-anomer ($\underline{1} \rightarrow \underline{2}$). There are downfield displacements of four of the ring proton resonances, which are accompanied by increased shielding of the corresponding $^{13}C$ nuclei and also of the hydroxyl protons.

These variations were originally attributed to non-bonded de-stabilizing interactions ---- e.g., resulting from the alteration of an equatorially-oriented OH-1 to an axial one -- by analogy with the well-known [4] "γ - gauche" effect. However, as the origin of the latter effect has

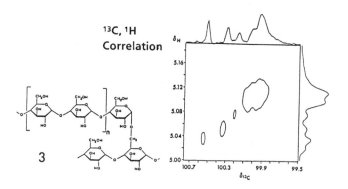

**1** ... OH, H

**2** ... OH, H

| $\Delta\delta^a$ | 1 | 2 | 3 | 5 | 4,6 |
|---|---|---|---|---|---|
| $^1H$ | +0.60 | +0.27 | +0.16 | +0.29 | no |
| $^{13}C$ | -3.9 | -2.7 | -2.9 | -4.6 | significant |
| $O^1H$ | -0.36 | -0.34 | -0.14 | - | change |

[a] + downfield shift;  – upfield shift

been a subject of substantial disagreement [5-7], the overall status of the question today remains uncertain from the theoretical point of view.

Steric interactions are invoked widely, nevertheless, to rationalize a host of similar kinds of chemical shift observations on carbohydrate molecules. A seemingly – related case, involving solution contributions to shielding is found [8] in spectra of amylopectin (3), the branched component of starch. When the polymer is dissolved in $Me_2SO$, resonances due to carbons and protons of residues associated with branching are much more broadly dispersed than when the solvent is water. The 2D $^{13}C$ – $^1H$ correlation spectrum below shows that the <u>more</u> strongly shielded of the $^1H$ nuclei are appended to the <u>less</u> strongly shielded $^{13}C$ nuclei. As these converse effects on the chemical shifts are analogous to those [3] that accompany the introduction, or removal, of a steric interaction, they imply that, in the $Me_2SO$ – rich environment, the glycosidic bonds of a number of residues associated with branching experience less strain ($^{13}C$ shift downfield, $^1H$ shift upfield) than in water.

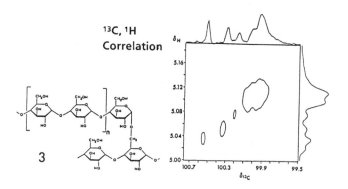

**$^{13}C, ^1H$ Correlation**

**3**

8

Furanose sugars and derivatives are characterized [9] by strikingly distinctive chemical shifts, in that both the $^{13}$C and $^{1}$H nuclei of their 5-membered rings (5) are strongly deshielded relative to those of the configurationally – related pyranoses (4), as the following data show:

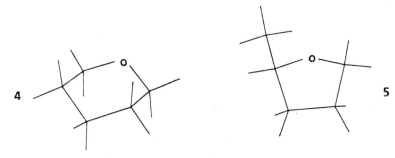

OVERALL DIFFERENCES ($\Sigma\Delta\delta$) IN THE SHIELDING OF FIVE- AND SIX-MEMBERED RING COMPOUNDS

| Pair of structural isomers | $\Sigma\Delta\delta$ (ppm) $^{13}$C | $^{1}$H |
|---|---|---|
| β–D–fructofuranose – β–D–fructopyranose | 25.9 | |
| Methyl β–D–xylofuranoside – methyl β–D–xylopyranoside | 19.6 | 3.8 |
| Methyl β–D–ribofuranoside – methyl β–D–ribopyranoside | 18.8 | 1.4 |
| Methyl α–D–glucofuranoside – methyl α–D–glucopyranoside | 19.9 | 3.0 |
| Methyl α–D–mannofuranoside – methyl β–D–mannopyranoside | 33.4 | 2.2 |

Incorporated into these differences are chemical shift variations among the furanoses that, as for the pyranoses, are ascribable [9] to configurational changes and their associated steric interactions; e.g., vicinal cis hydroxyl groups, which must engage in steric compression, give rise to increased shielding relative to that for the trans 1,2–diol, or 1,3–diols. The diamagnetic ($\sigma^d$) term of the chemical shift, may be a more important factor for a furanose, than a pyranose, because it shows [9] a strong dependence on furanose conformation. Otherwise, little attention has been given to this intriguing influence of ring size on chemical shift.

Another distinctive type of chemical shift pattern is found among bicyclic anhydride derivatives. In response to configurational inversion, the C-2, C-3 and H-2, H-3 nuclei of 2,3-anhydropyranoses (6, 7) show [10] chemical shift changes in the same direction, rather than the opposite ones that characterize such pyranoses as glucose (above). Possibly, this distinction between an oxirane and vic-diol represents a balance between contributing electronic and steric influences, whereby the latter is more important in the diol, and the former (e.g., the ring current) the overriding factor in determining the chemical shift of the $^{13}$C and $^{1}$H nuclei of the oxirane ring.

6 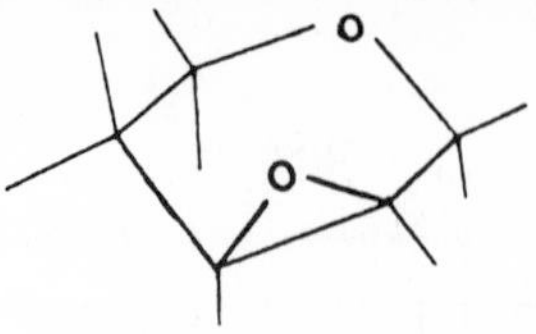 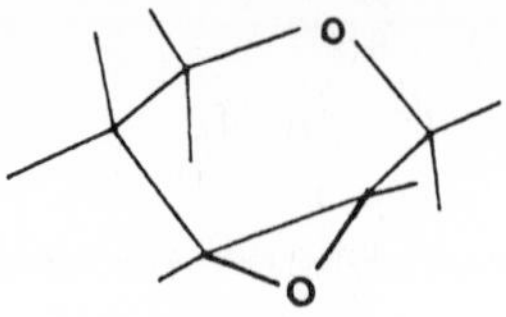                                                                          7

## Rotamers of Exocyclic Hydroxymethyl Groups

Analyses of the rotational isomerism of a $-CH_2OR$ group appended to a sugar ring (see, e.g., ref. 11) have utilized the geometry - dependence of vic-$^1H-^1H$ coupling. The g,g and g,t rotamers are prominent in solutions of pyranoses (e.g. glucose) having an eq. O-4 substituent because, in contrast to the t,g rotamer, this substituent entails minimal interactions with the OR-6 group; by analogy, the g,g rotamer is disfavored when O-4 is axial. Values of $J_{5,6S}$ and $J_{5,6R}$ for 1-deoxynojirimycin, a D-glucose analog that functions as an antibiotic, correspond to a g,t (8, R=R'=H): g,g (9, R=R'=H) ratio of ~1:1, the anticipated sort of value based on the characteristics of glucose and its derivatives. A unique opportunity for checking the rationale for this equilibrium population is provided [12] by the N-methyl derivative, because an equatorial substituent as in 8(R=Me, R'=H) would incorporate a syn-periplanar interaction with OH-6 that should destabilize the g,t rotamer of this derivative, whereas the g,g rotamer should be destabilized if the methyl group is axial, as in 9(R=H, R'=Me). Overall, the N-methyl diastereomers (10,11) were found in an eq.: ax. ratio of ~11:1, and in accord with this there was a strong preponderance (>90%) of the g,g rotamer. Hence, formula 10 is taken to be a good configurational representation of N-methyl-1-deoxynojirimycin in aqueous solution.

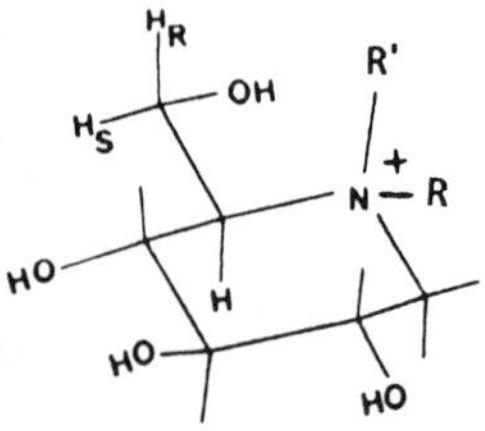

(g,t)          8 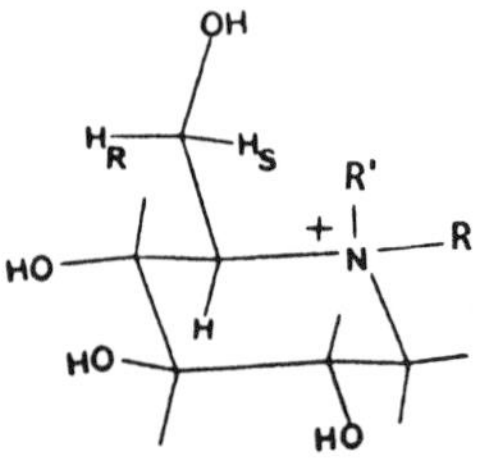                (g,g)          9 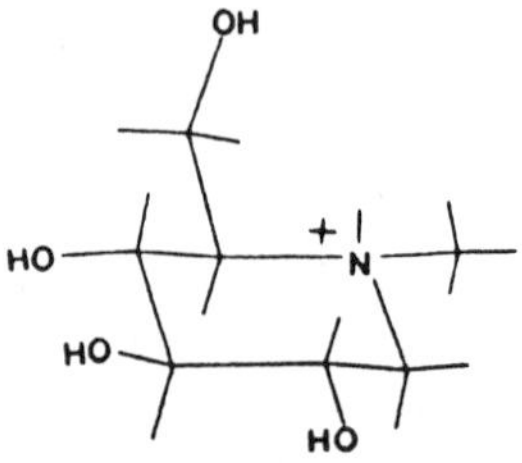                10

(11, axial epimer)

## Comparing Solution and Solid State $^{13}C$ Chemical Shifts

Important advances in solid state NMR have opened up excellent opportunities, such as the examination of poorly soluble carbohydrates, or long - standing questions as to how similar is stereochemistry in solution and the solid state?

10

The identification of 8, 9, and 10 as equatorial or axial N-methyl derivatives, included the fact that their respective methyl-$^{13}$C chemical shifts are δ 43.7 (eq.) and 38.4 (ax.); the latter may reasonably be attributed to the increased shielding effect of a γ-gauche interaction, referred to earlier. Analogous $\delta_c$ values are observed [13] for the N-methyl diastereomers of atropine (12) and scopolamine (13), not only from spectra of their solutions but also, using [13] the CP/MAS technique, of crystalline specimens of known (X-ray) N-methyl stereochemistry. As the latter affords an unequivocal basis for relating an N-methyl group orientation to its $^{13}$C-chemical shift, the overall data for 8-13 fully validate the proposition that the more strongly shielded N-methyl-$^{13}$C of an eq.-ax. pair is that of the axial epimer.

Atropine · HBr
eq 41.1 (D$_2$O), 39.6 (solid)
ax 33.9 (D$_2$O)

Scopolamine · HBr
eq 46.9 (D$_2$O)
ax 32.7 (D$_2$O), 33.2 (solid)

N-methyl epimerization (18:1 eq:ax)

## Coupling, $^{13}$C-$^1$H and Stereochemistry

As spin-spin coupling between $^{13}$C and $^1$H-nuclei across one, two, and three bonds is geometry dependent, these parameters find valuable applications to carbohydrate stereochemistry. Anomers of sugars and derivatives are differentiated by the fact [14] that when 0-1 is axially oriented (14), $J_{C-1,H-1}$ is larger (generally by ~ 10 Hz) than that of its equatorial counterpart (15). This characteristic is utilized commonly for the determination of anomeric configurations in oligosaccharides and polysaccharides, particularly when $^3J_{H-1, H-2}$ values are not readily extracted from the $^1$H spectrum. Possibly, the orientational influence on $^1J_{CH}$ is stereoelectronic, stemming [15] from contributions by the lone pairs on the ring oxygen atom. Among support for this proposal is the observation [16] that when the oxygen is replaced by a sulfur heteroatom (as in 16 and 17) the corresponding interaction between the $^{13}$C-1 and $^1$H-1 nuclei appears to lack an orientational component.

The magnitude of $^2J_{CH}^{17}$ is affected strongly by the orientation of adjacent C-0 bonds. Whereas, for example, a value of about -5 Hz is

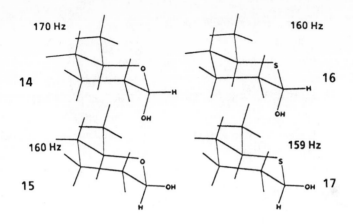

characteristic when the H atom is _gauche_ with respect to geminal
O-substituents (left), the $^2J_{CH}$ value becomes strongly positive (~+5 Hz)
when a C-O bond is _anti_ (right). Relatively little effort has been made to
gain stereochemical information from $^2J_{CH}$ data, probably because such data
have been difficult to obtain experimentally. This neglect will likely
diminish with the present, increasing, usage of 2D J-resolved methodology.

Easier access to coupling information should also facilitate the application of three-bond $^{13}C-^1H$ coupling to stereochemical problems. Among
possibilities meriting examination, are $^3J_{CH}$ data for the molecular segment
$^{13}C-C-O-^1H$ and, accordingly, the rotational isomerism of hydroxyl groups
that those couplings should reflect. Data readily acquired [18] from a 2D
J-resolved spectrum of α-D-glucose (18) in $Me_2SO-d6^{19}$ are presented as an
example: the coupling pattern shown was exhibited on selective irradiation
of OH-2, and a set of $^3J$ values was obtained (another set is listed for
OH-4). Previously, comparable information was available only by the
synthesis of selectively $^{13}C$-enriched compounds. With such data as these, a
meaningful analysis of hydroxyl-group orientation (in this solvent) should
be feasible.

Orientations of Glycosidic Bonds

The magnitude of coupling ($^3J_{CH}$) across the glycosidic bond that links
one sugar residue with another, is of interest as a descriptor [20] of the
molecular conformation of oligosaccharides and polysaccharides in solution.
$^3J_{CH}$ values compiled from the spectra of a variety of model compounds generate a "Karplus-type" curve from which the bond-torsion angles, Φ and Ψ,
may be estimated. However, the relative sign of these angles, i.e. + or -,

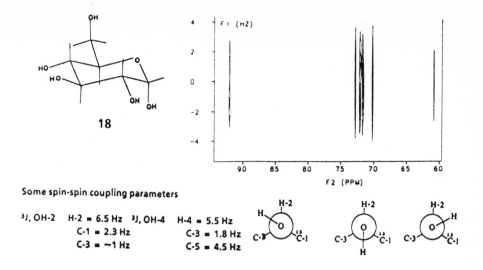

**Some spin-spin coupling parameters**

| $^3J$, OH-2 | H-2 = 6.5 Hz | $^3J$, OH-4 | H-4 = 5.5 Hz |
|---|---|---|---|
| | C-1 = 2.3 Hz | | C-3 = 1.8 Hz |
| | C-3 = ~1 Hz | | C-5 = 4.5 Hz |

must be obtained in some other way. Molecular-dynamics methods, now used intensively for evaluating conformational possibilities, should be of help in choosing appropriate signs for the torsional angles given by $^3J_{CH}$. Combined in this fashion, as shown below, the experimental [21] and computed [22] data for the disaccharide derivative, 1,6-anhydro-β-cellobiose hexaacetate (19), suggest that the observed couplings correspond to Φ and Ψ values of +20° and –55°, respectively. These, however, must represent a time–averaged description, rather than an explicit one, of the orientations adopted by the glycosidic linkage, a fundamental question that is addressed in detail by other participants in this symposium.

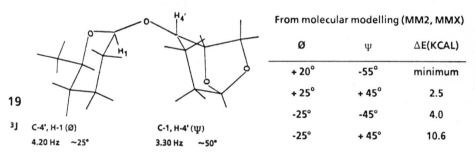

**From molecular modelling (MM2, MMX)**

| Ø | Ψ | ΔE(KCAL) |
|---|---|---|
| + 20° | -55° | minimum |
| + 25° | + 45° | 2.5 |
| -25° | -45° | 4.0 |
| -25° | + 45° | 10.6 |

**19**

| $^3J$ | C-4', H-1 (Ø) | | C-1, H-4' (Ψ) | |
|---|---|---|---|---|
| | 4.20 Hz | ~25° | 3.30 Hz | ~50° |

Much can be learned about the conformations of large molecules by molecular-modelling, as well as from correlations with crystal structures (as determined with X-rays, or by solid-state NMR spectroscopy). Hence, as the inter-residue $^{13}C-^1H$ couplings can now be measured with great facility by 2D methods, more than ever they constitute practical experimental parameters with which computed data may be compared.

NOE measurements provide [21] complementary information about how a glycosidic linkage may be oriented. In this approach the NOE values are translated into distances between protons of adjacent residues. Although a qualitative description is easily obtained by the use of newly-available 2D

NOE methods, and is frequently given, there have been relatively few quantitative studies. One example [23] of the latter type deals with 1,6-anhydro-β-maltose hexaacetate (20). From the NOE values associated with the interactions among H-1, H-3′ and H-4′ illustrated in the correlation spectrum shown below, disaccharide derivative (20) was determined to have inter-proton distances between H-1 and H-4′ and H-1 and H-3′ of 2.18±0.12 and 2.91±0.24Å, respectively. As with the inter-residue $^3\underline{J}_{CH}$ data, these values are commensurate with more than one set of torsional angles ($\Phi$, $\Psi$), and define a limiting region of allowed solution conformations for (20). Also in common with the coupling data, NOE's help to provide an experimental framework within which calculations of molecular dynamics can be assessed.

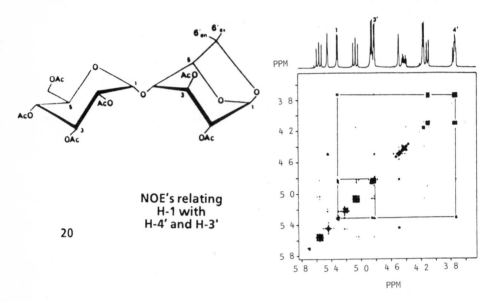

20

NOE's relating
H-1 with
H-4′ and H-3′

ACKNOWLEDGMENTS

The excellent cooperation of my colleagues in our studies cited here, is acknowledged with deep appreciation.

REFERENCES

1. A. S. Perlin, B. Casu, and H. J. Koch, 1970, Can. J. Chem., 48:2596;
   H. J. Koch, and A. S. Perlin, 1970, Carbohydr. Res., 15:403;
   A. S. Perlin, 1976, MTP Int. Rev. Sci. Org. Chem. Ser., One, 7:1.
2. R. U. Lemieux, and J. D. Stevens, 1966, Can. J. Chem., 44:249.
3. A. S. Perlin, and H. J. Koch, 1970, Can. J. Chem., 48:2639.
4. D. K. Dalling, and D. M. Grant, 1967, J. Am. Chem. Soc., 86:6612;
   D. M. Grant, and B. V. Cheney, 1967, J. Am. Chem. Soc., 89:5315.
5. H. J. Schneider, and V. Hoppen, 1974, Tetrahedron Lett., 579.
6. H. Beierbeck, and J. K. Saunders, 1976, Can. J. Chem., 54:2985.
7. D. G. Gorenstein, 1977, J. Am. Chem. Soc., 99:2254.
8. Q. -J. Peng, and A. S. Perlin, 1987, Carbohydr. Res., 160:57;
   P. Dais, and A. S. Perlin, 1982, Carbohydr. Res., 100:103.

9.   A. S. Perlin, N. Cyr, H. J. Koch, and B. Korsch, 1973, Ann. N.Y.
     Acad. Sci., 222:935; R. G. S. Ritchie, N. Cyr, B. Korsch,
     H. J. Koch, and A. S. Perlin, 1975, Can. J. Chem., 53:1424;
     A. S. Perlin, 1977, Isotopes in Org. Chem., E. Buncel, and
     C. C. Lee, ed., Elsevier, 3:229; N. Cyr, and A. S. Perlin, 1979,
     Can. J. Chem., 57:2504.
10.  M. M. Abdel-Malik, Q. -J. Peng, and A. S. Perlin, 1987, Carbohydr.
     Res., 159:11.
11.  D. M. Mackie, A. Maradufu, and A. S. Perlin, 1986, Carbohydr. Res.,
     150:23.
12.  R. Glaser, and A. S. Perlin, 1988, Carbohydr. Res., 182:169.
13.  R. Glaser, Q. -J. Peng, and A. S. Perlin, 1988, J. Org. Chem.,
     53:2172.
14.  A. S. Perlin, and B. Casu, 1969, Tetrahedron, Lett., 2921;
     J. A. Schwarcz, and A. S. Perlin, 1972, Can. J. Chem., 50:3667;
     K. Bock, J. Lundt, and C. Pederson, 1974, J. Chem. Soc. Perkin,
     Trans., 2:293.
15.  J. Augé, and S. David, 1976, Nouv. J. Chim., 1:57.
16.  J. E. N. Shin, and A. S. Perlin, 1979, Carbohydr. Res., 76:165;
     V. S. Rao, and A. S. Perlin, 1981, Carbohydr. Res., 92:141.
17.  J. A. Schwarcz, N. Cyr, and A. S. Perlin, 1975, Can. J. Chem.,
     53:1872; N. Cyr, G. K. Hamer, and A. S. Perlin, 1978, Can. J.
     Chem., 56:297.
18.  F. Sauriol, and A. S. Perlin, unpublished.
19.  A. S. Perlin, 1966, Can. J. Chem., 44:539; B. Casu, M. Reggiani,
     G. G. Gallo, and A. Vigevani, 1964, Tetrahedron Lett., 2839.
20.  A. S. Perlin, R. G. S. Ritchie, and A. Parfondry, 1974, Carbohydr.
     Res., 37:C1; A. Parfondry, N. Cyr, and A. S. Perlin, 1977,
     Carbohydr. Res., 59:299; G. K. Hamer, F. Balza, N. Cyr, and
     A. S. Perlin, 1987, Can. J. Chem., 56:3109.
21.  P. Dais, T. K. M. Shing, and A. S. Perlin, 1984, J. Am. Chem. Soc.,
     106:3082.
22.  R. Glaser, and A. S. Perlin, unpublished.
23.  P. Dais, and A. S. Perlin, 1988, Magn. Reson. Chem., 26:373.

# APPLICATIONS OF 2D NMR SPECTROSCOPY TO CARBOHYDRATES

L. Lerner

Department of Chemistry
University of Wisconsin-Madison
Madison, WI  53706

Two-dimensional high resolution nuclear magnetic resonance spectroscopy (2D NMR) is a powerful tool for determining the structure and
conformation of carbohydrates in solution.  Applications to carbohydrate
problems have followed closely on the heels of development of sophisticated two dimensional -- and now three dimensional -- pulse sequences
(Hoffman and Davies, 1988; Fesik et al., 1989; Vuister et al., 1989).

The first task in using NMR to extract conformational information is
assignment of resonances:  which crosspeak goes with which pair of inter-
acting spins.  Carbohydrates are usually difficult to assign because most of
their protons have overlapping chemical shifts in the 3-4 ppm region.  Until
the introduction of two-dimensional techniques, only a few carbohydrate
resonances such as the anomeric protons and methyl protons could be readily
assigned.  Vliegenthart and coworkers (1983) pioneered the use of the
"structural-reporter group" to identify primary structures of oligosaccharides by characteristic chemical shifts of those protons with distin-
guishable chemical shifts.  To extend analysis beyond identification of
residues, more complete assignment is necessary to determine solution
conformation by analysis of coupling constants and NOE's.  Recent advances
in the application of 2D methods for studying oligosaccharides have been
reviewed by Dabrowski (1988).

This presentation is a brief description of some further developments
in two-dimensional techniques, both homonuclear and heteronuclear,
especially useful for assigning $^{1}$H and $^{13}$C resonances in carbohydrates when
only limited amounts of sample are available.  For carbohydrates with many
identical subunits, it is also possible that no known method will yield
complete assignments without derivatization or isotopic labelling.  It is

*NMR Applications in Biopolymers,* Edited by
J. W. Finley *et al.,* Plenum Press, New York, 1990

important to note that there are numerous alternate approaches to assignment.  For a clear and thorough review of the entire scope of 2D methods,
the reader is referred to Kessler et al. (1988).

## Homonuclear Correlation

Correlation experiments that rely on scalar coupling establish connectivities among protons within each ring of an oligosaccharide.  The original
COSY experiment has been refined and modified in numerous ways.  The basic
COSY pulse sequence can be modified further to select for double or triple
(or higher) quantum coherence transfers which can simplify a connectivity
pattern (Piantini et al., 1982; Homans et al., 1986).  Such multiple quantum
filtered COSY experiments improve resolution near the diagonal and produce
well-resolved multiplets.  E. COSY (Exclusive COSY, Griesinger et al., 1985)
is a combination of multiple quantum filtered COSY spectra of different
orders.  The E. COSY and related P. E. COSY methods (primitive E. COSY,
Müller, 1987) allow accurate measurement of small coupling constants between
nuclei with a shared coupling partner.  The practical result is the ability
to measure small passive couplings accurately in the presence of larger
active couplings.  Bax and Lerner (1988) used a modified P. E. COSY sequence
and a modified double quantum filtered COSY sequence to measure all the
three-bond proton-proton coupling constants in the deoxyribose rings of an
oligonucleotide (except for those involving H5', H5").  These coupling
constants can be used to estimate dihedral angles, and hence provide
information about local conformation, provided a suitable Karplus relation
is available.

An alternative 2D experiment for obtaining complete scalar coupling
networks is TOCSY (total correlation spectroscopy, Braunschweiler and Ernst,
1983) in its most widely used form, HOHAHA (Bax and Davis, 1985).  A slice
through a HOHAHA spectrum will show peaks for all protons scalar coupled to
one another within a sugar residue.  However, the distribution of magnetization around the spin system can be impeded by a small coupling, such as
typically found between H4 and H5 in galactosyl residues.  To circumvent the
bottleneck of a small coupling, Inagaki et al. (1989) described one-
dimensional and two-dimensional versions of a relayed HOHAHA pulse sequence.
In this method, magnetization is transferred first by application of MLEV-17
mixing and then by the conventional homonuclear relay sequence described by
Eich et al. (1982).

With their characteristic chemical shifts (~4.5-4.7 ppm for $\beta$ and
5.0-5.2 for $\alpha$ at ambient temperatures) and $^3J_{H1,H2}$ values (~8 for $\beta$ and
~4 Hz for $\alpha$), anomeric protons usually provide the toehold required to begin
tracing connectivities in carbohydrates.  When the anomeric protons are
well-resolved, it may be more efficient to use one-dimensional versions of

two-dimensional experiments by applying selective pulses. Selectivity can be improved by using shaped (Gaussian) pulses (Bauer et al., 1984; Kessler et al., 1986).

For example, the basic HOHAHA sequence can be modified to suppress signals from all but one spin system (i.e., one sugar residue) in an oligosaccharide (Davis and Bax, 1985). The addition of a z-filter and appropriate phase cycling to this sequence yields subspectra in pure phase, so that $^1$H–$^1$H scalar coupling constants can be determined (Subramanian and Bax, 1987; Lerner and Bax, 1987). An example of this method is shown in Fig. 1, which is a $^1$H subspectrum from the galactose residue of the tri-saccharide α-Neu5Ac-(2→3)-β-Gal-(1→4)-Glc. The dramatic simplification achieved by generating subspectra can be appreciated by comparing the sub-spectrum with the original one-dimensional $^1$H spectrum of this sample, drawn along one axis in Fig. 2. The assignment of the ring protons and carbons of this trisaccharide, by a combination of three inverse-detected $^1$H–$^{13}$C chemical shift correlation experiments, is described by Lerner and Bax (1978).

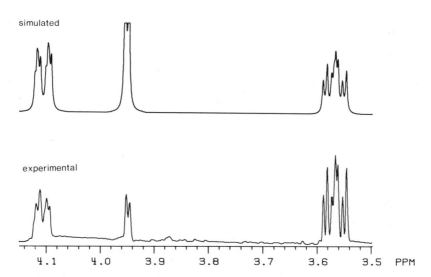

Fig. 1.  $^1$H subspectra at 500 MHz, 20°C, of α-Neu5Ac-(2→3)-β-Gal-(1→4)-Glc. Top, simulated subspectrum; bottom, experimental subspectrum obtained by selective irradiation of the Gal H-3 resonance at 4.105 ppm in the pulse sequence described by Bax and Subramanian, 1987. The sample contained 3.5 mg of the trisaccharide (purchased from BioCarb, Lund, Sweden) in 350 μl $^2$H$_2$O. Simulation was based on assuming two sets of Gal H-2 and H-3 atoms, corresponding to 2:1 β:α anomers of glucose. This subspectrum was simulated with the NMRSIM program provided by GE-Nicolet in its 1280 software package.

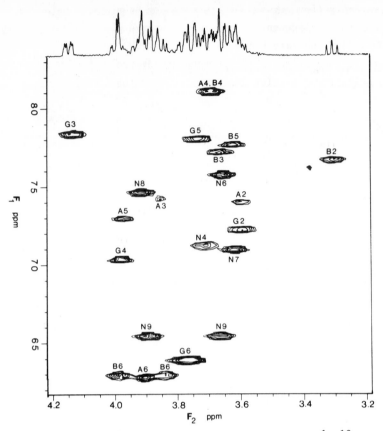

Fig. 2. Portion of chemical shift correlation map for $^1$H–$^{13}$C pairs
directly bonded in the trisaccharide, α–Neu5Ac–(2→3)–β–Gal–
(1→4)–Glc (same sample as in Fig. 1), generated using the
pulse sequence described by Subramanian and Bax, 1986.
Total measuring time was 2.4 hours. Each crosspeak
corresponds to a $^1$H–$^{13}$C pair, labelled N for α–Neu5Ac,
G for Gal, A for α–Glc, B for β–Glc. The corresponding
portion of the one–dimensional $^1$H spectrum at 500 MHz is
displayed along the horizontal axis. $F_1 = ^{13}$C chemical
shift, $F_2 = ^1$H chemical shift, relative to TSP.

## Heteronuclear Correlation

Sometimes the proton resonances of an oligosaccharide are too overlapping to be disentangled by homonuclear correlation alone. In such cases,
two–dimensional $^1$H–$^{13}$C chemical shift correlation maps may enable assignment
because of the greater spread in $^{13}$C chemical shifts. Heteronuclear
chemical shift correlation by means of coherent transfer of transverse
magnetization between scalar–coupled protons and low–gamma (gyromagnetic
ratio) nuclei was first proposed by Maudsley and Ernst (1977). Morris and
Hall (1981) provided an early example of the utility of this experiment for
assigning carbohydrates. The initial proposal has since spawned many

important variants (for reviews, see Bax, 1982; Ernst et al., 1987). One of
the most important refinements has been to observe protons, the abundant
nuclei, instead of the rare heteronuclei ($^{13}$C, $^{15}$N, $^{31}$P, among others).
Inverse detection experiments (also known as reverse or indirect or proton
detection) exploit the greater sensitivity of protons. General aspects of
inverse detection are reviewed by Ernst et al., 1987; Griffey and Redfield,
1987; and Kessler et al., 1988. The schemes rely on the generation of
multiple quantum coherence between coupled $^1$H and $^{13}$C energy levels. There
are several possible paths for coherence transfer between coupled protons
and carbon-13 nuclei, but the greatest potential gain in sensitivity can be
achieved by transferring coherence from protons to carbon-13 and then
converting the multiple quantum coherences back into detectable proton
single quantum coherence. The basic sequence to achieve this double
transfer was proposed by Müller (1979). The increase in sensitivity arises
from the higher gyromagnetic ratio of protons and the shorter longitudinal
relaxation times of protons relative to carbon-13.

Successful inverse-detection pulse schemes must suppress the unwanted
signal from protons that are not coupled to the rare spins (in this case,
$^1$H–$^{12}$C pairs). Several methods have been developed, each appropriate for a
particular application. The unwanted $^1$H signals can be reduced by presaturation, caused by application of a series of 180° pulses before the start of
the experiment. For low molecular weight compounds, insertion of a BIRD
pulse sandwich (bilinear rotation, Garbow et al., 1982):

$^1$H  $(90_x - \Delta - \quad 180_y - \Delta - 90_x)$

$^{13}$C  $\quad\quad\quad\quad (180_x)$

inverts protons bonded to carbon-12, leaving those bonded to carbon-13
unchanged (Bax and Subramanian, 1986). The delay $\Delta$ is optimized for the
average zero-crossing time of the inverted protons, to minimize their
signal. Choosing a single best value for this delay may not be possible in
molecules with several functional groups, such as glycopeptides or lipopoly-
saccharides, which usually contain protons having different $T_1$ values. Then
a compromise delay value must be used. Insertion of a BIRD pulse does not
work well for samples of larger molecular weights because of the loss of
signal during the delay due to the negative NOE effect when $\omega\tau_c > 1$. In
that case, the unwanted signals are removed by phase cycling. The inverse-
detected HMQC sequence optimized for larger molecules is described by
Sklenář and Bax (1986).

Relayed coherence transfer experiments can be used to establish a
connection between nuclei with a common coupling partner, even if they are
not directly coupled to each other. These remote $^1$H–$^{13}$C connectivities can

be studied by combining heteronuclear correlation with homonuclear relay, or vice versa, in various possible coherence transfer pathways, such as $^1$H – $^1$H– X. Different paths offer different advantages, but the path X – $^1$H – $^1$H, described by Field and Messerle (1986) for $^{31}$P – $^1$H – $^1$H correlation in phosphorylated monosaccharides, offers the sensitivity of inverse detection. The sensitivity is further increased when the final $^1$H-$^1$H transfer step is via an MLEV–17–based HOHAHA method (Lerner and Bax, 1986).

For most monosaccharides, the combination of a one–bond and a relay $^{13}$C–$^1$H map will suffice for complete $^1$H and $^{13}$C assignment. But for oligosaccharides, and for non–protonated carbons, correlations via long–range scalar coupling are necessary. A long–range $^1$H–$^{13}$C map provides information that may be redundant with the relay map, with lower sensitivity. However, it can be used to identify glycosidic linkages via $^3J_{COCH}$. Also, assignment of non–protonated carbons such as carbonyls can be obtained from a long–range map. Bax and Summers (1986) developed the HMBC method (Heteronuclear Multiple Bond Correlation) for obtaining chemical shift correlations between long–range scalar coupled $^1$H–$^{13}$C pairs. Signals arising from directly–bonded $^1$H–$^{13}$C pairs can be suppressed by insertion of a low–pass J–filter (Kogler et al., 1983). An example of its application is shown in Fig. 3.

To improve the resolution and sensitivity of the HMBC experiment, Bax and Marion (1988) proposed a "mixed–mode" data–processing procedure which yields a phase–sensitive (absorptive) spectrum in the $F_1$($^{13}$C) dimension and an absolute magnitude spectrum in the $F_2$ ($^1$H) dimension. Recently, Williamson et al. (1989) modified the HMBC sequence to obtain spectra that were phase–sensitive in both dimensions for molecules with small (~2 Hz) $^1$H–$^1$H scalar couplings.

There are other heteronuclear methods that can be used to sequence oligosaccharides. Bax et al. (1984) demonstrated the use of a selective INEPT sequence to show the position of linkages between sugar residues in a disaccharide, trisaccharide, and polysaccharide. In the same reference, they demonstrated a one–dimensional heteronuclear shift correlation experiment, the decoupled SPT experiment, for completing assignment of a partially–assigned $^{13}$C spectrum. Batta and Liptak (1985) used a 2D DEPT experiment to highlight $^3J_{COCH}$ couplings for identification of the interglycosidic linkages in a trisaccharide.

Inverse–detected experiments are an important tool for analysis of oligosaccharides, because many oligosaccharides of biological importance are not available in the amounts required for conventional chemical shift correlation spectroscopy. There is a theoretical gain of thirty–two in terms of experimental time if magnetization is transferred from $^1$H to $^{13}$C and thence to $^1$H again (Ernst et al., 1987). In practical terms, this means

22

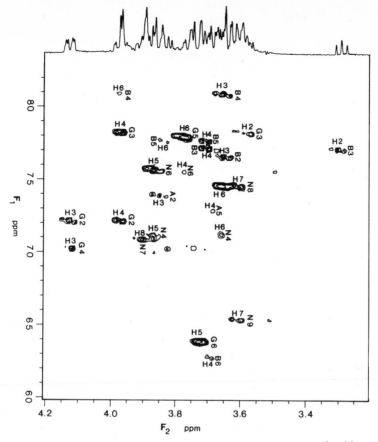

Fig. 3. Portion of chemical shift correlation map for $^1$H–$^{13}$C pairs remotely bonded in the trisaccharide, α–Neu5Ac–(2→3)–β–Gal–(1→4)–Glc (same sample as in Fig. 1), generated using the pulse sequence described by Bax and Summers, 1986. Acquisition parameters are described by Lerner and Bax, 1987. Each crosspeak arises from long-range (multiple bond) scalar coupling between $^1$H and $^{13}$C. Crosspeaks are labelled horizontally with the $^{13}$C assignment and vertically with the $^1$H assignment. Note the excellent suppression of one-bond crosspeaks (shown in Fig. 2).

being able to obtain a one-bond $^1$H–$^{13}$C correlation map, a relay map, and a long-range $^1$H–$^{13}$C map in less than a twenty-four hour period on a few mM sample. (The author makes no estimate of the time necessary to analyze the data!) To achieve maximal sensitivity, it is important to optimize pulse lengths and delays during the pulse sequence. It is also important to have excellent spectrometer stability and clean high power heteronuclear de-coupling during acquisition. The hardware and pulse sequences for this class of experiments are now available with commercial high field spectrometers.

ACKNOWLEDGMENTS

The author thanks the Arthritis Foundation for its continued financial support and A. Bax for his invaluable expertise and advice.

REFERENCES

Batta, G., and Liptak, A., 1985, J. Chem. Soc., Chem. Commun., 1985:368–370.
Bauer, C., Freeman, R., Frenkiel, T., Keeler, J., and Shaka, A. J., 1984, J. Magn. Reson., 58:442–457.
Bax, A., 1982, "Two-dimensional Nuclear Magnetic Resonance in Liquids," D. Reidel Publishing Company, Boston.
Bax, A., Griffey, R. H., and Hawkins, B. L., 1983, J. Magn. Reson., 55:301–305.
Bax, A., Egan, W., and Kovac, P., 1984, J. Carbohydr. Chem., 3:593–611.
Bax, A., and Davis, D. G., 1985, J. Magn. Reson., 65:355–360.
Bax, A., and Subramanian, S., 1986, J. Magn. Reson., 67:565–569.
Bax, A., and Lerner, L., 1988, J. Magn. Reson., 79:429–438.
Bax, A., and Marion, D., 1988, J. Magn. Reson., 78:186–191.
Braunschweiler, L., and Ernst, R. R., 1983, J. Magn. Reson., 53:521–528.
Dabrowski, J., 1987, Ch. 6 in W. R. Croasmun and R. M. K. Carlson (eds.), 1978, "Two-dimensional NMR spectroscopy for chemists and biochemists, Methods in Stereochemical Analysis," v. 9, VCH Publishers, Inc., New York.
Davis, D. G., and Bax, A., 1985, J. Am. Chem. Soc., 107:7197–7198.
Eich, G., Bodenhausen, G., and Ernst, R. R., 1982, J. Am. Chem. Soc., 104:3732–372.
Ernst, R. R., Bodenhausen, G., and Wokaun, A., 1987, "Principles of Nuclear Magnetic Resonance in One and Two Dimensions," Oxford University Press, Oxford.
Fesik, S. W., Gampe, Jr., R. T., and Zuiderweg, E. R. P., 1989, J. Am. Chem. Soc., 111:770–772.
Field, L. D., and Messerle, B. A., 1986, J. Magn. Reson., 66:483–490.
Garbow, J. R., Weitekamp, D. P., and Pines, A., 1982, Chem. Phys. Lett., 93:504–509.
Griesinger, C., Sorensen, O. W., and Ernst, R. R., 1985, J. Am. Chem. Soc., 107:6394–6396.
Griffey, R. H., and Redfield, A., 1987, Quart. Rev. Biophys., 19:51–82.
Hoffman, R. E., and Davies, D. B., 1988, J. Magn. Reson., 80:337–339.
Homans, S. W., Dwek, R. A., Boyd, J., Mahmoudian, M., Richards, W. G., and Rademacher, T. W., 1986, Biochemistry, 25:6342–6350.
Inagaki, F., Shimada, I., Kohda, D., Suzuki, A., and Bax, A., 1989, J. Magn. Reson., 81:186–190.
Kessler, H., Oschkinat, H., Griesinger, C., and Bermel, W., 1986, J. Magn. Reson., 70:106–133.
Kessler, H., Gehrke, M., and Griesinger, C., 1988, Angew. Chem. Int. Ed. Engl., 27:490–536.
Kogler, H., Sorensen, O. W., Bodenhausen, G., and Ernst, R. R., 1983, J. Magn. Reson., 55:157–163.
Lerner, L., and Bax, A., 1986, J. Magn. Reson., 69:375–380.
Lerner, L., and Bax, A., 1987, Carbohydr. Res., 166:35–46.
Maudsley, A. A., and Ernst, R. R., 1977, Chem. Phys. Lett., 50:368–372.
Morris, G. A., and Hall, L. D., 1981, J. Am. Chem. Soc., 103:4703–4711.
Müller, L., 1979, J. Am. Chem. Soc., 101:4481–4484.
Piantini, U., Sorensen, O. W., and Ernst, R. R., 1982, J. Am. Chem. Soc., 104:6800–6801.
States, D. J., Haberkorn, R. A., and Ruben, D. J., 1982, J. Magn. Reson., 48:286–292.
Sklenář, V., and Bax, A., 1987, J. Magn. Reson., 71:379–383.
Subramanian, S., and Bax, A., 1987, J. Magn. Reson., 71:325–330.

Vliegenthart, J. F. G., Dorland, L., and van Halbeek, H., 1983, <u>Adv. Carbohydr. Chem. Biochem.</u>, 41:209-374.

Vuister, G. W., de Waard, P., Boelens, R., Vliegenthart, J. F. G., and Kaptein, R., 1989, <u>J. Am. Chem. Soc.</u>, 111:772-774.

Williamson, D. S., Smith, R. A., Nagel, D. L., and Cohen, S. M., 1989, <u>J. Magn. Reson.</u>, 82:605-612.

TWO-DIMENSIONAL NMR SPECTRUM EDITING OF CARBOHYDRATES

Bruce Coxon

Center for Analytical Chemistry
National Institute of Standards and Technology
Gaithersburg, MD 20899

INTRODUCTION

Expansion of the application of multiple-pulse NMR methods to chemical analysis has been accompanied by the development of many different NMR spectrum editing techniques that are directed towards selective display of the spectra of various molecular features. One of the most popular one-dimensional (1D) methods of this type has been the Distortionless Enhancement by Polarization Transfer (DEPT) technique (Doddrell et. al., 1982), which is often used to generate carbon-hydrogen multiplicity information for use in automated $^{13}$C database and spectral search programs. In its most common form, (Doddrell et al., 1982; Bendall et al., 1982) the 1D DEPT method involves the acquisition of three spectra at three polarization transfer pulse flip angles ($\theta$) of 45°, 90°, and 135°, followed by the construction of three linear combinations of the spectra which form the $\underline{X}H$, $\underline{X}H_2$, and $\underline{X}H_3$ heteronuclear subspectra ($X = {}^{13}C$ or $^{15}N$, or other nuclei).

A related polarization transfer technique is the Phase Oscillations to MaxiMIze Editing (POMMIE) method (Bulsing et al., 1984; Bulsing and Doddrell, 1985), for which certain advantages over the DEPT technique have been claimed (Bulsing et al., 1984; Bulsing and Doddrell, 1985). The pulse sequence for the 1D POMMIE method is derived from that for the DEPT method by replacement of the polarization transfer pulse ($\theta$) by a pair of contiguous 90° pulses. Variation of the radiofrequency (RF) phase difference ($\phi$) between these latter pulses is used to acquire three different raw spectra, and, because the theoretical dependencies of the DEPT and POMMIE methods are identical, the phase angle $\phi$ may take the same set of values that are used for $\theta$ in the DEPT experiment (Bulsing et al., 1984; Bulsing and Doddrell, 1985).

*NMR Applications in Biopolymers,* Edited by
J. W. Finley *et al.,* Plenum Press, New York, 1990

We have been interested in the experimental development and optimization of two-dimensional (2D) NMR spectrum editing methods that offer the possibility of simplification of the 2D NMR spectra of complex biomolecules by generation of separate 2D subspectra for different chemical groups, for example, the CH, $CH_2$, and $CH_3$ structural types. The display of separate 2D subspectra may facilitate the interpretation of complex 2D NMR spectra, particularly in cases where the 2D resonances of different structural types are coincident, or overlapped. Three-dimensional (3D) NMR spectroscopy (Bodenhausen and Ernst, 1981; Griesinger et al., 1987) is a recently available alternative for the simplification of 2D NMR spectra, but with the penalty of substantially increased hardware and software requirements, and data acquisition and processing times.

Selective 2D DEPT heteronuclear shift correlation spectroscopy (Levitt et al., 1983; Nakashima et al., 1984a, 1984b) is an early example of a 2D DEPT spectrum editing technique, which was demonstrated by generation of separate 2D $^{1}$H-$^{13}$C chemical shift correlation maps for the C̲H, C̲H$_2$, and C̲H$_3$ groups in 2-butanol (Nakashima et al., 1984a), and cholesterol (Nakashima et al., 1984b). The production of these separate 2D subspectra basically requires the construction of linear combinations of three 2D data sets, a process that under certain conditions may be achieved during data acquisition (Nakashima et al., 1984a), or failing that, by post-acquisition data processing, using either the resident NMR program (Nakashima et al., 1984b; Coxon, 1985), or specially written software (Coxon, 1985, 1986). An important facet of the selective 2D DEPT CH shift correlation experiments (Nakashima et al., 1984a, 1984b) and similar work in our own laboratories (Coxon, 1983, 1985, 1986) was the recognition that the use of a modified set of values θ = 30°, 90°, and 150° permits the construction of 2D subspectra as simple 1:1 combinations of the 2D raw data sets, whereas the combination of 2D (or 1D) data sets acquired by use of θ = 45°, 90°, and 135° requires a minimum of one nonunit coefficient (Doddrell et al., 1982; Bendall et al., 1982).

For 1D DEPT NMR spectrum editing, linear combinations of the three raw spectra have commonly been constructed by use of several nonunit coefficients, in order to compensate for imperfections in pulse timing and homogeneity, probehead tuning, pulse width calibration, and solvent dielectric (Doddrell et al., 1982; Bendall et al., 1982). For 2D NMR spectrum editing, this approach was perceived to be more of a problem, because typical data acquisition systems, high level NMR microprograms, and the earlier Pascal compilers were not well suited to performing arithmetic computations on fractions of 2D data matrices.

Our initial studies of 2D NMR spectrum editing methods focused on the 2D DEPT J̲(CH)-resolved technique (Coxon, 1983, 1985, 1986). Methods were developed for the automated acquisition of three phase-comparable, data matrices, using the values $\theta$ = 30°, 90°, and 150° (Coxon, 1985, 1986). Two software methods were investigated for the automated construction of J-resolved subspectra from 1:1 combinations of the matrices, using either a sequence of high-level microprograms to combine the matrices by sequential handling of files, or a more efficient Pascal program that computes the $\underline{CH}_2$ and $\underline{CH}_3$ 2D subspectra within a single compound statement (Coxon, 1985, 1986). The methods were tested by application to methyl 2,3-anhydro-4,6-$\underline{O}$-benzylidene-$\alpha$-$\underline{\underline{D}}$-mannopyranoside (Coxon, 1986).

In subsequent work, a pulse sequence for 2D POMMIE $\underline{J}$(CH)-resolved $^{13}$C NMR spectrum editing was used to explore three methods for the automated acquisition of data for 2D spectrum editing (Coxon, 1988). Our interest in the POMMIE technique was stimulated by the reported advantages of this technique over the DEPT method, including (a) less sensitivity to RF pulse inhomogeneity because of the more extensive phase cycling that is possible with POMMIE, (b) better $^{13}$CH suppression, i.e. reduced error signals in POMMIE subspectral editing, (c) better ability to edit in a number of different ways, and (d) more independence of the pulse phase in POMMIE (to a first approximation) from both RF homogeneity and pulse power (Bulsing et al., 1984; Bulsing and Doddrell, 1985).

The first two 2D POMMIE $\underline{J}$(CH)-resolved methods involved the acquisition of sets of three 2D data matrices in either sequential or interleaved modes, by use of the three values $\phi$ = 30°, 90°, and 150° (Coxon, 1988). The required 2D subspectra were computed as 1:1 linear combinations of the three raw data matrices by use of the same Pascal program (written in fixed point arithmetic) that had been used (Coxon, 1985, 1986) for the 2D DEPT analog of the technique. In a third method, the subspectra were constructed directly during acquisition, by rotation of the phase shifts of the POMMIE data read pulse and the spectrometer receiver (Coxon, 1988). The latter method is expected to be the least sensitive to long-term variations in sample or spectrometer stability, but has the disadvantage that the $\underline{CH}$ components of the 2D data are acquired twice, and, as will be seen, the method does not readily allow the use of nonunit coefficients. The three methods were tested by application to a series of carbohydrate derivatives and small peptides (Coxon, 1988).

While the use of 1:1 linear combinations provided a convenient entry into the computation of 2D NMR subspectra, either by use of high-level NMR microprograms (Nakashima et al., 1984b; Coxon, 1985, 1986) or Pascal programs (Coxon, 1985, 1986, 1988), further work with a broader range of compounds has revealed many cases where the residual (edited) $^{13}$C signals in the subspectra are unacceptably large, even when a 5 mm probehead of relatively good RF homogeneity is used. This was particularly evident when the

2D POMMIE carbon-proton shift correlation spectrum editing technique was studied. For this reason, our recent work has included investigations of the use of nonunit coefficients in 2D NMR spectrum editing computations, and modified Pascal programs that include either two, four, or six floating point coefficients have been written for this purpose. In this work, the number of floating point coefficients in the NMR spectrum editing computation has been increased gradually, to assess the effect on the computation time. The use of such coefficients resembles techniques that were developed earlier for 1D DEPT spectrum editing. However, the algorithms for 2D spectrum editing have now been made slightly more general, and recent improvements in NMR hardware and software have facilitated the use of floating point computations.

During this work, certain deviations of the POMMIE method from theoretical behavior have been encountered, and these have been studied by the generation of $^{13}$C intensity-phase shift $\phi$ dependence curves under various experimental conditions. The main thrust of this work has been in NMR spectrum editing for the 2D POMMIE J(CH)-resolved, CH chemical shift correlation, and CH chemical shift correlation with BIlinear Rotation Decoupling (BIRD) techniques. Carbohydrate derivatives have been used as model compounds for these investigations because of their well known advantages in NMR studies.

METHODS

The pulse sequence used for 2D POMMIE heteronuclear J(CH)-resolved $^{13}$C NMR spectrum editing is shown in Fig. 1. The sequence is similar to that for 2D DEPT J(CH)-resolved $^{13}$C NMR spectroscopy (Coxon, 1983, 1985, 1986), except that the $\theta$ pulse in the DEPT sequence is replaced by a pair of adjacent 90° pulses. The second 90° pulse in this pulse sandwich [known as the multiple quantum (MQ) read pulse] is phase shifted by an angle $\phi$ from the first 90° pulse, which has been described as the MQ formation pulse (Bulsing et al., 1984; Bulsing and Doddrell, 1985). This sequence contains a delay, D, that is set equal to the combined widths of the MQ formation and read pulses, i.e. equal to the $^1$H 180° pulse width, in order to equalize the $^{13}$C dephasing and refocusing periods. All of the spectra shown in the Figures were acquired by use of a standard $^1$H decoupler (power $\simeq$ 10 W, $^1$H 90° pulse width $\simeq$ 26 μs) at 400 MHz. "Standard" here refers to a commercial decoupler unit delivered in 1980. Recent commercial decoupler models have more power, but were not available for this work. Studies of the dependence of the POMMIE $^{13}$C intensities of methyl 2,3-di-O-methanesulfonyl-α-D-glucopyrano-side on the phase shift $\phi$ have been conducted for several different 2D POMMIE pulse sequences. This was done by converting the 2D sequences to 1D sequences by replacing the incremented delays ($t_1$/2) in the 2D sequences

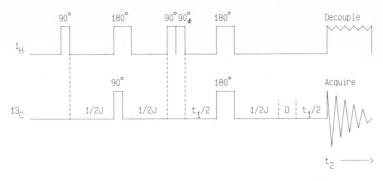

Fig. 1.   Pulse sequence for 2D POMMIE J(CH)-resolved $^{13}$C NMR
          spectroscopy.  An average value of the one-bond $^{13}$CH
          coupling constant J was used to calculate the 1/2J delay.
          The delay D is equal to the width of the $^{1}$H 180° pulse.


with fixed delays.  A series of 97 1D POMMIE $^{13}$C NMR spectra was acquired by
incrementing the value of $\phi$ by 2° from 0° to 192°.  The data from the phase
incremented spectra were used to generate the $^{13}$C intensity-$\phi$ dependence
curves.  A similar set of data was collected for the DEPT method by
incrementing the pulse width $\theta$ in a 1D DEPT experiment by 2.1° (0.6 µs)
from 2.1° to $\simeq$ 205°.

    A number of different strategies may be considered for the optimization
of 2D POMMIE spectrum editing experiments.  During data acquisition, the set
of three values of $\phi$ can be adjusted to alternative values, as might also
the relative widths of the MQ formation and read pulses.  The latter possibility arises because the amplitudes and phases of signals obtained by use
of values of $\phi$ in the first and second quadrants have some dependence on
these pulse widths.  Proton or $^{13}$C pulse power might also be varied.

    Other techniques may be applied during data processing.  For example,
linear combinations of the 2D data matrices could be constructed with
nonunit coefficients, to give more exact subtraction of signals.  Iterative
adjustments of these coefficients [the Varian[*] approach to 1D DEPT spectrum
editing (Richarz et al., 1982)] could also be made, but for computational
reasons, this method has not yet been applied to 2D spectrum editing.

    One particular problem that we have encountered in 2D editing studies
of carbohydrate derivatives, is that isolated methyl $^{13}$C nuclei, such as

[*] Certain commercial equipment, instruments, or materials are identified
    in this paper to specify adequately the experimental procedure.  Such
    identification does not imply recommendation by the National Institute
    of Standards and Technology, nor does it imply that the materials or
    equipment are necessarily the best available for the purpose.

those in acetyl, methanesulfonyl, and methoxyl substituents, tend to have $T_2$ values that are longer than those of the CH and $CH_2$ carbons. In this situation, residues of the narrow $\underline{C}H_3$ resonances in the edited 2D subspectra tend to be more obtrusive than those of the $\underline{C}H$ and $\underline{C}H_2$ signals. To counter this effect, one can discriminate against $CH_3$ carbons with long $T_2$ values by (a) acquiring data for a shorter time in the $\underline{t}_1$ domain by collecting fewer spectra in the 2D data set, and (b) filtering the data through a window function that suppresses the latter part of the free induction decay (FID), for example, a sine-bell squared function with an offset of $\pi/2$ rad. Method (a) is of limited utility, because it results in loss of resolution in the $\underline{F}_1$ dimension, and, therefore, we have used method (b) exclusively.

Our observation of $^{13}C$ intensity-$\phi$ dependence curves that were shifted to higher values of $\phi$ under certain conditions (see RESULTS AND DISCUSSION) led us to investigate 2D POMMIE data acquisitions in which the three experimental values of $\phi$ were varied, in an effort to compensate for the rightward displacement of the curves. In these experiments, the chief criterion used for the selection of a new set of values for $\phi$ was equality of the methyl $^{13}C$ intensities in spectra obtained by use of values of $\phi$ in the two quadrants $0° < \phi < 90°$ and $90° < \phi < 180°$. This procedure requires a pulse programmer equipped with 1° (or less) phase shifting of the $^1H$ decoupler frequency. Inequalities of signal amplitudes measured by use of the corresponding symmetrically placed $\theta$ or $\phi$ angles in the two quadrants are a common problem in both DEPT (Bendall et al., 1982) and POMMIE experiments, that can (hopefully) be corrected by the more elaborate computation schemes given hereafter. However, inequalities of the phases of signals for the two quadrants in POMMIE experiments have also been observed, and these are more difficult to correct.

The original computation method used for 2D DEPT and 2D POMMIE spectrum editing based on unit coefficients (fixed point arithmetic) in Pascal programs (Coxon, 1985, 1986, 1988) is shown in Scheme 1.

| Scheme 1. | CH | = | Data(90) |
|---|---|---|---|
| Unit coefficients | $CH_2$ | = | Data(30) − Data(150) |
| | $CH_3$ | = | Data(30) + Data(150) − Data(90) |

Modified computation methods for 2D spectrum editing using either two, four, or six floating point coefficients are shown in Schemes 2, 3, and 4, respectively.

| Scheme 2. | CH | = | Data(90) |
|---|---|---|---|
| Two coefficients | $CH_2$ | = | Data(30) − Data(150) − a.Data(90) |
| | $CH_3$ | = | Data(30) + Data(150) − $\underline{b}$.Data(90) |

| Scheme 3. | CH | = | Data(90) |
|---|---|---|---|
| Four coefficients | $CH_2$ | = | Data(30) − a.Data(150) − b.Data(90) |
| | $CH_3$ | = | Data(30) + $\underline{c}$.Data(150) − $\underline{d}$.Data(90) |

Scheme 4.           $CH_2$  =  Data(30) − $\underline{a}$.Data(150) − $\underline{b}$.Data(90)
Six coefficients     $CH_3$  =  Data(30) + $\underline{c}$.Data(150) − $\underline{d}$.Data(90)
                     $CH$    =  Data(90) − $\underline{e}$.$CH_2$ − $\underline{f}$.$CH_3$

In these schemes, Data($\phi$) refers to a 2D POMMIE data matrix acquired by use of an MQ read pulse phase shift angle of $\phi$ (in degrees). However, the values of $\phi$ shown in the expressions are only generic labels, meaning that in practice, values in the general vicinity of $\phi$ were used; e.g., $\phi$ = 40°, 96°, and 152°.

In Scheme 4, the $\underline{C}H_2$ and $\underline{C}H_3$ residues in the $\underline{C}H$ subspectrum are minimized by subtraction of small fractions of the $\underline{C}H_2$ and $\underline{C}H_3$ subspectra that have just been computed. For computational efficiency, this correction is done while the appropriate segments of the subspectra are still in memory. The values of up to six floating point coefficients ($\underline{a}$-$\underline{f}$) were determined by a separate 1D calibration experiment that was usually run immediately after the acquisition of the three 2D data matrices. For calibration of the constants, only the first row of each matrix is acquired, but with the number of scans increased typically by a factor of 16, in order to obtain good signal:noise ratio.

Since the presentation of our work at the symposium, a number of improvements have been incorporated in the Pascal program for the floating point computations. These changes include (a) the six coefficients are entered by keyboard at program start, (b) the 2D data matrix dimensions (size of FID x number of spectra acquired) are read automatically from the −1 sector of the (Bruker) data set, and are printed for inspection, and (c) the compile time option :H (use the Bruker firmware floating point package) is used to speed the floating point computations.

Two different pulse sequences that we have used for 2D POMMIE carbon-proton chemical shift correlation spectrum editing are shown in Fig. 2 and Fig. 3. Like the sequence for the $\underline{J}$(CH)-resolved technique (see Fig. 1),

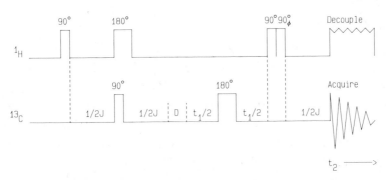

Fig. 2.  Pulse sequence for 2D POMMIE carbon-proton chemical shift correlation [13]C NMR spectroscopy. An average value of the 1/2$\underline{J}$ delay was used. The delay $\underline{D}$ is equal to the width of the [1]H 180° pulse.

33

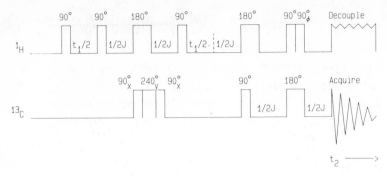

Fig. 3.  Pulse sequence for 2D POMMIE carbon–proton chemical shift
correlation $^{13}$C NMR spectroscopy with BIRD $^1$H–$^1$H
decoupling.  An average value of the 1/2J delay was used.
The phases of the components of the composite $^{13}$C 180°
pulse are $\underline{x}$ = 0° and $\underline{y}$ = 90°.


the sequence in Fig. 2 contains a delay $\underline{D}$ equal to the width of the $^1$H 180°
pulse, in order to equalize the $^{13}$C dephasing and refocusing periods.  This
sequence produces CH chemical shift correlation spectra in which all of the
$^1$H–$^1$H coupling constants are present in the $\underline{F}_1$ dimension, although limited
digital resolution in this dimension may cause them not to be resolved.

The sequence in Fig. 3 contains a BIlinear Rotation Decoupling (BIRD)
unit for $^1$H–$^1$H decoupling in the $\underline{F}_1$ dimension (Bax, 1983), and also a
composite 180° $^{13}$C pulse for improvement of $^{13}$C spin inversion.  This
sequence yields CH chemical shift correlation spectra in which the $^1$H–$^1$H
coupling constants of nongeminal protons are removed in the $\underline{F}_1$ dimension.
This type of sequence is less suitable for larger molecules than for smaller
ones, because it contains five 1/2J delays, which for molecules with short
$\underline{T}_2$ values offer the opportunity for the transverse magnetization to decay
significantly, before the sequence is finished.


RESULTS AND DISCUSSION

Some results of a POMMIE $^{13}$C intensity-phase shift $\phi$ dependence study
for selected carbon atoms of methyl 2,3-di-$\underline{O}$-methanesulfonyl-$\alpha$-$\underline{D}$-gluco-
pyranoside at standard decoupler pulse power ($\simeq$ 10 W) are shown in Fig. 4,
in which the intensity dependence curves for C-3, C-6, and an SCH$_3$ are all
displaced to the right.  The results for a series of such experiments in
terms of the phase angles $\phi$ at which absolute value maximum and minimum $^{13}$C
intensities were observed for $\underline{C}$H, $\underline{C}$H$_2$, and $\underline{C}$H$_3$ signals are summarized in
Table 1, and may be compared with the results of a pulse width incremented
1D DEPT experiment, and with the turning points expected from theory
(Doddrell et al., 1982; Bendall et al., 1982; Bulsing et al., 1984; Bulsing
and Doddrell, 1985). Within experimental error, the DEPT results do not

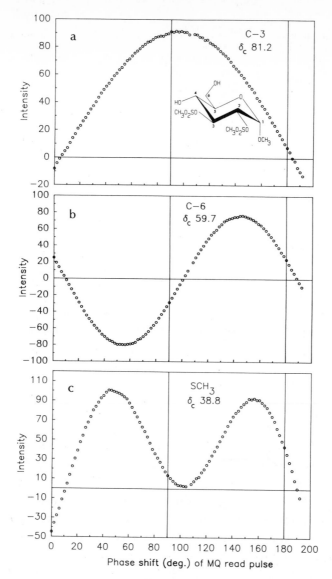

Fig. 4.  Dependence of POMMIE $^{13}$C intensities on the phase shift $\phi$
of the multiple quantum read pulse, for selected resonances
of methyl 2,3-di-O-methanesulfonyl-α-D-glucopyranoside,
(a) CH, (b) CH$_2$, and (c) CH$_3$.  Standard decoupler pulse
power ($\simeq$ 10 W) was used (see text).

deviate significantly from theory.  However, for the POMMIE experiments, a
significant range of deviations was observed.

2D POMMIE J(CH)-Resolved Spectrum Editing

Contour plots of the 2D POMMIE J(CH)-resolved CH, CH$_2$, and CH$_3$ $^{13}$C
subspectra that were generated for an artificial mixture of seven methyl
hexosides in dimethylsulfoxide-d$_6$ (DMSO-d$_6$) solution are displayed in

Table 1. Dependence of $^{13}$C Intensity Maxima and Minima on Pulse Flip
Angle ($\theta$) and Multiple Quantum Read Pulse Phase Angle ($\phi$)
in DEPT and POMMIE Spectra of Methyl 2,3-di-O-methane-
sulfonyl-$\alpha$-D-glucopyranoside

|  | CH max.[a] | $CH_2$ max. | $CH_2$ max. | $CH_2$ min. | $CH_3$ max. | $CH_3$ max. | $CH_3$ min. |
|---|---|---|---|---|---|---|---|
| Theory[b] ($\theta, \phi$) | 90 | 45 | 135 | 90 | 35.3 | 144.7 | 90 |
| DEPT[c] ($\theta$) | 91-97 | 46.6 | 135.5 | 91 | 36-38 | 148 | 91-93 |
| POMMIE[d] ($\phi$) | 86-104 | 50-56 | 140-146 | 95-100 | 38-50 | 148-158 | 95-104 |

[a] Angles (in degrees) for absolute value maximum and minimum $^{13}$C
intensities. Minimum values (zero crossing points) near to 0° or
180° are not reported, because these are of less interest in the
editing experiments. The solvent was dimethylsulfoxide-$d_6$.

[b] Calculated by taking derivatives of the functions CH = $k \cdot \sin x$, $CH_2$ =
$k \cdot \sin 2x$, $CH_3$ = $3k(\sin x + \sin 3x)/4$, where $k$ = $\gamma_H/\gamma_C$, and $x$ = $\theta$ or $\phi$.

[c] From pulse width incremented 1D DEPT experiment ($\delta\theta$ = 2.12°).

[d] From phase incremented 1D POMMIE experiments ($\delta\phi$ = 2°), using
standard proton decoupler, $^1$H 90° ~ 26 µs. The shift in the curves
depends on the pulse power at the proton frequency.

Figs. 5a, 5b, and 5c, respectively, and may be compared with the total
(unedited) spectrum shown in Fig. 5d. The raw data were acquired by use of
the phase shift set $\phi$ = 40°/96°/152°, and the linear combinations were
constructed according to Scheme 2 (two floating point coefficients).
This level of computation was sufficient in this case to give good
suppression of residual signals in the subspectra. The types of carbon
atoms may be identified readily from the multiplicities of the J-spectra,
and although not relevant to this particular example, the method should be
able to separate overlapped J-spectra of different carbon types, i.e., even
those that have the same chemical shift.

2D POMMIE J(CH)-resolved $^{13}$C subspectra for the more complex oleandomycin molecule are shown in Figs. 6a-c. In this case, the raw data were
measured with the phase shift set $\phi$ = 45°/105°/145°, and the linear combinations were computed by means of Scheme 4, with six floating point
coefficients. For this experiment, typical values of the coefficients were
$a$ = 1.02, $b$ = 0.15, $c$ = 0.88, $d$ = 1.22, $e$ = -0.1, and $f$ = 0.12. The almost
coincident CH and $CH_3$ J-spectra at ~ 40 ppm in the total spectrum ($\phi$ = 150°)
shown in Fig. 6d are distinguished in the CH and $CH_3$ subspectra displayed in
Figs. 6a and 6c, respectively.

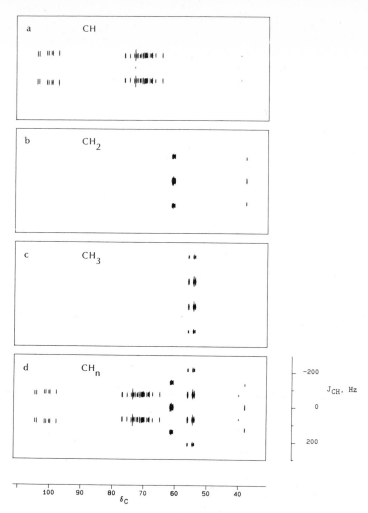

Fig. 5.  Contour plots of the C̲H, C̲H$_2$, and C̲H$_3$ 2D POMMIE
         J̲(CH)-resolved $^{13}$C NMR subspectra of a mixture of methyl
         α-D-glucopyranoside, methyl β-D-glucopyranoside, methyl
         α-D̄-mannopyranoside, methyl α-D̄-galactopyranoside, methyl
         β-D̄-galactopyranoside, methyl ᾱ-D-altropyranoside, and
         meᵗhyl 2-deoxy-α-D-arabino-hexopȳranoside (0.1 mole/L each
         in DMSO-d$_6$).  The data were acquired by use of MQ read
         pulse phase shifts φ = 40°/96°/152° and were combined with
         two floating point coefficients.  The C̲H$_n$ spectrum in (d)
         represents the data from φ = 152°.

## 2D POMMIE CH Chemical Shift Correlation Spectrum Editing

     This technique was initially studied by using a solution of the simple
model compound: methyl α-D-glucopyranoside tetraacetate in DMSO-d$_6$.  The 2D
POMMIE shift correlation sequence shown in Fig. 3 was used with the phase
shift values φ = 30°/90°/150° to acquire three raw data sets, from which 1:1
linear combinations were computed according to Scheme 1.  Contour plots of

37

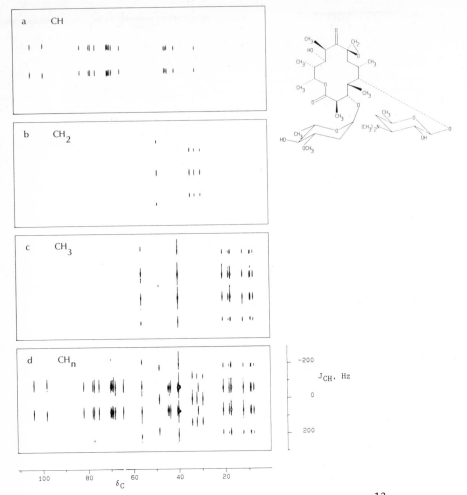

Fig. 6.   Contour plots of the 2D POMMIE J(CH)-resolved $^{13}$C NMR
subspectra of a 0.2 mole/L solution of oleandomycin in
DMSO-$d_6$.   The data were acquired by using $\phi$ = 45°/105°/
145°, and were combined with six floating point coefficients.   The complete spectrum in (d) represents the data
from $\phi$ = 145°.

the resulting 2D POMMIE CH chemical shift correlated CH, CH$_2$, and CH$_3$
subspectra are shown in Figs. 7a, 7b, and 7c, respectively, and may be
compared with the complete spectrum shown in Fig. 7d, which represents the
data for $\phi$ = 150°.   Because of the limited digital resolution (6.25
Hz/point) available in the $F_1$ ($^1$H) dimension and the fact that the chemical
shifts of H-6 and H-6′ are quite similar, these protons are correlated with
C-6 by a single crosspeak in the CH$_2$ subspectrum shown in Fig. 7b.   For this
example, a measure of the quality of the spectral editing is indicated by
the low residues of edited signals in the $F_2$ projections of the 2D POMMIE CH

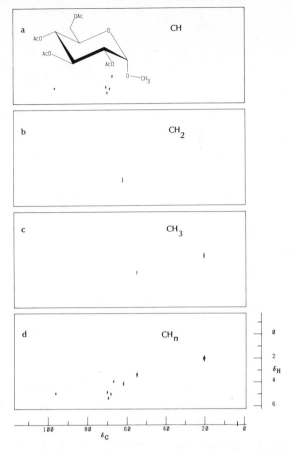

Fig. 7.  Contour plots of the $\underline{C}H$, $\underline{C}H_2$, and $\underline{C}H_3$ 2D POMMIE carbon-proton chemical shift correlation $^{13}C$ NMR subspectra of methyl $\alpha$-D-glucopyranoside tetraacetate (1 mole/L in DMSO-$\underline{d}_6$). The data were obtained with $\phi$ = 30°/90°/150°, and were combined with unit coefficients.  The $\underline{C}H_n$ spectrum in (d) represents the data from $\phi$ = 150°.

chemical shift correlated $\underline{C}H$, $\underline{C}H_2$, and $\underline{C}H_3$ subspectra (see Figs. 8a, 8b, and 8c, respectively).  In the complete correlation spectrum projection shown in Fig. 8d (obtained with $\phi$ = 150°), it may be seen that the methyl carbon signals are substantially more intense than the methine and methylene carbon resonances.  The signal at highest field is undoubtedly the most intense, because it contains four, almost coincident acetyl methyl carbon resonances.  However, the strong intensity of the methoxyl carbon resonance at $\delta_c$ 55 ppm deserves comment.

According to the theoretical dependences (Bulsing et al., 1984; Bulsing and Doddrell, 1985) $\underline{C}H = \underline{k} \cdot \sin \phi$, $\underline{C}H_2 = \underline{k} \cdot \sin 2\phi$, and $\underline{C}H_3 = 3\underline{k}(\sin \phi + \sin 3\phi)/4$ ($\underline{k} = \gamma_H/\gamma_C$), methyl carbon signals measured by the POMMIE method at $\phi$

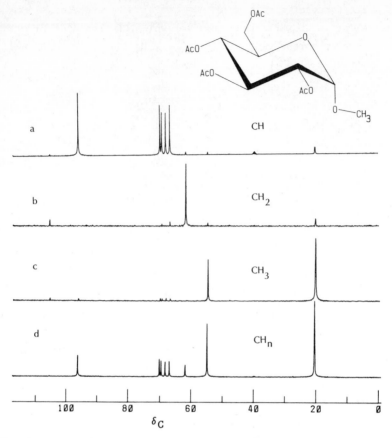

Fig. 8. $\underline{F}_2$ projections of the 2D POMMIE CH chemical shift correlation $^{13}C$ subspectra of methyl $\alpha$-D-glucopyranoside tetraacetate shown in Fig. 7.

= 150° should have intensities 1.125/0.5 = 2.25 times greater than those of the methine carbon resonances, and 1.125/0.866 = 1.30 times greater than the methylene carbon intensities. The experimental peak height ratios for the data in Fig. 8d range from $OCH_3/CH$ = 2.6–4.4 to $OCH_3/CH_2$ = 5.0. Whilst the low intensity of the $CH_2$ projection is explicable in terms of slight non-equivalence of the H-6 and H-6' protons, the greater than expected dominance of the methoxyl carbon signal may be attributed to a longer $\underline{T}_2$ (narrower resonance) of this carbon nucleus, in spite of the suppressive filtering that was applied.

Unfortunately, the CH shift correlated editing results obtained for methyl $\alpha$-D-glucopyranoside tetraacetate by the simple $\phi$ = 30°/90°/150° --Scheme 1 method turned out to be atypically good. For other more complex molecules, it became apparent that more elaborate methods based on modification of the $\phi$ angle set and/or the use of nonunit coefficients in the linear combinations of the data sets were especially needed for 2D POMMIE CH

40

chemical shift correlated editing, and might also be expected to enhance the analogous selective 2D DEPT technique, for which $CH_3$ residues in $CH_2$ subspectra have been reported (Nakashima et al., 1984).

Some results from more elaborate data acquisition and processing are shown in Fig. 9, which displays the 2D POMMIE CH shift correlated $CH$, $CH_2$, and $CH_3$ subspectra of oleandomycin that were obtained by using the pulse sequence shown in Fig. 3 with $\phi$ = 48°/98°/158°, followed by computation of linear combinations according to Scheme 3 (four floating point coefficients). Inspection of the $CH_2$ subspectrum (Fig. 9b) reveals that three out

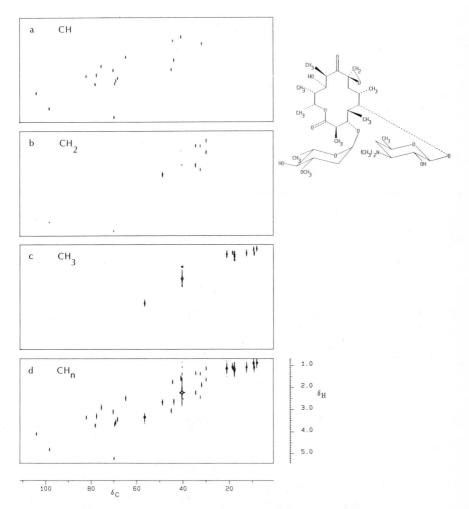

Fig. 9.  Contour plots of the 2D POMMIE CH chemical shift correlation subspectra of oleandomycin (0.2 mole/L in DMSO-$d_6$). The raw data were acquired by use of $\phi$ = 48°/98°/158°, and were combined with four floating point coefficients. The complete spectrum in (d) represents the data from $\phi$ = 158°.

of four of the methylene carbon atoms of oleandomycin (the three at higher field) display their correlation signals as apparent doublets (more lines may be seen under expansion, higher resolution, or proton decoupling in the $F_1$ dimension) (Nakashima et al., 1984; Bain et al., 1988).

2D POMMIE CH chemical shift correlation CH, $CH_2$, and $CH_3$ subspectra of oleandomycin that were obtained with BIRD $^1H$–$^1H$ decoupling in the $F_1$ dimension [as well as the usual WALTZ–16 composite pulse $^1H$ decoupling (Shaka et al., 1983) in the $F_2$ dimension] are shown in Figs. 10a, 10b, and 10c, respectively, and may be compared with the total spectrum in Fig. 10d

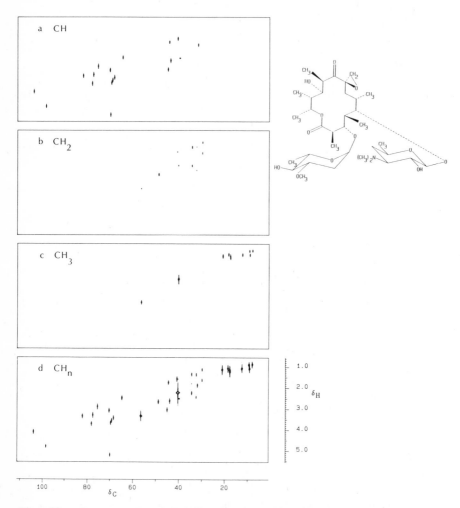

Fig. 10. Contour plots of the 2D POMMIE CH chemical shift correlation subspectra of oleandomycin (0.2 mole/L in DMSO-$d_6$) acquired by use of $\phi$ = 42°/95°/150° with BIRD $^1H$–$^1H$ decoupling, and processing with four floating point coefficients. The $CH_n$ spectrum in (d) is the data from $\phi$ = 150°.

that was measured with $\phi = 150°$. These data were acquired by using the pulse sequence in Fig. 3 with the phase angle set $\phi = 42°/95°/150°$, with editing computations according to Scheme 3 with four floating point coefficients. The BIRD subsequence gives homonuclear decoupling in the $F_1$ dimension of only the nongeminal protons, so that the $CH_2$ correlation signals of oleandomycin still appear as three doublets and one singlet in the $CH_2$ subspectrum shown in Fig. 10b. However, comparison of the linewidths of the $CH$ cross peaks in the BIRD subspectrum shown in Fig. 10a, with those in the nondecoupled subspectrum displayed in Fig. 9a revealed significant narrowing of the former cross-peaks.

The N-methyl $^{13}C$ signal of the dimethylamino group of the D-desosamine moiety (Celmer and Hobbs, 1965) of oleandomycin at ~ 40 ppm is exceptionally strong in the total spectrum shown in Fig. 10d, and survives the editing procedure to give noticeable residues in the $CH$ and $CH_2$ subspectra (see Figs. 10a and 10b). Moreover, because of the particular problems associated with the methylene group, the $CH_2$ subspectrum (Fig. 10b) also contains two $CH$ residues and a $CH_3$ residue of the methoxyl group of the L-oleandrose moiety (Celmar and Hobbs, 1965) of oleandomycin. The complete $CH_n$ spectrum shown in Fig. 10d contains an example of a $CH_2$ correlation multiplet with a central peak (Nakashima et al., 1984b; Bain et al., 1988).

The observation of $CH_2$ correlation multiplets for oleandomycin illustrates a general problem in attempts to edit the CH chemical shift correlation spectra of chiral biomolecules. The methylene carbon atoms in these molecules often bear nonequivalent protons that split the $^{13}CH_2$ correlation cross peak into a multiplet. For heteronuclear correlation pulse sequences of the type shown in Fig. 2, this multiplet is commonly an AB quartet. However, sequences that contain the BIRD subsequence (for example, that shown in Fig. 3) may yield an AB quartet, either with, or without one or more additional central lines (Nakashima et al., 1984b; Bain et al., 1988).

It has been proposed that such artifacts in the center of the $CH_2$ shift correlation quartet arise because the BIRD pulse transfers some spin coherence between nonequivalent methylene protons (Bain et al., 1988). Apparently, these artifacts may or may not be observed because of a marked dependence on the chemical shift difference of the methylene protons (Nakashima et al., 1984b), a dependence that has been verified experimentally for a cyclophane derivative, and confirmed by a theoretical analysis (Bain et al., 1988). This analysis indicates that the dependence on the chemical shift difference arises because the artifact signals originate from two sources which can interfere either constructively, or destructively. As the frequency separation of the methylene proton signals increases, the intensities of the artifact peaks pass through a series of alternating maxima and zero points, so that whether or not these peaks are observed depends on the numerical value of the relative shift of the particular pair of methylene protons (Bain et al., 1988). Simulations have shown that the artifact will have fine structure (perhaps as complex as a quartet) due not only to the geminal $^1H$-$^1H$ coupling, but also to any differences between the $^1J_{CH}$ values of the methylene protons (Bain et al., 1988).

Splitting of the $\underline{C}H_2$ correlation signal by these various mechanisms tends to exaggerate the _relative_ intensities of the $\underline{C}H$ and $\underline{C}H_3$ residues in the $\underline{C}H_2$ subspectra by factors of two to five, or more, leading to aesthetically less pleasing edited spectra (see Figs. 9b and 10b).

Generation of a good $\underline{C}H_2$ subspectrum may be especially useful for structural analysis, because inspection of the complete CH shift correlation spectrum (for example, that shown in Fig. 9d) does not immediately indicate whether two cross peaks that happen to have the same $^{13}C$ chemical shift, but different $^1H$ shifts are due to different $^{13}C$ nuclei of any type, or to one $\underline{C}H_2$ group with nonequivalent protons.

Since the symposium, a more detailed study has been made of the dependence of POMMIE $^{13}C$ signal intensities and phases on several different factors, including pulse power at the proton resonance frequency, probe size, and the relative widths of the MQ formation and read pulses. Two notable observations have been (a) the POMMIE intensity-phase shift $\phi$ dependence curves returned to their symmetrical, theoretical positions (about $\phi = 90°$) at high proton pulse power (110 W, $^1H$ 90° pulse width = 8 µs) and (b) phase matching of signals acquired by use of values of $\phi$ in the two quadrants was achieved by lengthening the MQ formation pulse by 1.2 µs and shortening the MQ read pulse by 1.2 µs.

Another optimization strategem investigated recently involves the rationale that since the $\underline{C}H_2$ crosspeaks tend to be the least intense, for the reasons given already, then perhaps the $\phi$ values in the two quadrants should be chosen so as to maximize the $\underline{C}H_2$ intensities. If the intensity-$\phi$ dependence curves exhibit theoretical behavior (as at high proton pulse power), then calculations based on the theory (Bulsing et al., 1984; Bulsing and Doddrell, 1985) indicate that $\underline{C}H_2$ signals are maximal at $\phi = 45°$ and 135°, whereas $\underline{C}H_3$ signals have maxima at $\phi = 35.3°$ and 144.7° (see Table 1). Because the $\underline{C}H_3$ signals acquired by the DEPT and POMMIE methods are normally quite strong, loss of $\underline{C}H_3$ intensity can readily be tolerated.

CONCLUSIONS

Adjustment of the phase shift $\phi$ angle set to values other than 30°/90°/150° or 45°/90°/135° appears to be an appropriate technique for improvement of POMMIE spectrum editing if only standard decoupler power ($\leq 10$ W) is available (as in most unmodified NMR spectrometers manufactured prior to the late 1980s). However, we have found that this adjustment technique does not offer a complete solution to the problem of displaced $^{13}$C intensity-$\phi$ dependence curves, because the displacements appear to be offset dependent. For example, the intensity dependence curves for the various chemically shifted $\underline{C}$H signals of methyl 2,3-di-$\underline{O}$-methanesulfonyl-$\alpha$-$\underline{\underline{D}}$-glucopyranoside have been found to be displaced to different extents. Correction of this problem might require the use of different sets of floating point coefficients for different resonances, which would be computationally difficult.

The author's currently preferred data acquisition protocol for both 2D POMMIE $\underline{J}$(CH)-resolved and 2D POMMIE CH chemical shift correlation spectrum editing is to use (a) high proton pulse power (generated from a booster amplifier with 10.5 db gain after filtering at the $^1$H frequency) to symmetrize the $^{13}$C intensity-$\phi$ dependence curves, (b) the values $\phi = 45°$ and $135°$ to maximize the $CH_2$ intensities, and (c) adjustment of the MQ formation pulse width by +1.2 μs and the MQ read pulse width by −1.2 μs, to equalize signal phases for the two quadrants of $\phi$. Our preferred data processing procedure is to use (a) construction of linear combinations of the three raw data sets by means of floating point computations with six coefficients (Scheme 4), and (b) a sine-bell squared window function with an offset of π/2.

## REFERENCES

Bain, A. D., Hughes, D. W., and Hunter, H. N., 1988, Magn. Reson. Chem., 26:1058.

Bax, A.. 1983, J. Magn. Reson., 53:517.

Bendall, M. R., Doddrell, D. M., Pegg, D. T., and Hull, W. E., 1982, Bruker brochure: "DEPT".

Bodenhausen G., and Ernst, R. R., 1981, J. Magn. Reson., 45:367.

Bulsing, J. M., Brooks, W. M., Field, J., and Doddrell, D. M., 1984, J. Magn. Reson., 56:167.

Bulsing, J. M., and Doddrell, D. M., 1985, J. Magn. Reson., 61:197.

Celmer, W. D., and Hobbs, D. C., 1965, Carbohydr. Res., 1:137.

Coxon, B., 1983, Poster B-33, 24th Experimental NMR Conference, Asilomar, CA, April 10-14.

Coxon, B., 1985, Abstract B-17, 26th Experimental NMR Conference, Asilomar, CA, April 21-25.

Coxon, B., 1986, J. Magn. Reson., 66:230.

Coxon, B., 1988, Magn. Reson. Chem., 26:449.

Doddrell, D. M., Pegg, D. T., and Bendall, M. R., 1982, J. Magn. Reson., 48:323.

Griesinger, C., Sorensen, O. W., and Ernst, R. R., 1987, J. Am. Chem. Soc., 109:7227.

Levitt, M. H., Sorensen, O. W., and Ernst, R. R., 1983, Chem. Phys. Lett., 94:540.

Nakashima, T. T., John, B. K., and McClung, R. E. D., 1984a, J. Magn. Reson., 57:149.

Nakashima, T. T., John, B. K., and McClung, R. E. D., 1984b, J. Magn. Reson., 59:124.

Richarz, R., Ammann, W., and Wirthlin, T., 1982, Varian Application Note: "DEPT", no. Z-15.

Shaka, A. J., Keeler, J., and Freeman, R., 1983, J. Magn. Reson., 53:313.

DETERMINATION OF COMPLEX CARBOHYDRATE STRUCTURE USING CARBONYL CARBON
RESONANCES OF PERACETYLATED DERIVATIVES

Warren J. Goux

Department of Chemistry
University of Texas at Dallas
2601 North Floyd Road-Chem B2.502
P.O. Box 830688
Richardson, TX 75083-0688

Oligosaccharides constitute the most abundant and diverse group of
compounds present in living systems. Their functions range from that of
antigenic determinants, such as the ABO blood-group determinants of man, to
the purported regulation of gene expression in plants (Watkins, 1972;
Darvill and Albersheim, 1984). Their physiological roles appear to be influenced by their tertiary and, ultimately, their primary structure (Brisson
and Carver, 1983; Homans et al., 1986; Carver, 1984; Montreuil, 1980).
Their structural complexity arises from the large number of structurally
unique residues present and from the variety of ring configurations and of
glycosidic linkages that can be made to and from neighboring residues.

In recent years, $^{1}$H NMR spectroscopy has been used to determine the
primary structures of oligosaccharides isolated, usually in limited quantities, from glycolipids, glycoproteins, yeast cell-walls, and the urine of
patients suffering form glycoproteinosis (Vliegenthart et al., 1981; 1983).
As a basis for structural identification, "reporter-group" resonances, most
arising from anomeric protons between 4.0 and 5.5 ppm, are compared with
those observed in spectra of model compounds. Structural details, including
pyranose or furanose ring-structure and glycosidic linkages made to and from
neighboring residues, can be gleaned from proton chemical shifts and coupling constants. Possible complications inherent in the method include
overlap of these reporter-group resonances among themselves or with other
resonances in the spectrum, particularly those arising from the solvent (HOD
or $H_2O$).

$^{13}$C NMR spectroscopy has also been used to determine primary structures
of oligosaccharides and oligosaccharides covalently linked to peptides, proteins, and lipids (Allerhand and Berman, 1984; Dill et al., 1985; Sillerud

et al., 1982).  In a manner analogous to the $^1$H NMR spectral method, structural details are inferred from the shifts of pyranose or furanose ring-carbon atoms.  However, carbon spectra are inherently easier to interpret, owing to the much greater chemical-shift range over which resonances occur and the lack of complexities arising from spin-spin coupling and overlap of resonances with those arising from the solvent.  Unfortunately, these advantages are usually overshadowed by the low sensitivity of the $^{13}$C nucleus, making acquisition of spectra of samples isolated in limited quantities (1-5 mg) difficult or impossible.

Initially our goal was to develop a novel NMR method for structural analysis of complex carbohydrates which shared some advantages of both the $^{13}$C and $^1$H NMR methods.  We decided to look first at the carbonyl carbon resonances of peracetylated forms of mono- and oligosaccharides (Goux, 1988; Goux and Unkefer, 1987).  Our thinking was if $^{13}$C-enriched reagents were used in the preparation of the derivatives, a $^{13}$C NMR spectrum would result which had the simplicity of a carbon spectrum with enhanced sensitivity. The peracetylation reaction can be easily carried out on a few milligrams of sample using [1-$^{13}$C]acetic pivalic anhydride in dry pyridine (Fig. 1). Extraction with chloroform and evaporation of the solvent yields the peracetylated derivative.  The sample can then be redissolved in a deuterated organic solvent for NMR spectral analysis.  Typically such solvents are of low viscosity and rotational diffusion of the sample is faster than in more viscous aqueous solvents.  As a result, spin-spin relaxation times of the molecule of interest are long and resonance linewidths are correspondingly narrower than might be expected for the underivatized sample in an aqueous solvent (Allerhand et al., 1971).

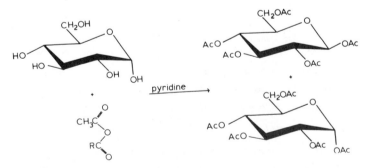

Fig. 1.  Base catalyzed acetylation of mono- and oligo-saccharides generally yields an anomeric mixture of peracetylated derivatives.

Figure 2 shows spectra of peracetylated forms of glucose di- and trisaccharides. The many resonances present in each of the spectra arise from carbonyl carbons substituted at various positions of the pyranoid ring structures. Although these resonances are spread only over about a 2.5 ppm range, there is little overlap of signals. For example, 20 and 22 carbonyl resonances of a peracetylated isomaltotriose anomeric mixture can be resolved (Fig. 2E). Notice that the pattern of resonances changes upon changing the anomeric configuration of one of the two disaccharide residues (compare the spectrum of β-cellobiose, which has a β1->4 glycosidic linkage, to that of β-maltose, which has a α1->4 glycosidic linkage) or upon changing the position of the glycosidic linkage to a neighboring residue (compare the spectrum of β-cellobiose, where the previously reducing residue is substituted at position 4, to β-gentiobiose, where the corresponding substitution is at position 6). Also notice that molecules having similar structural elements appear to have groups of resonances with common chemical shifts. For example, resonances 3-5 in the spectrum of peracetylated isomaltose have shifts nearly identical to resonances 3-5 in the spectrum of peracetylated isomaltotriose.

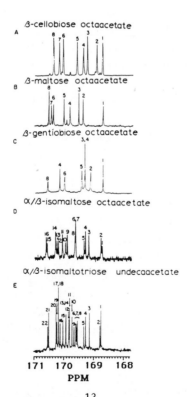

Fig. 2.  The proton-decoupled $^{13}$C NMR spectra of peracetylated derivatives of oligosaccharides of glucose acquired at 9.4 T.

By assigning each of the resonances in these and similar spectra to specific acetoxy substituents, the effect of different types of structural changes upon resonance chemical shift could be studied. Ultimately, if such correlations could be revealed, the carbon spectrum of a particular compound could be predicted. Conversely, and more importantly, a particular spectrum could be used to infer a possible structure. Figure 3 summarizes the strategy used in assigning each of the carbonyl carbon resonances for a octa-0-acetyl-α-lactose sample. Pyranosyl proton resonances were first assigned by using homonuclear correlated spectroscopy (COSY). In the COSY contour plot shown in Fig. 3A resonances on either of the two axes also lie along the diagonal running from the lower left to the upper right. Off-diagonal contours, lying either above or below the diagonal, arise from sets of coupled spins. Having assigned the doublets at 6.26 and 4.55 ppm to H-1 of the α-glucose and β-galactose residues on the basis of their unique chemical shifts (Gagnaire et al., 1976), all sets of resonances can be assigned by tracing out the coupling network. Figure 3B shows the $^{13}C-^{1}H$ shift correlation spectrum of the compound, where fixed delay-times before and following the final mixing pulse have been adjusted to emphasize long-range $^{13}C-^{1}H$ coupling. Carbonyl carbon resonances can be assigned to specific substituents from cross-peaks arising from the coupling to nearest neighboring pyranosyl protons. For example, the carbonyl carbon resonance lying farthest upfield can be assigned to C1-Ac[*], as it is seen to correlate with a proton doublet at 6.26 ppm, previously assigned to the anomeric proton of the α-glucose residue. Carbonyl carbon resonances also are seen to correlate with proton resonances arising from acetyl methyl protons (1.95-2.25 ppm). Chemical shifts of these protons have been shown to depend on pyranosyl ring structure (Goux, 1988; Goux and Unkefer, 1987; Appleton et al., 1986).

To date carbonyl carbon resonances have been assigned in over 30 compounds (Goux, 1988; Goux and Unkefer, 1987). Many of these compounds are structural fragments of larger, more complex structures which have been isolated from naturally occurring glycans. Recall that one of our goals was to investigate whether the pattern of carbonyl resonances observed in the spectrum of a peracetylated oligosaccharide of unknown structure could be used to yield information as to (a) what types of peracetylated residues are present, and (b) how they are glycosidically linked to other residues. Figure 4 shows the ranges of assigned carbonyl carbon resonance shifts of all compounds studied to date. In a few cases, there are carbonyl

---

[*] Abbreviations used to denote atom types are C-1, for the anomeric carbon of the pyranose ring, C1-Ac, for the carbonyl carbon of the acetoxy group substituted at C-1, and H1-AcMe, for the methyl protons of the acetoxy substituted at C-1.

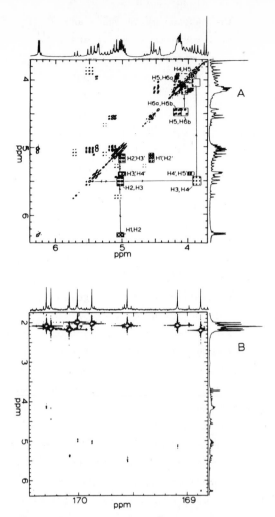

Fig. 3. (A) The COSY spectrum (contour plot) of octa-O-acetyl-α-
lactose. (The normal [1]H NMR spectrum is shown along both of
the chemical-shift axes. Arrows in the plot illustrate how
the D-glucopyranosyl-ring proton resonances can be assigned
by tracing out their coupling network. Unlabeled, off-
diagonal contours arise from pyranosyl-ring protons of
octa-O-β-lactose.) (B) The [13]C-[1]H shift-correlation
spectrum. (Normal [13]C and [1]H spectra are shown along the
horizontal and vertical axes.)

resonances which are characteristic of specific residue types. For example,
carbonyl carbon resonances present between 171.3 and 171.5 ppm in the spectrum of an acetylated unknown can be assigned to the OAc-3-group of
acetylated N-acetyl-α-D-glucosamine residues. A single resonance between
169.1 and 169.3 ppm and two resonances between 169.8 and 170.2 ppm appear to
be a characteristic spectral pattern of acetylated β-galactose residues.

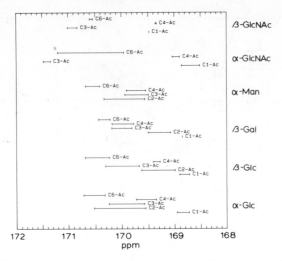

Fig. 4.  Shift ranges of carbonyl resonances in peracetylated residues contained in a variety of parent structures (over 30) (Goux, 1988; Goux and Unkefer, 1987).

Although the acquisition of NMR data on a greater number of peracetylated compounds may help further to confirm the uniqueness of these spectral patterns, it should be apparent from Fig. 4 that ambiguities will persist. For example, three carbonyl carbon resonances between 169.5 and 169.9 ppm might be indicative of the presence of either acetylated α-glucose and α-mannose residues.  These types of degeneracies can, however, be removed if the shifts of resonances arising from pyranosyl-ring and acetyl methyl protons coupled to each carbonyl carbon atom are also considered.  Rather than considering each of the resonances in Fig. 4 as a data point in one-dimensional "shift space", they now become data points in three-dimensional shift space.  The comprehensive carbonyl carbon chemical-shift data shown in Fig. 4 can then be plotted as shown in Fig. 5.  Each of the pseudo-three-dimensional cubes shown contains shift data for the acetyl groups of four different types of residues occurring in a variety of parent compounds. Different types of data points represent the assignments made from $^{13}C-^{1}H$ shift-correlation spectra.  Open circles, filled circles, open triangles, filled triangles, and open diamonds represent shift data for acetyl groups substituted at O-1, O-2, O-3, O-4, and O-6.

Close inspection of the data in Fig. 5 reveals two general conclusions which can be made as well as some interesting anomalies.  For each residue type there is a clustering of different types of data points.  The physical interpretation of this is that a particular acetyl group of a particular residue type has a limited range of shifts associated with it, regardless of the parent compound in which it occurs.  Whether this is due to the

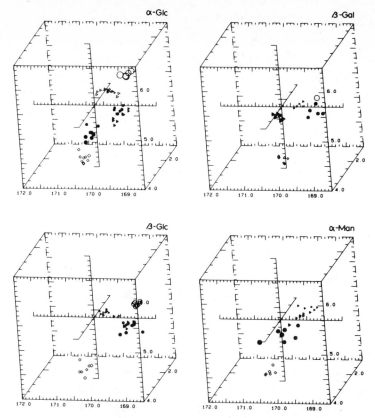

Fig. 5.   Resonance shifts of carbonyl carbon atoms and nearest-
          neighboring acetyl methyl protons and pyranosyl-ring protons,
          taken from the data contained in Goux, 1988; Goux and Unkefer,
          1987. [Different data-point types represent assigned substi-
          tuents.  Key: (O) 1-O-acetyl, (●) 2-O-acetyl, (∇) 3-O-acetyl,
          (▼) 4-O-acetyl, (◊) 6-O-acetyl.  Points are plotted in
          pseudo-three-dimensional shift-space with larger points lying
          nearer towards the front of the cube.]

limited number of compounds studied or as a result of the insensitivity of
shifts associated with a particular acetyl group to structural variations is
a question which cannot be answered completely at present.  It may also be
seen that when all of the clusters are considered together the pattern
formed is unique to each type of residue.  For example, the shift data for
the acetyl group substituted at C-2 in β-glucose (filled circles) appears to
overlap corresponding data of β-galactose.  However, the region of shift
space occupied by the acetyl group substituted at C-4 (filled triangles) is
clearly different in the two types of residues.

In some cases there appear to be data points in Fig. 5 that lie in a
region of shift space apart from a cluster of data points of the same type.
These anomalies usually have as a basis the unique structures of parent
compounds from which the data was taken.  For example, one of the data

points representing the C-2 acetyl group of α-mannose (filled circle) appears to be removed from the region of space occupied by similar data points. Reference to the original data shows that the parent compound from which the anomalous data was taken is glycosidically linked to a neighboring residue through the nearby C-3 hydroxyl (compound 10). Apparently the proximity of this linkage perturbs the shifts of C2-Ac and H2-AcMe. Data points representing shifts of C-2 acetoxy groups of acetylated α-glucose residues fall into two regions of space. Reference back to the raw data shows that one of the subgroups was taken from shift data of C-2 acetoxy groups having neighboring C-1 acetoxy groups, while the other subgroup arises from compounds glycosidically linked at C-1.

## ALTERNATIVE METHODS FOR CORRELATING PERACETYLATED OLIGOSACCHARIDE STRUCTURE AND NMR CHEMICAL SHIFTS

By presenting shift data for each peracetylated oligosaccharide residue in the manner shown in Fig. 5, one can visually recognize regions of shift space occupied by any given acetyl group on any given type of residue. However, the method also has some serious disadvantages. One of these is the experimenter's inability to discriminate subtle trends in the data arising from combinations of shift variables. In addition, some of the data shown may not be of any use in helping the experimenter discriminate between types of residues or glycosidic linkages. In Fig. 5 this appears to be the case for the shifts associated with C6-Ac. Data like these only add noise and obscure patterns which may be present. An alternate method of treating the data is to arrange the shifts associated with each residue into vector form. Most unsubstituted pyranosides then have 12 components corresponding to the proton and carbon NMR shifts associated with four acetoxy substituents and one component corresponding to the anomeric proton shift. In order that each of shift variables have equal weight, they can be scaled to unit variance (autoscaling). Each of the objects in the resulting data matrix can then be presented as a point in 13-dimensional hyperspace. One of the simplest of "pattern recognition" methods, the Nearest Neighbor approach, has been used to assign each of the objects to a specific residue class. The method works by polling for residue type the Nearest Neighbors closest to an object, as measured by their Euclidean distance in n-space.

$$ dij = \sqrt{\left[ \sum_{k=1}^{n} \left( X_{ik} - X_{jk} \right)^2 \right]} \tag{1} $$

This method has previously been applied to classifying objects according to their $^{13}$C NMR or mass spectra and it is probably the best of the pattern recognition methods to use on small data sets (Sharaf et al., 1986; Kowalski and Bender, 1972a; Lowry et al., 1977). Initially, four classes were defined: (1) all peracetylated α-Glc residues (2) peracetylated β-Gal

residues (3) peracetylated α–Man residues and (4) peracetylated β–Glc residues. The objects included in the data set, their class and their assigned
class using the KNN method are summarized in Table 1. Residues 1-17 are all
pyranosides having no glycosidic linkage substitutions. The remaining residues are either pentaacetylated monosaccharides or have glycosidic linkages
at positions 3, 4 or 6. When all 13 variables (features) are included,
there are 9 and 8 out of 38 (20%) objects misclassified using the 1-Nearest
Neighbor and 3-Nearest Neighbor methods, respectively. Although these
results are not remarkable, they do illustrate that different residue types
are separated by some distance in 13-dimensional feature space.

Distances between objects in 13-dimensional hyperspace can be viewed by
the experimenter by picking the two directions in hyperspace for which there
is the greatest variance in the data (Sharaf et al., 1986; Kowalski and
Bender, 1972b; Albano et al., 1979; Kowalski, 1975; Jurs, 1986; Johnels et
al., 1983; Edlund and Sjostrom, 1977; Edlund and Wold, 1980; Wold, 1976;
Albano et al., 1978; Wold and Sjostrom, 1986; Wold, 1978). Points in
n-space are then projected onto the surface created by these two "principal
components". Mathematically, these directions or eigenvectors are obtained
by diagonalization of the covariance matrix. Such a principal component
analysis is shown in Fig. 6 for all of the objects in the data set and in
Fig. 7 for subdata sets formed from peracetylated α–Glc residues. Figure 6
shows that all of the β–Gal residues (filled diamonds) are well separated
from the other three classes of residues. Notice that object 19, the only
pentaacetylated β–Gal residue lies slightly apart from the other β–Gal
residues, almost forming a class of its own. Although there is some
clustering among the members of each of the remaining three classes, there
appears to be significant overlap between classes. For example, objects 25,
26, 29 and 30, all α–Glc residues lie apart from many of the other α–Glc
residues and appear near to many of the β–Glc residues. The principal
component plots for the α–Glc data shown in Fig. 7 emphasize the inhomogeneity of each of these classes. The plot contain 3 separate clusters of
points. Previously reducing residues 1->4 linked form a cluster of points,
as do previously reducing 1->6 linked residues and previously nonreducing or
internal 1->6 linked residues. Taken together, the PC plots in Figs. 6 and
7 suggest that objects may be classified according to the linkages they make
with neighboring residues as well as with respect to their residue types.

MODELING CLASS STRUCTURE:  SUPERVISED LEARNING

Using the aforementioned pattern recognition methods, it is largely
up to the experimenter to define classes. For example, an object near a
1->4 linked α–Glc residue may be arbitrarily classified as belonging to a
class containing only α–Glc residues or alternatively, as belonging to a

Table 1. Results of KNN and SIMCA Calculations.

| Object[a] | Class | Assigned Class | | |
|---|---|---|---|---|
| | | 1–KNN | 3–KNN | SIMCA |
| 1 AMG[b] | 1 | 1 | 1 | 1 |
| 2 BMA2B[b] | 1 | 1 | 1 | 1 |
| 3 AIMA2B[b] | 1 | 1 | 1 | 1 |
| 4 BIMA2B | 1 | 1 | 1 | 1 |
| 5 AIMA3C | 1 | 1 | 1 | 1 |
| 6 BIMA3C[b] | 1 | 1 | 1 | 1 |
| 7 BMGAL | 2 | 2 | 2 | 2 |
| 8 BGAL3B | 2 | 2 | 2 | 2 |
| 9 ALACTB | 2 | 2 | 2 | 2 |
| 10 ALACNB | 2 | 2 | 2 | 2 |
| 11 AMM | 2 | 2 | 2 | 2 |
| 12 MAN13B | 3 | 3 | 3 | 3 |
| 13 MAN14B | 3 | 3 | 3 | 3 |
| 14 MAN16B | 3 | 3 | 3 | 3 |
| 15 BMG[b] | 4 | 4 | 4 | 4 |
| 16 BCELB[b] | 4 | 4 | 4 | 4 |
| 17 BGENTB | 4 | 4 | 4 | 4 |
| 18 AGLC[b] | 1 | 4* | 4* | 1 |
| 19 BGAL[c] | 2 | 2 | 2 | 2 |
| 20 MAN13A | 3 | 3 | 3 | 3 |
| 21 MAN16A | 3 | 3 | 3 | 3 |
| 22 MAN14A[c] | 3 | 4* | 4* | 3 |
| 23 BGLC[b] | 4 | 4 | 4 | 4 |
| 24 BCELA | 4 | 1* | ?* | 4 |
| 25 AIMA2A[b] | 1 | 4* | 4* | 1 |
| 26 AIMA3A[b] | 1 | 4* | 4* | 1 |
| 27 AIMA3B | 1 | 1 | 1 | 1 |
| 28 BIMA3B | 1 | 1 | 1 | 1 |
| 29 ALACTA[a] | 1 | 4* | 4* | 4* |
| 30 ANLACA[a] | 1 | 4* | 4* | 1 |
| 31 BNLACA[b] | 4 | 1* | ?* | 4 |
| 32 BMAL2A | 4 | 4 | 4 | 4 |
| 33 BIMA2A | 4 | 4 | 4 | 4 |
| 34 BGENTA | 4 | 4 | 4 | 4 |
| 35 BIMA3A[b] | 4 | 1* | 4 | 4 |
| 36 ANLACB | 2 | 2 | 2 | 2 |
| 37 BNLACB | 2 | 2 | 2 | 2 |
| TOTAL MISASSIGNMENTS | | 9 | 8 | 1 |

[a] The first letter at the beginning of the object abbreviation indicates the anomeric form; the letter at the end of the name indicates the sequence number ("A" is at the reducing end). Misassignments are marked with an asterisk.

[b] Used in modeling class structure in SIMCA. There were 11 test objects. All members except objects 19 and 22 were used in modeling classes 2 and 3.

[c] Outliers not used in SIMCA modeling.

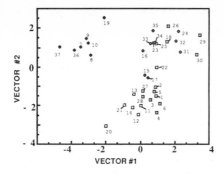

Fig. 6.   Principal component plot of residues listed in Table 1.
Coded symbols refer to α–glucose (▣), β–galactose (◈),
α–mannose (□), and β–glucose (◇) residues.

class consisting of 1–>4 glycosidically linked residues, regardless of its
residue type.   An alternative approach is for the experimenter to arbitrarily define a class consisting of a representative sampling of objects
having some portions of their structures in common.   The class can be
modeled using the principal components best describing the data according to
the expression (Sharaf et al., 1986; Kowalski, 1975; Jurs, 1986; Johnels et
al., 1983; Edlund and Sjostrom, 1977; Edlund and Wold, 1980; Wold, 1976;
Albano et al., 1978; Wold and Sjostrom, 1986)

$$y_{ik}^{(q)} = a_i^{(q)} + \sum_{a=1}^{A_q} b_{ia}^{(q)} \theta_{ak}^{(q)} + e_{ik}^{(q)} \tag{2}$$

where $y_{ik}$ are the components contained in the initial data matrix, $a_i$ are
means for variable i, $b_{ia}$ are the loading of the $A_q$ principal components,

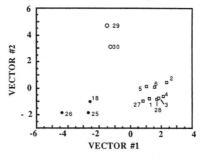

Fig. 7.   Principal component plot of the peracetylated α–glucose
residues listed in Table 1 (class 1).   Residues 1–6, 27 and
28 are α–methyl pyranoside and other previous nonreducing
residues 1–>6 linked.   Residues 18, 25 and 26 are
α–glucopyranose or other previous reducing residues 1–>6
linked.   Residues 29 and 30 are previous reducing residues
1–>4 linked.

$\theta_{ak}$ are the coordinates of the transformed points (scores) and $e_{ik}$ are residuals or differences between the actual components of the data matrix and the sum of the first two terms on the right. The objects contained in class q are mathematically described by the expression as lying along a line ($A_q=1$) or contained in a plane ($A_q=2$) or bounded volume ($A_q=3$). The most significant principal components are selected by crossvalidation (Sharaf et al., 1986, Wold, 1978), a method similar to those for determining the statistically most significant terms in a binomial expansion. Outliers in the class may be eliminated using statistical criteria (F-test). This method for independently modeling each class according to its most significant components was developed by Wold et al. (Sharaf et al., 1986; Albano et al., 1979; Johnels et al., 1983; Edlund and Sjostrom, 1977; Edlund and Wold, 1980; Wold, 1976; Albano et al., 1978; Wold and Sjostrom, 1986; Wold, 1978) and is known by the acronym SIMCA (Soft Independent Modeling of Class Analogy). Typically modeling is carried out using some of the objects contained in a class training set and the validity of the model can be tested on the remaining objects belonging to the same class (test set). Again, classification of the test objects is determined on a statistical basis.

In addition to providing a means of classifying objects, the residuals in Eq. (2) provide valuable information regarding class homogeneity, separation between classes (interclass distance) and the relative strength of any given variable to model the structure of a class or to discriminate between two classes. Variable modeling power (Mp (i)) is defined such that the smaller the residuals associated with a variable in modeling its own class members, the larger Mp (i) will be, where Mp (i) can have a maximum value of unity. Variable discriminating power ($\theta i$) is a measure of a variable's contribution to the interclass distance (Wold and Sjostrom, 1986).

The modeling power and discriminating power of variables may be used to eliminate noise from the data set. By eliminating those variables only with low discriminating power, one is falsely emphasizing interclass distances, making it easier to correctly classify test objects by chance. On the other hand, variables having both low discriminating power and modeling power can be eliminated (Sharaf et al., 1986; Albano et al., 1979; Kowalski, 1975; Jurs, 1986; Wold and Sjostrom, 1986).

The SIMCA routine was initially run using all 13 chemical shift variables and four classes corresponding to the four different residue types previously defined. All classes could be modeled using only one significant principal component. Objects 29 and 30, the α-Glc residues of peracetylated Gal(1->4)Glc and NeuAc(2->3)Gal( 1->4)Glc were found to be outliers in class 1; object 29, pentaacetylated β-Gal was found to be an outlier in class 2; object 22, the 1-0-methyl peracetylated α-mannoside residue of Man( 1->4)Man-OMe was found to be an outlier in class 3; there were no

Table 2.  Interclass Distances and Variable Modeling Power and
Discriminatory Power Determined from Seven Variable
SIMCA Calculation.

A.  Interclass Distances

| Class/Class | 1($\alpha$–Glc) | 2($\beta$–Gal) | 3($\alpha$–Man) | 4($\beta$–Glc) |
|---|---|---|---|---|
| 1($\alpha$–Glc) | 1.0 | 11.0 | 5.6 | 2.7 |
| 2($\beta$–Gal) | | 1.0 | 7.3 | 9.3 |
| 3($\alpha$–Man) | | | 1.0 | 7.7 |
| 4($\beta$–Glc) | | | | 1.0 |

B.  Variable Discriminatory Power[b]

| Classes | Chemical Shift Variable | | | | | | |
|---|---|---|---|---|---|---|---|
| | C2–Ac | H2–AcMe | H3 | H3–AcMe | H4 | H4–AcMe | H1 |
| (1,2) | 6.2 | 9.9 | 25.7* | 9.8 | 15.8 | 6.7 | 20.3 |
| (1,3) | 2.9 | 1.8 | 12.3* | 7.9 | 7.2 | 1.5 | 6.0 |
| (1,4) | 5.2 | 2.8 | 4.6 | 2.6 | 6.6* | 0.6 | 1.5 |
| (2,3) | 5.4 | 22.7* | 7.6 | 1.4 | 3.5 | 12.9 | 5.4 |
| (2,4) | 1.6 | 6.4 | 22.0* | 9.2 | 19.1 | 11.2 | 18.0 |
| (3,4) | 5.6 | 5.4 | 7.9 | 13.6 | 16.7* | 3.9 | 5.0 |
| Total | 27.0 | 49.0 | 80.1 | 44.4 | 68.8 | 36.8 | 56.2 |

[b] The variable which is most discriminating between each pair
of classes is marked with an asterisk.

C. Variable Modeling Power[c]

| Modeling Power | Chemical Shift Variable | | | | | | |
|---|---|---|---|---|---|---|---|
| | C2–Ac | H2–AcMe | H3 | H3–AcMe | H4 | H4–AcMe | H1 |
| Mp(1) | 0.6* | 0.2 | 0.1 | 0.4 | 0.6* | 0.1 | 0.5 |
| Mp(2) | 0.1 | 0.1 | 0.8* | 0.1 | 0.5 | 0.8* | 0.1 |
| Mp(3) | 0.5 | 0.7* | 0.4 | 0.6 | 0.1 | 0.7* | 0.1 |
| Mp(4) | 0.2 | 0.1 | 0.4 | 0.8* | 0.1 | 0.3 | 0.3 |
| Totals | 1.4 | 1.2 | 1.7 | 1.9 | 1.3 | 1.9 | 1.0 |

[c] The variable(s) with greatest modeling power in each class are
marked with an asterisk.

outliers in class 4. On the basis of their low discriminating power and modeling power, variables corresponding to the H2, C3-Ac, C4-AC, C6-Ac, H6 and H6-AcMe chemical shifts were eliminated. Class 1 was divided into 6 training objects and 5 test objects while class 4 consisted of 5 training objects and 5 test objects. Classes 2 and 3, each consisting of 6 objects, were not divided into test and training sets because of the small number of objects contained in each.

The results of test object classification, interclass distances and modeling and discriminatory powers of the seven variables used in modeling are listed in Tables 1 and 2. On the basis of classification of test objects, the SIMCA method does quite well. All of the 11 test objects in class 1 and 4 were classified correctly. However, object 29, the per-acetylated α-Glc residue of Gal( 1->4) Glc, which was previously determined to be an outlier of class 1, was incorrectly assigned to class 4, which consisted of peracetylated β-Glc residues. Given the heterogeneity of classes 1 and 4 and their proximity to one another (small interclass distance as shown in Table 2), this result is not too surprising. The classes in general would undoubtedly be better modeled and have fewer outliers if more data were available. As shown by the interclass distances in Table 2, class 2, consisting of peracetylated β-Gal residues, is well separated from the other classes. According to variable discriminating powers, class 1 and class 2 (peracetylated α-Glc and β-Gal residues) can be distinguished from one another largely on the basis of H1 and H3 chemical shifts. This result makes sense structurally since these protons are near carbons of opposite chirality. Similarly, class 2, β-Gal residues, differs from class 3, α-Man residues, primarily as a result of the C2-Ac chemical shift. It is interesting that the H-1 anomeric proton shift is only the third most discriminating variable overall. This speaks in favor of an integrated approach where one makes use of all chemical shift information in order to distinguish between residue types, rather than using only the anomeric protons as spectroscopic reporters of residue types.


REFERENCES

Albano, C., Blomquist, G., Dunn, III, W., Edlund, U., Eliasson, B., Johansson, E., Norden, B., Sjostrom, M., Soderstrom, B., and Wold, S., 1979, Characterization and Classification Based on Multivariable Analysis, in: "27th Inter. Congress Pure and Appl. Chem.," A. Vermavwori, ed., Pergamon Press, New York, NY.
Albano, C., Dunn, III, W., Edlund, U., Johansson, E., Norden, B., Sjostrom, M., and Wold, S., 1978, Four Levels of Pattern Recognition, Anal. Chim. Acta, 103:429.
Allerhand, A., and Berman, E., 1984, A Systematic Approach to the Analysis of $^{13}$C NMR of Complex Carbohydrates. α-D-Mannopyranosyl Residues in Oligosaccharides and their Implications to Studies of Glycoproteins and Glycopeptides, J. Am. Chem. Soc., 106:2400.

Allerhand, A., Doddrell, D., and Komoroski, R., 1971, Natural Abundance
    Carbon-13 Partially Relaxed Fourier Transform Nuclear Magnetic
    Resonance Spectra of Complex Molecules, J. Chem. Phys., 55:189.
Appleton, M. L., Cottrell, C. E., and Behrman, E. J., 1986, NMR
    Assignments of Acetyl and Trityl Groups in Derivatized Carbohydrates
    via Proton-Carbon Long-Range Couplings, Carbohydr. Res., 158:227.
Brisson, J. R., and Carver, J. P., 1983, Conformation of $\alpha$-D(1->3) and
    $\alpha$-D(1->6)-Linked Oligomannosides Using Proton NMR, Biochemistry,
    22:1362.
Carver, J. P., 1984, The Role of 3-D Structure in the Control of N-Linked
    Oligosaccharide Biosynthesis, Biochem. Soc. Trans., 12:517.
Darvill, A. G., and Albersheim, P., 1984, Phytoalexins and Their
    Elicitors, Annu. Rev. Plant Physiol., 35:243.
Dill, K., Berman, E., and Pravia, A. A., 1985, Natural-Abundance $^{13}$C NMR
    Spectral Studies of Carbohydrates Linked to Amino Acids and
    Proteins, Adv. Carbohydr. Chem. Biochem., 43:1.
Edlund, U., and Sjostrom, M., 1977, Analysis of $^{13}$C NMR Data by Means of
    Pattern Recognition Methodology, J. Magn. Res., 25:285.
Edlund, U., and Wold, S., 1980, Interpretation of NMR Substituent
    Parameters by Use of a Pattern Recognition Approach, J. Magn. Res.,
    37:183.
Gagnaire, D. Y., Taravel, F. R., and Vignon, M. R., 1976, Assignment of
    $^{13}$C NMR Signals of Peracetylated Disaccharides Containing Glucose,
    Carbohydr. Res., 51:157.
Goux, W. J., 1988, The Determination of Complex Carbohydrate Structure by
    Using Carbonyl Carbon Resonances of Peracetylated Derivatives,
    Carbohydr. Res., 184:47.
Goux, W. J., and Unkefer, C. J., 1987, The Assignment of Carbonyl
    Resonances in $^{13}$C NMR Spectra of Peracetylated Mono- and
    Oligosaccharides Containing D-Glucose and D-Mannose:  An Alternative
    Method for Structural Determination of Complex Carbohydrates,
    Carbohydr. Res., 159:191.
Homans, S. W., Dwek, R. A., Boyd, J., Mahmoudian, M., Richards, W. G.,
    and Rademacher, T. W., 1986, Conformational Transitions in N-Linked
    Oligosaccharides, Biochemistry, 25:6342.
Johnels, D., Edlund, U., Grahn, H., Hellberg, S., Sjostrom, M., Wold, S.,
    Clementi, S., and Dunn, III, W., 1983, Clustering of Aryl Carbon-13
    NMR Substituent Chemical Shifts.  A Multivariable Data Analysis
    Using Principal Components, J. Chem. Soc. Perkin Trans. II, 863.
Jurs, P. C., 1986, Pattern Recognition Used to Investigate Multivariable
    Data in Analytical Chemistry, Science, 232:1219.
Kowalski, B. R., 1975, Measurement Analysis by Pattern Recognition.
    Survey of Computer Aided Methods for Mass Spectral Analysis, Anal.
    Chem., 47:1152A.
Kowalski, B. R., and Bender, C. F., 1972a, The K-Nearest Neighbor
    Classification Rule Applied to NMR Spectral Analysis, Anal. Chem.,
    44:1405.
Kowalski, B. R., and Bender, C. F., 1972b, Pattern Recognition.  A
    Powerful Approach to Interpreting Chemical Data, J. Amer. Chem.
    Soc., 94:5632.
Lowry, S. R., Isenhour, T. L., Justice, Jr., J. B., McLafferty, F. W.,
    Dayringer, H. E., and Venkataraghavan, R., 1977, Comparison of
    Various K-Nearest Neighbor Voting Schemes with Self-Training
    Interpretive and Retrieval Systems for Identifying Molecular
    Substructures from Mass Spectral Data, Anal. Chem., 49:1720.
Montreuil, J., 1980, Primary Structure of Glycoprotein Glycans, Adv.
    Carbohydr. Chem. Biochem., 37:157.
Sharaf, M. A., Illman, D. L., and Kowalski, B. R., 1986, Chapter 6:
    Exploratory Data Analysis, in:  "Chemometrics" (Chemical Analysis,
    V. 82), John Wiley and Sons, New York, NY.
Sillerud, L. O., Yu, R. K., and Schafer, D. E., 1982, Assignment of $^{13}$C
    NMR Spectra of Gangliosides, Biochemistry, 21:1260.

Vliegenthart, J. F. G., Dorland, L., van Halbeek, H., 1983, High
Resolution [1]H NMR Spectroscopy as a Tool in Structural Analysis of
Carbohydrates Related to Glycoproteins, Adv. Carbohydr. Chem.
Biochem., 41:209.

Vliegenthart, J. F. G., van Halbeek, H., and Dorland, L., 1981,
Applicability of 500 MHz [1]H NMR for the Structure Determination of
Carbohydrates Derived from Glycoproteins, Pure Appl. Chem., 53:45.

Watkins, W. M., 1972, Blood Group Specific Substances, in:
"Glycoproteins, Part B," 2nd edition, A. Gottschalk, ed., Elsevier,
Amsterdam.

Wold, S., 1976, Pattern Recognition by Means of Disjoint Principal
Component Models, Pattern Recog., 8:127.

Wold, S., 1978, Cross-Validatory Estimation of the Number of Components
in Factor and Principal Component Models, Technometrics, 20:397.

Wold, S. and Sjostrom, M., 1986, SIMCA: A Method for Analyzing Chemical
Data in Terms of Similarity and Analogy, in: "Chemometrics: Theory
and Application," A. Kowalski, ed., American Chemical Society,
Washington, DC.

SELECTIVE EXCITATION TECHNIQUES FOR WATER SUPPRESSION IN ONE- AND
TWO-DIMENSIONAL NMR SPECTROSCOPY

Vladimír Sklenár

Institute of Scientific Instruments
Czechoslovak Academy of Sciences
Kralovopolska 147
CS 612 64 BRNO, CZECHOSLOVAKIA

INTRODUCTION

The measurement in non-deuterated water solutions is a long-standing
problem in the pulse Fourier transform NMR spectroscopy. Very large signal
from the water protons (c ~100 M) introduces serious problems which arise
from the limited ADC (analog-to-digital converter) resolution, nonlinear-
ities in the detection channel and from computer data manipulations. As a
result, special experimental techniques which suppress the intense water
resonance are required to overcome the NMR system limitations. History of
the development of special experimental tools began in the middle seventies
with the introduction of the well known 214 pulse sequence by A.G. Redfield
et al. (1975). Despite the great effort and substantial contributions in
the following years the field remained open for further development.

The techniques which have been applied for this purpose can be divided
into two groups. With methods of the first group the elimination of water
resonance is achieved by applying the acquisition rf pulse at the time when
both longitudinal and transverse components of water magnetization are zero.
This can be accomplished by selectively saturating the water resonance
(Schaefer, 1972; Jesson et al., 1973; Bleich and Glasel, 1975; Hoult, 1976;
Krischna, 1976; Campbell et al., 1977) or by taking advantage of the difference between spin-lattice relaxation rates of the water and the measured
resonances (Patt and Sykes, 1972; Benz et al., 1972; Moobery and Krugh,
1973; Krugh and Schaefer, 1975; Gupta, 1976; Gupta and Gupta, 1979).
Techniques of this class have revived recently with the development of the
phase coherent presaturation (Zuiderweg et al., 1986) and the spin-echo
water eliminated spectroscopy (Bryant and Eads, 1985; Rabenstein et al.,
1985; Eads et al., 1986; Rabenstein et al., 1986; Connor et al., 1987) which
exploits the difference between the transverse relaxation times of water and

solute protons. Methods of the second group are based on the application of
rf excitation which is made frequency selective, having a zero spectral
density at the water resonance position. This approach which leaves the
water magnetization unperturbed in the equilibrium position is much more
convenient for studies of exchangeable protons since their resonances are
not affected by magnetization transfer, cross-relaxation, or intermolecular
interactions with the excited water molecules. Many interesting techniques
of selective excitation have been introduced recently, ranging from time
shared hard pulse sequences (Sklenar and Starcuk, 1982; Plateau and Gueron,
1982; Hore, 1983a, 1983b; Plateau et al., 1983; Turner, 1983; Clore et al.,
1983; Starcuk and Sklenar, 1985, 1986; Morris et al., 1986; Hall and Hore,
1986; Wang and Pardi, 1987; Levitt and Roberts, 1987; Sklenar and Bax,
1987a; Sklenar et al., 1987; Levitt et al., 1987; Bax et al., 1987; von
Kienlin et al., 1988) and combinations of soft and hard pulses (Redfield and
Kunz, 1979; Sklenar et al., 1987a; Sklenar and Bax, 1987b) to applications
of shaped rf pulses with a time dependent amplitude (Gutow et al., 1985;
McCoy and Waren, 1987). The present contribution is devoted to the description of selective pulse excitation techniques which produce spectra that are
free of phase distortions. This type of excitation can be advantageously
applied in the pure phase 2D NMR experiments. Two types of two-stage selective excitation techniques are discussed. Methods of the first group apply
the spin-echo type pulse sequences to remove the frequency dependent phase
shift which leads to the baseline undulations. The second group of techniques takes advantage of a combination of soft and hard rf pulses to obtain
excitation profiles with a very narrow non-excitation region close to the
water resonance and uniform, full excitation over a wide frequency bandwidth
of interest. Examples of applications in various one- and two-dimensional
NMR experiments are presented.

PHASE DISTORTIONS IN SELECTIVE EXCITATION

A common problem in all selective excitation techniques is an existence
of a phase gradient across the excited region which stems from the offset
dependent precession during the excitation period. Although this gradient
can be in principle removed by applying the linear or higher order phase
corrections, the baseline roll that results from such a manipulation can
lead to intense distortions in the measured spectra (Plateau et al., 1983).

In spectra obtained by applying the selective excitation, the time
origin for Fourier transform which coincides with the beginning of the data
acquisition differs substantially from the actual start of a spin system
evolution. During this time delay, the magnetization vectors rotate around
the z-axis of the rotating frame of reference acquiring offset dependent
phase shifts. Consequently, the frequency function $F(\omega)$ of each line in a

spectrum which describes the resonance lineshape must be multiplied by the complex phase factor:

$$F(\omega) = [A(\Omega) + iD(\Omega)].\exp (i.\omega.T),$$

where T specifies the time delay and $A(\Omega)$ and $D(\Omega)$ are absorptive and dispersive components of a Lorentzian line with a resonance offset $\Omega$, respectively. To remove this phase distortion, the linear phase correction must be applied and the resulting lineshape is characterized by a modified complex function of the form:

$$F_m(\omega) = A(\Omega).\cos[(\omega-\Omega).T] + i.D(\Omega).\sin[(\omega-\Omega).T].$$

This lineshape is no longer Lorentzian and the multiplication of the real part by the cosine function leads to the familiar baseline roll.

There are two kinds of baseline undulations in spectra produced by selective excitation methods for water suppression. Their effect is demonstrated using a computer simulation and the results are shown in Fig. 1. Distortions of the first kind stem from the interference of modified lineshapes in the measured part of a spectrum (Fig. 1a). The extent of these undulations is dependent on the relative positions of the individual resonances, their linewidths and on the duration of the excitation sequence. The undulations of the second kind originate from the intense residual water resonance (Fig. 1b). The distorted lineshape of its signal extends far off resonance, rolling a baseline in a large range of resonance offsets. Very high water suppression can partially remove the second kind of baseline roll from a spectrum but the undulations of the first kind can remain. This is a reason why the linear phase correction cannot compensate for the delayed acquisition and therefore cannot produce a perfect spectrum.

PURE PHASE SELECTIVE EXCITATION TECHNIQUES

The best remedy for baseline distortions is an application of excitation techniques that generate spectra without a phase gradient across the measured region. The first selective train with this property was the Jump and Return (JR) pulse sequence ($90°_x$-t-$90°_{-x}$) developed in 1982 by Plateau and Gueron. The main disadvantage of this simple pulse train is its high sensitivity to pulse imperfections, and consequently, rather pure water suppression. Recently, a number of new techniques has appeared in literature. Although relying on different principles these methods have been designed with the same intention: to produce NMR spectra with pure phase across the large frequency bandwidth (Levitt and Roberts, 1987; Sklenar and Bax, 1987a; Sklenar et al., 1987b; Levitt et al., 1987; Bax et al., 1987; von Kienlin et al., 1988; Redfield and Kunz, 1979; Sklenar et al., 1987a; Sklenar and Bax, 1987b).

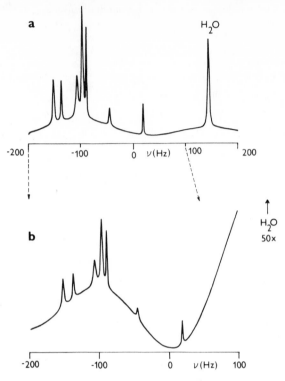

Fig. 1. Computer simulation of rolling baseline in NMR spectra obtained by applying the selective excitation.
a) Undulations arising from the interference of modified lineshapes when all signals in the spectrum have comparable intensities.
b) Rolling baseline that arises from the residual water resonance which has an intensity 50 times larger than the signals of interest. In both cases the linear phase gradient of 180°/100 Hz was removed by applying the linear phase correction.

## SELECTIVE SPIN-ECHO WATER SUPPRESSION PULSE SEQUENCE

Performance of the JR excitation can be substantially improved by forming the spin-echo sequence and applying the EXORCYCLE phase cycling (Sklenar and Bax, 1987a; von Kienlin et al., 1988).

A refocusing pulse ideally is a 180° rotation about an axis in the transverse plane. Imperfections in such a pulse can result in severe amplitude and phase distortions. In 1977, Bodenhausen et al. demonstrated that spurious effects of this pulse can be effectively removed by using the four step phase cycle which they named EXORCYCLE. In this procedure the phase of the refocusing pulse is incremented in 90° steps in consecutive scans and data are alternatively added and substracted from a computer

memory.  A Cartesian rotation matrix R of an arbitrary excitation sequence
is given by:

$$R = \begin{vmatrix} 11 & 12 & 13 \\ 21 & 22 & 23 \\ 31 & 32 & 33 \end{vmatrix} \qquad (1)$$

Applying the symmetry properties of the rotational matrices (Yan and Gore,
1987) it can be shown that for excitation sequences with the overall duration $t_R$, which are amplitude symmetrical with respect to $t_R/2$, and are
either phase symmetrical or antisymmetrical with respect to $t_R/2$, the effect
of the pulse after completion of a EXORCYCLE is described by the matrix:

$$R^E = \sum_{k=0}^{3} R_z^{-1}\left(k\,\Pi/2\right) \cdot R \cdot R_z\left(k \cdot \Pi/2\right) \qquad (2)$$

which acquires a simple diagonal form:

$$R^E = \begin{vmatrix} (11-22) & 0 & 0 \\ 0 & (22-11) & 0 \\ 0 & 0 & 0 \end{vmatrix} \qquad (3)$$

As a result, the magnetization vector with coordinates $[x,y,z]$ prior to the
180° pulse is rotated to a new position $A.[x,-y,0]$.  This transform represents rotation of transverse magnetization by 180° about an axis in the
transverse plane.  In addition, the transverse magnetization is modulated in
amplitude by a factor $A=(11-22)$ and the z-component contributions are completely substracted from the spectrum.  The implicit result of the EXORCYCLE
phase cycling procedure follows immediately from the form of the rotational
matrix (3):  all composite water suppression sequences with symmetry properties mentioned above can be applied to form the phase reversal echo
without introducing the offset dependent phase shift.

If the selective phase-cycled refocusing sequence is combined with
another selective excitation pulse, then a sequence results that produces
excellent water suppression (Fig. 2).  To avoid existence of any linear or
higher order phase distortion in the final spectrum, the selective excitation applied prior to the refocusing pulse must produce spectra with either
pure phase or linear phase gradient.  This linear gradient can be effectively removed by adjusting the delays in the echo sequence asymmetrically.
If two JR sequences with evolution delays $t_1$ and $t_2$ are combined, the
intensity profile depends on $\sin\omega t_1 \cdot (1-\cos\omega t_2)$ which for ratio $t_1/t_2=1/2$
reduces to $\sin^3\omega t_1$.  The excitation profile of this 1-1 ECHO pulse train has
a much broader near zero excitation region and the sequence gives substantially better water suppression.  Excitation profiles for 1-1 and other
composite water suppression sequences used for echo generation in
combination with the Jump and Return pulse train are shown in Fig. 3.

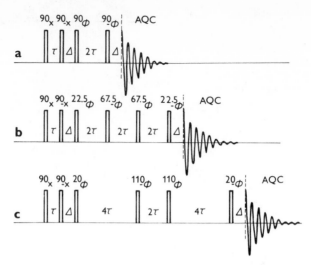

Fig. 2.  The spin-echo selective excitation pulse sequences.  In all
cases the Jump and Return pulse train is applied for the
selective excitation.  The 1-$\bar{1}$ (a), 1-$\bar{3}$-3-$\bar{1}$ (b) and
1--$\bar{5}$.$\bar{4}$-5.4--$\bar{1}$ (c) pulse sequences are used as a refocusing
pulse.  The phase shift by 180° is denoted by the overbar.
In all schemes the phase $\phi$ is cycled along the x,y,-x,-y
axis in the EXORCYCLE manner and data are alternatively
added and subtracted from the computer memory.  Maximum
excitation is achieved for offsets $\pm$ (2k+1)/4$\tau$, where
k=0,1,2,3....  The delay $\Delta$ is set to a short value 1-50 μs
and serves only for switching over the different rf phases.

In these sequences the effective water elimination is achieved in two
stages.  In the first stage the water suppression per individual scan is not
very high, being of the same order as that obtained with the regular Jump
and Return sequence, but this is sufficient to overcome the dynamic range
problems.  Very high suppression of the order 5000-10000 is obtained only in
the second stage after combining the results of four scans in the EXORCYCLE
manner.  There are two reasons for very high degree of water suppression.
The first one is the form of the frequency excitation profiles with broad
non-excitation regions.  The second one is a result of the EXORCYCLE selectivity as far as the variation of rf field intensity is concerned (von
Kienlin et al., 1988).  In the regular Jump and Return sequence the large
residual water signal originates from regions where both the static magnetic
and radio frequency fields are homogeneous.  In that case tilt of the effective rf field from the transversal plane already occurs very close to the
water resonance frequency and the water magnetization trajectory does not
return to the z direction at the end of the excitation sequence.  The broad
humps underneath the water resonance that may extend over several hundred
Hertz and which are difficult to shim out stem from these inhomogenous
regions.  As can be seen in Fig. 4 where excitation profiles of the 1-1 and

68

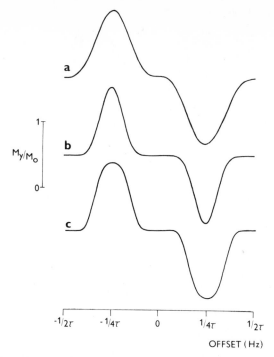

Fig. 3.  Excitation profiles of the spin-echo selective excitation
sequences calculated by the exact numerical solution of
the Bloch equation.  Pulse trains: a) 1-Ī.(1-Ī); b) 1-Ī.
(1-3̄-3-Ī) and c) 1-Ī. (1--5̄.4̄-5.4--Ī).  After completion of
the four step EXORCYCLE, spectra contain only in-phase
y-components; the x-components are effectively subtracted
from a spectrum.

1-1 ECHO pulse trains are computed as a function of the resonance offset and
of the intensity of rf field, the EXORCYCLE removes effectively the corresponding water signal from the spectrum.  The attainable water suppression
can be very high.  Due to the $\sin^3$ flip angle dependence of the signal
intensity in regions with full excitation, lineshapes of the excited
resonances can be substantially improved, increasing the final resolution.

To demonstrate the experimental performance of the selective spin echo
pulse train, one-dimensional spectra of decapeptide LH-RH (pGlu-His-Trp-
Ser-Tyr-Gly-Leu-Arg-Pro-Gly-$NH_2$) measured in 90% $H_2O$/10% $D_2O$ and obtained
with the 1-1 ECHO and non-optimized 1Z̄1 pulse sequences are shown in Fig. 5.
The experiments were carried out at 500 MHz on a modified NT-500 spectrometer.  As compared to the spectra obtained with 1Z̄1 pulse train
(Fig. 5 a,b) the 1-1 ECHO gives much higher water suppression and the
spectrum is free of baseline undulations (Fig. 5c).

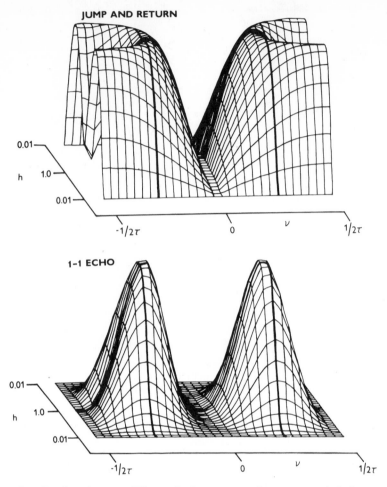

JUMP AND RETURN

0.01

h   1.0

0.01

$-1/2\tau$          0          $\nu$          $1/2\tau$

1-1 ECHO

0.01

h   1.0

0.01

$-1/2\tau$          0          $\nu$          $1/2\tau$

Fig. 4.   Excitation profiles of the Jump and Return and 1-1 ECHO
pulse sequences computed as a function of two parameters:
$\underline{\nu}$ – the resonance offset and $\underline{h}$ – the parameter of the rf
field homogeneity $(1.0 > h > 0)$.


WATER SUPPRESSION USING A COMBINATION OF HARD AND SOFT PULSES

Another possibility for obtaining the pure phase spectra is the use of
a hard/soft pulse combination first suggested by Gupta and Redfield in 1979.
In principle, this type of excitation can give very good intensity profile
with narrow non-excitation window close to the water resonance and supply
the spectra without any phase gradient.

Conceptually the method is extremely simple (Fig. 6a).  With the
carrier placed on water, a selective $90°_{-x}$ pulse rotates its magnetization
into the transversal plane, while leaving all other resonances unperturbed.
A subsequent nonselective $90°_{-x}$ pulse returns water back to the equilibrium
position and rotates the resonances of interest into the xy plane.

70

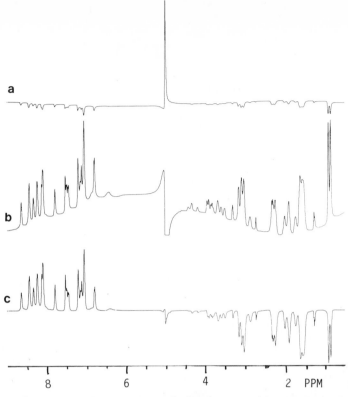

Fig. 5.  The proton NMR spectra of the decapeptide LH-RH (pGlu–His–
Trp–Ser–Tyr–Gly–Leu–Arg–Pro–NH$_2$) measured at 500 MHz on a
modified NT–500 spectrometer. 10 mg of the hormone were
dissolved in 0.4 ml of 90% H$_2$O/10% D$_2$O with 100 mM NaCl,
pH=5.8 and the measurements were done at 1°C.   Spectra a)
and b), differing in the vertical scale only, were obtained
applying the symmetrized 1–$\bar{2}$–1 pulse train (Sklenar and
Starcuk, 1982; Starcuk and Sklenar, 1985).   The attained
water suppression amounted to 350.   The baseline roll stems
mainly from the residual water signal.   Spectrum c) was
generated using the 1–1 ECHO excitation sequence.   After 64
acquisitions, combining the EXORCYCLE and CYCLOPS phase
cycles, the water suppression higher than 10000 was
obtained.   The spectrum does not require any linear or
higher order phase correction and is free of baseline
undulations.

    The simplest approach relies on the application of the long selective
soft pulse with a rectangular envelope.  This pulse gives an excitation and
phase profile shown in Fig. 7a.  A wider near-zero excitation region at the
cost of deteriorating phase profile can be obtained by positioning the hard
90°$_{-x}$ pulse in the center of the soft pulse (Fig. 6b).  The phase profile of
this pulse (Fig. 7b) is very poor because the excitation of the resonances
of interest by the hard 90°$_{-x}$ pulse occurs at a time t/2 before data acquisition is started.  The much smaller oscillations in phase observed far

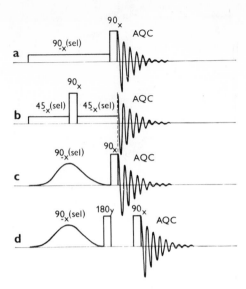

Fig. 6. The hard/soft rf pulse combinations for the selective water
suppression. The carrier of the soft pulse is set exactly
to the water resonance frequency.

off-resonance with the sequence a are due to the small amount of excitation
at these offsets by the rectangular soft pulse. Although in 1D NMR spectra
these small variations in phase are difficult to observe, they cause serious
problems when this type of excitation is incorporated in phase-sensitive 2D
experiments.

Better excitation and phase profiles (Fig. 7c) can be obtained by
shaping the soft pulse, which provides minimal excitation far off-resonance
(Fig. 6c) and optimizes the excitation profile close to the water resonance.
Different pulse shapes can be applied for this purpose. The basic criteria
imposed on the excitation profile of the selective shaped pulse are as
follows:

i)  very small variations of the z component in the relatively wide
range of excitation frequencies close to the water resonance;

ii) very small phase gradient in the excitation region.

These criteria are extremely difficult to meet applying a single shaped
pulse. Using amplitude modulation it is relatively easy to develop pulse
shapes which produce selective excitation with a very good performance with
respect to criterion i). Unfortunately, the phase gradient across the
excitation region is usually quite large and the components which are not
aligned with the y axis after application of a soft pulse do not return to
the +z-axis by using the non-selective $90°_{-x}$ pulse. In principle, this
defect can be corrected by using the sequence d (Fig. 6d). This scheme
employs a shaped $90°_{-x}$ pulse followed by non-selective $180°_{-y}$ pulse with a

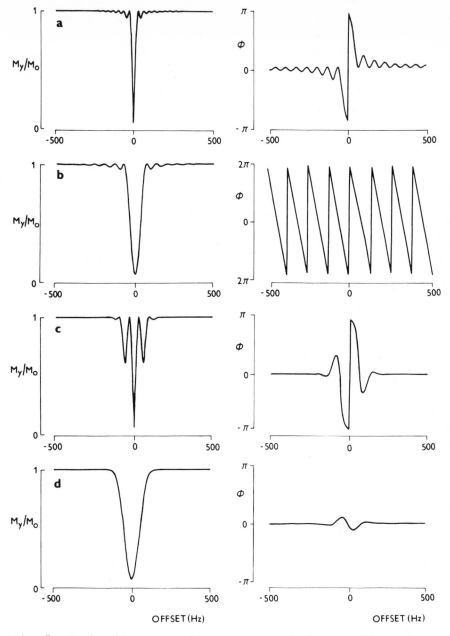

Fig. 7.   Excitation spectra (y-components and phase profiles) of
various hard/soft rf pulse combinations shown in Fig. 6a
(a), 6b (b), 6c (c) and 6d (d).   The spectra were computed
using the exact solution of the Bloch equation. Length of
the selective soft pulses was 15 ms in all cases.   The
delay of 7.5 ms was used to refocus the water signal after
the selective shaped pulse in the sequence d.

subsequent short delay, needed for all transverse water magnetization
components to refocus to the –y axis.   Then a nonselective 90°$_{-x}$ pulse

follows which brings water back to the z axis and excites the rest of the spectrum. The amplitude and phase profiles of this approach are shown in Fig. 7d. However, this sequence is quite sensitive to inhomogeneity of the rf field and experimentally obtainable water suppression per single scan is significantly lower than with schemes a–c. Also, a successful application of the self-refocused selective 90° pulses in combination with non-selective excitation developed recently by Warren et al. (1988) is very sensitive to experimental imperfections.

The experimental schemes which give in practice pure phase NMR spectra even in the region of water resonance are shown in Fig. 8. These sequences combine selective and nonselective excitation pulses with a short spin-lock period which destroys the water magnetization by rotating its components in an inhomogeneous rf field. Using this approach the water resonance is effectively saturated within a very short period of time. The two-stage sequences start with a soft $90°_{-y}$ pulse. In the simplest approach (sequence a) the rectangular long pulse is applied that rotates the water magnetiza-tion to the x axis. A subsequent non-selective 90° pulse followed by a short spin-lock period is used to measure the remaining z magnetizations. Appropriate phase cycling of these two pulses ($\phi$=x,-x,x,-x; $\psi$=y,y,-y,-y; receiver: +,-,+,-) leads to pure phase spectra which are obtained even when applying only very short spin-lock periods – of the order 2–3 ms. The water magnetization and out-off phase components of the measured resonances are defocused by the inhomogeneity of the rf field during the spin-lock pulse which is applied in the perpendicular direction to the water magnetization. Phase cycling ensures that good water suppression does not require poor rf homogeneity or very long spin-lock pulses. The degree of $H_2O$ suppression per individual scan is highly dependent on the inhomogeneity of rf field. On our 500 MHz system using a 3 ms soft pulse, suppression by a factor 30–50 is usually obtained. The residual water and x components of signals which are not destroyed applying the spin lock are effectively subtracted from the spectrum during the phase cycling procedure. This phase cycling constitutes the second stage of the water suppression scheme. The excitation profile, when using the rectangular soft pulse, is shown in Fig. 9a. Typically, the water signal is suppressed by a factor of about 300 with a soft pulse of 60 Hz rf field strength, which is more than sufficient to avoid any inter-ference of the water resonance with the rest of the spectrum. More importantly, only a very narrow region of resonances located about $\pm 0.25$ ppm around the water resonance is attenuated by more than 30%. A wider near zero excitation region is obtained by using the Gaussian soft pulse. Application of this pulse (Fig. 8b) improves the amplitude profile and provides more uniform excitation close to the water resonance (Fig. 9b). Although the water suppression is not as high as can be obtained with some other schemes, the phase of the residual water signal is absorptive and

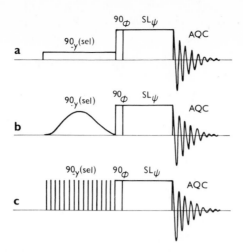

Fig. 8. Excitation schemes for the spin-lock water suppression. The rectangular soft pulse (a), shaped soft pulse (b), or the hard pulse sequence approximating the shaped soft pulse (c) are used for selective water excitation. Duration of the spin-lock period is adjusted to 2-2.5 ms. The four step phase cycle is applied to remove the residual components in the x-axis direction from the spectrum ($\phi$ = x,-x,x,-x; $\psi$ = y,y,-y,-y; receiver:+.-.+.-).

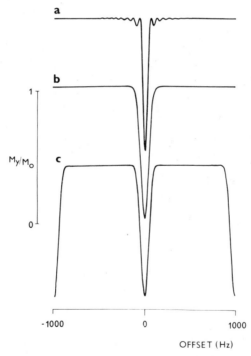

Fig. 9. Excitation profiles of the spin-lock water suppression sequences shown in Fig. 8a (a), 8b (b) and 8c (c). 15 ms soft pulses were applied with a rectangular (a) and a Gaussian envelope (b). In case of the spectrum (c) a Gaussian envelope was approximated by the DANTE pulse train with the component flip angles of 1.5, 3, 6, 9, 12, 13.5, 13.5, 12, 9, 6, 3 and 1.5 degrees. The time delay of 1 ms between the component rf pulses generates non-excitation sidebands at frequencies $\pm$ k.1000 Hz (k=1,2,3...).

therefore it does not interfere with the rest of the spectrum. Of course, this scheme is of saturation type but since the saturation occurs in a few milliseconds, the usual disadvantages of presaturation do not apply.

PULSE SHAPING FOR WATER SUPPRESSION

Even better excitation profiles can be obtained by shaping the rectangular envelope of the soft pulse in order to compress the excitation close
to the water resonance frequency. Such shaping functions can be designed
using different procedures. Very good excitation profiles result when
applying the approach developed recently by Z. Starcuk and coworkers (1988).
Their procedure is based on a straightforward idea. An analytically defined
waveform in the frequency domain is generated from a suitable set of base
functions. This frequency waveform is in the second step Fourier transformed to give the first order approximation of the pulse shaping function
in the time domain. The corresponding excitation spectrum is subsequently
computed from the transient solution of the Bloch equations in order to take
into account the non-linear behavior of a spin system. The whole procedure
is iteratively repeated varying the adjustable parameters in the analytical
expressions of the frequency waveform until the target excitation profile is
obtained. This enables one to compensate for the non-linear properties of
the spin system. In order to keep the procedure as simple as possible and
without a need of computerized optimization, the choice of base functions
plays a crucial role. Set of infinitely differentiable Gaussian functions
is one of the best solutions since all their derivatives are continuous and
the resulting excitation falls off at the fastest possible rate.

For example, two Gaussian functions can be used to derive a pulse shape
with a very good performance. The frequency spectrum was synthesized from
two Gaussians of the identical halfwidth of 150 Hz. In this case the
optimized separation amounted to 128 Hz (Fig. 10a). The resulting pulse was
16ms long. Its shape is presented in Fig. 10b. The computed excitation
profile when this shaped pulse is applied in combination with the spin-lock
water suppression sequence (Fig. 8b) is shown in Fig. 11. As can be seen
the double Gaussian pulse shaping gives an almost ideal region with near-
zero excitation. The width of this region is adjustable by varying the
length of the shaped pulse.

A combination of soft and hard pulses requires a system which enables
fast switching between the high and low power and a possibility to adjust
and keep the phase relationship between these two modes of operation
absolutely constant. These technical problems can be circumvented by
substituting a regular train of hard rf pulses of identical rf field
strength for an amplitude modulated soft pulse. An appropriate definition
of individual flip angles and phases makes it possible to approximate an

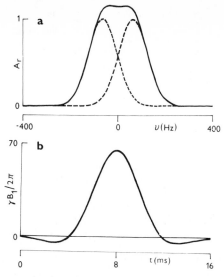

Fig. 10.   The frequency waveform (a) composed of two Gaussian
functions with halfwidths of 150 Hz and identical intensities.   Fourier transform of this waveform generates the 16
ms pulse envelope (b) with the maximum rf field strength
of 66 Hz.

arbitrary pulse shape.   This can be understood as a generalization of the
DANTE pulse sequence (Morris and Freeman, 1978).   Application of this
approach is shown in Fig. 12a where the double Gaussian shaped pulse is used
to derive the corresponding selective excitation pulse train.   Although the
component flip angles at the beginning and at the end of the sequence are

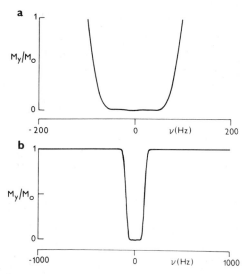

Fig. 11.   The calculated excitation spectra of the spin-lock water
suppression pulse sequence with the 16 ms double Gaussian
shaped soft pulse (scheme of Fig. 8b).   a) Expanded region
of a zero-excitation; b) the whole spectrum.

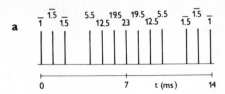

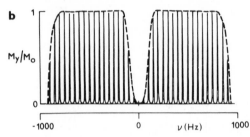

Fig. 12. The hard pulse sequence approximating the 16 ms double Gaussian shaped pulse (a) and the corresponding excitation spectrum measured using the spin-lock water suppression scheme of Fig. 8c (b). Numbers specify the flip angles of the component rf pulses; the phase shift by 180° is denoted by the overbar. The rf field was adjusted to 5.5 kHz giving the 1° pulse width of 0.5 μs; 2.5 ms spin-lock period was applied. The experimental result is compared with the computer simulation (dashed line).

quite small with properly adjusted electronics it is possible to achieve a very good experimental performance. A comparison of the theoretically computed and experimentally measured spectra is presented in Fig. 12b. As a result of the time sharing approximation, excitation sidebands are observed in the computed and measured spectra. Their frequency is given as $1/t$, where t denotes the time interval between the hard pulses in the excitation train. Both computed and measured spectra were obtained applying the spin-lock water suppression scheme shown in Fig. 8c.

INCORPORATION OF SELECTIVE EXCITATION IN TWO-DIMENSIONAL NMR EXPERIMENTS

In principle, there are two possibilities for incorporating the water suppression schemes, which apply the selective excitation in 2D NMR experiments. All proton pulses in a particular pulse sequence can be substituted by selective excitation trains, or the water suppression sequence can be applied as a read pulse for $t_2$ acquisition only. The first approach leaves the water magnetization unperturbed in the course of the whole experiment but modifies the excitation profiles in both $t_1$ and $t_2$ dimension; the second one with a nonselective excitation in the $t_1$ domain does not restrict the excitation in this dimension but brings water into the nonequilibrium state with a large transverse magnetization. For maximum water suppression it is important to minimize the presence of its transverse magnetization before a read pulse, otherwise the suppression in a single scan is much poorer and the dynamic range of the spectrometer becomes a problem.

To demonstrate, 2D NOE spectroscopy is chosen as an example. Water suppression in 2D NOE spectroscopy is most easily accomplished by replacing the last pulse of the NOESY sequence by a selective read pulse (Fig. 13). A homospoil pulse (HSP) is used during the mixing period. This gives an additional advantage of removing coherent transfers through single-quantum and higher orders of coherence. On our system an 8 ms homospoil pulse is sufficient to eliminate the water resonance almost completely. An equal delay following the HSP is needed for recovery of the deuterium lock and for the decay of eddy currents.

An application of the 1–1 ECHO sequence to the study of exchangeable protons in the double stranded DNA fragment formed by the Dickerson dodecamer d(CGCGAATTCGCG) using 2D NOE spectroscopy is taken as an example. It is possible to choose a single $\tau$ value in the 1–1 sequence that ECHO provides near-optimal excitation of both imino and amino protons in this DNA (Sklenar et al., 1987b). Chemical shift ranges of the amino protons (6.2–8.5 ppm) and imino protons (12.2–14.5 ppm) in double stranded DNA allow to specify the optimal delay as:

$$\tau = 1/(11.2*f_o) \qquad\qquad (4)$$

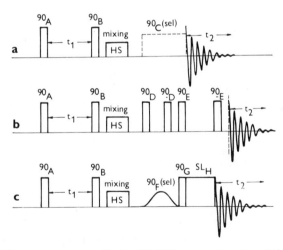

Fig. 13.   Pulse sequences of the 2D NOE experiments with a selective 90° read pulse for $t_2$-acquisition. a) general scheme; b) a version with the 1–1 ECHO excitation read pulse; c) a version with a spin-lock water suppression scheme. Phases A and B of the non-selective 90° pulses are cycled as in the regular experiment without water suppression; the phases of selective read pulses follow the schemes of 1D versions appropriately shifted with respect to the phase cycles A and B.

where $f_0$ stands for the working frequency of the spectrometer specified in MHz. Using this setting, a maximum excitation is achieved at 7.5 and 13.1 ppm with the carrier positioned to the water resonance at 4.7 ppm. A 2D NOE spectrum obtained at 35°C in 90% $H_2O$ with 500 optical density units of the dodecamer duplex is displayed in Fig. 14. This spectrum shows the connectivities of imino to amino protons. Cross sections taken through the 2D data matrix at the resonance positions of $GH_1$ and $TH_3$ imino protons are shown in Fig. 15. Using these 2D NOE data measured in $H_2O$ and data previously published by D. Hare et al. (1983), almost all crosspeaks have been assigned.

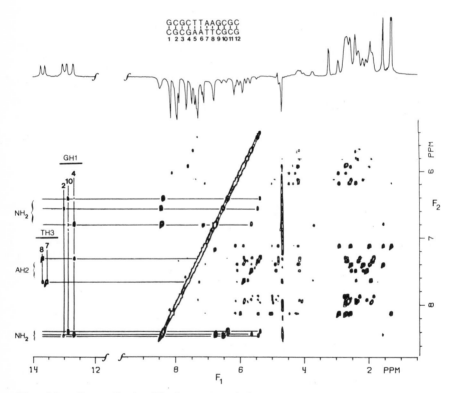

Fig. 14.   Part of the 2D phase sensitive NOE spectrum measured at 500 MHz displaying the interactions with the imino resonances of d(CGCGAATTCGCG)$_2$. The data were obtained using 500 O.D. of DNA dissolved in 0.5 ml 90% $H_2O$/10% $D_2O$ with 10 mM phosphate buffer, 100 mM NaCl, pH=7.0 at 30°C.

The 1–1 ECHO selective read pulse with $\tau$ = 179 μs and the mixing time of 100 ms was applied. The spectrum on top of the 2D matrix was obtained using the 1D version of the 1–1 ECHO pulse sequence.

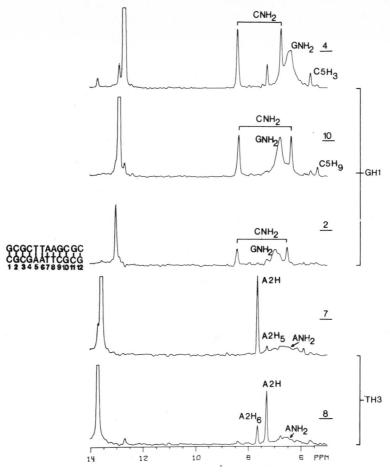

Fig. 15.  Cross sections through the 2D phase sensitive NOE spectrum of d(CGCGAATTCGCG)$_2$ shown in Fig. 14.  The sections have been taken at $F_2$ frequency of the GH (2,4,10) and TH (7,8) imino protons and show connectivities to the nearby protons.  No baseline correction has been applied.

As the second example, the 2D NOE spectrum of 11 mM small globular protein BPTI is shown in Fig. 16.  The spectrum was measured using the spin-lock water suppression sequence as a read pulse.  The mixing time was 125 ms and spectra were obtained at 42°C.  The spectrum on the top of the 2D matrix was measured by applying a 1D version of this water suppression scheme.  It demonstrates that the residual water signal is indeed in phase with the measured resonances.  As can be seen, this water suppression sequence supplies a 2D data matrix which is highly symmetrical, having a narrow gap with near-zero excitation close to the water resonance along the $F_2$ frequency axis.  This allows an identification of connectivities between the amino and CH protons resonating in the immediate vicinity of water.

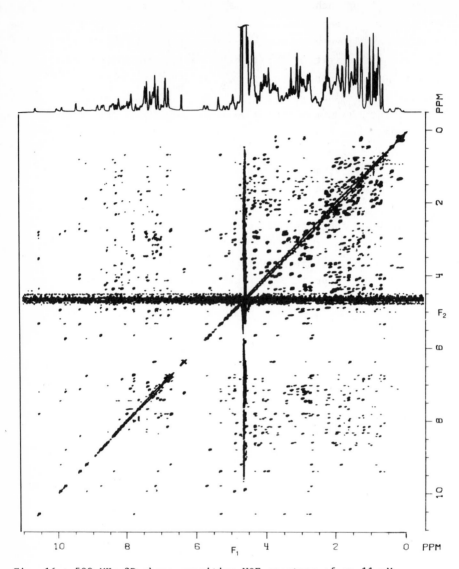

Fig. 16.   500 MHz 2D phase sensitive NOE spectrum of an 11 mM
solution of BPTI obtained with a 125 ms mixing time in 90%
$H_2O/10\%$ $D_2O$ (pH=6.6, 42°C, 100 mM NaCl). The spectrum was
measured by applying the spin-lock water suppression
scheme with a rectangular soft pulse.   For non-selective
90° pulses and the spin-lock period ($\tau$ = 2.42 ms) a 7.9
kHz rf field strength was applied; a low power rf pulse
with a 61 Hz rf field strength was used. The spectrum
results from 2x350x1024 data matrices with 64 scans per $t_1$
value, recycle time of 1.5 s, and the total measuring time of
9.5 hours.

CONCLUSIONS

Many different approaches for eliminating the intense water resonance from proton NMR spectra have been proposed in the past. Selective excitation techniques that leave water magnetization unperturbed at the end of the excitation train constitute an attractive variant for studies of exchangeable protons. Their resonances form an indispensable source of information for the elucidation of biomolecular structure in solution. For measurements of absorption mode 2D NMR spectra the selective excitation techniques which produce spectra free of phase distortions are required. Two-stage selective excitation methods which were described in this contribution offer several advantages: they produce pure phase NMR spectra that are free of baseline undulations and the water suppression attained in the second stage can be very high. Spin echo selective excitation sequences are easy to adjust and the EXORCYCLE phase cycling effectively eliminates the broad humps underneath the water signal. The excitation schemes which couple a combination of soft and hard rf pulses with a short spin-lock period produce excitation profiles with narrow nonexcitation regions where resonances very close to the water signal can be observed. These sequences require special hardware modifications and are more demanding on instrumental stability.

ACKNOWLEDGEMENT

I am indebted to Ad Bax for many stimulating discussions, cooperation and warm hospitality during my stay in his lab at National Institutes of Health in Bethesda, MD. I also wish to express my thanks to Zenon Starcuk for making available his manuscript (Starcuk et al., 1988) prior to publication and for a longtime cooperation. Stimulating discussions with L. Pucek, R. Fiala, and P. Kessler and their help with simulations are greatly acknowledged.

REFERENCES

Bax, A., Sklenar, V., Clore, M., and Gronenborn, A., 1987, J. Amer. Chem. Soc., 109:6511.
Benz, F. W., Feeney, J., and Roberts, G. C. K., 1972, J. Magn. Reson., 8:114.
Bleich, H. E., and Glasel, J. A., 1975, J. Magn. Reson., 18:401.
Bodenhausen, G., Freeman, R., and Turner, D. L., 1977, J. Magn. Reson., 27:511.
Bryant, R. G., and Eads, T. M., 1985, J. Magn. Reson., 64:312.
Campbell, I. D., Dobson, C. M., and Ratcliffe, R. G., 1977, J. Magn. Reson., 27:455.
Clore, G. M., Kimber, B. J., and Gronenborn, A. M., 1983, J. Magn. Reson., 54:170.
Connor, S., Nicholson, J. K., and Everett, J. R., 1987, Anal. Chem., 59:2885.
Eads, T. M., Kennedy, S. D., and Bryant, R. G., 1986, Anal. Chem., 58:1752.

Gupta, R. K., 1976, J. Magn. Reson., 24:461.

Gupta, R. K., and Gupta, P., 1979, J. Magn. Reson., 34:657.

Gutow, J. H., McCoy, M., Spano, F., and Waren, W. S., 1985, Phys. Rev. Lett., 55:1090.

Hall, M. P., and Hore, P., 1986, J. Magn. Reson., 70:350.

Hare, D. R., Wemmer, D. E., Chou, S. H., Drobny, G., and Reid, B. R., 1983, J. Mol. Biol., 171:319.

Hore, P. J., 1983a, J. Magn. Reson., 54:539.

Hore, P. J., 1983b, J. Magn. Reson., 55:283.

Hoult, D. I., 1976, J. Magn. Reson., 21:337.

Jesson, J. P., Meakin, P., and Kneissel, G., 1973, J. Amer. Chem. Soc., 95:618.

Krischna, N. R., 1976, J. Magn. Reson., 22:555.

Krugh, T. R., and Schaefer, W. C., 1975, J. Magn. Reson., 19:99.

Levitt, M. H., and Roberts, M. F., 1987, J. Magn. Reson., 71:576.

Levitt, M. H., Sudmeier, J. L., and Bachovchin, W. W., 1987, J. Amer. Chem. Soc., 109:6540.

McCoy, M., and Waren, W. S., 1987, Chem. Phys. Lett., 133:165.

Moobery, E. S., and Krugh, T. R., 1973, J. Magn. Reson., 17:128.

Morris, G., and Freeman, R., 1978, J. Magn. Reson., 29:433.

Morris, G. A., and Smith, K. I., 1986, J. Magn. Reson., 68:350.

Patt, S., and Sykes, B. D., 1972, J. Chem. Phys., 56:3182.

Plateau, P., Dumas, C., and Gueron, M., 1983, J. Magn. Reson., 54:46.

Plateau, P., and Gueron, M., 1982, J. Amer. Chem. Soc., 104:7310.

Rabenstein, D. L., Fan, S., and Nakashima, T. T., 1985, J. Magn. Reson., 64:541.

Rabenstein, D. L., and Fan, S., 1986, Anal. Chem., 58:3178.

Redfield, A. G., Kunz, S. D., and Ralph, E. K., 1975, J. Magn. Reson., 19:275.

Redfield, A. G., and Kunz, S., in: "NMR and Biochemistry," (S. J. Opella and P. Lu, Eds.), pp. 225-239, Dekker, New York, 1979.

Schaefer, J., 1972, J. Magn. Reson., 6:670.

Sklenar, V., and Bax, A., 1987a, J. Magn. Reson., 74:469.

Sklenar, V., and Bax, A., 1987b, J. Magn. Reson., 75:378.

Sklenar, V., Brooks, B., Zon, G., and Bax, A., 1987b, FEBS Lett., 216:249.

Sklenar, V., and Starcuk, Z., 1982, J. Magn. Reson., 50:495.

Sklenar, V., Tschudin, R., and Bax, A., 1987a, J. Magn. Reson., 75:352.

Starcuk, Z., Pucek, L., and Starcuk, Jr., Z., 1988, J. Magn. Reson., 80:352.

Starcuk, Z., and Sklenar, V., 1985, J. Magn. Reson., 61:567.

Starcuk, Z., and Sklenar, V., 1986, J. Magn. Reson., 66:391.

Turner, D. L., 1983, J. Magn. Reson., 54:146.

von Kienlin, M., DeCorps, M., Albrand, J. P., Foray, M. F., and Blondet, P., 1988, J. Magn. Reson., 76:169.

Wang, C., and Pardi, A., 1987, J. Magn. Reson., 71:154.

Warren, W. S., McCoy, M., and Hasenfeld, A., Presentation at 29th Experimental NMR Conference, Rochester, New York, 1988.

Yan, H., and Gore, J. C., 1987, J. Magn. Reson., 71:116.

Zuiderweg, E. R. P., Hallenga, K., and Olejniczak, E. T., 1986, J. Magn. Reson., 70:336.

OXYGEN EXCHANGE AND BOND CLEAVAGE REACTIONS OF CARBOHYDRATES STUDIED
USING THE $^{18}O$ ISOTOPE SHIFT IN $^{13}C$ NMR SPECTROSCOPY

Tony L. Mega and Robert L. Van Etten*

* Purdue University
Chemistry Department
W. Lafayette, IN  47907

Isotopic labeling, particularly through the use of $^{18}O$-enriched water,
has been important in defining our present understanding of the mutarotation
process and reactions involving carbohydrate bond cleavage (Rittenberg and
Graff, 1958; BeMiller, 1967; Capon, 1969). A related reaction, the exchange
of oxygen at the anomeric carbon atom of reducing sugars, may also be observed by the use of $^{18}O$-water, and an understanding of the kinetics of this
exchange process is an important prerequisite to carbohydrate studies such
as bond cleavage reactions that make use of the $^{18}O$ label. Unfortunately,
there are relatively few data in the literature that characterize the oxygen
exchange reactions of even the most biologically important monosaccharides.
No comparable investigation has appeared since Rittenberg and Graff's study
of glucose oxygen exchange in 1958. As a result, meaningful comparisons of
oxygen exchange rates between different sugars have been quite difficult due
to the wide variety of experimental conditions that have been used (Risley
and Van Etten, 1982).

Noting that the lack of data on the rates of oxygen exchange in sugars
was to a large extent the result of the cumbersome nature of the analytical
techniques that had been required for their measurement, we demonstrated
that a technique pioneered in our laboratory, the $^{18}O$ isotope shift in $^{13}C$
NMR spectroscopy (Risley and Van Etten, 1979), could be employed to acquire
a wealth of oxygen exchange data for sugars in a nearly continuous assay
mode (Risley and Van Etten, 1982). This advancement in technique thus
opened up the possibility that a wide range of oxygen exchange data might be
obtained and in this way could facilitate careful comparisons between
structure and reactivity among the biologically important monosaccharides.

The possibility to study oxygen exchange events at carbon using $^{13}C$ NMR
is a result of the small upfield shifts in the $^{13}C$ NMR signals when $^{16}O$

*NMR Applications in Biopolymers*, Edited by
J. W. Finley *et al.*, Plenum Press, New York, 1990

atoms directly bonded to the $^{13}$C nuclei under observation are replaced by $^{18}$O atoms. For example, the $^{13}$C NMR signal for the carbonyl carbon of [$^{18}$O]acetone is shifted upfield by 0.050 ppm with respect to the carbonyl signal for [$^{16}$O]acetone (Fig. 1). By measuring the relative intensities of the two signals, the percentage incorporation of $^{18}$O in the sample may be readily estimated.

The magnitude of the isotope shift at the anomeric carbon atoms of several sugars have been measured in this laboratory and the results are listed in Table 1. In general, the magnitude of the isotope shift at the anomeric carbon is approximately 0.02 ppm, but some variation is seen among the different anomers of a given sugar. Since it has been found that the isotope shift is to some extent influenced by the hybridization of the carbon atom (Risley and Van Etten, 1980; Vederas, 1980; Diakur et al., 1980), it seemed possible that variations in its magnitude among the different anomers might arise from the influence of the "anomeric effect." However, this interpretation does not appear to be consistent with all of the available data on the preferred conformations of the pyranose sugars listed in Table 1. For example, although the α-pyranose forms of D-glucose and 2-deoxy-D-ribose each exhibit larger isotope shifts than the β-pyranose forms, it is the α-pyranose form of glucose and the β-pyranose form of 2-deoxy-D-ribose that would be expected to exhibit the anomeric effect

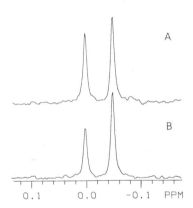

$^{13}$C NMR Spectra of $^{18}$O-Labeled Acetone

Fig. 1.   Shown are the carbonyl regions of $^{13}$C NMR spectra of (A) 56% $^{18}$O-labeled and (B) 64% $^{18}$O-labeled acetone. The presence of an $^{18}$O attached to the carbonyl carbon causes a small (0.050 ppm) upfield shift in its $^{13}$C NMR signal with respect to the $^{16}$O species (referenced to 0 ppm in the figure). Thus, by measuring the relative intensities of the two signals one may determine the $^{18}$O content of the sample in a nondestructive mode.

Table 1. $^{18}O$ Isotope-induced Shifts for the Anomeric Carbons of Various Sugars[a]

| Sugar | Isotope shift | | (ppm) | |
|---|---|---|---|---|
| | α–pyranose | β–pyranose | α–furanose | β–furanose |
| D–Erythrose[b] | | | 0.017 | 0.019 |
| D–Ribose | 0.015 | 0.019 | 0.017 | 0.016 |
| 2–Deoxy–D–ribose | 0.020 | 0.017 | 0.017 | 0.017 |
| D–Glucose | 0.019 | 0.017 | | |
| D–Mannose | 0.018 | 0.017 | | |
| D–Fructose | 0.021 | 0.023 | 0.022 | 0.021 |

[a] All isotope shifts are upfield with respect to the unlabeled ($^{16}O$) sugar (Table 4 from [Mega, 1989]).

[b] In unbuffered water containing 25% $D_2O$ (Risley and Van Etten, 1982).

(Stoddart, 1969; Durette and Horton, 1971; Lemieux et al., 1971). It appears that some additional rationale must be sought to explain all of the trends noted in Table 1.

The utilization of the $^{18}O$ isotope–induced shift in $^{13}C$ NMR to follow the exchange of oxygen at the anomeric carbon atom of a monosaccharide is illustrated for the case of D–fructose (Fig. 2). In this experiment, a highly $^{18}O$–labeled sample of D–fructose was placed into a solution of ordinary water and the loss of the $^{18}O$ label from β–fructofuranose and β–fructopyranose was observed over a period of several hours. (The α–furanose and α–pyranose forms are not shown because they comprise less than 10% of the fructose in solution [Angyal, 1984]). An alternative oxygen exchange experiment was also conducted in which the unlabeled sugar incorporates the label from $^{18}O$–enriched water, although this latter approach is more costly due to the larger amounts of $^{18}O$–water that are generally required. By either of these experimental approaches an oxygen exchange rate constant may be obtained by making the appropriate logarithmic plot of the change in the percentage incorporation of $^{18}O$ into the sugar with time (Model et al., 1968; but see also Röhm and Van Etten, 1986). The resulting kinetic plots are found to be linear. Furthermore, the rate constants obtained from both types of experiments are found to be the same regardless of which anomer is studied. This can be seen qualitatively for the different anomers of D–fructose in Fig. 2. The identity in exchange rates for different anomers has also been found for the other sugars which

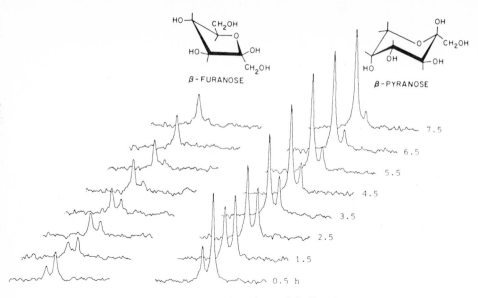

β-FURANOSE

β-PYRANOSE

7.5
6.5
5.5
4.5
3.5
2.5
1.5
0.5 h

Oxygen Exchange Kinetics of D-Fructose

Fig. 2.   $^{13}C$ NMR spectra illustrating the loss of the $^{18}O$ label from the predominant anomers in a 20 mM solution of 70% $^{18}O$-labeled [2-$^{13}C$]fructose in 20% $D_2O$ at pH 1.7 and 25°C. The signals for the anomeric carbon atoms of the β-furanose and β-pyranose ring forms appear at 102.31 ppm and 98.87 ppm, respectively. The signals for the $^{18}O$-labeled anomers are shifted upfield relative to the $^{16}O$ species by 0.021 ppm for β-fructofuranose and 0.023 ppm for β-fructopyranose. A logarithmic plot of the peak intensities gave an oxygen exchange rate constant ($k_{ex}$) of 6.4 x $10^{-5}$ $s^{-1}$.

have been studied (Mega, 1989). This result is consistent with an oxygen exchange mechanism in which the open chain carbonyl intermediate undergoes a relatively slow reversible hydration reaction in competition with a rapid ring-closure reaction that produces the various ring forms.

The ease of data collection afforded by the present $^{13}C$ NMR technique allowed a careful comparison of the pH dependence of the oxygen exchange reactions of several of the most biologically important sugars. In contrast to previous results based on more limited data (Rittenberg and Graff, 1958), it was found that the monosaccharide oxygen-exchange reaction depends linearly on both the hydronium and the hydroxide ions. Moreover, distinct regions of water-catalysis are observed, particularly for the aldoses glucose, mannose and ribose, which exhibit pH-independent regions extending over approximately 3 pH units. These results are summarized in Fig. 3, which shows pH-rate profiles constructed for various sugars. These

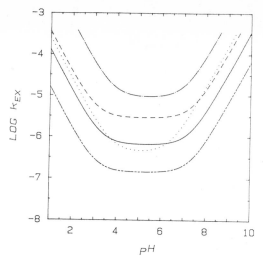

pH-Rate Profiles for Oxygen Exchange of Various Sugars

Fig. 3.   The pH-dependence of the oxygen exchange kinetics of
(·————·) 2-deoxy-D-ribose, (— — —) D-ribose, (·····)
D-fructose, (————) D-mannose, and (- - ——) D-glucose at
25-26°C (Mega, 1989; Mega, Cortes and Van Etten, 1990).

curves are based upon a non-linear least-squares fit of the oxygen exchange
data to Eq. (1) using the program KINFIT (Knack and Röhm, 1981).

$$k_{ex} = k_w + k_{H+}[H+] + k_{OH-}[OH-] \tag{1}$$

In Eq. (1), $k_{ex}$ is the observed pseudo-first-order oxygen exchange rate
constant and $k_w$, $k_{H+}$, and $k_{OH-}$ represent the pH-independent, hydronium
ion-dependent and hydroxide ion-dependent reactions, respectively.

The relative rates of oxygen exchange between the various sugars that
we have studied can be rationalized providing that one takes into account
the relative proportions of the open chain species for the different sugars.
For example, the ketohexose D-fructose exhibits exchange rates that range
from 3- to 20-fold faster than the rates observed for the aldohexose
D-glucose. At first this appears surprising, since simple aldehydes are
known to exchange oxygen at much faster rates than simple ketones (Cohn and
Urey, 1938). However, the proportion of D-fructose that is present in the
open chain form in aqueous solution at 25°C is over 2 orders of magnitude
greater than is the case for D-glucose (Hayward and Angyal, 1977). Thus,
although the actual oxygen exchange event at the carbonyl carbon of the open
chain form of glucose is much faster than that at the carbonyl carbon of the
open chain form of fructose, a more rapid exchange rate is observed in
fructose because such a relatively large proportion of the reactive species
is available.

Wertz et al. (1981) derived the equation that relates the rate of oxygen exchange to the proportion $N_c$ of the sugar that is present in the open chain form.

$$k_{ex} = \frac{1}{2} N_c k_{hy} \tag{2}$$

Given values of $N_c$, one may then use this equation to calculate the pseudo-first-order rate constant, $k_{hy}$, for the hydration reaction of the open chain species which underlies the oxygen exchange process. When such calculations are made it is found that the kinetics of monosaccharide hydration can be interpreted in terms of steric and inductive effects by drawing analogies with the hydration kinetics of simple aldehydes and ketones. For example, it has been found that the introduction of electron withdrawing groups adjacent to the carbonyl carbon in simple aldehydes and ketones decreases the rate of hydration in the region of acid catalysis and increases the rate of hydration in the region of water catalysis (Buschman et al., 1982). This is the trend that is observed when the hydration kinetics of the polyhydroxy aldehyde glucose (calculated from Eq. (2) together with the experimental oxygen exchange results) are compared with those of acetaldehyde (Mega, 1989).

The straightforward interpretation of monosaccharide oxygen exchange kinetics in terms of the hydration of the small percentage of the open chain species that is available is significant because it is at odds with a proposal that had been advanced earlier to rationalize other oxygen exchange results. Because of the seemingly slow rate of glucose oxygen exchange relative to mutarotation, Isbell et al. (1969) proposed the existence of "pseudo-acyclic" intermediates that would allow interconversion between the various ring forms without passage through a free aldehyde which would be subject to oxygen exchange. Such novel intermediates are unnecessary under the present interpretation since the hydration events underlying the oxygen exchange kinetics are characterized by normal reaction rates. In fact, as noted above, the calculated rate of hydration of glucose in the region of water catalysis exceeds that observed in acetaldehyde. Furthermore, the relatively fast rate of mutarotation may be explained by estimating the effective molarities of the backbone hydroxyl groups of the sugars in the ring-closure reaction (Mega, 1989). In glucose, for example, we estimate that the effective molarity of the C-5 hydroxyl group is on the order of $10^5$ M--a value that is consistent with other intramolecular reactions involving the hydroxyl group. Thus, the increased rate of mutarotation over oxygen exchange may be reasonably explained in terms of the large entropic advantage enjoyed by the backbone hydroxyl groups in the intramolecular ring-closure reaction as compared to the intermolecular hydration reaction (Kirby, 1980; Mega et al., 1990).

In addition to their intrinsic interest, monosaccharide oxygen exchange rates are often critical to the interpretation of studies involving carbohydrate bond cleavage reactions that employ $^{18}O$ labeling. The position of bond cleavage in the acid-catalyzed hydrolysis of the disaccharide sucrose, for example, remained unresolved for decades, in part because of uncertainties associated with experimental techniques, but also because of the lack of adequate kinetic data about the oxygen exchange processes of the reaction products (Oon and Kubler, 1982). If the rates of oxygen exchange in the hydrolysis products were very fast relative to the rate of bond cleavage then isotopic labeling with $^{18}O$-water would be of little use in establishing the position of bond cleavage. The availability of kinetic data such as that illustrated in Fig. 3 made possible the selection of reaction conditions under which the position of bond cleavage in sucrose could be determined.

A series of $^{13}C$ NMR spectra that illustrate the incorporation of $^{18}O$ into the products upon the hydrolysis of sucrose in 60% $^{18}O$-water are shown in Fig. 4. (A more complete series of spectra may be found in Mega and Van Etten [1988]). In the portions of the spectra that are shown, the disappearance of the signal corresponding to the glucosyl C-1 carbon of sucrose is accompanied by the appearance of the signals for the anomeric carbon atoms of the predominant forms of fructose and glucose. For the glucose anomers that are formed upon hydrolysis, it can be seen that the $^{18}O$ content slowly increases until it reaches the level of 60%. In contrast, the fructose that is formed appears to be 60% $^{18}O$-labeled from the very early stages of the reaction. In fact, the fructose which is formed has an $^{18}O$ content that averages $60 \pm 2\%$ throughout the reaction, while the extent of incorporation of $^{18}O$ into the glucose anomers is entirely consistent with the kinetics of glucose oxygen exchange under the particular reaction conditions that were employed in the experiment.

These observations strongly support a sucrose hydrolysis mechanism in which the glucose that is initially formed retains the sucrose bridge oxygen atom. The latter is then only slowly replaced by the $^{18}O$ present in the solvent as a result of the oxygen exchange process. Under this fructosyl-oxygen bond cleavage mechanism, the fructose that is produced would be expected to be labeled at the level of the $^{18}O$ present in the solvent. The spectra that were obtained support this expectation. It was also possible to make independent measurements of fructose oxygen exchange under identical reaction conditions in order to rule out the possibility that initially unlabeled fructose had rapidly become labeled as a result of the oxygen exchange process.

Analysis of the data does not support the presence of any significant amount of glucosyl-oxygen cleavage. Instead, the results are most simply

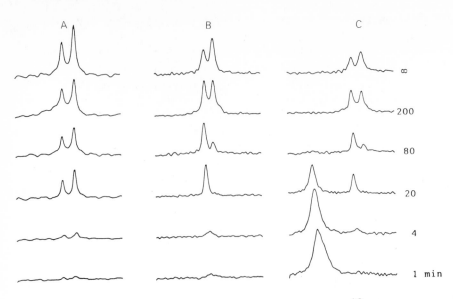

| | | |
|---|---|---|
| A | B | C |

∞

200

80

20

4

1 min

The Acid-Catalyzed Hydrolysis of Sucrose in 60% $^{18}$O-Water

Fig. 4.   A solution of 0.9 M sucrose plus 2 M NaCl was hydrolyzed in 60% $^{18}$O-water (60% $H_2^{18}O$, 20% $D_2O$, 20% $H_2O$) at pH 2.9 and 80°C. The spectra show the incorporation of $^{18}$O into the hydrolysis products. The presence of $^{18}$O causes an upfield shift in the $^{13}$C NMR signals of 0.023 ppm for β-fructopyranose (A), 0.017 ppm for β-glucopyranose (B), and 0.019 ppm for α-glucopyranose (C). The disappearance of the sucrose (glucosyl) C-1 peak can also be seen in (C). Upon hydrolysis, the fructose that is produced is $^{18}$O-labeled to the extent of 60% (A), however the glucose that is produced becomes $^{18}$O-labeled only slowly as a result of the oxygen exchange reaction (B and C) (Mega and Van Etten, 1988).

interpreted in terms of a hydrolysis mechanism in which the initial products are α-D-glucopyranose and a fructosyl carboxonium ion. Independent experimental evidence using $^{1}$H NMR under similar reaction conditions has also demonstrated the initial formation of α-D-glucopyranose upon hydrolysis, thus providing important confirmation of these results (Mega and Van Etten, 1988).

Thus it can be seen that the use of the $^{18}$O isotope shift in $^{13}$C NMR spectroscopy provides a convenient and powerful technique for the study of carbohydrates in solution. The ability to simultaneously observe several different anomeric species as well as their $^{18}$O-isotopomers under identical reaction conditions allows for valuable insights into the tautomeric equilibria of monosaccharides as well as the straightforward interpretation of carbohydrate bond cleavage processes.

ACKNOWLEDGEMENTS

This research was supported by U.S. Public Health Service Research Grant GM27003 from the National Institute of General Medical Sciences and by instrumentation grants from the National Institutes of Health (RR01077) and the NSF (BBS-8714258).

REFERENCES

Angyal, S. J., 1984, Adv. Carbohydr. Chem. Biochem., 42:15-68.
BeMiller, J. N., 1967, Adv. Carbohydr. Chem., 22:25-108.
Buschmann, H.-J., Dutkiewicz, E., and Knoche, W., 1982, Ber. Bunsenges. Phys. Chem., 86:129-134.
Capon, B., 1969, Chem. Rev., 69:407-498.
Cohn, M., and Urey, H. C., 1938, J. Am. Chem. Soc., 60:679-687.
Diakur, J., Nakashima, T. T., and Vederas, J. C., 1980, Can. J. Chem., 58:1311-1315.
Durette, P. L., and Horton, D., 1971, Adv. Carbohydr. Chem. Biochem., 26:49-125.
Hayward, D. L., and Angyal, S. J., 1977, Carbohydr. Res., 53:13-20.
Isbell, H. S., Frush, H. L., Wade, C. W. R., and Hunter, C. E., 1969, Carbohydr. Res., 9:163-175.
Kirby, A. J., 1980, Adv. Phys. Org. Chem., 17:183-278.
Knack, I., and Röhm, K.-H., 1981, Hope-Seyler's Z. Physiol. Chem., 362:1119-1130.
Lemieux, R. U., Anderson, L., and Conner, A. H., 1971, Carbohydr. Res., 20:59-72.
Model, P., Ponticorvo, L., and Rittenberg, D., 1968, Biochemistry, 7:1339-1347.
Mega, T. L., and Van Etten, R. L., 1988, J. Am. Chem. Soc., 110:6372-6376.
Mega, T. L., 1989, Ph. D. Thesis, Purdue University, West Lafayette, IN.
Mega, T. L., Cortes, S., and Van Etten, R. L., 1990, J. Org. Chem., 55:522-528.
Oon, S. M., and Kubler, D. G., 1982, J. Org. Chem., 47:1166-1171.
Risley, J. M., and Van Etten, R. L., 1979, J. Am. Chem. Soc., 101:252-253.
Risley, J. M., and Van Etten, R. L., 1980a, J. Am. Chem. Soc., 102:4609-4614.
Risley, J. M., and Van Etten, R. L., 1980b, J. Am. Chem. Soc., 102:6699-6702.
Risley, J. M., and Van Etten, R. L., 1982, Biochemistry, 21:6360-6365.
Rittenberg, D., and Graff, C., 1958, J. Am. Chem. Soc., 80: 3370-3372.
Röhm, K. H., and Van Etten, R. L., 1986, Arch. Bioch. Bioph., 244:128-136.
Stoddart, 1969, "Stereochemistry of Carbohydrates," R. N. Castle, Ed.; Wiley-Interscience: New York, pp. 35-55.
Vederas, J. C., 1980, J. Am. Chem. Soc., 102:374-376.
Wertz, P. W., Garver, J. C., and Anderson, L., 1981, J. Am. Chem. Soc., 103:3916-3922.

COMPUTER-AIDED CONFORMATIONAL ANALYSIS BASED ON NOESY SIGNAL INTENSITIES

Niels H. Andersen, Xiaonian Lai, Philip K. Hammen and
Thomas M. Marschner

Department of Chemistry
University of Washington
Seattle, WA  98195

ABSTRACT

The basis for the development of a suite of programs that allow the
user to determine motional features (the correlation time and the significance of segmental motion) and the optimum conditions for future experiments from a NOESY signal matrix is presented.  This automated evaluation
of NOESY data serves as the initial step of an iterative conformational
analysis which uses the molecular model manipulation capabilities of modern
graphics workstations.  Incorporated in these programs is NOESYSIM, a
calculation subroutine which uses a set of molecular coordinates (and a
correlation time estimate) together with user entered experimental parameters (acquisition time, sweep width, mixing time and cycle repetition
time) to generate an accurately calculated NOESY signal matrix reflecting
those conditions and the specified conformational model.  Conformational
refinement then consists of iterative comparisons of the experimental
signal matrix with a series (or systematically sampled set) of model
coordinates corresponding to a dynamics' course, driven-minimization or
torsional grid search.  These procedures and developments are illustrated
with examples including: solution conformations of prostanoids; studies of
the folding preferences and media-dependent changes in conformation for
peptide hormones; and the structure elucidation of a novel undecapeptide
macrolide antibiotic (lysobactin).  For larger molecules, even constrained
grid searches have too high a dimensionality and one must resort to
distance-constraint based minimizations.  A novel procedure for deriving
more accurate distance constraints (corrected for secondary NOEs) is
detailed and a new strategy for conformation elucidation, based on this
procedure, is outlined.

*NMR Applications in Biopolymers*, Edited by
J. W. Finley *et al.*, Plenum Press, New York, 1990

# INTRODUCTION

The question of how to properly convert NMR data into a dynamic or static structure hypothesis is being explored in many laboratories now that 2D methods and modern high-field instruments make the collection of extensive data sets for complex biopolymers nearly routine [1]. Three types of NMR data are available for structure evaluation: chemical shifts, scalar coupling constants, and relaxation data. Intramolecular relaxation is dominated by distance-dependent dipolar coupling, leading to the well known nuclear Overhauser effect (NOE) [2]. The local and intermediate-range geometric dependence of chemical shifts is very complex and, at present, does not appear to provide a means for structure elucidation. Scalar coupling through three-bond pathways can provide conformational insights. Typically the coupling constant has a definable dependence on the proton dihedral angle ($\theta$) and thus the values observed, if they can be attributed to a single static structure, can be used to derive torsional constraints or target values for modeling. However the typical Karplus relationships between $\theta$ and $^3J$ often allow for two or more indistinguishable ranges of $\theta$ that can explain any specific measured J-value. Thus, NOEs remain as the single most useful structure-dependent NMR measurement.

It has been recognized for some time that a collection of NOEs (or intensities from a 2D NOESY spectrum) should, in principal, provide the conformation of a substance as a solution to a distance-geometry problem. In the case of typical 1D NOE experiments, it is well known that the magnetization transfer rate (which is proportional to $r^{-6}$) corresponds to the initial growth rate of the fractional enhancement.[#]

$$-\sigma_{ij} = \left[ \frac{1}{\tau_m} \left( f_j[i] \right) \right]_{\tau_m \rightarrow 0}$$

Extensions of this relationship to NOESY data can be made if the data lie in the linear growth region. If the isolated spin-pair approximation (ISPA) [3, 4] is valid, one obtains--

$$\frac{r_{ij}}{r_{ref}} = \left( \frac{S_{ref}}{S_{ij}} \right)^{\frac{1}{6}}_{\tau_m \rightarrow 0} \quad \text{and} \quad S_{ij} \simeq n_j \sigma_{ij} \tau_m S^0_{ii} \qquad (1a, 1b)$$

---

[#] We will use $\tau_m$ as the symbol for the period during which magnetization transfer (or cross-relaxation) is allowed for both 1D and 2D experiments. Sigma ($\sigma$) is defined as a magnetization transfer rate (rather than another decay rate [6]), following the lead of Kalk and Barendsen [5]. The full details of this formalism appear in the next section.

As noted by Borgias and James [4c] distances from an ISPA calculation work
quite well for protein structural studies. "This success...is largely due
to the fact that accurate interproton distances are not required. For
tertiary structure, ...the judicious use of 'long-range distances' [gives]
overall structures reasonably well-constrained by important loop-closing
inter-residue contacts. Similarly, secondary structure classification into
$\alpha$-helix and $\beta$-sheet can be made on the basis of existence or absence of 2D
NOE cross-peaks without need for accurate distances." Significant, though
somewhat less spectacular, progress has also been made in solving oligonucleotide structures. More accurate treatments indicate that the NOE
cross-peak intensity depends in a more complex way on all of the distances,
not just the specific interproton vector [7, 4a], and these considerations
are increasingly being incorporated in data analysis for biomacromolecules
[4, 8, 9].

In some respects, small molecules are not as well suited for NOESY
studies [10, 11]. Independent of whether they display Overhauser enhancements or weak cross-saturation, accurate NOE data is more difficult to
collect. In small molecules, magnetization at individual spin sites decays
slowly, largely by leakage rather than transfer to other sites. As can be
seen in Fig. 1, even at the small molecule optimum, NOE cross-peaks are only
25% of the relative intensity observed for macromolecules [11]. The long
relaxation times observed usually exceed the NOESY experiment recycle times
that are dictated by instrument availability and resolution requirements.
Experiments must therefore be performed under non-ideal conditions. When an
effectively infinite magnetization recovery period is not included, the
fundamental symmetry [12] of the NOESY experiment is no longer observed
thereby complicating the analysis of small molecule NOE data [13, 14].

The Theoretical Basis for Conformational Analysis from NOESY Data

The spectral progression during $\tau_m$ of a NOESY sequence can be shown
schematically as in Fig. 2.

The intensity data from a NOESY experiment thus corresponds to a signal
matrix, as shown below, left side of Eq. (2).

$$\begin{bmatrix} S_{ii} & S_{ij} & S_{ik} & S_{il} \\ S_{ji} & S_{jj} & S_{jk} & \cdots \\ S_{ki} & \cdots\cdots\cdots \end{bmatrix} \rightarrow \begin{bmatrix} 1 & (S_{ij} \ S_{ii}) & (S_{ik} \ S_{ii}) & (S_{il} \ S_{ii}) \\ (S_{ji} \ S_{jj}) & 1 & (S_{jk} \ S_{jj}) & \cdots \\ (S_{ki} \ S_{kk}) & \cdots & \cdots & \cdots \end{bmatrix} \qquad (2)$$

We have recently shown [14] that the row-normalized form of the matrix
(right side, above) has particularly useful properties for small molecules
and non-ideal experiments. It is much less influenced by the experimental

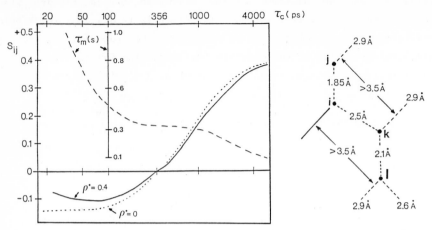

Fig. 1. Simulation of cross-peak intensities at $S_{ii}(\tau_m) = (1/2)S_{ii}^0$ for
an effectively infinite preparatory delay. $S_{ij}$ and the mixing
time used for its observation are plotted versus the correlation time in a semilog plot. The ij spin-pair corresponds to
a relatively well isolated geminal pair at $r_{ij}$ = 1.85 A. The
spin system used in the simulation is shown to the right of
the plot. The simulation was carried-out at two values of $\rho^*$
(non-dipolar contributions to relaxation) using the method of
reference 13b.

NOESY sequence: $PD-90^\circ-t_1-90^\circ-\tau_m-90^\circ$ Acq. $(t_2)$

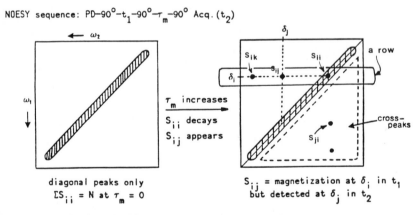

Fig. 2. An idealized NOESY. The axes and a selection of signal matrix
elements are labeled. A slice at a single value of $\omega_1$ corresponds to a row. The projection of a row or limited range of
rows produces an $\omega_2$-spectral cross-section. The peak heights
within a $\omega_2$-spectra correspond to the fate of $S_{ii}(\tau_m=0)$ and
includes cross-peaks designated as $S_{ij}$, $S_{ik}$, etc.

conditions and provides better approximations of the effective cross-relaxation rates. This residual-driver normalized matrix is diagonal asymmetric under all conditions. However, as we will show, the cross-diagonal averages are excellent estimates of the effective net magnetization transfer rates between sites.

## NOESY Theory, Spectral Calculations and Experiment Simulation

For an ideal NOESY experiment the signal matrix can be equated with the mixing coefficients (A) which in turn are related to the relaxation rate matrix (R) as shown below. The elements of R can be calculated directly from molecular coordinates and spectral densities.

$$
A = \begin{pmatrix} a_{ii} & a_{ij} & a_{ik} & \cdots \\ a_{ji} & a_{jj} & a_{jk} & \cdots \\ \cdots\cdots\cdots\cdots \end{pmatrix} = \exp\left(-R\tau_m\right); \quad R = \begin{pmatrix} R_{ii} & R_{ij} & R_{ik} & \cdots \\ R_{ji} & R_{jj} & R_{jk} & \cdots \\ \cdots\cdots\cdots\cdots \end{pmatrix} \tag{3}
$$

$$
R_{ii} = \rho^* + \rho_i \quad ; \quad \rho_i = 0.057 \left[ 6J(2\omega) + 3J(\omega) + \tau_c \right] \sum_j \frac{1}{r_{ij}^6} \tag{4a}
$$

$$
R_{ij} = -\sigma_{ij} \quad ; \quad \sigma_{ij} = \frac{0.057}{r_{ij}^6} \left[ \tau_c - 6J\left(\omega_i + \omega_j\right) \right]
$$

$$
= \frac{0.057}{r_{ij}^6} \left[ \tau_c - \frac{6\tau_c}{1 + 4\omega^2\tau_c^2} \right] \tag{4b}
$$

In the classic formulation of Macura and Ernst [7], recovery toward equilibrium is described by

$$
m\left(\tau_m\right) = \left[\exp\left(-R\tau_m\right)\right] m(0)
$$

and all of the equations are based on $M_o$, the total equilibrium magnetization, and when possible intensity dependence on $\tau_m$ is factored into leakage terms ($R_L$) and cross-relaxation ($R_C$). They derive

$$
R_C = 2\left|\sigma_{ij}\right| \quad ; \quad a_{ij}\left(\tau_m\right) = \left[\exp\left(-R\tau_m\right)\right]_{ij} \frac{n_j}{N} M_o \quad ; \quad N = \sum_k n_k
$$

and indicate an expansion of the exponential for short $\tau_m$ values.

$$
a_{ij} \sim \left[ \delta_{ij} - R_{ij}\tau_m + \frac{1}{2} \sum_k R_{ik}R_{kj}\tau_m^2 - \cdots \right] \frac{n_j}{N} M_o \tag{5}
$$

where $\delta_{ij}$ is the Kronecker delta function ($\delta_{ij} = 0$ when $i \neq j$). The first term in the expansion ($-R_{ij}\tau_m$) corresponds to the ISPA analysis.

In the alternative to this matrix formulation, NOESY signal intensities can be obtained by numeric integration [9, 13b] of the classic Solomon equations [6] through the time course of the NOE experiment. Our approach, which now utilizes matrix methods, is conceptually (and historically) related to this alternative since it arose from earlier studies of the simulation of 1D transient NOE experiments [13b]. In such experiments one begins with a set amount of non-equilibrium magnetization at the perturbation site i ($S_{ii}^o$ using the 2D terminology) which decays during the experiment, $S_{ii}(\tau_m) \simeq [\exp(-R_{ii}\tau_m)]S_{ii}^o$ if all spins had been perturbed -- $S_{ii}(\tau_m) = a_{ii} S_{ii}^o$. It is convenient to view the NOE as having a growth and decay component. At any instant some effective growth rate applies to the remaining $S_ij$. For convenience the overall experiment can be similarly conceptualized.

$$S_{ij} = a_{ij}S_{ii}^o \simeq \sigma_{ij}^{eff}\tau_m S_{ii}^o \left[\exp\left(-R_L\tau_m\right)\right] \tag{6}$$

What we found experimentally was that $\sigma_{ij}^{eff}$ was very well approximated by Eq. (7);

$$\sigma_{ij}^{eff} = \frac{1}{2\tau_m}\left(\frac{S_{ij}}{S_{ii}} + \frac{S_{ji}}{S_{jj}}\right) \tag{7}$$

where the effective rate could be obtained from peak ratios in the two symmetry-related 1D experiments [13a]. As it turns out, the same relationships apply to the NOESY experiment [11, 14].

The Effects of Preparatory Delay (PD) Truncation

In an ideal NOESY experiment with perfect pulses, all $S_{ii}^o$ values correspond to populations, $S_{ii}^o = n_i$. The time allowed for recovery to equilibrium during the NOESY sequence (see Fig. 2) is $t_2$ + PD. For small molecules in practical experiments, this period is much less than three times the selective relaxation times, and incomplete recovery occurs. As a result, less frequency-labeled magnetization is available for generating signal intensity in subsequent cycles. When the decay of non-equilibrium magnetization from individual spin sites is dominated by slow leakage, rather than transfers among the protons, a wide range of leakage rates is present and differential truncation of signal intensities occurs at short recycle times.

$$S_{jj}^o \neq \frac{n_j}{n_i} S_{ii}^o$$

This has profound effects on both the 1D and NOESY experiment. These transient experiments are no longer symmetric ($S_{ij} \neq S_{ji}$). The 2D data matrix fundamentally lacks diagonal symmetry.[≠] For this reason, and generally signal truncation, the experimental data is no longer proportional to the mixing coefficients.

---

[≠] See the Appendix for further details.

Some examples which illustrate the severity of these problems are shown in Figs. 3, 4 and in Table 1, which were generated using experiment simulation using numeric integration [13b]. Figure 3 specifically indicates the effects of PD-truncation on NOE growth as revealed by simulated experiments for a small molecule. The observed NOEs ($S_{ij}$ and $S_{ji}$) are smaller at PD = 1.5s and the two symmetrically related NOEs become increasingly different. In contrast the residual driver normalized [13] data is independent of PD. Also plotted is the calculated curve based on a two-term effective transfer rate. This curve falls almost exactly halfway between the two normalized curves. Equation (8) therefore appears to be an excellent approximation.

$$\frac{1}{2\tau_m}\left(\frac{S_{ij}}{S_{ii}} + \frac{S_{ji}}{S_{jj}}\right) \simeq \sigma_{ij} + \frac{\tau_m}{2}\sum_{k \neq i,j}\sigma_{ik}\sigma_{kj} \tag{8}$$

In the spin system illustrated in Fig. 3, secondary NOEs are relatively insignificant as seen in the linearity of the calculated curve ($— + — + —$).

Simulations for systems showing increasing secondary influences appear in Fig. 4 where the ISPA is acceptable only to $\tau_m \lesssim 100$ ms. As in the previous case, Eq. (8) remains a good approximation ($\pm 10\%$) out to $\tau_m = 600$ ms.

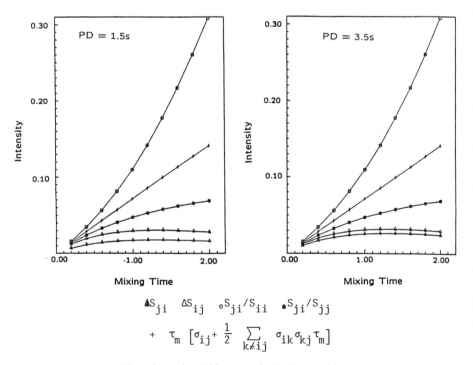

$$\blacktriangle S_{ji} \quad \triangle S_{ij} \quad \circ S_{ji}/S_{ii} \quad \bullet S_{ji}/S_{jj}$$

$$+ \quad \tau_m \left[\sigma_{ij} + \frac{1}{2}\sum_{k \neq ij}\sigma_{ik}\sigma_{kj}\tau_m\right]$$

Fig. 3. The Effects of PD Truncation.

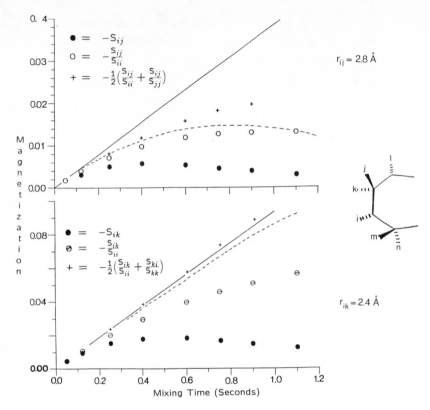

Fig. 4.  Cross-peak growth (•) during $\tau_m$ for NOESY spectra collected
with truncated delay: $\tau_c$ = 160 ps at 500 MHz $^1$H observation,
with a cycle time of 2.1 s.  Data is taken from experiment
simulations based on the illustrated model with $\rho^*$ = 0.2s$^{-1}$.
The solid line corresponds to the linear limit isolated spin-
pair assumption.  The broken line shows the expected curve
based on the two term expansion, Eq. (8).  The symbols for
cross/auto peak ratios are indicated on the figure.  (Taken
from reference 11.)

Before closing this section, we would like to present a more quantitative estimate of the errors associated with the ISPA and with ignoring
leakage.  We have therefore simulated the complete NOESYs for four distinctive conformers of tryptophan under two sets of conditions:

1)  as a free solution state molecule -- $\omega\tau_c$ = 0.25 ($\tau_m$ = 600 ms) --
with a 2 s preparatory delay (and also an infinite one for
comparison),

2)  as a component of a large peptide -- $\omega\tau_c$ = 12.6 ($\tau_m$ = 100 ms) --
with a 1.2 s preparatory delay.

Table 1.   Deviation of Approximate Methods from Exact Solution:
over all observable[a] $S_{ij}$ for 4 conformers of TRP.

R.M.S. % Deviation at $\omega\tau_c = 0.25$, $\tau_m = 600$ ms

| Approx. method # → | PD = 2 sec | | | infinite PD | | |
|---|---|---|---|---|---|---|
| quant. calculated | 1 | 2 | 3 | 1 | 2 | 3 |
| $S_{ii}$ | ... | 62.5 ... | 2.1 | ...... | 2.1 | ..... |
| $S_{ij}$ | 129.4 | 55.4 | 1.4 | 77.6 | 5.9 | 1.4 |
| $S_{ij}/S_{jj}$ | 117.2 | 47.7 | 2.9 | | | |
| $S_{ij}/S_{ii}$ | 81.9 | 7.0 | 2.9 | 82.0 | 7.3 | 2.8 |

R. M. S. % Deviation at $\omega\tau_c = 12.6$, $\tau_m = 100$ ms

| | 1 | 2 | 3 | 1 | 2 | 3 |
|---|---|---|---|---|---|---|
| $S_{ii}$ | ... | 43.6 ... | 8.9 | ...... | 8.9 | ..... |
| $S_{ij}$ | 111.7 | 45.9 | 6.9 | 56.3 | 43.2 | 6.5 |
| $S_{ij}/S_{ii}$ | 73.0 | 37.0 | 12.1 | 68.8 | 41.6 | 11.2 |

Maximum/Minimum % Deviation[b] at truncated PDs

| | @ $\omega\tau_c = 0.25$ | | | @ $\omega\tau_c = 12.6$ | | |
|---|---|---|---|---|---|---|
| $S_{ii}$ | ... 123/−10 ... | | 0/−4 | ... 55/16[b] ... | | 0/−18 |
| $S_{ij}$ | 262/62[b] | 136/−7 | 4/−3 | 221/−72 | 55/−84[c] | 12/−30 |
| $S_{ij}/S_{jj}$ | 292/−27 | 144/−53 | 8/−1 | 175/−81 | 23/−89[c] | 30/−18 |
| $S_{ij}/S_{ii}$ | 148/22[b] | 21/0 | 8/0 | 173/−80 | 18/−88[c] | 30/−17 |

a) Observable cross-peaks are defined as having an exactly calculated
value of $|S_{ij}| > 0.005$.

b) The sign of deviation is defined as $\Delta$ = approximate method minus exact
method.  A positive minimum deviation indicates that the approximate
method overestimates the value for all spins or spin-pairs.

c) These very large negative values indicate that some of the observable
NOEs are predominantly (up to 89%) secondary in nature.

Table 1 shows the rms % deviations between approximate calculations of auto-peaks and $S_{ij}$ values or ratios and those obtained using an exact method.[≠] The comparison is for only those $S_{ij}$ that are judged observable ($S_{ij} \gtrsim 0.005$).

The three approximate methods examined were:

method #1 – linear limit ISPA: $\quad S_{ii} \simeq e^{-R_{ii}\tau_m}$ ; $\quad S_{ij} \simeq \sigma_{ij}\tau_m$

method #2 – leakage corrected: $\quad S_{ii} \simeq e^{-R_{ii}\tau_m}$ ;

$$S_{ij} \simeq \sigma_{ij}\tau_m\left[\tfrac{1}{2}\left(S_{ii} + S_{jj}\right)\right]$$

method #3 – 2 term $\sigma^{eff}$: $\quad S_{ij} = \sigma_{ij}^{eff}\tau_m S_{ii}^o\left[\tfrac{1}{2}\left(\dfrac{S_{ii}}{S_{ii}^o} + \dfrac{S_{jj}}{S_{jj}^o}\right)\right]$

By comparing the errors associated with each method at infinite versus truncated PDs and comparing the errors between methods one can assess the relative importance of each error class for the small and large molecule domains. For small molecules, leakage is very significant and at truncated delays row normalized data are much more accurately estimated (7 vs 48 rms-% error). In the large molecule time domain, PD-truncation effects are smaller but spin diffusion is a more severe problem. A two-term expansion of $\sigma^{eff}$ is only accurate to circa 10% at a field strength of 500 MHz when the molecular correlation time is 4 ns. For relatively small NOEs ($r_{ij} > 2.8$ Å), the ISPA introduces systematic errors. This can be seen when the deviations are given on a max/min rather than on a root-mean-square basis.

The approximate methods calculations were formulated to put the ISPA in the best possible light. The leakage correction corresponds to that which was suggested by Macura et al. [15]. This leakage correction is an obvious improvement over the linear growth approximation: it provides almost as good a correction for PD-truncation effects as method #3. It, however, fails to account for secondary NOEs which are indeed significant for both the large and small molecules based on these simulations.

## Conformation Elucidation Strategies

The numerous reported strategies for utilizing NOESY data in conformational refinement can be divided into two classes. The first class includes as the primary data reduction the derivation of distance constraints from NOESY intensities; the second relies on the comparison of model dependent calculated spectral intensities with the experimentally observed values. To our knowledge, the effects of PD-truncation,

---

[≠] The exact calculations were performed using NOESYSIM (vide infra): the same results can be obtained with a greater cost in cpu time using numeric integration.

as previously described, have not been considered for either method in other laboratories. However significant progress has been made in accounting for secondary NOEs by both methods. The derivations of distance constraints and their utilization in conformational refinement will be considered first.

Deriving Distance Constraints

Just as the mixing coefficients can be obtained from a diagonalized relaxation matrix; the reverse process [16], known as a "direct solution" can be accomplished <u>if a complete high precision signal matrix is available</u>. The full set of cross relaxation rates, and thus a set of distance constraints free of spin diffusion artifacts, are obtained as an eigenvalue solution to $R = -(1/\tau_m)X(\ln \Lambda)X^{-1}$, assuming that the signal matrix, $S = X\Lambda X^{-1}$, is identical to the mixing coefficients. In practice, we find that a sufficiently complete signal matrix is rarely, if ever, available for even moderately-sized systems. Although this "direct solution method" is frequently mentioned, to our knowledge it is only used by the Abbott group [8]. For systems in which some intensities are unmeasurable due to peak overlap, it is necessary to place estimated intensities into the signal matrix prior to the "direct" solution. Several approaches have been suggested including the use of average diagonals and estimates based on relaxation data from models. Borgias and James [4c] have examined error propagation due to incorrect values and random errors in "the direct solution method" and based on that analysis they prefer the second approach. They have developed the program CORMA to generate the theoretical spectra. Boelens, Koning and Kaptein [4d] recommend "back transformation of a mixed NOE matrix with experimental and calculated elements" for an iterative derivation of distance constraints. It should be noted that PD-truncation effects preclude the derivation of accurate distance constraints by any "direct solution" method.[≠] The signal truncation and diagonal asymmetry combine to yield estimates of $R_{ij}$ that are inaccurate since errors in diagonal elements propagate into the off-diagonal.

Once distance estimates have been generated, they can be used to generate structure hypotheses by a number of distinct procedures. Distance-geometry algorithms based either on metric matrix embeds in 3-space [17] or on a multi-dimensional space of torsion angle dimensions [18] can be used. "Experimental distances" can also be used as pseudobond potentials added to a variety of force-fields such as AMBER [19] or CHARMm [20]. The latter combination, alternating constrained dynamic and minimizations using CHARMm, appears to be the most common method at present and is available in commer-

---

[≠] The specific nature of the diagonal asymmetry due to PD-truncations and implications that this has upon data scaling for direct solutions appear in the Appendix.

cial molecular modeling and graphics software packages. A hybrid program with "dynamical simulated annealing" has recently been reported [21]. The authors indicate that it samples a larger region of conformational space in a more computationally efficient manner. Nevertheless, the quality of the output will be largely determined by the accuracy of the distance constraints used. As noted previously the accuracy for macromolecular studies appears to be available from current data sets even when the ISPA is used. In our experience this is not the case for small molecule problems and peptides.

In order to provide an indication of the errors in distance constraints that can be expected in ISPA analyses (including those utilizing the leakage correction of Macura et al. [15]), we have searched through accurate NOESY simulations for in excess of 25 peptide and drug systems performed in this laboratory over the last two years. We selected NOESY cross-peaks corresponding ($\pm$20%) in intensity to a 3.2 $\overset{o}{A}$ isolated spin-pair under the same conditions. In principle, a $\pm$20% error in an NOE cross-peak corresponds to $\pm$0.1 $\overset{o}{A}$ at a nominal distance of 3.2 $\overset{o}{A}$. Table 2 shows the actual interproton separations that gave these NOE intensities. These studies were based on calculated spectral simulations without added random noise. Additional error sources, and their estimated contributions, are also given in the table. The systematic error associated with the distance estimates by the ISPA, is due largely to accumulated secondary NOEs. For the large molecule cases, these are of the same sign as the primary NOE and this leads to a 0.4–0.8 $\overset{o}{A}$ underestimation of the actual distance in many cases. For small molecules, NOEs from proton-pairs at $\gtrsim$ 3.2 $\overset{o}{A}$ are frequently unobservable due to accumulated secondary NOEs of opposite sign. The errors seen in Table 2 obviously limit the precision of any conformational refinement based on distance constraints obtained by an ISPA analysis. In the 2.4–2.8 $\overset{o}{A}$ region,

Table 2. Distance Estimation Errors at 3.2 $\overset{o}{A}$ Nominal. Model distances yielding NOEs equivalent to a 3.2 $\overset{o}{A}$ ISP.

|  | $\omega\tau_c = 0.36$ | $\omega\tau_c = 6.3$ | $\omega\tau_c = 12.6$ |
|---|---|---|---|
| $r_{ij}$ range | 2.96–3.14$\overset{o}{A}$ | 3.27–4.05$\overset{o}{A}$ | 3.29–4.27$\overset{o}{A}$ |
| typical systematic error | +0.15$\overset{o}{A}$ | (−0.5$\overset{o}{A}$) | (−0.6$\overset{o}{A}$) |
| Other Errors |  |  |  |
| intensity measure | $\pm$ 0.25$\overset{o}{A}$ | $\pm$ 0.2$\overset{o}{A}$ | $\pm$ 0.16$\overset{o}{A}$ |
| PD-truncation | $\pm$ 0.25$\overset{o}{A}$ | $\pm$ 0.15$\overset{o}{A}$ | $\pm$ 0.12$\overset{o}{A}$ |

estimation errors are less severe (see Appendix, Table 3) but still substantial. (This analysis prejudiced us against refinement based on distance constraints, until very recently. We will return to this subject in the concluding remarks.) Obviously, even rather poor direct solutions from incomplete matrices will be a significant improvement and the increasing attention to this problem [4, 8, 9] bodes well for NOESY-based conformational refinement.

## Structure Refinement by Comparison of Calculated Model Spectra and Experimental Data

This approach has been pioneered by the U. C. San Francisco group [4a, c] using the program CORMA for calculating the NOESY expected from the refining structure [22]. The primary concern about this approach is its strong initial model bias [8a]. It is very difficult to prove that an improvement in fit, as the initial structure is modified, represents a movement toward the global rather than a local minimum. For oligonucleotide structures, this approach has been automated based on random variation of torsional and other angular parameters in the COMATOSE program [4c]. It remains to be established, however, whether this program escapes local minima. One additional shortcoming, particularly for smaller molecules, is the failure to account for PD-truncation effects. In a later section we describe the coupling of NOESYSIM with comprehensive conformational scans as a means for overcoming both of these problems in small molecule conforma tional refinement.

## Convergence Characteristics of Equations (5) and (8)

Equations (5) and (8) bear a formal similarity. Equation (5) is a standard Taylor expansion of an exponential and, with limited terms, is a good approximation of $a_{ij}$ only at rather small values of $\tau_m$. The expansion of the exponential implicitly includes decay components: k in the summation includes i and j. Borgias and James [23], in support of the use of CORMA, have noted that two and three term expansions do not approach the values of $a_{ij}$ in "practical" cases. Based on the examples they illustrated one would conclude that the addition of the second and third power terms actually produced much worse estimates.

In contrast our Eq. (8) is a phenomenological summation over magneti- zation transfer pathways and corresponds to an "effective" cross-peak growth rate. As stated in Eq. (8), leakage is accounted for by using the appropri- ate cross/auto-peak ratios. In order to make this distinction clear we have compared the convergence characteristics of these two equations using simu- lated data for a peptide structure model. These simulations were carried out at $\tau_c$ = 2 ns ($\tau_m$ = 60 and 160 ms), values that correspond to those seen in lysobactin (an undecapeptide from which the fragment was taken, vide infra). They were also repeated at values that would apply in small molecule studies. Figure 5 illustrates some of these results. For an unsymmetric data set the comparison has the form:

$$S_{ij} \text{ and/or } S_{ij} \simeq a_{ij} = -R_{ij}\tau_m + \frac{1}{2}\sum_K R_{ik}R_{kj}\tau_m^2 - \frac{1}{6}\sum_{K,L} R_{ik}R_{kl}R_{lj}\tau_m^3 + \cdots$$

versus

$$\frac{1}{2\tau_m}\left(\frac{S_{ij}}{S_{ii}} + \frac{S_{ji}}{S_{jj}}\right) \simeq \sigma_{ij} + \frac{\tau_m}{2}\sum_{k\neq i,j}\sigma_{ik}\sigma_{kj} + \frac{\tau_m^2}{6}\sum_{\substack{k\neq i,j,1 \\ 1\neq i,j,k}}\sigma_{ik}\sigma_{kl}\sigma_{lj} + \cdots$$

The superiority of Eq. (8) is immediately apparent from Fig. 5: only the second term is required for either the small or large molecule case. As $\tau_c$

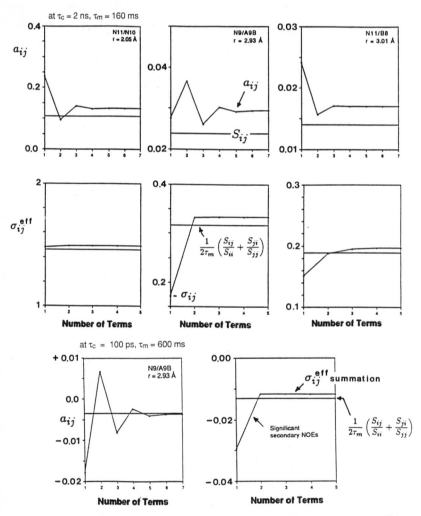

Fig. 5. The convergence characteristics of Eq. (8) <u>versus</u> Eq. (5). Each panel shows the calculated values as <u>additional</u> terms are added to the expansion. Each panel is labeled with the applicable interproton distance. Calculations for both a small molecule ($\omega\tau_c = 0.31$) and a moderate-sized peptide ($\omega\tau_c = 6.3$) are included, each using appropriate mixing times.

approaches 5 ns or more, tertiary NOEs may need to be considered.  In
contrast Eq. (5) approaches $a_{ij}$ rather slowly, and even with sufficient
terms the inequality of $a_{ij}$ and $S_{ij}$ remains a problem.  At the shorter $\tau_m$
value, the Taylor series converges more rapidly (Fig. 6) but the non-
equivalence of $a_{ij}$ and $S_{ij}$ is still a problem.

## The NOESYSIM Suite of Programs

As previously reported [13b], we have done experiment simulations, one
row at a time, to generate NOESY matrices that include the effects of
non-ideal experimental conditions.  In contrast to some reports [9], we
found numerical integration procedures much less efficient than matrix
diagonalization even for the ideal one-cycle experiments.  We therefore
sought a means for including PD-truncation effects in a matrix solution
for the NOESY intensities.

Numerous full simulations revealed that $S_{ij} = a_{ij}S_{ii}^o$ was indeed an
equality for both cross- and auto-peaks (i=j).  Thus, our task was
reduced to finding a matrix solution for $[S]\tau_m = 0$ at truncated PD-values.
Having already diagonalized $R$, it is an easy matter to calculate a
"control" matrix $C$

$$C = \exp\left[-R\left(t_2 + PD\right)\right]$$

The summation of any row (or column) of $C$ yields the magnetization that
is "lost" in a truncated-PD experiment.  Thus $S_{ii}^o = n_i$ for PD = $\infty$ becomes

$$S_{ii}^o = n_i\left(1 - \sum_j c_{ij}\right)$$

where $c_{ij}$ are elements of matrix $C$.  Incorporating this feature allows us
to calculate exact NOESY spectra for 40-spin systems in less than a
second
at 10 MIPs.

## The Operation of NOESYSIM

NOESYSIM is a program that, among other features, calculates the exact
theoretical NOESY spectrum that corresponds to a particular structural model
and experimental conditions.  To accomplish this the following input is
required:
    1)  experimental parameters – acquisition time, sweep width, mixing
time and cycle repetition time.
    2)  an initial structural model (which can be in an arbitrary
conformation[≠]) in the form of a protein data bank file such as those

---

[≠] For peptides, as an example we routinely utilize a loose "random" coil
($\phi = -80$, $\psi = +90°$) and an approximate $\alpha$-helix ($\phi = \psi = -55°$) as
alternative arbitrary conformation inputs.  These computer models are
generated using ALEX (Squibb, see note § next page) or QUANTA
(Polygen).

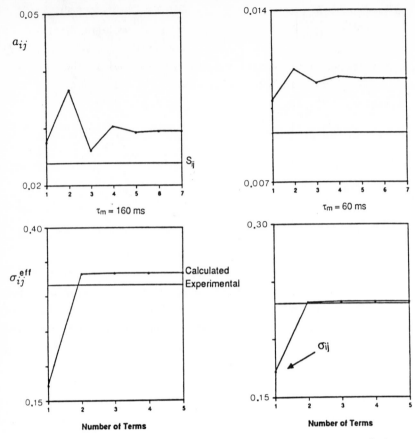

Fig. 6. Convergence $\tau_m$ = 160 ms versus 60 ms: $\omega\tau_c$ = 5.6, $r_{ij}$ = 2.93 $\overset{o}{A}$.

generated by most commercially available molecular graphics programs.

3) a user-selected list of proton pairs that are relatively isolated and whose $r_{ij}$-values are not dependent on the conformation. (These are typically geminal pairs and cis-olefinic or vicinal protons on aromatic rings.)

4) the experimental signal matrix (to the extent that it can be measured), including unambiguous instances of unobserved cross-peaks as zero values.

When the above data are available, a subroutine of NOESYSIM determines the best fit value of the correlation time ($\tau_c$) and the scale factor that should be applied to the experimental signal matrix to make the intensities comparable to those calculated by NOESYSIM. The fit that provides $\tau_c$ is based on a minimization of

$$\left(\frac{S_{ij}}{S_{ii}}\right)^{calc} \quad \underline{versus} \quad \left(\frac{S_{ij}}{S_{ii}}\right)^{exp}$$

over the user-selected "isolated" ij values. At this point, the rms deviation over $S_{ij}$, $S_{ii}$ and $S_{jj}$ is calculated and the best fit value of $\rho^*$ [see Eq. (4a)] is determined. The experimental data is then rescaled on that basis.

The primary function of NOESYSIM is the rapid calculation of a complete theoretical signal matrix from molecular coordinates and a given set of experimental parameters. After each such calculation the theoretical spectrum is compared to the experimental data and a measure of fit is calculated. This measure of fit is a weighted r.m.s. residual with the weighting favoring large cross-peaks and accounting for a user-entered lower cut-off for intensity measurements. These features are designed for conformational refinement. As the structural model is altered – by conformational scanning, a dynamics course, or some form of constrained minimization – the measure of fit can be plotted or otherwise examined as a function of the model-alteration para-meters. At present, NOESYSIM has not been integrated with any molecular graphics application during automated conformation searches, but this feature should soon be implemented. In our laboratory, we expect to make NOESYSIM minimizations an option of the conformational scan capabilities of ALEX.[§]

Another recent enhancement of NOESYSIM allows the theoretical spectrum to be displayed and plotted, as a 2D contour plot. This option is illustrated in Fig. 10 (vide infra). In order to use this option the operator must enter all of the chemical shifts, line widths, spectral width, and the number of $t_1$ and $t_2$ values. The resulting data set (an 'smx' file) can also be manipulated and plotted (as contour, stacked, or difference plots) using the commands in FT-NMR (Hare Research).

At first glance, the option of producing a theoretical spectrum in the same format as the experimental data may seem of only cosmetic value. However we have found it invaluable. Of necessity the experimental signal matrix contains many elements that are indeterminate: no precise value can be assigned due to peak overlaps or spectral artifacts. But the simulated contour plots can be compared to the experimental data in all regions and provide thereby additional bases for interpreting crowded spectral regions and for eliminating structure hypotheses that predict significant peaks which are not observed in the experimental spectrum.

---

[§] ALEX is a model manipulation and display program developed at the Squibb Institute for Medical Research (by Dr. J. Gougoutas).

ILLUSTRATIVE APPLICATIONS

All of the NOESY spectra used in these examples are from pure-phase
absorptive experiments at 500 MHz. The details of data collection and
initial data reduction are presented elsewhere [11, 24, 25]. For both
comparisons to calculated theoretical spectra and for distance estimation
routines, the signal intensity values should be peak volume integrals.
To date we have not been completely satisfied with those volume reporting
options (activated by peak picking) that we have examined. The intensity
data used in this work is taken from peak areas (or heights) in
projection spectra corresponding to the summation of an inclusive set of
columns or rows for each resolved resonance [11, 24a]. This procedure
can be applied across the entire spectrum when the diagonal peak is
resolved, or through spectral fragments when only cross-peaks are fully
resolved.

In the sections that follow, the initial stages of analysis and
resonance assignment are not shown, nor do we attempt to present fully
annotated spectra for each case. Our emphasis is on the conformation
refinement stage of the studies. In a number of cases, refinement is
still in progress as more accurate distance constraints and, in the case
of peptides, stereospecific assignments become available.

Prostanoid Solution Conformations

The differing stereoisomeric forms of prostanoids with an oxobicy-
cloheptane core mimic many of the distinct classes of prostaglandin
pharmacology [26]. Three analogs that figure in our discussion are
illustrated; they represent different stereochemical series.

SQ-28,989

$R_\omega = C_5H_{11}$ : SQ-26,271   R = $C_4H_9$, X = H : SQ-26,536
$R_\omega = $ cyclo-$C_6H_{11}$ : SQ-27,986   R = $CH_3$, X = $C_6H_5$ : SQ-28,668

Molecules of this size in typical organic or aqueous media have correlations times of 80–120 ps with some additional segmental motion in the
far ends of the two side-chains. The molecular complexity is well-suited
for torsion-scan conformation search procedures. Typically there are
four to six torsional degrees of freedom that must be scanned, indicated
here by the carbon atoms in the bonds: 12/13, 14/15, 8/7, 7/6, 5/4, and
15/16 (or 15/1′).

112

Figure 7 shows the typical complexity and resolution present in a 1K
by 1K NOESY spectrum that can be recorded in an overnight run for a 10mM
solution.  The following features deserve note:  (1) the superior
resolution in the $\omega_2$ dimension which allows for the clear distinction of
the 11/12 and 11/exo-10a crosspeaks, (2) the cross peaks between H-15 and

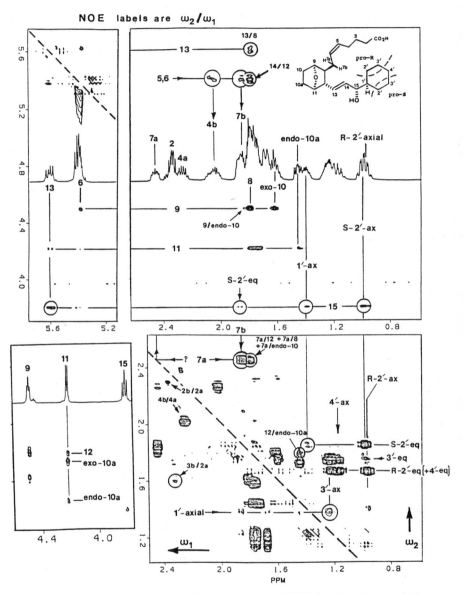

Fig. 7.  Segments of a 1K x 1K NOESY on SQ-27986 (10mM in CDCl$_3$):
PD = 2s, $\tau_m$ = 800 ± 80ms, 64 transients for each 383 t$_1$
values.  This is an unsymmetrized representation with the
contour level cut-off selected so that some of the t$_1$
streak echoes appear for slowly relaxing auto-peaks.
Portions of the oppositely signed diagonal are indicated by
dashed lines.

the 2'-methylene appear for on only one side of the cyclohexyl ring (indicated
as S-2'-eq/15 and S-2'-ax/15 in the figure; however the pro-R/pro-S assignments are not absolute), and (3) the non-vicinal NOEs - 13/8, 14/12, 9/6,
15/13, and 7a/12 - that define the conformation.  Specific rotameric preferences are observed for C4 → C8 and C12 → C1'.  In this regard, the observation
of NOEs between olefinic protons 5 and 6 and only the upfield members of the
C-4 and C-7 methylenes is particularly significant.  The NOE peaks observed
are a clear indication of the strong preference for the conformation in which
H-12/H-13 and H-14/H-15 are nearly antiparallel.  In the absence of an a
priori assignment of the pro-R/S sides of the cyclohexyl substituents,
however, a complete conformational assignment cannot be made for this system.

Prostanoid analogs provide ideal examples of conformational analysis
by torsion scans directed toward locating regions of conformational space
that fulfill observed distance constraints or predict the observed values
of NOE cross-peak ratios.  These two scan options have been incorporated

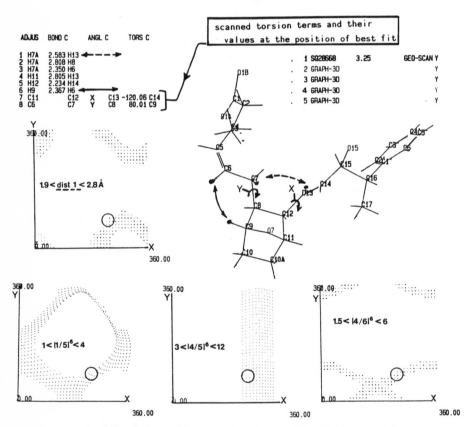

Fig. 8.   Defining the torsion angles exo to the bicyclo core of
          SQ-28668 by two dimensional scans using ALEX.  The figure is a
          screen dump from ALEX run on an Evans & Sutherland MPS.

into ALEX for some time. Figure 8 shows the results for four such scans for analog SQ-28668. In this type of scan procedure, two or three torsion angles are selected for comprehensive scans at 10° resolution while monitoring a selected distance or NOE ratio. Each scan result can be shown as a graph of the region of conformational space that fit the constraints. In Fig. 8 only one region (encircled) of a map with C7/C8 and C12/C13 torsion coordinates fits the NOEs observed between protons 6, 7A, 9, 11, 12, 13, and 14.

If the data could be appropriately analyzed (for small molecules) using distance derived by the ISPA, one would expect, except for experimental scatter, a linear correlation between $S_{ij}$ and $k(r_{ij})^{-6}$. If Eq. (8) represents a significant improvement in data analysis the same data set, examined as comparisons of $\sigma_{ij}^{eff}$ values (experimental <u>versus</u> those calculated from a model conformation) should show a better linear correlation with less out-liers at a two-term expansion for the calculated values of $\sigma^{eff}$. Figure 9 (next page) establishes that this is true for experimental data. In this case, the NOESY data for prostanoid analog SQ-28989 are used to illustrate the effect. The correlation in panel B is more nearly linear and further demonstrates the superiority of the cross-diagonal average of row normalized data for obtaining accurate estimates of effective cross-relaxation rates (Panel B, - filled versus open symbols).

The points shown in Fig. 9 include only well-resolved, quantifiable cross-peaks whose corresponding auto-peaks can also be measured. This precludes the use of some of the data present in the more crowded upfield region. The NOESYSIM plot function provides a means for checking the validity of the conformational model in this spectral region. Figure 10 illustrates a comparison of the simulated spectrum (and the conformation upon which it is based) and the corresponding segment of the experimental matrix. We believe that the remaining minor differences reflect a combination of factors: incomplete definition of the C5-C4 torsion angle(s); and, most notably, segmental motion in the extreme ends of the two side-chains. An effectively isotropic model was used for the simulation.

## Linear Peptide Studies

Linear peptides composed of less than 20 residues are typically viewed as lacking secondary structure [27]. The near-consensus opinion considers these systems as "random coils" or largely extended "structures" with much segmental motion and numerous populated conformers. Thus it is believed that NOESY studies will either yield low intensity data (due to segmental motion) or data which will refine to nonsensical "average" structures.

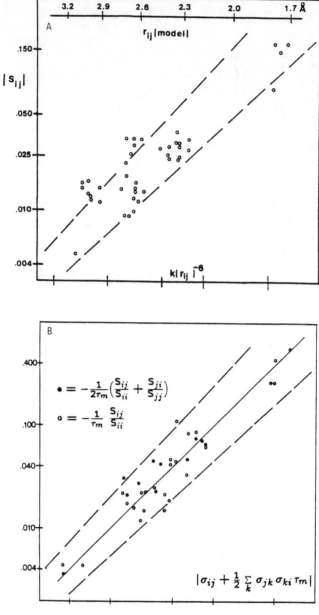

Fig. 9. Improved fit between model distances (A) or effective
magnetization transfer rates (B) and experimental
cross-peak intensities for a small molecule ($\tau_c$ = 95 ps).
The data set is for prostanoid SQ-28989 (10 mM, $\tau_m$ =
800 ms). The dashed line limits correspond to a comparable
ultimate distance error (± 0.20 Å). In order to show the
full data set (experimental $\sigma$ values from 0.003 →
0.62 s$^{-1}$), a log–log scale is used. For A, the ISPA
analysis, k is chosen to yield the best linear fit of unit
slope.

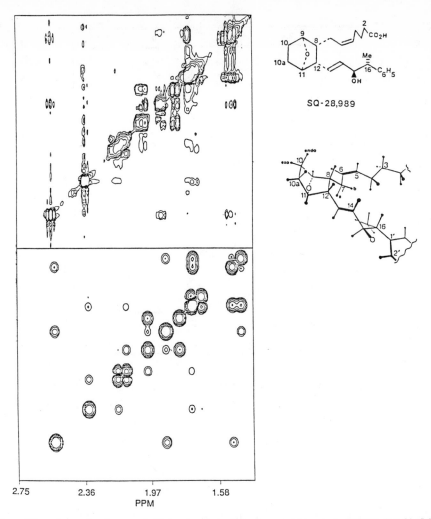

Fig. 10. A comparison of the simulated and experimental NOESY (upfield segment) for SQ-28989 (structural inserts include the current conformation in our refinement procedure).

During the last year we have applied NOESY to a variety of linear peptides including neuropeptides, hormones (GnRH; and napharelin, an analog), and pheromones (yeast $\alpha$–factor) in aqueous, mixed aqueous and organic media. High precision, easily quantitated NOESY data, indicating $\omega\tau_c$ = 3-8, has been obtained for 9–13mers at just below ambient temperatures. Changes in relative NOESY cross-peak intensities are observed upon both solvent changes and residue substitution, reflecting changes in the conformation preferences within the applicable conformational equilibria. Figure 11 shows the NH/$\alpha$–methine regions (labeled by residue numbers) for GnRH in two mixed aqueous solvent compositions, and shows a number of readily apparent shifts in NOE ratios of the type – NH$^i$/$\alpha^i$ vs NH/$\alpha^{i-1}$ and NH$^i$/$\alpha^i$ vs NH$^{i+1}$/$\alpha^i$ – which reflects changes in $\phi_i/\psi_{i-1}$ and $\phi_i/\psi_i$ respectively.

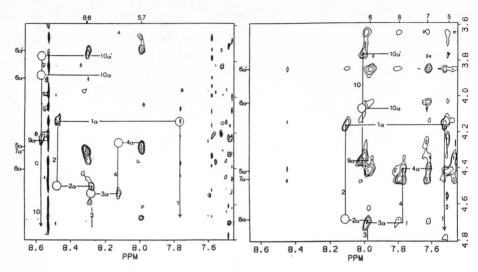

Fig. 11. NOESY spectral fragments for 4 mM GnRH in two compositions of HFIP:$H_2O$. Left Panel, 12.5% HFIP, $\tau_m$ = 180 ms, 600 TPPI increments collected in 40 h; right panel, 50% HFIP, $\tau_m$ = 120 ms, 220 $t_1$ increments (hypercomplex) in 25 h. In each panel the location of unobserved or weak intraresidue NH/αH cross-peaks is shown by an empty circle. The most notable change (in the annotated portion of the sequential assignment included) is the change in inter- and intra-residue peaks associated with the S4-NH.

In the more aqueous media, at 87.5% $H_2O$, many of the intra-residue NH/α cross-peaks are below the level of detection. Simulations reveal that this result implicates a significant amplitude of high frequency φ variation. For certain regions of the structure, however, this segmental motion does not contain a comparable variation in ψ since strong inter-residue cross-peaks remain. At high hexafluoroisopropanol (HFIP) concentrations, a nearly complete set of sequential and intra-residue connectivities is apparent. Several of the inter-residue cross-peaks are actually of diminished intensity due to the lesser contribution of more extended conformers.

118

Similar comparisons also demonstrate substitution-induced conformational changes. These studies utilized GnRH and napharelin ([D-naphthy-lalanine⁶]-GnRH) in several mixed aqueous and fully organic media. Figure 12 shows a more detailed picture of a comparable region for native GnRH dissolved in DMSO. Even in this highly solvating (and thus internal-structure disrupting) medium, small intermediate-range NH/$\alpha$ NOEs are observed. These must reflect significant populations of specific turns as shown on the figure. In attempts to model GnRH, we were unable to locate any single conformer that could explain these signals: if the $\phi/\psi$ values are scanned to place S4-$\alpha$ and R8-NH within $\lesssim$ 3.5 Å; the H2-$\alpha$/G6-NH distance is always far too large to produce the observed peak and no secondary pathways exist for this spin transfer. These results stand in notable contrast to recently published studies of GnRH which were taken to imply highly random structures in a variety of media [28].

We believe that an excellent case can now be made for quantitative NOESY analysis of peptide conformational equilibria. The major distinction between proteins and peptides is a dynamic one. In proteins, tertiary structure elements in effect restrict local $\phi/\psi$ conformational spaces into different regions of rigid secondary structure. Smaller peptides are unlikely to be "rigid" under any circumstance other than specific receptor complexation, but nonetheless distinct conformational preferences can exist, and be altered by environment. Quantitative analysis of NOESY data should serve to elucidate these conformations just as qualitative NOESY analysis has provided insights into protein secondary and tertiary structure features.

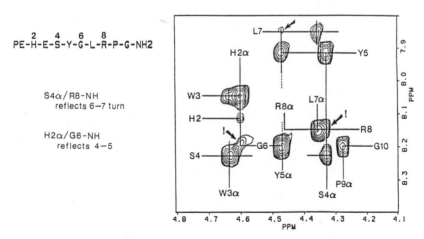

Fig. 12.   NH/$\alpha$-methine NOESY details for GnRH (7mM in DMSO), $\tau_m$ = 220 ms (to accommodate a co-collected TPPI-COSY). Intermediate-range NH/$\alpha$ connectivity are indicated by bold arrows.

## Conformational Refinement of Lysobactin

In contrast to the conformational ambiguities of small linear
peptides, many bioactive cyclic peptides, although still lacking
"tertiary" structure have, due to the covalent loop constraints, single
definable conformations.  In order to elucidate these with the resolution
required for understanding biorecognition phenomena, high precision NOESY
data is required.  If distance constraints are to be used in refinement,
they should be corrected for indirect spin-transfer pathways.  Our recent
work on lysobactin (1) provides illustrations of the use torsion scan
strategies and comparisons with constrained CHARMm dynamics and
minimization.

We recently completed a structure elucidation for lysobactin (1) and
its benzylaminolysis product (2) exclusively from NMR data.  Each residue
was identified and the sequence was determined from the 2D NMR spectra.
With the exception of the D,L,L,L sequence for residues 1→4 and the
L,-,L,L-configuration of 8-11, which are supported by NOE analysis, the
absolute configurational assignments are those that resulted from a
parallel degradative structure elucidation [29].[§]

In this report, we will take up the story at the stage in which
subtle regiochemical and stereochemical questions were being addressed,
leaving the assignment and sequencing problem for other accounts
[25].  The NMR structural information at that point was conveniently
summarized in Fig. 13 (taken from reference 25b), which collects intra-
and vicinal inter-residue interactions that are probed either by dipolar
or scalar coupling data.  The NOEs are expressed on an intensity scale
(values 0 to 8) with  corresponding distance constraints given in the
footnote.  These deserve some comment.  The distance ranges assigned are
in some cases more narrowly defined than is typical of protein NOESY

---

[§] Essentially de novo NMR structure elucidation for peptides of this size
and complexity are rare.  Many NOESY studies of peptides (and proteins)
have been reported; but in these studies, the residues and their
sequence and relative configurations were known at the onset.

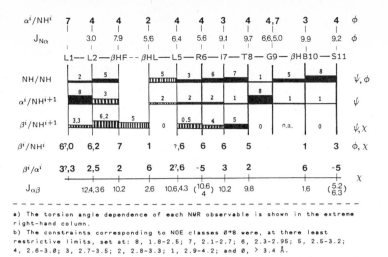

Fig. 13.  a) The torsion angle dependence of each NMR observable is
shown in the extreme right-hand column.
b) The constraints corresponding to NOE classes 0→8 were,
at their least restrictive limits, set at:  8, 1.8-2.5; 7,
2.1-2.7; 6, 2.3-2.95; 5, 2.5-3.2; 4, 2.6-3.0; 3, 2.7-3.5;
2, 2.8-3.3; 1, 2.9-4.2; and 0, > 3.4 Å.

studies [3b, 21, 30].  This was possible for some categories by virtue of
the application of Eqs. (7) and (8).  When a specific NOE can be measured
in all possible NOESY cross-sections and similar quantitation is
available for the NOEs that correspond to individual steps in the
anticipated two-step pathways, these equations can be used to correct
$\sigma_{ij}^{eff}$ to more accurately reflect $\sigma_{ij}$.  This point is considered in detail
under Future Prospects at the end of this account.  In excess of 120 net
spin transfer pathways were quantitated for this undecapeptide; thus many
of the constraints could be defined to $\pm$ 0.3 Å (or better resolution).

To apply these constraints to structure refinement, it must be
demonstrated that lysobactin is in a single "conformational state."  The
best evidence for this comes from the constant values of scalar couplings
that depend on $\phi_i$ and $\chi_i^1$ with variation in temperature and solvent
composition.  Macrolide 1 displays, to the extent they are resolved in 1D
spectra, the same ($\pm$ 0.4 Hz) $J_{N\alpha}$ and $J_{\alpha\beta}$ values whether dissolved in
water, $CD_3CN$, DMSO, or $C_6D_6$-DMSO mixtures.

In addition to the constraints implied by Fig. 13, a number of
intermediate-range constraints could also be defined from uniquely
assigned NOEs between non-vicinal residues.  Some of these are
highlighted in Fig. 14 which also serves to illustrate the general
features of the NOESY data set.

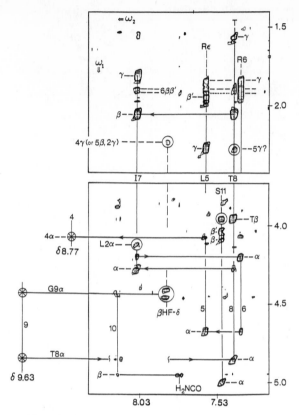

Fig. 14.   Details of the crowded NH/α and NH/β NOESY cross-peak
regions for lysobactin:  10 mM in 4:1 $C_6D_6/d_6$DMSO ($\tau_m$ =
100 ms) at 283K.  Inter-residue cross-peak which are not
directly sequential (i, j≠i±1) are encircled and labeled.
These become the most significant constraints in conformational refinement.  Key sequential connectivities that
appear off-scale of this spectral fragment are added at
the left margin. (Taken from reference 25b.)

Turning to the use of this data set for defining regiochemical and
conformational details, our first example is the linkage isomerism
possible in the vicinity of the lactone closure.  In principle, either
carboxyl of the β-OH-Asx or that of Ser could be assigned as the sight of
lactonization and the aspartate could be functioning as either an α- and
β-amino acid in the continuing peptide chain.  A NOESY sequencing study
[25b] on ring-opened derivative 2, served to establish that serine was
the C-terminus and the sight of lactonization, however it was still not
apparent whether the α- or β-carboxyl of Asn was the free carboxamide.
In the following discussion we refer to β-OH-Asn with a free
α-carboxamide as the iso-form (βHI, or iso) to distinguish it from the
normal free β-carboxamide form (βHN).

D-Leu–Leu–NH

'βHL'—C $\underset{O}{\overset{3}{|}}$

OH

H₂N≡----C=O

O

'Gly' —N—$\overset{10}{|}$ OH
H

We adopted a conformation scan strategy for distinquishing between the iso and normal forms (see Fig. 15). The most important constraints in the β-OH-Asn$^{10}$ region were: $J_{\alpha\beta} = 1.6$, $J_{N\alpha} = 9.9$ Hz, and d(11–NH/10–NH) = 2.0–2.35 Å. The first listed constrains $\chi^1_{10}$ to the two narrow ranges of values for the erythro- and threo-form of β-OH-Asn; the latter depends largely on two (for normal) or three (iso) torsion angles. A total of eight models (two $\chi^1$ values for each erythro and threo form of the iso and normal structures) had to be evaluated using a series of conformational scans of two and three torsional dimensions. Additional constraints (used as target values for the scans) that were evaluated in these searches were: d(10–NH/β10) = 3.1–4.2, d(NE/10–NH) = 3.0–3.8, d(NE/β10) = 2.6–3.0, d(NE/α10) = 2.8–3.45, d(NE/11–NH) > 2.75, d(11–NH/α10) = 3.1–4.2, d(11–NH/β10) > 3.3 Å. The method is illustrated (in Fig. 15) by two scans yielding conformations displaying d(11NH/10–NH) = 2.0–2.35 Å. When L-Ser was assumed, a normally linked L-threo-β-hydroxyAsn unit was clearly the best fit to the data set.

In order to extend the structural hypothesis, the backbone angles of the best fit model of fragment (10–11) were fixed (±15°) and residues 7→9 were sprouted onto the structure. Ten NOE-derived constraints, of which d(11–NH/β8) = 2.75–3.35 Å was the most restrictive, were then used in a series of additional scans of two to three dimensions until an acceptable fragment conformation was obtained. The Ile unit was assumed to be the L-form. Due to a report concerning katanosin B [31], both L-allo- and D-allo-Thr were examined. The resulting conformational model [$\phi_8$ = −110°, $\psi_8$ = +75°, $\phi_9$ = −65°, $\psi_9$ = −10°, $\phi_{10}$ = −125°, $\psi_{10}$ = −30°, $\phi_{11}$ = −115°], which appears in later figures (Figs. 17 and 18) and contains the L-allo-Thr form, gave a superior fit. With somewhat more generous constraint error limits the D-allo-Thr model refined to: $\phi_8$ = +130°, $\psi_8$ = −85°, $\phi_9$ = −80°, $\psi_9$ = +20°, $\phi_{10}$ = −105°, $\psi_{10}$ = −40°, $\phi_{11}$ = −105°.

In a similar manner a model of residues 1→5 was constructed, eight NOE-derived constraints and one torsion constraint (10:$J_{\alpha\beta}$ = 10.2 Hz) were selected, and sequential conformation scans yielded a conformational model. These two fragments were linked by an ester bond, the NHCα unit

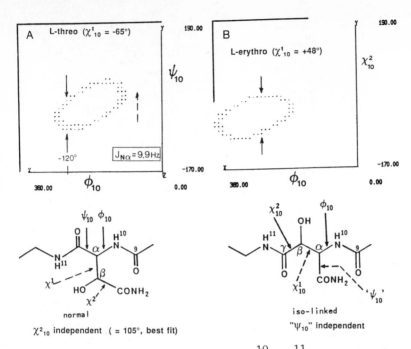

Fig. 15.   Conformation scans for β-OH-Asn[10]-Ser[11] of lysobactin, the
normal and iso-forms are illustrated.  In each graph the
region of conformational space yielding a 10-NH→11-NH
distance of 2.0-2.35 Å is stippled, each point
corresponding to a conformation at 10° torsion angle
resolution.  Other torsion and stereochemical constraints
are given in the figure.  The arrows indicate regions
(±20°) fitting the scalar coupling constraint.  The best
fits (over the indicated independent variable) over eight
NOE-derived constraints within and between residues 10 and
11 were (rms deviation):  A, normal form L-threo-βHN
(0.12 Å); B, iso form L-erythro-βHI (0.56 Å).

of R6 was added to L5, the R6 carbonyl was added to I7, and the torsion
angles of the newly "formed" bonds were scanned to place Cα6 and C'6 in
proximity (~3Å).  In this manner a crude cyclic model, with chemically
unreasonable bonding in R6, but retaining the previously refined backbone
angles in residues 1→5 and 7→11, was obtained.  With the newly added loop
constraint, further use of torsion scans was precluded.  We therefore
turned to CHARMm for further refinement.  However we wished to retain, to
the extent possible, the refinement results for the fragments.  The
further stages of refinement are outlined below.

A short steepest descent CHARMm minimization (with ten key NOE
constraints) served to fix the geometry in R6 without appreciable alteration of the experimentally-derived conformations in fragments 1→5 and
7→11.  Figure 16 shows the peptide backbone changes during this process.

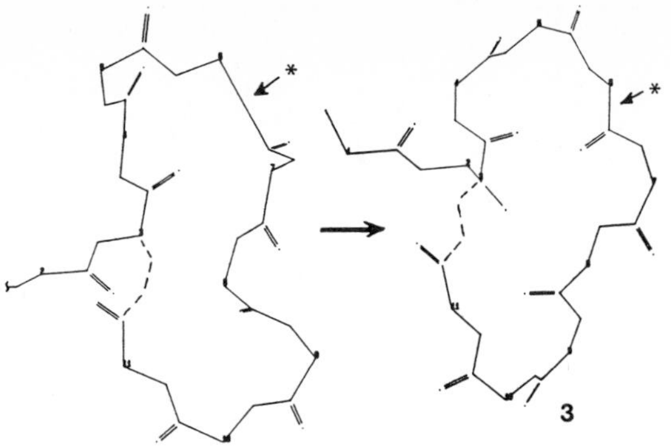

Fig. 16. Backbone changes during a steepest descent CHARMm
minimization. The ring closure lactone unit is sketched
in, since the "backbone only" program does not recognize
this as a backbone feature.

Model 3 [32] was the basis for further refinement.  In order to
assess the significance of the NOE constraints, the stages of refinement
were carried out alternately with:  no constraints, the limited set of 10
constraints, and full set of 26 constraints.  Figure 17 shows a
particularly revealing sequence which serves to establish:  a) that the
NOE constraints do define a unique geometry, b) that the torsion scan
method gives the same results, and that c) the unconstrained CHARMm
minimization suggests a structure (6) which does not fit the NMR data.
The sequence illustrated is:  3→4 (2 ps of constrained r.t. dynamics
followed by a similarly constrained steepest descent minimization, 10
constraints), 4→5 (constrained conjugate gradient minimization with 26
constraints), 4→6 (a similar unconstrained minimization following a 2 ps
dynamics course).  At each stage, a least squares fit of the original
torsion scan model of fragment 7→11 is superimposed on the structure and
the backbone r.m.s. difference over this portion of the structures is
reported.  Points a and b below, are strongly supported by the
observation that model 4, which deviates significantly (0.85 Å r.m.s.
over 7→11) from its precursor fragment, re-refines to the same
conformational family (0.27 Å r.m.s.) when the full set of constraints is
imposed on the CHARMm potentials.

The excellent correspondence of the CHARMm structure and our earlier
fragment conformations from torsion scans also applies to residues 2→5.
Figure 18 shows this graphically and also depicts the fragment (7→11)
conformation in greater detail.

To this stage, no constraints  based on prodiastereotopic $\beta CH_2$
protons had been used; thus only two constraints involving R6 protons
figured in the refinement.  R6 had been introduced with an S-C$\alpha$ center

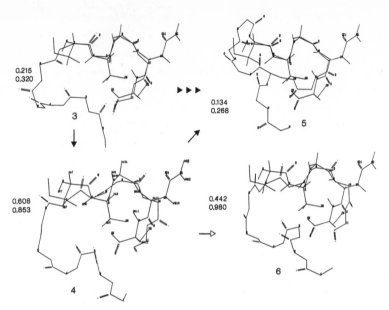

Fig. 17.  A comparison of the NOE-distance constrained CHARMm
          refinement sequence (backbone structures) with:  the
          torsion scan derived conformer at the (7→11)-fragment, and
          the results of unconstrained minimization.  At each stage
          the backbone atom rms deviations (Å) over residues 8→10
          (upper value) and 7→11 (lower value) are listed.

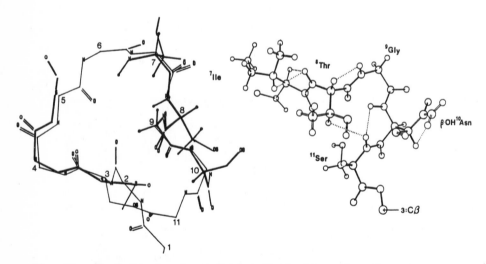

Fig. 18.  The bold structures depict the backbone (+ selected
          side-chains and protons) conformations of fragments
          corresponding to residues 1→5 and 7→11 individually
          refined ($\phi, \psi, \chi$) to fit NOE-derived distance constraints
          using torsion scans.  These are superimposed upon a cyclic
          conformer of the [L-Arg[6], L-allo-Thr[8]]-diastereomeric
          model of lysobactin which was produced by short con-
          strained minimizations and low-temperature dynamics using
          the CHARMm program.

based on the then available results of the degradative structure elucidation [29]. As we begin the process of examining bases for $\beta CH_2$ pro-R/S assignment using NOESYSIM, we found that inversion at $C\alpha6$ improved the fit. A model of [D-Arg$^6$]-lysobactin was obtained from model 5 by following that C$\alpha$ inversion by another refinement cycle: 2 ps dynamics and conjugate gradient minimization with 33 constraints afforded a low energy conformer which is shown in two views in Fig. 19. This is our current working model for lysobactin. We expect to obtain a precise definition of structure (and, assess the significance of segmental motion) as we continue to add constraints based on stereospecific assignments obtained by HETCOR and $^{13}C_{=O}$-edited 2D-J methods [33].

FUTURE PROSPECTS

In the case of prostanoids, a comprehensive scan of all significant degrees of torsional freedom would be in only 5 to 6 dimensions. Similar dimensionalities would be required for many small molecule problems of interest in pharmaceutical, food, and agricultural chemistry. At present, ALEX scans can include more than three dimensions, but human visual perception which is limited to observing intensity values distributed in three dimensions often limits graphical analysis. Scans of up to 6-7 dimensions are reasonably rapid and provide an absolute assurance that the global minimum has been located for small molecules. For larger systems such as peptides, it will be possible to complete comprehensive scans for fragments that are large enough to show a substantial number of intra-, short-range sequential, and intermediate-range through space interactions.

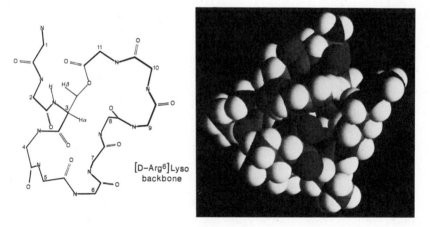

[D-Arg$^6$]Lyso backbone

Fig. 19.  Two views of the current model of the [D-Arg$^6$, L-allo-Thr$^8$]-lysobactin structure. The backbone is shown in the left panel. The right panel shows a space-filling model of the same conformation from a different perspective.

Another area of current effort is modeling based on comparison with simulation data displayed in the same format as the experimental matrix (as illustrated in Fig. 10). As previously noted this utilizes data in the diagonal and crowded cross-peak regions which is otherwise not available for either "constraint definition" or direct intensity comparison. In order to make these comparisons in a manner which lacks user-bias it will be necessary to develop automated evaluations of fit.

More Precise Experimental Distance Constraints

As shown in the lysobactin conformational refinement, torsion scan methods cannot be applied when covalent loop constraints are present and will also be of untenable magnitude for large linear systems. Constrained minimizations and dynamics, for example those available within CHARMm, can handle such systems quite efficiently. Although the nagging local minimum question will remain, it is apparent that much improvement in conformation refinement would be gained if distance constraints of $\pm 0.2$ Å accuracy were available in the 2.4–4.0 Å range. (At present typical constraints classes are 1.8–2.7, 1.8–3.5, and 2.7–5.0 Å.) The examples presented in this work clearly show that two-step spin transfer pathways and PD-truncation effects are the primary sources of error in experimentally-derived constraints. We are currently developing a series of programs (generically called DISCON) which should provide distance constraints free of these difficulties. The initial constraints are derived without assuming any starting structure hypothesis and thus should be free of user-selected model bias. The current working version of this procedure is outlined here:

1) A signal matrix with all quantifiable intensities (including observed zero values) is entered.

2) If any $S_{ij}$ was measured, but the corresponding $S_{ji}$ intensity was unmeasurable, the $S_{ij}$ value is reflected onto $S_{ji}$.

3) The "observed" intensities are flagged for further reference and the data set is scaled and a best fit $\tau_c$-value is calculated.

4) The missing auto-peak intensities are replaced by average values by category: a) amide-NHs, b) methines, c) methylenes, and d) methyls.

5) Missing $S_{ij}$ values are replaced by average values for the following classes: a) intraresidue NH/$\alpha$, NH/$\beta$ and methine/methyl interactions, b) geminal interactions, and c) $\alpha$-H$^i$/NH$^{i+1}$ connectivities.

6) Surrogate zero values are inserted for all remaining unmeasurable or undifferentiable cross-peaks. The resulting full matrix is designated as a "mixed signal matrix."

7) Row normalize the mixed signal matrix and calculate experimental

$\sigma^{eff}$ values. The latter being derived by Eq. (7), not by "direct solution."

8) For each flagged observed value of $S_{ij}$, distance constraints are obtained by an interactive fit to the two term expansion of $\sigma^{eff}$.

$$\sigma_{ij}^{(n)} = \sigma_{ij}^{eff} - \frac{\tau_m}{2} \sum_{k \neq i,j} \sigma_{jk}^{eff} \sigma_{ki}^{eff} \; ; \; \sigma_{ij}^{(n+1)} =$$

$$\sigma_{ij}^{eff} - \frac{\tau_m}{2} \sum_{k \neq i,j} \sigma_{jk}^{(n)} \sigma_{ki}^{(n)} \; ; \; \sigma_{ij}^{(n+2)} = \cdots$$

where (n, n+1, etc.) are the sequential estimates. In practice only 2-4 interactions are required. The final $\sigma_{ij}$ estimates from the experimental data are converted to distance constraints ($d_{ij}$) using the known field strength and the best fit $\tau_c$. In this manner a $d_{ij}$ value (corrected for known large secondary contributions and average small secondary contributions to the NOE) is obtained for each resolved cross-peak.

9) Use the $d_{ij}$ values and applicable torsion constraints with torsion scans, CHARMm or D-SPACE [34] to arrive at an initial conformational model.

10) Apply NOESYSIM to the initial conformational model and replace all averaged or surrogate-zero values with the model-dependent calculated values. Repeat determination of best fit $\tau_c$ and scale factor. Scale the new mixed signal matrix and determine the residual over the "flagged" experimental values.

11) Iterate through steps 7) $\rightarrow$ 10) to convergence.

The procedure bears some resemblance to that suggested by Kaptein [4d] in that a mixed signal matrix is constructed. It differs in two notable respects: 1) no initial conformational hypothesis is required to derive the first set of $d_{ij}$ values; and 2) the direct matrix solution is replaced by a two-term $\sigma^{eff}$ expansion [derived from Eq. (8)] which is less computationally intensive and less subject to error propagation due to the surrogate values. These added average values are used only to correct for two step pathway contributions and none of them are used to directly calculate $d_{ij}$s for later use in refinement.

Preliminary evaluations of this strategy outline using simulated experimental data suggest that $d_{ij}$ estimates will fall within the $\pm 0.2$ Å accuracy limits required for high resolution refinement of small molecules. The improved accuracy of distance estimates by this procedure (=method #3B) in these studies is shown below.

maximum $|d_{ij} - r_{ij}|$

| $r_{ij}$ | #1B | #2 | #3B |
|---|---|---|---|
| $\lesssim 2.9Å$ | 0.42 | 0.37 | 0.05 |
| 2.9–3.3Å | 0.60 | 0.54 | 0.09 |
| 3.3–4.0Å | 0.88 | 0.72 | 0.18 |

Method #1B is direct use of Eq. (1a) (ISPA); method #2 has its previous meaning.


APPENDIX

An Example of PD–Truncation Induced Diagonal Asymmetry and its Implications in Data Analysis

The nature of the deviation from diagonal symmetry that results from short PD periods is best illustrated by specific example. In this appendix, we present simulated signal matrices calculated for a specific relaxation matrix (corresponding to a small molecule case). The relaxation rate matrix employed here is given, in full, below; and followed by Table 3 which shows the major portion of the signal matrix calculated for $\tau_m$ = 0.9s, with a recycle time of 2.1s at 500 MHz observation. A $t_c$ = 20s signal matrix was also calculated for comparison. The latter is diagonal symmetric and the individual peak intensities can be calculated by multiplying the parenthetic values in each second row by 0.01.

The relaxation rate matrix (1/sec), off-diagonal elements are $W_0$–$W_2$

| | | | | | | |
|---|---|---|---|---|---|---|
| 0.8042 | −0.1662 | −0.0986 | −0.0080 | −0.0260 | −0.0050 | −0.0080 |
| −0.1662 | 2.2878 | −0.0260 | −0.0986 | −0.4700 | −0.1662 | −0.0080 |
| −0.0986 | −0.0260 | 0.5260 | −0.0020 | −0.0610 | −0.0040 | −0.0010 |
| −0.0080 | −0.0986 | −0.0020 | 1.8596 | −0.0040 | −0.1662 | −0.4700 |
| −0.0260 | −0.4700 | −0.0610 | −0.0040 | 1.4713 | −0.0260 | −0.0020 |
| −0.0050 | −0.1662 | −0.0040 | −0.1662 | −0.0260 | 1.1122 | −0.0080 |
| −0.0080 | −0.0080 | −0.0010 | −0.4700 | −0.0020 | −0.0080 | 1.4077 |

In the short PD theoretical spectrum illustrated (at $\tau_m$ = 900 ms), "symmetry-related" cross-peak intensities can differ by as much as 50% due to truncation effects. However, residual-driver normalized intensities, $S_{ij}/S_{ii}$ are identical to the values that would be observed with an infinite recovery period. It has been suggested [16c, 15, 8c] that column normalized NOESY data be used for the derivation of rates (and thus distances) by the "direct solution" procedure using matrix diagonalization. Such a normalization procedure is the 2D equivalent of classical fractional enhancement - % NOE = $-S_{ij}/S_{jj}^o$. As can be seen in the middle entries in Table 3, <u>this procedure produces even greater distortions from diagonal symmetry</u> for the short PD experiment. We therefore suggest that this is not the appropriate data reduction strategy for small molecule problems. Fractional enhancement matrices are symmetric about the diagonal only when all initial ($\tau_m$ = 0)

Table 3. The corresponding signal matrix, at $\tau_m$ = 0.9s with $t_c$ = 2.1s, expressed in a variety of commonly used formats.(a)

(parenthetic values are those calculated at $t_c$ = 20s) $S_{jj}^o$ at $t_c$ = 2.1s (b)

```
 0.3428        -0.0277        -0.0338        -0.0002        +0.0005
39.3 (49.0)    -2.67 (-3.97)  -4.63 (-4.84)  -0.02 (-0.03)  +0.05 (+0.07) 0.8726
   ——          -8.08 (-8.10)  -9.86 (-9.87)  -0.06 (-0.06)  +0.15 (+0.14)

-0.0417         0.1546        -0.0017        -0.0127        -0.0860
-4.78 (-3.97)  14.9 (14.7)    -0.23 (-0.17)  -1.24 (-1.21)  -8.83 (-8.18) 1.0360
-27.0 (-27.0)    ——           -1.10 (-1.16)  -8.21 (-8.23)  -55.6 (-55.6)

-0.0262        -0.0009         0.3388        -0.0001        -0.0119
-3.00 (-4.84)  -0.09 (-0.17)  46.4 (62.6)    -0.01 (-0.03)  -1.22 (-2.19) 0.7304
-7.73 (-7.74)  -0.27 (-0.27)    ——           -0.03 (-0.05)  -3.51 (-3.50)

-0.0003        -0.0120        -0.0003         0.2091        +0.0029
-0.03 (-0.03)  -1.16 (-1.21)  -0.04 (-0.03)  20.5 (21.0)    +0.3 (+0.3)   1.0201
-0.14 (-0.14)  -5.74 (-5.76)  -0.14 (-0.15)    ——           +1.4 (+1.4)

+0.0007        -0.0728        -0.0195        +0.0026         0.2540
+0.08 (+0.07)  -7.03 (-8.18)  -2.67 (-2.20)  +0.3 (+0.3)    26.1 (28.5)   0.9744
+0.28 (+0.25)  -28.7 (-28.7)  -7.68 (-7.71)  +1.0 (+1.0)      ——

+0.0012        -0.0255        -0.0009        -0.0310         0.0000
+0.14 (+0.15)  -2.46 (-3.26)  -0.12 (-0.11)  -3.04 (-3.96)  0.0 (0.0)     0.9168
+0.41 (+0.40)  -8.72 (-8.73)  -0.31 (-0.29)  -10.6 (-10.6)  0.0 (0.0)
```

(a) For each i,j matrix entry: the top value is the peak intensity (at $t_c$ = 2.1s) expressed as a fraction of a unit proton intensity, the middle entry is the percent fractional cross-relaxation (= $S_{ij}/S_{jj}^o$), and the bottom set are the residual-driver normalized (= row normalized) intensities (= $S_{ij}/S_{ii}$) multiplied by 100 to be comparable to the values in the second entry position.

(b) The last column lists the relative intensities that would be observed in a control spectrum with the same recycle time. These correspond to diagonal intensities at $\tau_m$ = 0.

auto-peaks have intensities in proportion to their equilibrium populations. (The $S^o_{jj}$ values for the short PD NOESY are also given in Table 3).

The non-equivalence of $S_{ij}$ and $S_{ji}$ is reminiscent of the SKEWSY experiment [16a] and has a similar source: the frequency labeled magnetizations of spins i and j are not identical if $R_{ii} \neq R_{jj}$. When spin i is the more rapidly recovering member of the pair, $|S_{ij}| > |S_{ji}|$ due to the larger value of $M_i$ (more nearly recovered to its full equilibrium value prior to the first 90° pulse of the NOESY sequence) which governs all net transfers that are initiated from site i [see Eq. (6)]. Fortunately these effects are fully accounted for by either row-normalization [as in Eq. (8)] or by the use of NOESYSIM with explicit inclusion of the cycle time in the calculation of signal matrices.

ACKNOWLEDGMENTS

This work would have been impossible without the assistance of J. Gougoutas (Squibb Institute) who incorporated a variety of NOE-related options into the ALEX modeling and molecular display software. The present studies were supported by grants from the National Science Foundation Biophysics Program (#DMB-85-02737) and from the Office of Naval Research (Molecular Recognition Program, #N00014-88-K-0202). A grant from E. R. Squibb and Sons, Inc. covered some of the costs associated with data collection for prostanoid and lysobactin samples.

REFERENCES

1.  K. Wüthrich, 1986, NMR of Proteins and Nucleic Acids, John Wiley & Sons, New York; R. Kaptein, R. Boelens, R. M. Scheek, and W. F. van Gunsteren, 1988, Biochemistry, 27:5389.
2.  R. E. Schirmer, and J. H. Noggle, 1972, J. Am. Chem. Soc., 94:2947, and references cited therein.
3.  a) G. M. Clore, and A. M. Gronenborn, 1985, J. Magn. Reson., 61:158; b) A. M. Gronenborn, and G. M. Clore, 1985, Investigation of the solution structure of short nucleic acid fragments by means of NOE measurements, in: "Prog. in Nuclear Magnetic Resonance Spectroscopy," (J. W. Emsley, J. Feeney, and L. H. Sutcliffe, eds.), Pergamon, Oxford.
4.  a) J. W. Keepers, and T. L. James, 1984, J. Magn. Reson., 57:404; b) E. T. Olejniczak, R. T. Gampe, Jr., and S. W. Fesik, 1986, J. Magn. Reson., 67:28; c) B. A. Borgias, and T. L. James, J. Mag. Reson., in press, d) R. Boelens, T. M. G. Koning, and R. Kaptein, 1988, J. Molecular Struc., 173:299.
5.  A. Kalk, and H. J. C. Berendsen, 1976, J. Magn. Reson., 24:343.
6.  I. Solomon, 1955, Phys. Rev., 99:559.
7.  S. Macura, and R. R. Ernst, 1980, Mol. Phys., 41:95.
8.  a) S. W. Fesik, G. Bolis, H. L. Sham, and E. T. Olejniczak, 1987, Biochemistry, 26:1851; b) ANF - E. T. Olejniczak, R. T. Gampe, Jr., T. W. Rockway, and S. W. Fesik, 1988, Biochemistry, 27:7124; c) P. A. Mirau, 1986, 27th Experimental NMR Conference, Asilomar CA, poster WK12.

9.  a) M. P. Williamson, 1987, <u>Magn. Reson. in Chem.</u>, 25:356; b) J. F. Lefevre, A. N. Lane, and O. Jardetzky, 1987, <u>Biochemistry</u> 26:5076; c) M. Madrid, and O. Jardetzky, 1988, <u>Biochimica et Biophysica Acta</u>, 953:61; see also ref. 3a.
10. This distinction between small and large molecule systems has long been recognized, but the 2D disadvantage for small molecules was mistakenly overstated:  J. K. M. Sanders, and J. D. Mersh, 1982, <u>Prog. in NMR Spetroscopy</u>, 15:353.
11. N. H. Andersen, H. L. Eaton, and C. Lai, "Quantitative Small Molecule NOESY:  A Practical Guide for Derivation of Cross-Relaxation Rates and Internuclear Distances," <u>Magn. Res. in Chem.</u>, in press.
12. M. P. Williamson, and D. Neuhaus, 1987, <u>J. Magn. Reson.</u>, 72:369; O. W. Sorensen, C. Griesinger, and R. R. Ernst, 1987, <u>Chemical Phys. Letters</u>, 135:313.
13. a) N. H. Andersen, K. T. Nguyen, C. J. Hartzell, and H. L. Eaton, 1987, <u>J. Magn. Reson.</u>, 74:195; b) H. L. Eaton, and N. H. Andersen, 1987, <u>J. Magn. Reson.</u>, 74:212.
14. H. L. Eaton, N. H. Anderson, and X. Lai, "Recent Extensions of NOESYSIM, A Program for Rapid Computation of NOESY Intensity Matrices from Atomic Coordinates and Experimental Conditions," paper #112, 29th ENC (4/88; Rochester, N.Y.).
15. S. Macura, B. T. Farmer, II, and L. R. Brown, 1986, <u>J. Magn. Reson.</u>, 70:493.
16. a) J. Bremer, G. L. Mendz, and W. J. Moore, 1984, <u>J. Amer. Chem. Soc.</u>, 106:4691; b) C. L. Perrin, and R. F. Gipe, 1984, <u>J. Amer. Chem. Soc.</u>, 106:4036; c) E. R. Johnston, M. T. Dellwo, and J. Hendrix, 1986, <u>J. Magn. Reson.</u>, 66:399; d) S. W. Fesik, T. J. O'Donnell, R. T. Gampe, Jr., and E. T. Olehniczak, 1986, <u>J. Amer. Chem. Soc.</u>, 108:3165.
17. DISGEO is a metric matrix distance geometry program [T. F. Havel, I. D. Kuntz, and G. M. Crippen, 1979, <u>Biopolymers</u>, 18:73; 1983, <u>Bull. Math. Biol.</u>, 45:665] which has been modified for applications in protein structure refinement - T. F. Havel, and K. Wüthrich, 1984, <u>Bull. Math. Biol.</u>, 46:673; 1985, <u>J. Mol. Biol.</u> 182:281.
18. a) W. Braun, and Gō, N., 1985, <u>J. Mol. Biol.</u>, 186:611; b) program DISMAN - A. D. Kline, W. Braun, and K. Wüthrich, 1986, <u>J. Mol. Biol.</u>, 189:377.
19. T. A. Holak, J. H. Prestegard, and J. D. Forman, 1987, <u>Biochemistry</u>, 26:4652.
20. The CHARMm MM and dynamics program was developed by M. Karplus, and associates [Brooks et al., 1983, <u>J. Comput. Chem</u>, 4:187]; it now contains NOE pseudopotentials, see G. M. Clore, A. M. Gronenborn, A. T. Brunger, and M. Karplus, 1985, <u>J. Mol. Biol.</u>, 186:435; G. M. Clore, A. M. Gronenborn, G. Carlson, and E. F. Meyer, 1986, <u>J. Mol. Biol.</u>, 190:259.
21. M. Nilges, G. M. Clore, and A. M. Gronenborn, 1988, <u>FEBS Letters</u>, 229:317.
22. E. Suzuki, N. Pattabiraman, G. Zon, and T. L. James, 1986, <u>Biochemistry</u>, 25:6854; N. Zhou, A. M. Bianucci, N. Pattabiraman, and T. L. James, 1987, <u>ibid</u>, 26:7905; V. J. Basus, M. Billeter, R. A. Love, R. M. Stroud, and I. D. Kuntz, 1988, <u>Biochemistry</u>, 27:2763.
23. B. A. Borgias, and T. L. James, Methods in Enzymology in press.
24. a) N. H. Andersen, H. L. Eaton, and K. T. Nguyen, 1987, <u>Magn. Reson. in Chem.</u>, 25:1025; b) N. H. Andersen, H. L. Eaton, K. T. Nguyen, and P. Hammen, 1987, <u>Adv. Prostaglandin Thromboxane and Leukotriene Res.</u>, 17:794.
25. a) N. H. Andersen, P. Hammen, K. Banks, M. A. Porubcan, and A. A. Tymiak, ACS Nat'l Mtg, Denver (4/87), Abstracts, paper ORGN-57; b) N. H. Andersen, P. Hammen, K. Banks, T. Pratum, M. A. Porubcan, and A. A. Tymiak, <u>J. Am. Chem. Soc.</u>, submitted.

26. P. W. Sprague, J. E. Heikes, D. N. Harris, and R. Greenberg, 1983, Adv. Prostaglandin Thromboxane and Leukotriene Res., 11:337.

27. a) O. Jardetzky, 1980, Biochem. Biophys. Acta, 621:227; A. Marqusee, and R. L. Baldwin, 1987, Proc. Natl. Acad. Sci., USA, 84:8898, and references cited therein; see also - b) S. Mammi, N. J. Mammi, and E. Peggion, 1988, Biochemistry, 27:1374; c) N. Zhou, and W. A. Gibbons, 1986, J. Chem Soc. Perkin Traus. II, 637.

28. K. V. Chary, S. Srivastava, R. V. Hosur, K. B. Roy, and G. Govil, 1986, Eur. J. Biochem., 158:323.

29. a) A. A. Tymiak, T. J. McCormick, and S. E. Unger, 193rd Nat'l Mtg. of Amer. Chem. Soc. (Denver, 4/87), Abstracts, paper ORGN-56; b) A. A. Tymiak, T. J. McCormick, and S. E. Unger, J. Org. Chem., submitted.

30. M. P. Williamson, T. F. Havel, and K. Wüthrich, 1985, J. Mol. Biol., 182:295; J. M. Moore, D. A. Case, W. J. Chazin, G. P. Gippert, T. F. Havel, R. Powls, and P. E. Wright, 1988, Science, 240:314; G. M. Clore, A. M. Gronenborn, M. Nilges, and C. A. Ryan, 1987, Biochemistry, 26:8012.

31. Katanosin B appears to be identical to lysobactin, it was recently assigned a similar structure, differing in being the [D-Arg, D-allo-Thr]-form: T. Kato, H. Hinoo, Y. Terui, J. Kikuchi, and J. Shoji, 1988, J. Antibiotics, XLI:719.

32. Model 3 is the L-allo-Thr form. To date, our less extensive attempts to carry-out a similar sequence of refinement using a [D-allo-Thr]-fragment(7-11) unit have failed to afford cyclic models which fit the NMR data.

33. a) T. K. Pratum, P. K. Hammen, and N. H. Andersen, 1988, paper #140, 29th E.N.C., Rochester, NY (4/88); b) T. K. Pratum, P. K. Hammen, and N. H. Andersen, 1988, J. Magn. Reson., 78:376.

34. A distance geometry program from Hare Research, Inc. - for recent applications see: A. Pardi, D. R. Hare, M. E. Selsted, R. D. Morrison, D. A. Bassolino, and A. C. Bach, II, 1988, J. Mol. Biol., 201:625; W. Nerdal, D. R. Hare, and B. R. Reid, 1988, ibid, 717; J. H. B. Pease, and D. E. Wemmer, 1988, Biochemistry, 27:8491, - the last gives a more detailed description of the program.

DETERMINATION OF DNA AND PROTEIN STRUCTURES IN SOLUTION VIA COMPLETE
RELAXATION MATRIX ANALYSIS OF 2D NOE SPECTRA

Thomas L. James(a), Brandan A. Borgias(a),
Anna Maria Bianucci(b), and Ning Zhou(c)

(a) Departments of Pharmaceutical Chemistry and Radiology
    University of California, San Francisco, CA  94143
(b) Istituto di Chimica Farmaceutica e Tossicologica
    Dell'Universita di Pisa, 56100 Pisa, Italy
(c) Biological Sciences Department, University of Calgary
    Calgary, Alberta, Canada T2N 1N4

INTRODUCTION

A major objective of scientists for years has been the determination of
molecular structures in non-crystalline environments.  In general, re-
searchers were forced to accept limited structural information due to the
techniques available; certainly high-resolution structures such as those
derived from x-ray diffraction (XRD) on crystals could not be remotely
attained.  However, recent developments provide us with the means to obtain
considerable insight into solution structures (with resolution approaching,
but not equaling, that of XRD on crystals).

Figure 1 shows a proposed scheme which combines experimentally observable structural constraints with various calculational strategies to
generate structures.  Although the scheme in its entirety has not been
accomplished, its component parts have been or are being developed.  The
scheme relies upon two recent developments applied in conjunction to provide
a direct route to molecular structure in noncrystalline phases:  2D NMR and
calculational strategies, e.g., the distance geometry (DG) algorithm (Havel
et al., 1983; Braun and Go, 1985; Havel and Wüthrich, 1985), molecular mechanics (MM) (Weiner and Kollman, 1981; Singh et al., 1986), and molecular
dynamics (MD) (Singh et al., 1986; Karplus and McCammon, 1979; Gunsteren et
al., 1983).  The important ideas can be described quite simply.  The structure of any molecule can be determined if one can obtain a sufficient number
of structural constraints, e.g., internuclear distances and bond torsion
angles, to use in conjunction with holonomic constraints of bond lengths,
bond angles, and steric limitations.  A high-resolution structure requires a

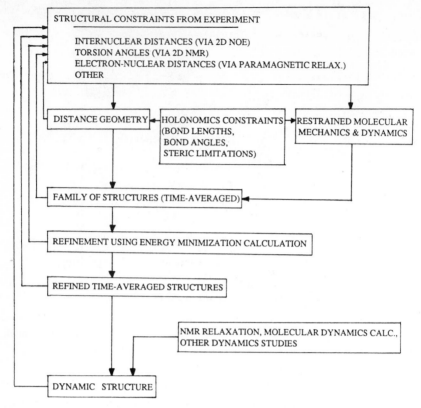

```
┌─────────────────────────────────────────────────────────────────┐
│ STRUCTURAL CONSTRAINTS FROM EXPERIMENT                            │
│ ───────────────────────────────────────                          │
│ INTERNUCLEAR DISTANCES (VIA 2D NOE)                               │
│ TORSION ANGLES (VIA 2D NMR)                                       │
│ ELECTRON-NUCLEAR DISTANCES (VIA PARAMAGNETIC RELAX.)              │
│ OTHER                                                             │
└─────────────────────────────────────────────────────────────────┘
```

DISTANCE GEOMETRY ← HOLONOMICS CONSTRAINTS (BOND LENGTHS, BOND ANGLES, STERIC LIMITATIONS) → RESTRAINED MOLECULAR MECHANICS & DYNAMICS

FAMILY OF STRUCTURES (TIME-AVERAGED)

REFINEMENT USING ENERGY MINIMIZATION CALCULATION

REFINED TIME-AVERAGED STRUCTURES

NMR RELAXATION, MOLECULAR DYNAMICS CALC., OTHER DYNAMICS STUDIES

DYNAMIC STRUCTURE

Fig. 1.  Scheme for deriving molecular structures in noncrystalline environments using experimental NMR data in conjunction with computational procedures.

large number of such constraints.  One can either include (MM, MD) or not include (DG) energetic considerations.

The 2D NMR experiment (Jeener, 1971) is now routinely used in a vast array of specialized experiments for spectrum assignment and structural characterization of nucleic acids and proteins.  Here we will principally examine the capability of using the two-dimensional nuclear Overhauser effect (2D NOE) experiment (Macura and Ernst, 1980); it has the potential for providing numerous internuclear distances.  The homonuclear 2D NOE experiment readily yields qualitative structural information.  This experiment relies on proton–proton dipolar relaxation to yield a correlation map between protons in close spatial proximity ($\leq 5$ Å).  The intensities of the cross–peaks in the 2D NOE spectrum are related to the distances between the protons and can therefore be used to obtain estimates of those distances. The use of 2D NOE in macromolecular structure determination is now becoming widespread. But, while considerable success has been achieved with this technique, there are certain limitations and precautions that are not always appreciated.

The 2D NOE experiment was first described and given a firm theoretical foundation in Ernst's laboratory (Macura and Ernst, 1980; Jeener et al., 1979). We previously described a matrix method for the explicit calculation of the homonuclear 2D NOE intensities expected for a known molecular structure (Keepers and James, 1984) and compared results so obtained for a rigid test molecule (proflavin) in solution with its x-ray crystal structure (Young and James, 1984). The essential concepts described in these seminal papers have been further elaborated by several other laboratories (Olejniczak et al., 1986; Macura et al., 1986; Kay et al., 1986a; Massefski and Bolton, 1985; Bull, 1987). We call this general approach to the calculation and interpretation of 2D NOE intensities CORMA for COmplete Relaxation Matrix Analysis. The critical feature in CORMA is the explicit treatment of the complete relaxation network and specifically spin diffusion in calculating the 2D NOE intensities. An alternative (and widely used) approach assumes that, with short experimental mixing times, a cross-peak intensity correlating a particular proton-proton pair depends on the distance separating that pair of protons alone. This assumption leads to the Isolated Spin Pair Approximation (ISPA). Most analyses of 2D NOE spectra to date have incorporated ISPA in the qualitative or semi-quantitative extraction of interproton distance limits based upon relative cross-peak intensities. Gronenborn and Clore (1985) provide a good review of this approach. Later we will demonstrate that the assumption of ISPA is often not valid in practical situations.

The problems addressed with DNA structure and with protein structure studies are often of a different nature. In general, we are interested in fairly subtle structural changes in the DNA helix which are sequence-dependent and, consequently, guide protein or drug recognition. These subtle variations demand detailed knowledge of the structure and, therefore, accurate internuclear distance determinations. But one can probably define a protein tertiary structure with moderate accuracy using distance geometry or restrained molecular dynamics calculations without accurately determining interproton distances. A qualitative assessment of the 2D NOE spectrum is often all that is needed to obtain the information necessary for calculation of a modestly high-resolution protein structure in solution. If (nearly) all proton resonances of a small protein are resolved and assigned and if the protein has a high content of $\alpha$-helix and $\beta$-sheet, the helices and $\beta$-sheet structures can be fairly easily defined by the qualitative presence or absence of certain cross-peaks. Then the observation of a (relatively) few "long-range" (in the primary sequence) cross-peaks will serve to orient the secondary structure relative to one another. Often one can get a rough idea of the structure immediately, but the use of distance geometry or restrained molecular dynamics will improve the picture and, not coincidentally, lend credence to the structure (or, more properly, family of acceptable structures) obtained. In the most favorable cases, RMS distance deviations of about 1.0 Å can be obtained for the protein backbone. The loop regions and terminal segments are usually the least defined moieties of these protein structures. So, the structures of proteins with low content of common secondary structural elements will be more difficult to pin down. In proteins possessing less common structural features, it may be especially valuable to have more accurate interproton distances for use with the computational techniques. But, more importantly, we will want better defined structures at ligand binding sites (with and without ligand bound). Use of a complete relaxation matrix approach offers the opportunity of determining protein solution structure with greater accuracy and resolution. And protein dynamics can potentially be incorporated into the analysis.

At the very least, a comparison of the experimental 2D NOE spectra with theoretical 2D NOE spectra calculated, using our program CORMA or a similar program, for any nucleic acid or protein structure proposed by distance geometry or restrained molecular dynamics calculations should be made. So far, no one has reported doing that for proteins. However, we have been doing that for our DNA structural studies for the past few years (Broido et al., 1985; Jamin et al., 1985; Suzuki et al., 1986; Zhou et al., 1987). Recently, we have been examining oligonucleotides containing alternating d(AT) duplexes since: a) such sequences are ubiquitous in the promoter region of genes being recognized by RNA polymerase; b) it has not been possible to crystallize oligomers with d(AT) segments longer than a tetramer; and c) at least seven structures have been proposed for alternating d(AT) sequences, largely on the basis of x-ray fiber diffraction data. Examples in this paper will be largely, but not exclusively, chosen from that interest.

## INTERPROTON DISTANCE DETERMINATION FROM 2D NOE EXPERIMENTS

The effect of cross-relaxation between two neighboring protons during the mixing time period $\tau_m$ of the 2D NOE experiment is to transfer magnetization between them. This results in cross-peak intensities in the spectrum that are roughly inversely proportional to the sixth power of the distance between the two neighboring protons. However, as illustrated in Fig. 2, the neighboring protons belong to an array of all protons in the molecular structure. So the cross-relaxation between the two is part of a coupled relaxation network. More rigorously, the whole relaxation network should be considered.

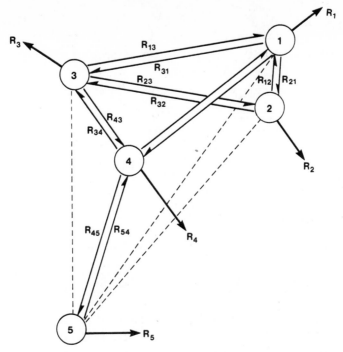

Fig. 2. Network of five protons in a molecule illustrating cross-relaxation rate effects arising from dipole-dipole inter-actions ($R_{ij}$) and non-dipolar relaxation ($R_i$).

Cross-relaxation during the mixing time $\tau_m$ is described by the system of equations (Macura and Ernst, 1980):

$$\frac{\partial M}{\partial t} = -RM \tag{1}$$

In Eq. (1), $M$ is the magnetization vector describing the deviation from thermal equilibrium ($M = M_z - M_0$), and $R$ is the matrix describing the complete dipole-dipole relaxation network. This is essentially an extension of the two-spin equations of Solomon (1955).

$$R_{ii} = 2\left(n_i - 1\right)\left(W_1^{ii} + W_2^{ii}\right) + \sum_{j \neq i} n_j \left(W_0^{ij} + 2W_1^{ij} + W_2^{ij}\right) + R_{1i} \tag{2a}$$

$$R_{ij} = n_i \left(W_2^{ij} - W_0^{ij}\right) \tag{2b}$$

Here, $n_i$ is the number of equivalent spins in a group such as a methyl rotor, and the zero, single, and double transition probabilities $W_n^{ij}$ are given (for isotropic random reorientation of the molecule) by:

$$W_0^{ij} = \frac{q\tau_c}{r_{ij}^6} \tag{3a}$$

$$W_1^{ij} = 1.5 \frac{q\tau_c}{r_{ij}^6} \frac{1}{1 + (\omega\tau_c)^2} \tag{3b}$$

$$W_2^{ij} = 6 \frac{q\tau_c}{r_{ij}^6} \frac{1}{1 + 4(\omega\tau_c)^2} \tag{3c}$$

where $q = 0.1\gamma^4 h^2$. The term $R_{1i}$ represents external sources of relaxation such as paramagnetic impurities and is generally ignored. The system of Eqs. (1-3) has the solution

$$M(\tau_m) = a(\tau_m)M(0) = e^{-R\tau_m}M(0) \tag{4}$$

where $a$ is the matrix of mixing coefficients which are proportional to the 2D NOE intensities. This matrix of mixing coefficients is what we wish to evaluate. The exponential dependence of the mixing coefficients on the cross-relaxation rates complicates the calculation of intensities (or the distances). Note that the expression for the above rate matrix is actually still an approximation in that it neglects cross-correlation terms between separate pairwise and higher order interactions (Bull, 1987; Werbelow and Grant, 1978). However, the importance of the cross-correlation terms appears to be small for $\tau_m \lesssim \tau_m^{opt}$, the optimal mixing time for maximum cross-peak intensity (Bull, 1987). The expressions given above also do not account for second-order effects due to strong scalar coupling (Kay et al., 1986a, 1986b). Explicit account of J coupling can be included in the analysis of the 2D NOE intensities. However, the magnitude of error due to neglect of scalar coupling should lead yield at most a 10% error, except in the case of very strong coupling ($J/\delta \gtrsim 0.5$) and short mixing times (50 ms) where the error could be ~30% (Kay, 1986b). The conclusion is that neglect of J coupling is probably satisfactory in most practical cases where cross-peaks arising from strongly coupled protons are generally not resolvable anyway (Kay, 1986b).

We will compare a few different methods of analyzing 2D NOE spectra for internuclear distance and structural information. This analysis will entail use of hypothetical data sets. This use of hypothetical data is a necessity, since we must know the structure and molecular dynamics exactly in order to understand the effects of any random or (systematic) errors in experimental spectral intensities or the limitations of the different methods being developed to determine structure. We can calculate the theoretical 2D NOE spectrum for the hypothetical structure using any motional model (Keepers and James, 1984). We can add random (or systematic) noise at any level desired. And we can consider any number of peaks to be overlapping. Furthermore, we can compare the various methods proposed in their abilities to handle realistic spectral limitations (Borgias and James, 1988). In other words, for a given dynamic structure we can create spectra with various realistic problems. Then we can see how well we are able to deduce the structure using the different methodologies without using our a priori knowledge of the structure.

Isolated Spin Pair Approximation (ISPA)

The exact evaluation of intensities (the mixing coefficients **a**) according to Eq. (4) is hindered by the absence of a general analytical expression for matrix exponentiation. This has led to the ISPA approach since the exponential can be recast into a Taylor series expansion:

$$\mathbf{a}\left(\tau_m\right) = e^{-\mathbf{R}\tau_m} \sim 1 - \mathbf{R}\tau_m + 1/2\mathbf{R}^2\tau_m^2 - \dots + \frac{(-1)^n}{n!}\mathbf{R}^n\tau_m^n + \dots \quad (5)$$

Truncating the series after the linear term results in a simple approximation for the mixing coefficients that is valid for short $(\tau_m \rightarrow 0)$ mixing times.

In some studies, NOE build-up curves are obtained to assess whether or not the short mixing time condition is achieved. The practical application of ISPA usually goes one step further to eliminate the dependence on correlation time by scaling all the distances with respect to a known reference distance which is assumed to have the same correlation time as the proton-proton pair of interest. The distance $r_{ij}$ between nuclei i and j is usually determined by comparing the cross-peak intensity $a_{ij}$ with that of a reference cross-peak $(a_{ref})$ which originates from two protons whose internuclear distance $r_{ref}$ is known (correlation time of ij pair and reference pair assumed to be equal). Then distances are calculated according to:

$$r_{ij} = r_{ref}\left(\frac{a_{ref}}{a_{ij}}\right)^{1/6} \quad (6)$$

Use of short mixing times definitely limits the signal-to-noise ratio obtainable with cross-peaks in the 2D NOE spectrum. One might also pose the question: how reliable are distances calculated using ISPA? This question was examined by carrying out the following calculations. 1) Simulate (using CORMA) proton 2D NOE spectra (500 MHz) for a DNA duplex octamer in B form, i.e., without assumptions of ISPA, for an effective isotropic correlation time of 4 nsec. at a few mixing times. 2) Assuming these simulated 2D NOE spectra were experimentally obtained, calculate interproton distances using ISPA with the following guidelines for oligonucleotides (Gronenborn and Clore, 1985): a) The correlation times for sugar-sugar, sugar-methyl, and sugar-base (with the exception of H1'-base) inter-proton vectors are the same as the intranucleotide H2'-H2" vector (reference distance 1.77 Å). b) The correlation times for base-base, base-methyl and H1'-base interproton vectors are the same as the intranucleotide H5-H6 vector (reference distance

2.46 Å). 3) Compare the internuclear distances calculated via ISPA with the distances from the model -- B DNA.

Figure 3 shows that comparison using the C2' geminal protons for reference. Clearly, analysis with ISPA introduces a systematic error; calculated distances are shorter than true distances. As the distances become larger, both the systematic deviation and the scatter increase. The amount of error between true and calculated distances depends on the exact environment of the proton pair. If a longer reference distance is used, the "data points" in the above figure effectively move up relative to the line of unit slope, so that shorter distances will be over-estimated while the longer distances will still be underestimated (but not as badly). It should be noted that the shortest mixing time illustrated, 50 ms, is actually shorter than that which is sometimes employed. But the deviation between true and calculated distances clearly becomes larger with longer mixing time. Deviations can be expected to be worse with slower motions and with proteins which possess higher proton density; note that on average there are 3.4 protons within 3 Å of any proton in DNA and 4.7 in proteins.

In summary, there are two problems with using ISPA. The requisite short mixing times give 2D NOE spectra with smaller signal-to-noise. Whenever at least one proton approaches either of the "isolated pair" (or both) at a distance less than the distance between the pair, the approxi-

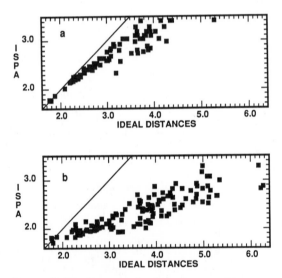

Fig. 3. Comparison of distances calculated according to the isolated spin pair approximation (ISPA) with actual distances. The ISPA distances were calculated from ideal intensities generated by CORMA for d(GGTATACC) in energy-minimized B-DNA conformation assuming a correlation time of 4 ns and a mixing time of a) 50 ms, and b) 250 ms.

142

mation breaks down for practical values of $\tau_m$ for molecules with an effective correlation time greater than a nanosecond.

The implication of using ISPA-derived distances with the distance geometry algorithm is that the bounds, in particular the upper bound, may need to be relaxed more than has been the case in calculations reported. There are also implications for restrained molecular dynamics calculations. If the distances are systematically underestimated, the protein may never be able to get to the vicinity of the global minimum during the MD simulation.

## Complete Relaxation Matrix Analysis (CORMA)

A more rigorous method of calculating intensities is to take advantage of linear algebra and the simplifications which arise from working with the characteristic eigenvalues and eigenvectors of a matrix. The rate matrix $\mathbf{R}$ can be represented by a product of matrices: $\mathbf{R} = \chi\lambda\chi^T$ where $\chi$ is the unitary matrix of orthonormal eigenvectors ($\chi^{-1} = \chi^T$), and $\lambda$ is the diagonal matrix of eigenvalues. The utility of making this transformation is that, since $\lambda$ is diagonal, the series expansion for its exponential (and consequently that of the mixing coefficient matrix) collapses:

$$a = \chi e^{-\lambda\tau_m}\chi^T \tag{7}$$

This calculation allows one to readily calculate all the cross-peak intensities for a proposed structural model. Comparison between calculated and measured intensities allows a determination as to the validity of a model structure. We have developed a program for performing this calculation, named CORMA, which is available upon request from the authors (Keepers and James, 1984; Borgias and James, 1988). Typically, CORMA is used to calculate the 2D NOE peak intensities for a feasible structural model and to compare the calculated intensities with experimental intensities numerically. However, it is also possible for the comparison to be made using contours and true chemical shift axes, as generated by the program SPHINX-LINSHA (Widmer and Wüthrich, 1987), as depicted in Fig. 4 for the aromatic-H1' region of [d(GGTATACC)]$_2$ in a B-D-B conformation in comparison with experimentally obtained intensities. Of course, we can compare 2D NOE spectra of the hypothetical structure with that of any other deduced structure. Another representation of the type of comparison that is possible using CORMA is shown in Fig. 5. This simple illustration, output directly from CORMA, is for a small portion of the protons of bovine pancreatic trypsin inhibitor (BPTI), comparing across the diagonal the 2D NOE peak intensities predicted for the two crystal structures (A and B) of BPTI which are available in the Brookhaven Protein Data Bank. This region was chosen because there are some differences between the two structures in the vicinity of glu 7; this is manifest in the spectra by intensity differences and by the absence of peaks in one form (indicated by shaded boxes) which are present in the other form.

143

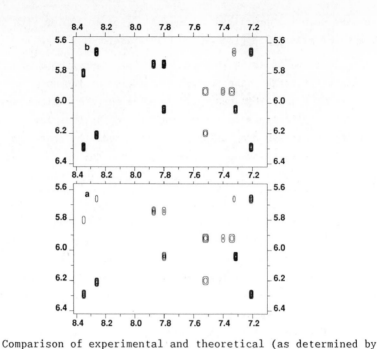

Fig. 4. Comparison of experimental and theoretical (as determined by CORMA) 2D NOE spectra for the aromatic-H1' region in [d(GGTATACC)]$_2$. Calculated intensities assume the octamer duplex has a correlation time of 4 ns, a B-D-B conformation, and that the spectra were obtained at 500 MHz with a mixing time of 250 ms (i.e., experimental conditions). Only those peaks which were unambiguously assigned and measured with high confidence are shown in these plots. Contour plots of the intensities were generated by software developed by Dr. Hans Widmer. CORMA-calculated intensities are shown in **a**; experimental intensities are shown in **b**. Line shapes for both plots are calculated assuming a natural linewidth of 5 Hz, and acquisition times of 0.15 sec. in both t$_1$ and t$_2$, a 45° phase-shifted sine-squared apodization was applied in both domains. Coupling constants were assigned to model the splitting in the experimentally obtained spectrum (not shown). Peak assignments are (in ppm): A4-H8 8.35, A6-H8 8.26, G1-H8 7.87, G2-H8 7.80, C8-H6 7.52, A6-H2 7.40, C7-H6 7.34, A4-H2 7.33, T3-H6 7.26, T5-H6 7.21, A4-H1' 6.29, A6-H1' 6.21, C8-H1' 6.20, G2-H1' 6.04, C7-H1' 5.92, T3-H1' 5.80, G1-H1' 5.74, T5-H1' 5.66 ppm. Displaying the intensities in this manner results in greater apparent contrast than the simple grey-scale representation of the following figure. However, the validity of this display is subject to the selection of appropriate coupling constants and the assumption that all peaks have the same natural linewidth.

It should be mentioned that an alternative approach to matrix diagonalization has been presented for dealing with the complete relaxation network (Lefevre et al., 1987). This method entails a numerical solution to the Bloch equations. Recent work has shown CORMA and the method used by Jardetzky and co-workers give the same results (M. Madrid, B. Borgias, O. Jardetzky, and T. James, unpublished results).

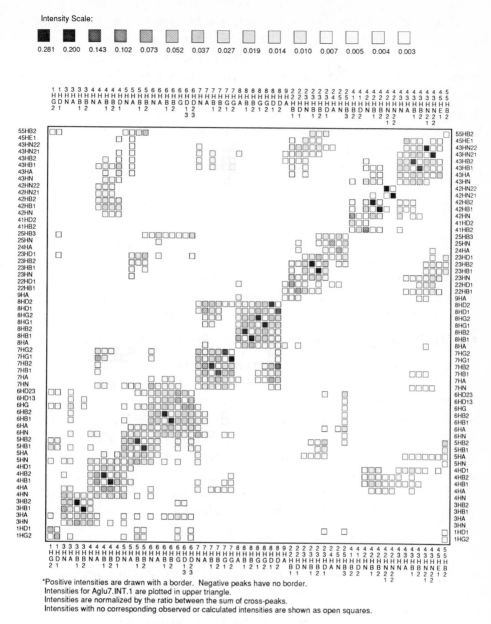

Intensity Scale:

| 0.281 | 0.200 | 0.143 | 0.102 | 0.073 | 0.052 | 0.037 | 0.027 | 0.019 | 0.014 | 0.010 | 0.007 | 0.005 | 0.004 | 0.003 |

*Positive intensities are drawn with a border. Negative peaks have no border.
Intensities for Aglu7.INT.1 are plotted in upper triangle.
Intensities are normalized by the ratio between the sum of cross-peaks.
Intensities with no corresponding observed or calculated intensities are shown as open squares.

SCHEMATIC PLOT OF 2D NOE INTENSITIES Calculated via CORMA
Comparison of A and B Crystal Forms of BPTI

Fig. 5.   Schematic intensities for some of the proton cross-peaks of
          bovine pancreatic trypsin inhibitor (BPTI) as determined by
          CORMA.  Experimental intensities were obtained at 500 MHz
          with a mixing time of 200 ms.  Calculated intensities
          assume an isotropic correlation time of 1.3 ns.  Crystal
          structures A and B, obtained from the Brookhaven Protein
          Data Bank, were used.  Protons were placed on the
          structures using standard bond angles and bond lengths.
          Resulting cross-peak intensities for A and B structures are
          shown in the upper and lower triangles, respectively.

145

The complete relaxation matrix approach is a) accurate, b) can accommodate any size spin system (computer size limitations only), c) is not
limited to any range of mixing times, d) incorporates spin diffusion
naturally, and e) can utilize any molecular motion model.  We have carried
out a number of theoretical calculations as well as experimental studies.
In summary (Keepers and James, 1984; Young and James, 1984; Borgias and
James, 1988):  a) Distances up to 5 Å (and possibly 6 Å) with an accuracy of
10% could be attainable with knowledge of individual relaxation times.
b) Detailed knowledge of the molecular motions is not required for this
distance accuracy; it is generally sufficient for a single "effective"
isotropic correlation time to be used in the spectral density expression for
any nucleus.  One can iteratively fit the simple 2D NOE spectra of small
molecules (Young and James, 1984), but that was initially not possible for
larger molecules.  Over the course of our studies, the sophistication of
analysis developed from evaluation of a few models on the basis of selected
intensities to a more detailed analysis of many closely related models using
all well resolved intensities (Broido et al., 1985; Zhou et al., 1987).

Direct Calculation of Distances From Experimental Spectra (DIRECT)

The limitation of the trial-and-error approach utilizing CORMA (vide
supra) is that it is governed by the choice of structural models.  It is
capable only of discriminating between the proposed structures and of
indicating regions of good (or bad) fit between the model and true solution
structure.  There are two possible ways to circumvent this limitation.  One
is the direct determination of distances without making the isolated spin
pair approximation -- essentially using the reverse of CORMA.  The other is
to automatically refine the structure based on the 2D NOE intensities.  We
will consider first the "reverse CORMA" approach.

An obvious solution to the problem of limited (and biased) trial
structures is to apply the same computational techniques used in CORMA to
the direct calculation of proton-proton distances from the experimental
intensities.  This approach has been discussed elsewhere (Olejniczak et
al., 1986; Borgias and James, 1988).  Rearrangement of Eq. (4) gives:

$$\frac{-\ln\left(\frac{a\left(\tau_m\right)}{a(0)}\right)}{\tau_m} = R \tag{8}$$

So, assuming isotropic tumbling and using Eqs. (2) and (3), distances
can be calculated directly from the rate matrix via diagonalization of the
experimental 2D NOE spectrum (matrix).

We have tested the performance of the direct calculation of distances
(DIRECT method) in the presence of several different types of data errors:

random noise, cross-peak overlap, and diagonal peak overlap (Borgias and James, 1988).  For relatively small molecules yielding spectra with very high signal-to-noise, in which most of the major peaks can be resolved and accurately estimated, this is clearly the ideal method of distance determination.  However, low resolution (peak overlap) and low (i.e., generally realistic protein or nucleic acid spectral) signal-to-noise hamper the accurate estimation of distances (Borgias and James, 1988).  These complications lead to a lack of knowledge about relaxation pathways.  This problem is exacerbated by spin diffusion which occurs at longer mixing times.  Under these circumstances, a significant amount of spin magnetization is distributed among a large number of cross-peaks, although the amount each may have could be so small that the intensity is less than the signal-to-noise ratio.  Consequently, the DIRECT approach has the same limitations as ISPA, namely, short mixing times are better, and longer distances will be underestimated.  Regardless, the DIRECT method is distinctly better than ISPA and has the additional feature that it is not explicitly model-dependent.

## Iterative Relaxation Matrix Analysis (IRMA)

An interesting and potentially quite useful extension of the DIRECT calculation of distances is iterative relaxation matrix analysis (IRMA) proposed by Prof. Robert Kaptein, University of Utrecht.  The general scheme of IRMA is shown in Fig. 6.  The critical feature of IRMA is the generation of the augmented intensity matrix which contains all the scaled experimental intensities and the intensities calculated from a suitable model structure (obtained from model building, distance geometry, etc.).  By solving the augmented intensity matrix for distances, one generates a distance set which will be in reasonable agreement with the model, but which is also partially restrained by the experimental intensities.  By iterating through the cycle of structure generation via distance geometry or molecular dynamics on one branch and the CORMA-type calculations on the other, one will eventually reach a self-consistent structure which incorporates all the structural information inherent in the 2D NOE intensities.  We have made some simple tests of the accuracy of the distances derived from the augmented intensity matrix.  For the initial model we generated a randomized structure in which the proton coordinates had a root-mean-square shift of 3 Å relative to the ideal structure from which "experimental" intensities were calculated by CORMA.  The short distances generated from the augmented intensity matrix were within a few percent of their ideal values.  Longer distances tended to reflect the round-off error of the weakest intensities, or the random distribution of the model.  Of course, a fundamental aspect of IRMA is iteratively cycling through the calculations such that the longer distances can also be improved.

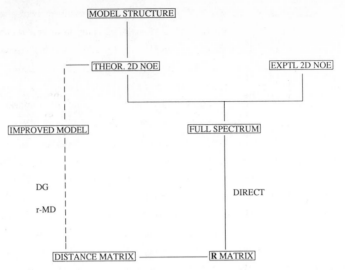

Fig. 6.  Schematic diagram of the iterative relaxation matrix analysis
(IRMA).  A model structure is used to generate a theoretical
2D NOE spectrum (using CORMA).  Wherever possible, experimental intensities are substituted into the theoretical
spectrum to yield a full spectrum suitable for direct solution
of the rate and distance matrices.  This process incorporates
all the effects of network relaxation and spin diffusion and
will be relatively accurate, depending on how close the model
is to the true structure.  To improve the structure, distance
geometry (DG) or restrained molecular dynamics (r-MD) can be
used to generate an improved model which incorporates the
distance restraints generated by the previous pass of the
intensity/distance calculation.

Refinement of Interproton Distances Using COMATOSE

The limitations of the ISPA and DIRECT methods for accurately determining interproton distances led us to pursue the less direct and more
time-consuming approach of iterative least-squares refinement of structure
based on the 2D NOE intensities.  With the increasing availability of more
powerful computers, it is becoming feasible to refine the structures of
modest-sized biopolymers (i.e., those with ca. 1000 protons) using their 2D
NOE spectra.  We have begun developing the program COMATOSE (COmplete Matrix
Analysis Torsion Optimized StructurE) for the refinement of molecular
structure based on 2D NOE intensities (Borgias and James, 1988).  The goal
is to optimize a trial structure while minimizing the error between
calculated and observed intensities.  We have incorporated CORMA into a
program with a Marquardt-Levenberg least squares optimizer.  We have chosen

to minimize the function:

$$E = \sum_i \left[ I_i^{obs} - k I_i^{calc} \right]^2 \tag{9}$$

where k is a constant scaling factor applied to all the calculated
intensities to normalize the sums of the calculated and observed intensities.
Consequently, we investigated the effects of experimental errors in peak
intensities and peak overlap on the refinement process.

The use of Cartesian coordinates to define the position of protons
provided too many parameters for refinement to proceed properly. To reduce
the number, one can take advantage of holonomic structural constraints.
Several alternative algorithms using different variable parameters were
tried, but the most promising one makes use of torsion angles and group
orientation parameters rather than the full set of coordinates. For nucleic
acids, the following parameters are defined for each nucleotide:  $\chi$, the
glycosidic bond torsion angle; $\gamma$, the C4′-C5′ bond torsion angle; sugar
pucker parameters P and $\theta_{max}$ (Altona and Sundaralingam, 1972) and group
orientation parameters $\alpha'$, $\beta'$, $\gamma'$, x, y, and z. The sugar pucker parameters
P and $\theta_{max}$ allow fewer parameters since the ring torsions are correlated and
constrained by the condition of ring closure, and are used to define each of
the furanose ring torsions (Altona and Sundaralingam, 1972). Each nucleotide is described as a freely floating entity, its location described by the
group orientation parameters, which is held in place only by the constraints
imposed by the fit between calculated and observed internucleotide NOE
intensities. The internucleotide torsion angles are not specifically
included in the description of the structure, although constraints such as
the O5′-O3′ angle could be included during refinement. This yields ten
structural parameters for each nucleotide.

Refinement against torsion angles as variable parameters has the
advantage that the number of parameters is considerably reduced from the
alternative choice of Cartesian coordinates. Furthermore, incorporation of
ancillary constraints such as bond lengths, bond angles and atom
connectivities, as well as others (vide infra) is easily accomplished.

We have found COMATOSE to work reasonably well for DNA fragments.
Figure 7 depicts the results from one study. This study was for a 250 ms
mixing time with experimentally realistic random noise. The other methods
(ISPA,DIRECT) compared were not very good at such a long mixing time.
Consequently, it is feasible that sufficient cross-peak intensity can be
measured that distances out to 6-7 Å can be determined as illustrated. Not
only does COMATOSE improve the distances, but the spectral density can be
refined as well. For the figure shown, the original hypothetical model for

149

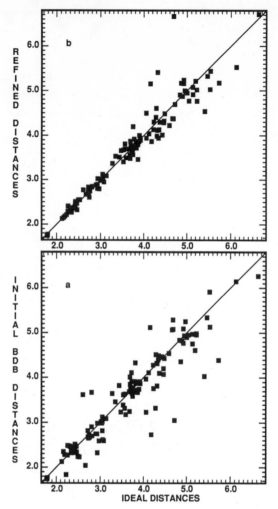

Fig. 7.   Comparison of distances obtained from COMATOSE refinement
with ideal distances.  Intensities were calculated by CORMA
for $[d(GGTATACC)]_2$ in energy-minimized BDB-DNA conformation
assuming an isotropic correlation time of 4 ns and a mixing
time of 250 ms.  The starting conformation was the BDB-DNA
conformation prior to energy minimization, and the initial
value chosen for a correlation time was 10 ns.  a) Prior to
COMATOSE refinement.  b) After COMATOSE refinement.

the study had an isotropic correlation time of 4.0 ns.  The COMATOSE calculations though were initiated with a correlation time of 10 ns, i.e., as a
rough guess one might choose in the absence of any other information.
COMATOSE refinement yielded a correlation time of 4.2 ns, simultaneously
yielding the distances plotted.  This level of performance has been found in
many cases now.

It is obvious that an initial structure is necessary for refinement via
COMATOSE.  In the case of DNA fragments, it is reasonable to start with

standard B form DNA and refine that structure using the experimental 2D NOE spectra. In the case of proteins, it may be appropriate to obtain an initial structure using ISPA (or, better, DIRECT) via analysis of the distances with either the methods of distance geometry or restrained molecular dynamics. COMATOSE could then be used to refine that structure. An iterative process involving the use of COMATOSE to arrive at the best available determination for a solution structure is shown in Fig. 8.

In short, the COMATOSE refinement process is much more cpu time-consuming than simpler methods for estimating the distances from 2D NOE intensities such as ISPA or the DIRECT method. However, it does not suffer from the systematic errors associated with such methods. No information is needed from the diagonal peak intensities (which are typically overlapping and difficult to measure), and overlapping or tiny cross-peaks pose no problem (other than loss of potential additional information) since these intensities are simply not included in the calculation of the error to be minimized. Major limitations of the COMATOSE approach are a) the neglect of energetic and steric constraints during the structure refinement process and b) ability of COMATOSE to find a local minimum rather than a global minimum. These occasionally allow conformations to be generated which satisfy the distance constraints imposed by the NOE intensities, but which are otherwise unacceptable. However, as shown in the scheme in Fig. 8, it may be possible to overcome both of these limitations by employing an iterative method in concert with molecular mechanics or molecular dynamics calculations.

DNA Fragments

Our development of the refinement procedures is quite recent. Pre-viously (Broido and James, 1985; Suzuki et al., 1986) and concomitantly (Zhou et al., 1987), we have derived DNA structural information by modeling DNA structures and comparing theoretical spectra calculated from the models with experimental 2D NOE spectra.

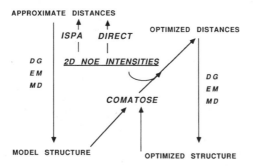

Fig. 8.  Scheme for the complete structure determination of a macromolecule.  DG, EM, and MD refer to distance geometry, energy minimization, and molecular dynamics, respectively.

More recently, the experimental 2D NOE data from [d(GGTATACC)]$_2$ were used with COMATOSE to see if the structure could be improved beyond that of the energy-minimized BDB model which we had previously (Zhou et al., 1987) found satisfactory from our model-building. As indicated in Fig. 9, optimization with COMATOSE did improve the fit of the resultant structure's theoretical 2D NOE spectra to the experimental spectra. We also could utilize the 225 optimized distances derived from COMATOSE as restraints using a molecular mechanics (AMBER) force field containing a harmonic NOE-distance restraint term. We have used such restrained AMBER calculations on [d(GGTATACC)]$_2$, which had already been subjected to COMATOSE optimization (A. M. Bianucci, B. Borgias, and T. L. James, unpublished results). The results (simplified) are summarized in Fig. 9. As illustrated, it was also possible to find essentially the same structure using standard B-DNA as the initial structure.

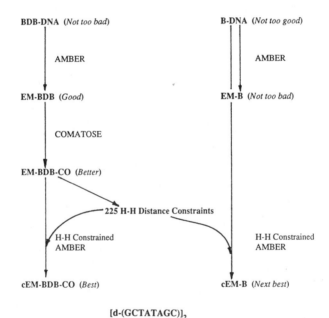

**BDB-DNA** (*Not too bad*)     **B-DNA** (*Not too good*)

AMBER     AMBER

**EM-BDB** (*Good*)     **EM-B** (*Not too bad*)

COMATOSE

**EM-BDB-CO** (*Better*)

225 H-H Distance Constraints

H-H Constrained     H-H Constrained
AMBER     AMBER

**cEM-BDB-CO** (*Best*)     **cEM-B** (*Next best*)

**[d-(GCTATAGC)]$_2$**

*Parenthetic comments refer to comparison of theoretical 2D NOE spectra for each model with experimental spectra.*

Fig. 9.   Refinement of [d(GGTATACC)]$_2$ structure via molecular energy refinement (AMBER), without and with COMATOSE-derived NOE-distance restraints, and optimization of theoretical 2D NOE spectra against experimental 2D NOE spectra using COMATOSE. EM and cEM prefixes to the model names denote energy-minimized, without and with NOE-distance constraints, respectively, and the CO suffix denotes COMATOSE-optimized.

CONCLUSIONS

The original scheme presented for structure determination (Fig. 1) indicated both distance geometry and restrained molecular dynamics could be usefully employed. Both should be used in conjunction with the complete relaxation matrix analysis of the 2D NOE data. It is not possible to know now which, if any, method will provide superior and will be the method of choice five years from now. In fact, it may induce greater confidence to derive essentially the same structure using a couple of different approaches. An important aspect of this development is that we can always compare theoretical spectra calculated from any interim refinement model with experimental spectra. (Notice the "feedback" arrows in Fig. 1.)

Not mentioned previously, but an important point, is that the structures will be improved by analysis of data from several 2D NOE spectra. The most useful data are obtained at mixing times where the contribution from spin diffusion is as important as the primary Overhauser effect, as the pattern of spin diffusion is quite sensitive to structure. The longer mixing times and complete relaxation matrix analysis used here will enable more distances and somewhat longer distances (5 Å or possibly 6 Å) to be measured. This should also result in better solution structures.

ACKNOWLEDGMENTS

The work described here was supported by the National Institutes of Health (Grants GM 39247 and CA 27343) and the National Science Foundation (Grant PCM 84-04198).

REFERENCES

Altona, C., and Sundaralingam, M., 1972, J. Amer. Chem. Soc., 94:8205.

Borgias, B. A., and James, T. L., 1988, "COMATOSE: A Method for Constrained Refinement of Macromolecular Structure Based on Two-Dimensional Nuclear Overhauser Effect Spectra," J. Magn. Reson., 79:493–512.

Braun, W., and Go, N., 1985, "Calculation of Protein Conformations by Proton-Proton Distance Constraints. A New Efficient Algorithm," J. Mol. Biol., 186:611–626.

Broido, M. S., James, T. L., Zon, G., and Keepers, J. W., 1985, "Investigation of the Solution Structure of a DNA Octamer [d-(GGAATTCC)]$_2$ Using Two-Dimensional Nuclear Overhauser Effect Spectroscopy," Eur. J. Biochem., 150:117–128.

Bull, T. E., 1987, "Cross-Correlation and 2D NOE Spectra," J. Magn. Reson., 72:397–413.

Gronenborn, A. M., and Clore, G. M., 1985, Investigation of the Solution Structures of Short Nucleic Acid Fragments by Means of NOE Measurements, in: "Progress in Nuclear Magnetic Resonance Spectroscopy," ed. Sutcliffe, L. H., Pergamon Press, Oxford, pp. 1–32.

Havel, T. F., Kuntz, I. D., and Crippen, G. M., 1983, "The Theory and Practice of Distance Geometry," Bull. Math. Biol., 45:665–720.

Havel, T. F., and Wüthrich, K., 1985, "An Evaluation of the Combined Use of Nuclear Magnetic Resonance and Distance Geometry for the Determination of Protein Conformations in Solution," J. Mol. Biol., 182:281–294.

Jamin, N., James, T. L., and Zon, G., 1985, "Two-Dimensional Nuclear Overhauser Effect Investigation of the Solution Structure and Dynamics of the DNA Octamer [d-(GGTATACC)]$_2$," Eur. J. Biochem., 152:157-166.

Jeener, J., 1971, AMPERE International Summer School, Basko Polje, Yugoslavia.

Jeener, J., Meier, B. H., Bachmann, P., and Ernst, R. R., 1979, "Investigation of Exchange Processes by 2D NMR Spectroscopy," J. Chem. Phys., 371:4546-4553.

Karplus, M., and McCammon, J. A., 1979, "Protein structural fluctuations during a period of 100 ps," Nature, 277:578.

Kay, L. E., Scarsdale, J. N., Hare, D. R., and Prestegard, J. H., 1986a, "Simulation of Two-Dimensional Cross-Relaxation Spectra in Strongly Coupled Spin Systems," J. Magn. Reson., 68:515-525.

Kay, L. E., Holak, T. A., Johnson, B. A., Armitage, I. M., and Prestegard, J. H., 1986b, "Second Order Effects in Two Dimensional Cross-Relaxation Spectra of Proteins: Investigation of Glycine Spin Systems," J. Am. Chem. Soc., 108:4242.

Keepers, J. W., and James, T. L., 1984, "A Theoretical Study of Distance Determinations from NMR. Two-Dimensional Nuclear Overhauser Effect Spectra," J. Magn. Reson., 57:404-426.

Lefevre, J. -F., Lane, A. N., and Jardetzky, O., 1987, "Solution Structure of the Trp Operator of Escherichia coli Determined by NMR," Biochemistry, 26:5076-5090.

Macura, S., and Ernst, R. R., 1980, "Elucidation of Cross Relaxation in Liquids by 2D NMR Spectroscopy," Mol. Phys., 41:95-117.

Macura, S., Farmer, B. T., and Brown, L. R., 1986, "An Improved Method for the Determination of Cross-Relaxation Rates From NOE Data," J. Magn. Reson., 70:493-499.

Massefski, Jr., W., and Bolton, P. H., 1985, "Quantitative Analysis of Nuclear Overhauser Effects," J. Magn. Reson., 65:526-530.

Olejniczak, E. T., Gampe, Jr., R. T., and Fesik, S. W., 1986, "Accounting for Spin Diffusion in the Analysis of 2D NOE Data," J. Magn. Reson., 67:28-41.

Singh, U. C., Weiner, P. K., Caldwell, J., and Kollman, P. A., 1986, AMBER 3.0, University of California, San Francisco.

Solomon, I., 1955, "Relaxation Processes in a System of Two Spins," Phys. Rev., 99:559.

Suzuki, E. -I., Pattabiraman, N., Zon, G., and James, T. L., 1986, "Solution Structure of [d-(AT)$_5$]$_2$ via Complete Relaxation Matrix Analysis of 2D NOE Spectra and Molecular Mechanics Calculations," Biochemistry, 25:6854-6865.

van Gunsteren, W. F., Boelens, R., Kaptein, R., and Zuiderweg, E. R. P., 1983, "Nuclear Acid Conformation and Dynamics, Report of NATO/CECAM Workshop," ed., Olson, W. K.

Weiner, P. K., and Kollman, P. A., 1981, "AMBER: Assisted Model Building with Energy Refinement. A General Program for Modeling Molecules and Their Interactions," J. Comp. Chem., 2:287-303.

Werbelow, L., and Grant, D. M., 1978, Adv. Magn. Reson., 9:189.

Widmer, H., and Wuthrich, K., 1987, "Simulated Two-Dimensional NMR Cross-Peak Fine Structures for [1]H Spin Systems in Polypeptides and Polydeoxynucleotides," J. Magn. Reson., 74:316-336.

Young, G. B., and James, T. L., 1984, "Determination of Molecular Structure in Solution via Two-Dimensional Nuclear Overhauser Effect Experiments: Proflavine as a Rigid Molecular Test Case," J. Am. Chem. Soc., 106:7986-7988.

Zhou, N., Bianucci, A. M., Pattabiraman, N., and James, T. L., 1987, "Solution Structure of [d-(GGTATACC)]$_2$: Wrinkled D Structure of the TATA Moiety," Biochemistry, 26:7905-7913.

# THE STRUCTURE AND BEHAVIOR OF THE STARCH GRANULE AS STUDIED BY NMR

J. M. V. Blanshard(a), E. M. Jaroszkiewicz(a) and
M. J. Gidley(b)

(a) Food Science Laboratories
    Department of Applied Biochemistry and Food Science
    Nottingham University, U. K.
(b) Unilever Research Laboratory, Colworth House,
    Sharnbrook, Bedford, UK MK44 ILQ

## INTRODUCTION

The purpose of this review will be to detail the contribution that NMR spectroscopy has afforded to our knowledge of the structure, gelatinization and retrogradation behavior of the starch granule. Those with only a cursory knowledge of starch should appreciate that other techniques have proved of enormous value in elucidating structure and behavior and the results of early NMR studies in many cases echoed the conclusions derived from other techniques. However, increasingly, NMR spectroscopy is making a distinctive contribution to our understanding of the starch granule and its physical chemistry.

By way of introduction, certain structural and behavioral character-istics of the starch granule may be listed. More detailed information is available in recent reviews (Blanshard, 1987; French, 1984). Each granule which, depending on its sources may vary from approximately 2-100 μ, is composed almost entirely of polyglucans. The two dominating macromolecules are the largely linear poly α-(1 → 4)-glucan amylose with a molecular weight as high as $1 \times 10^6$, and the branched polymer amylopectin with a molecular weight as high as $1 \times 10^9$. This is composed of poly α-(1 → 4)-glucan chains of variable length (depending on the starch source) but typically with a bimodal distribution of chain length, one population having a d.p. of 15-25, the other with a d.p. of 40-45; the branch points are α-(1 → 6) in character. The polymer chains are largely radially arranged and overall the structure is semicrystalline with two main crystalline forms identified by Katz and van Itallie as A and B (1930). Examination by both optical and electron microscopy reveals an annular structure with alternating regions of higher density and crystallinity. The structure is stable in cold water but the majority of plant starches, on heating in excess water at temperature regions characteristic of the starch, lose their order (as concluded by loss of x-ray crystallinity and optical birefringence) and swell. Continued heating may lead to further swelling and ultimate disruption of the granules. Cooling results in an increase in viscosity and usually the generation of a viscoelastic gel structure. Phase separation and recrystallization of the polymer chains result in an increase in gel rigidity, the recrystallization being responsible for the term retrogradation. A, B and V (the latter formed as a complex with lipids) polymorphic forms may be observed in retrograded material.

## THE STRUCTURE OF THE STARCH GRANULE

With such a structure it is not surprising that of all the NMR techniques presently available, the $^{13}$C cross polarization, magic angle spinning method has made a major contribution. For the sake of convenience the subject will be considered under two headings: (i) the structure of the poly $\alpha$-(1 → 4)-glucan chains and (ii) the structure of the granule itself. The primary emphasis will be placed on spectra and information derived from the solid state; reference to solution studies will be secondary and only considered where they illuminate the solid state condition.

### Structure of the Poly $\alpha$-(1 → 4)-Glucan Chains

Any consideration of the polymer must be built on a sound appreciation of the shifts associated with individual carbons and protons in the glucose monomer.

Assignment of $^{13}$C and $^{1}$H Shifts in $\alpha$-D-Glucose. Early results using $^{13}$C solution state NMR for glucose were reported by Pfeffer et al. (1984). Morris and Hall (1982) used heteronuclear 2D-NMR to determine both $^{13}$C and $^{1}$H chemical shifts of glucose and $\alpha$-D-glucopyranose oligomers in a 10% aqueous solution. However, Pfeffer et al. (1983a, 1983b) have elegantly established complete and unequivocal $^{13}$C resonance assignment of crystalline $\alpha$-D-glucose and $\alpha$-D-glucose monohydrate using the $\alpha$-D-[1 - $^{13}$C]-glucose and $\alpha$-D-[6 - $^{13}$C]-glucose to identify both the labeled sites themselves and those adjacent to them. Subsequent work by Hewitt et al. (1986) concurs with these results. The chemical shifts so determined are illustrated in Fig. 1.

### Conformation of $\alpha$-(1 → 4) Oligomers of $\alpha$-D-Glucopyranose

Maltose. Hewitt et al. (1986) have reported the almost perfectly resolved resonance of $\beta$-maltose monohydrate in the crystal state. The resonance of C-1 at the $\alpha$-D-(1 → 4) linkage occurs at 97.8 ppm whereas the C-1 atom of the reducing residue gives a peak at 105.2 ppm. The resonances

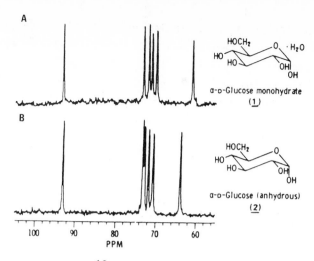

Fig. 1. A) 37.8 MHz $^{13}$C CP-MAS spectrum of α-D-glucose, H$_2$O
(1); B) α-D-glucose (2).

at 66.4 and 60.9 ppm can be attributed to the C-6 and C-6' atoms of the
primary hydroxymethyl groups but the rest of the spectrum can only be
assigned tentatively with the presumption that C-4' resonates at 80.9 ppm.

Morris and Hall (1981) have studied maltose in the solution state (10%
in D$_2$O) using heteronuclear 2D-NMR and have determined both $^{13}$C and $^1$H
chemical shifts in solution while Shashkov et al. (1986) have analyzed
n.o.e. data on α- and β-methyl maltosides and shown that the structure in
solution cannot be explained by only one conformer in solution but possibly
by four in equilibrium.

Maltotriose and Maltotetraose. In the solid state Saito and Tabeta
(1981) have examined the $^{13}$C chemical shifts of maltotriose by the CP-MAS
technique at 75.46 MHz. The C-1, C-4 and C-6 signals are well resolved, but
the C-2, C-3 and C-5 signals overlapped because of increased linewidth.
Conventional high resolution spectra have also been derived and the chemical
shifts determined. Morris and Hall (1982) have determined the $^{13}$C and $^1$H
chemical shifts in the same way as for maltose. Gidley (1988) has shown
that the $^3$J(H-4' → C-1) coupling constant determined by 2D-heteronuclear
J-resolved NMR increases from 4.5 for maltose to 5.3 for maltotriose and 6.1
HZ for maltotetraose. Although there is no accurate Karplus relationship
between 3-bond coupling constants and torsion angles across glycosidic
linkages optical rotation measurements concur with the NMR measurements to
suggest that maltotetraose is the lowest oligomer which models the
conformation of amylose in aqueous solution.

Cyclodextrins. Aqueous solutions of cyclodextrins show only one
resonance for each carbon site (i.e. C-1 to C-6) as chemical shift

157

differences between glucose units are averaged by fast molecular tumbling. However, such variations should be observed in solid state CP-MAS spectra as molecules are effectively immobile on the experimental timescale. Saito and Tabeta (1981) reported early studies of the $^{13}$C CP-MAS spectra of $\alpha$-, $\beta$-, and $\gamma$-cylcodextrins (cyclohexamer, -heptamer and -octamer respectively) but with relatively poor spectral resolution. More recently Gidley and Bociek (1986) have examined both $\alpha$- and $\beta$-cyclodextrin hydrates obtained by re-crystallization from water and found multiple splittings. For example, following resolution enhancement, Gidley and Bociek (1986) have demonstrated seven C-1 resonances for $\beta$-cyclodextrin and six C-4 resonances (Fig. 2) of which one has an unresolved shoulder and is of greater intensity suggesting it accounts for two sites. In fact discrete chemical shift values can be observed for all of the conformationally sensitive C-1 and C-4 sites of both $\alpha$- and $\beta$-cyclodextrin hydrates. Such results provide strong evidence that cyclodextrin solid state $^{13}$C chemical shifts are solely determined by intramolecular interactions.

The results for $\beta$-cyclodextrin in the solid state can be reconciled with those observed in aqueous solution since it has been shown (Gidley and Bociek, 1986) that appropriately weighted averages of the C-1 and C-4 sites of all monomers produce chemical shifts very close to those of the solution

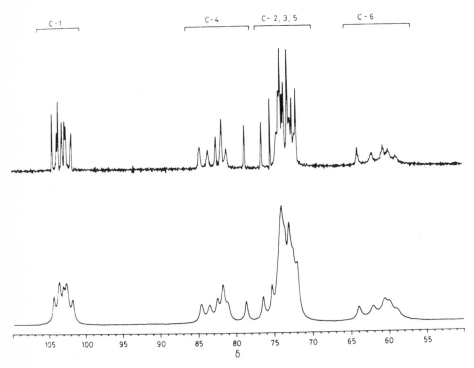

Fig. 2.   $^{13}$C CP-MAS spectrum of $\beta$-cyclodextrin hydrate; a line broadening of -20 Hz was used in the upper trace.

spectrum.  A similar comparison of the results for α-cyclodextrin also suggests that the partially collapsed structure of the α-cyclodextrin in the
solid state (Furo et al., 1987) expands in aqueous solution to relieve the
conformational strain.  This suggestion has recently been directly confirmed
through $^{13}$C CP-MAS studies of frozen solutions of α-cyclodextrin (Gidley and
Bociek, in press).

## Relation Between Chemical Shifts and Torsion Angles

At this stage, some mention should be made of the very detailed studies
attempting to correlate chemical shifts with the torsion angle associated
with the glycosidic linkage and the primary hydroxyl group at C-6.  These
angles are illustrated and defined in Fig. 3.

The Orientation of the CH$_2$OH Group.  The orientation of the primary
hydroxyl group at C-6 can be defined by two torsion angles referred to as
either gauche or trans.  The two possible situations are gauche-trans
(gt) or gauche-gauche (gg).  In this terminology the torsion angle
O[O-5-C-5-C-6-O-6] is stated first and then the torsion angle of
O[C-4-C-5-C-6-O-6].  Hewitt et al. (1986) suggested from the determination
of the $^{13}$C shifts of crystalline glucose, maltose, some of their derivatives
and a series of cyclodextrins that the gauche-gauche orientation at C-6 is
characterized by a shift of ~61.5 ppm whereas the gauche-trans orientation
occurs at ~65.0 ppm.  (A somewhat similar difference is observed between the
resonances of a C-1 atom in either the α- or β-glycosidic conformation).

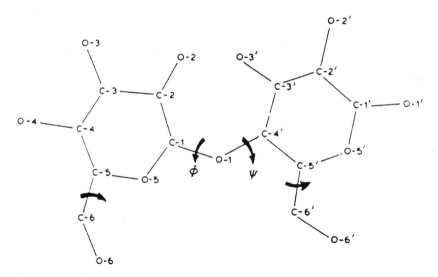

Fig. 3.  Schematic representation of the dimeric fragment of
         amylose, namely, α-maltose, along with the atom numbering
         and the torsion angles of interest.

Correlation Between Torsion Angles at $\phi$, $\psi$, $\chi$ and $^{13}$C Chemical Shifts.
Veregin et al. (1987) have proposed a correlation of the $^{13}$C CP-MAS NMR
chemical shifts of C-1 and C-4 with the X-ray determined torsion angles $\phi$
and $\psi$ of cyclodextrins which reflect the orientation of the D-glucosyl
residues about the $\alpha$-D-(1 $\to$ 4)-glycosidic linkage. They also correlated the
chemical shift of the C-6 with the torsion angle $\chi$ which describes the
orientation of the 0-6 about the (C-5 - C-6) bond. Having thus proposed a
relationship between the chemical shift of C-1 and the torsional angle $\phi$,
they then, as an example, computed the torsional angle associated with each
monomer in the maltotriose asymmetric unit of A starch or the dimer unit of
B starch.

Gidley and Bociek (1988) have examined in some detail the question of
the possible conformational origin of chemical shift effects and studied
possible correlations between C-1 shifts and the sum of the moduli of $\phi$ and
$\psi$ (/$\phi$/ + /$\psi$/) or /$\psi$/ (Fig. 4). The sum of the moduli expresses the extent
of non-coplanarity of H-1, C-1, 0-1, C-4' and H-4'. It proved possible to
compare the C-1 spectral line observed for amorphous starch samples over the

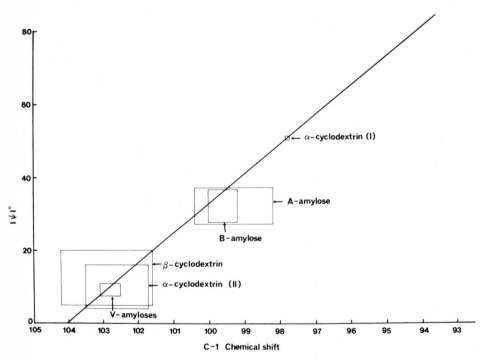

Fig. 4.  Representation of the correlation between C-1 chemical
shifts with the modulus of $\psi$. The range of chemical shifts
and torsion angles (determined by CP-MAS NMR and X-ray
diffraction, respectively), is represented by boxes for
each material. $\alpha$-Cyclodextrin I and II refer to the high
energy glycosidic linkage and the remainder of the molecule
respectively.

range 93–105 ppm with a calculated line which, employing the possible conformational space computed by Gagnaire et al. (1982), is derived by assessing the relative abundance of conformations have in $/\psi/$ or $/\phi/ + /\psi/$ values in 5–10° ranges. The solid line in Fig. 5 is the actual spectrum, the dashed line is generated by superimposing Lorentzian lines or 100 Hz width at half height corresponding to each 10° range of $/\phi/ + /\psi/$ or 5° range of $/\psi/$. The agreement between the observed amorphous C–1 signal and the two spectra predicted from conformational considerations is impressive.

<u>Poly $\alpha$-(1 → 4)-Glucans (DP 10 → 5,000)</u>

Early work by Saito and Tabeta (1981) showed that the $^{13}C$ signals of C–1, C–4, C–6 were well resolved whereas the C–2, C–3 and C–5 signal grossly overlapped. Similar results were observed by Hewitt et al. (1986) with amyloses of average DP 15 and 35. Subsequent work by Horii et al. (1987) with highly crystalline amyloses of DP 16–17 has shown that although considerable overlap may occur when the amyloses are dried, on hydration a triplet appears in the C–1 signal of A–amylose whereas a doublet appears in B–amylose and pronounced multiplicities are evident in the C–2, 3, 4, 5 region. Where high molecular weight amyloses were examined, multiplicities are largely obscured due, presumably, to overlap.

Gidley and Bociek (1988) have also observed (Fig. 6) multiple signals for C–2 to C–5 sites in A and B amyloses (their superior resolution was, in part, believed to be due to the use of monodisperse amyloses) and indeed

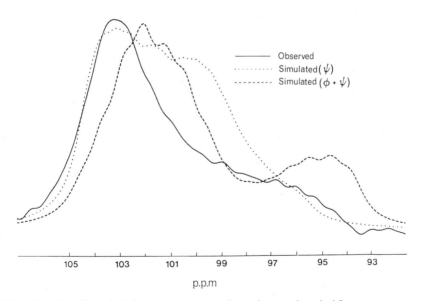

Fig. 5.   Predicted C–1 spectra assuming the equiprobable occurrence of all allowed conformations and based on correlations with $/\phi/ + /\psi/$ (dashed line) and $/\psi/$ (dotted line).

161

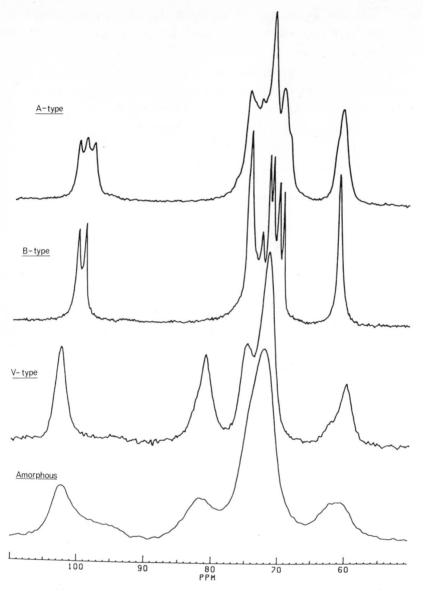

Fig. 6. $^{13}C$ CP-MAS NMR spectra of (a) crystalline A-type $\alpha$-(1 → 4)-glucan, (b) crystalline B-type $\alpha$-(1 → 4)-glucan, (c) $V_6$ amylose complex with sodium palmitate, and (d) amorphous starch prepared by ethanol precipitation of gelantinized maize starch.

doublet pairs at 72.5, 71.9 ppm and 71.0, 70.3 ppm were resolved in the B-type spectrum.  The observation of comparable splitting to that at C-1 suggests that the mechanisms responsible for the shift-effect (whether these be intra- or intermolecular) are the same in both instances.  However, dispersions of C-1 and C-4 chemical shifts are much greater than the total dispersion of C-2, C-3, and C-5 which suggests that conformational effects

162

are largely confined to C-1 and C-4 sites, while helix packing (inter-molecular) interactions may be responsible for producing the splittings of similar magnitude observed at different glucose sites.

It is also interesting that in the single helical $\alpha$-(1 → 4)-glucan, V-amylose (Fig. 6), there is no evidence of multiple resonances which equates with the X-ray diffraction evidence which points to six essentially equivalent glucose residues.

## The Structure of the Starch Granule

With the information already gained, considerable insight can now be obtained by a critical examination of $^{13}$C CP-MAS spectra of amorphous starch, crystalline V-, A- and B-type $\alpha$-(1 → 4)-glucans as illustrated in Fig. 6. There are significant variations in both signal widths and the number of resonances for each carbon site. It is assumed that the magic angle spinning and high power $^{1}$H decoupling remove chemical shift and dipole anisotropy and therefore the spectra reflect the distribution of the iso-tropic chemical shifts for each carbon resonance. It is also to be expected than amorphous materials will have broader resonances than the more crystal-line specimens. However, when multiplicities occur in situations where from solution studies only one resonance should be expected, this usually points to the existence of non-equivalent sites in a solid. It is therefore par-ticularly interesting that the A- and B-polymorphs display triplet and doublet signals respectively which accords well with their having asymmetric units with 3 and 2 glucose residues.

The information derived about the amylose polymers of low molecular weight permits the simulation of the $^{13}$C CP-MAS spectra of native starches by using an appropriate combination of model crystalline A or B and amor-phous materials (Fig. 7). In general the percentage crystallinity so determined is higher than that proposed by X-ray diffraction, the reason being that NMR investigates very local order in comparison to X-ray diffrac-tion. It is also clear from the similarity of amorphous phases of starches so diverse as waxy maize (100% amylopectin) and amylomaize (70% amylose) that the low (typically 4-5%) degree of branching in amylopectin does not greatly affect the distribution of local conformational features within the granule (Fig. 8). The results of the predicted C-1 resonance (Fig. 6) suggest that, all permitted conformations (according to Gagnaire et al., 1982) are present in amorphous starch phases. Such observations suggest a profound similarity of local conformational features within starch granules that has not been immediately evident. Supermolecular organization may, however, impose considerable functional differences.

A point that is worthy of note is the similarity between the $^{13}$C CP-MAS spectra of model amorphous materials and the spectra of single helical

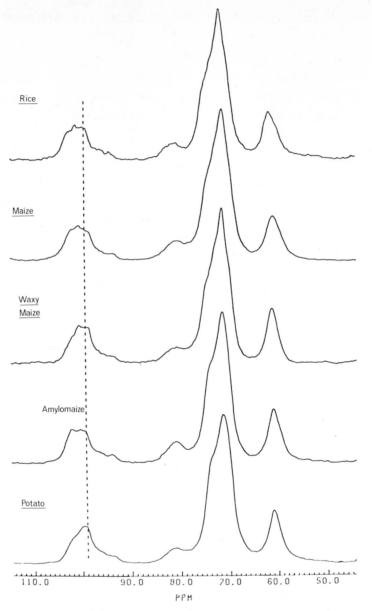

Fig. 7.    $^{13}$C CP-MAS NMR spectra of granular starches.

amyloses (but not of model double helical materials).  In particular C-4 chemical shifts are very similar and amorphous materials have a substantial (~40%) C-1 signal intensity in the range of the V-type C-1 signal (102-104 ppm).  Such information suggests that a substantial portion of the amorphous phase of starch granules consists of local conformations similar to those characteristic of V-type structures.

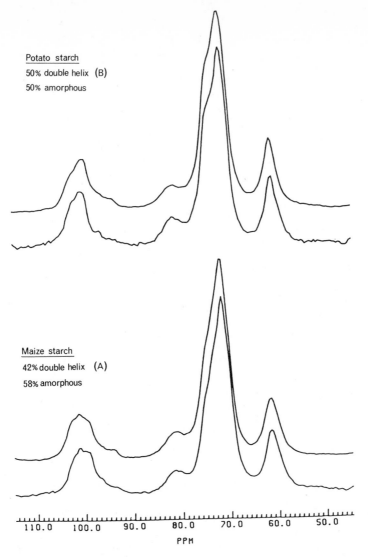

Fig. 8. Simulation of $^{13}C$ CP-MAS NMR spectra of potato starch and maize starch (upper traces) by selecting appropriate percentages of crystalline and amorphous components (lower traces).

THE GELATINIZATION OF THE STARCH GRANULE

Since the process of gelatinization is essentially a phase transition (Blanshard, 1987) and involves a sharp discontinuity in temperature dependent structural parameters, it is not surprising that the CP-MAS technique has thus far been largely confined to studies of the native starches and, as we shall see, the resultant, amorphous gelatinized and retrograded states. However, over the past two decades a number of less sophisticated techniques have been employed to probe the gelatinization process.

An early study of Jaska (1971) used steady state NMR and determined line widths of two suspensions of starches [20% and 40% (w/w) starch/ water] over a wide temperature range. The behavior of the two as a function of temperature was very different and it was difficult to satisfactorily rationalize the results.

Subsequent studies of the gelatinization process by Lelievre and Mitchell (1975) and Bhuiyan (1980) used pulsed NMR and followed $T_2$ relaxation behavior. A single exponential decay was observed under all conditions in which the concentration of the starch was varied between 5-40% (w/w) in the starch/water system and the temperature between ambient and 70°C. The general shape of the curves of $T_2$ vs. temperature for the various concentrations was the same, although the $T_2$ values obviously differed. The presence of a single $T_2$ was explained as a result of fast proton exchange between various proton sites. When the fast exchange occurs, the observed value of $T_2$ is given by the equation:

$$\frac{1}{T_{2(obs)}} = \sum_i \frac{f_i}{T_{2(i)}}$$

where $f_i$ is the fraction of protons associated with a site i with $T_{2(i)}$. Hence, the observed $T_{2(obs)}$ depends not only on the $T_{2(i)}$, but also on the fractional proton population. Before gelatinization, $T_{2(obs)}$ remained approximately constant or slightly decreased especially at the lower starch concentrations. At the gelatinization temperature which corresponds to the onset of melting of crystallites within the granule $T_{2(obs)}$ decreased and passed through a minimum. On further heating, granules completely disintegrated, resulting in increased chain mobility and therefore the $T_2$ increased.

Lechert (1981) has also employed $^2H$ NMR to investigate the loss of order and swelling attending the gelatinization process.

RETROGRADATION

Structural Changes

The phenomena of aggregation and gelation (which are more precise and descriptive terms than retrogradation) have been intensively studied in recent years by a variety of techniques, but particularly by rheological, calorimetric and X-ray diffraction methods. The outcome of recent studies with amylose gels has convincingly shown that in aqueous solution, the processes of aggregation/gelation result ultimately in the conversion of at least some random coils into highly-ordered crystalline structures. There is, however, less certainty about the molecular mechanisms.

Miles et al. (1985) envisage that gels are formed as a result of phase separation upon cooling molecularly entangled solutions with subsequent

166

crystallization.  In contrast, Sarko and Wu (1978) have proposed that gelation occurs because of the formation of double-helical junction zones which
produce the necessary cross-linking.  Recent studies with monodisperse
amyloses (DP 250-2800) have shown (Gidley, in press) that gelatin of such
amyloses can take place from dilute (i.e. non-entangled) aqueous solutions
which suggests that a specific cross-linking mechanism is operative.  For a
polysaccharide to gel, specific cross-linking involves the cooperative
interaction of many residues from each participating chain such that the
enthalpy gained from the sum of non-covalent inter-chain interactions is
sufficient to outweigh the entropic loss suffered by chains involved in
cross-linking.  These regions of inter-chain binding, so-called junction
zones, consist of conformationally-ordered chain segments.

## CP-MAS NMR Information

The exact nature of such interchain ordered structures can be examined
by X-ray diffraction (which is sensitive to long range ordering i.e. crystalline domains) and CP-MAS $^{13}$C NMR which will detect individual
molecularly-ordered species even where they are not present in extended
ordered arrays.

Figure 9 shows the $^{13}$C CP-MAS NMR spectra of amyloses of DP 40, 110 and
250 precipitated from aqueous solution.  The diffraction patterns were
recorded and identical both before and after recording the NMR spectra.  The
X-ray diffraction pattern for the precipitated DP 40 amylose is a well
resolved B-type pattern and suggests a major portion occurs in well-ordered
double-helical arrays.  Those for amyloses of DP 110 and 250 were more
diffuse suggesting both reduced absolute crystallinity and/or size of
crystalline arrays.  All three samples showed little or no evidence of the
presence of an amorphous component in either the NMR spectra (Fig. 9) or the
diffractograms.

The $^{13}$C CP-MAS NMR spectrum of the DP 40 amylose exhibits narrow
signals characteristic of a highly ordered solid while the doublet at the
C-1 site (100 ppm) points to the fact that there are two glucose residues in
the asymmetric unit.  Signals in the 70-75 ppm region are assigned to C-2,
3, 4, 5 sites and show more than one resonance for each carbon site.  For
the proposed (Gidley and Bociek, 1988) inter-helix mechanism of resonance
splitting, the multiplicities of such carbons should be the same as for C-1,
and on that basis two resolved doublets (70.3 and 71.0, 71.9 and 72.5 ppm),
one overlapping doublet (73.5 and ~75.4 ppm) and a single peak/unresolved
doublet (75.4 ppm) are proposed.

The spectrum of an amorphous $\alpha$-(1 → 4)-glucan is also shown in Fig. 9.
Not surprisingly, the signals are broader but there are marked chemical
shift differences with new significant signals at ~103 and 80-83 ppm both of

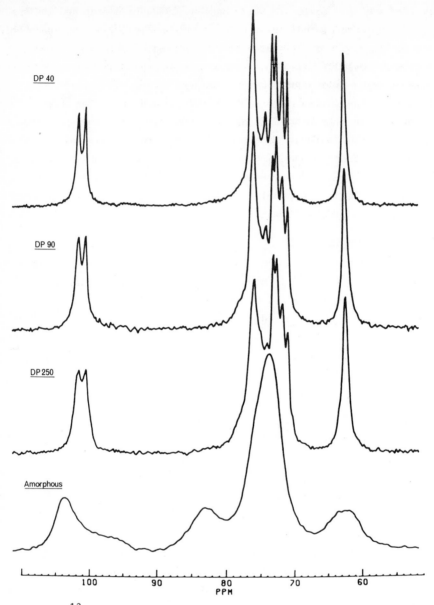

Fig. 9. $^{13}$C CP-MAS NMR spectra of retrograded amyloses of varying degrees of polymerization (DP); the amorphous sample is prepared by ethanol precipitation of gelatinized maize starch.

which, as we have seen already, are characteristic of the amorphous $\alpha$-(1 → 4)-glucan. On this basis, the other spectra in Fig. 9 have no detectable amorphous component. What then is the reason for the decrease in spectral resolution with increase in chain splitting? If multiple splitting is due to inter-helix interactions, then a decrease in resolution is explicable in terms of less perfect helix packing. It is therefore reasonable to suggest that aggregated DP 40 amylose is not cross-linked, DP 110 amylose is lightly cross-linked and DP 250 amylose is extensively cross-linked, the cross-linking hindering the growth of crystalline domains. With such an understanding Gidley (in press) has examined aqueous 10% w/v amylose gels with a weak B-type X-ray crystallinity, using synthetic monodisperse amyloses of DP 300 and 2800 and a polydisperse potato amylose of $DP_w$ ~3000. The extent of X-ray-determined crystallization (DP 300 ~5%; DP 2800 ~<1%) decreased with increasing DP. On examining these gels by NMR, Fig. 10.1 shows a conventional high resolution spectrum of a 10% amylose gel with typical broad resonances with either dispersion of chemical shifts or chemical shift anisotropy induced by network formation being the most probable culprits. The narrow resonances following magic angle spinning at 500 Hz (Fig. 10.2) suggests that chemical shift anisotropy is responsible for the width of the

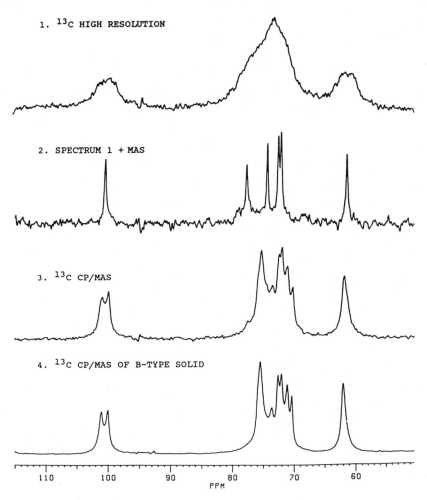

1. $^{13}$C HIGH RESOLUTION

2. SPECTRUM 1 + MAS

3. $^{13}$C CP/MAS

4. $^{13}$C CP/MAS OF B-TYPE SOLID

Fig. 10.   1) High resolution $^{13}$C spectrum of 10% w/v amylose gel;
2) High resolution spectrum + magic angle spinning (500 Hz); 3) $^{13}$C CP-MAS spectrum of 10% amylose gel;
4) $^{13}$C CP-MAS of solid B-type $\alpha$-(1 → 4)-glucan.

resonances. Amylose gels therefore contain some polysaccharide which has sufficient segmental motion for chemical shifts to be averaged on the NMR timescale but which experiences chemical shift anisotropy due to the network-based structure. Indeed the chemical shifts are identical to those found for an aqueous solution of amylose and other $\alpha$-(1 → 4)-glucans.

The more "solid like" regions can be studied by $^{13}C$ CP-MAS NMR using the same 10% w/v amylose gel, which is stable to the high spinning speeds (3 kHz) employed. The spectrum is shown in Fig. 10.3 and is essentially identical to that from crystalline B-type solid material (Fig. 10.4) showing that the "solid-like" polysaccharide in the gel was present as aggregated double helices. Examination of a 10% w/v amylose gel frozen to 233 K in which all the polysaccharide in the sample would be sufficiently "solid-like" to be detected by a CP-MAS experiment yielded a spectrum which can be simulated accurately by a combination of spectra due to B-type double helical amylose and amorphous material. The $^{13}C$ CP-MAS spectra of frozen aqueous $\alpha$-(1 → 4)-glucan solutions was found to be essentially identical to that of amorphous solid material which suggests that $\alpha$-(1 → 4) glucan solutions contain the same range of conformations as amorphous solids. This technique also provided information on the double helix content of such gels, and in this way values of 67, 67 and 83 ($\pm$ 2%) double helix content were estimated for the potato, DP 2800 and DP 300 amylose gels, respectively.

## $^{1}H$ Relaxation NMR Information

Investigation of the $^{1}H$ $T_2$ values of amylose gels in $D_2O$ by monitoring the decay of total magnetization following a Carr-Purcell-Meiboom-Gill pulse sequence provided data about the amylose protons which could be accurately modeled by assuming two ranges of relaxation behavior corresponding to $T_2$ values of ~10 usec and 1-10 msec. The relative proportions of the total decay attributable to the two relaxation timescales could then be calculated from the observed magnitudes. The proportion of protons exhibiting 10 usec $T_2$ values (reasonably the more solid component) was found to be 70, 71 and 88 ($\pm$ 2%) for the 10% w/v potato, DP 2800 and DP 300 amylose gels, respectively. These values did not vary following storage for 2h, 2d or 8d at 293 K and agree reasonably well with those from the frozen $H_2O$ gels. Gidley (in press) therefore proposed that amylose gels contain rigid, double helical segments ($^{1}H$ $T_2$ 10 usec for 10% gels) which act as junction zones and more mobile inter-junction segments ($^{1}H$ $T_2$ 1-10 msec for 10% gels).

## Time-Dependant Phenomena

Lechert (1981), and more recently Wynne-Jones and Blanshard (1986) have investigated changes in hydration of starch gels and bread by $^{1}H$ NMR. In the latters' work a 35% (w/w) water/starch gel showed a significant increase in the unfreezable water (Fig. 11) over a period of 180 h when determined by

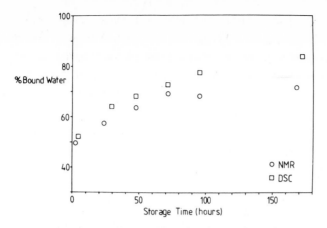

Fig. 11.   Changes in the bound water (as a percentage of the total
           water) in a 35% (w/w) water/starch gel.

both DSC and NMR.   However, a study of the $T_2$ relaxation behavior established that although a fairly rapid change occurred over the first 5 hours in $T_2$
for a 30% (w/w) water/amylopectin gel (Fig. 12) little time dependence was
observed in both 50 and 60% (w/w) water/amylose gels and, in the latter,
there was no evidence of the retrogradation process.   It therefore appears
that in a concentrated starch (i.e. mixed amylopectin/amylose) system
approximating to the level of hydration in bread, the change in state of
water is due solely to the amylopectin fraction with the rate of change
possibly moderated by the amylose.

   An important derivative aspect of these results was their implications
for the staling of bread where there is the complication of gluten as an
additional component.   An examination of the effects of storage of a heated
gluten/water system showed no change in $T_2$ over a period of 2 weeks.   When
these studies were extended to a standard Chorleywood bread, again no
changes were observed in $T_2$ on storage either at 2°C or 37°C.   However, a

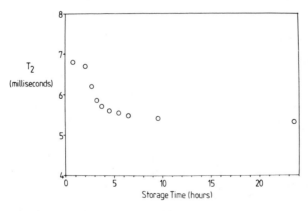

Fig. 12.   Time evolution of the NMR spin-spin relaxation time ($T_2$)
           for a 30% (w/w) water/amylopectin gel.

171

more detailed investigation of the fractions of "unfreezable" (frequently loosely termed "bound") water by DSC and NMR showed significant change. These need to be interpreted in the light of changes of so called "free" or "freezable" water. The authors concluded that in the bread system, the water binding by starch increases with time as in the simple starch system, but concurrently water is also transferred from the gluten phase and enters the starch phase as free water. The quantities of water and the rates involved in these processes are comparable and thus the overall effect is to produce a minimal change in the fraction of "unfreezable" water and consequently little change in the relaxation behavior of the system. This change, as shown by WillHoft (1971), is non-heat reversible and "refreshening" of the bread produces no change in the system as seen by NMR.

## REFERENCES

Bhuiyan, M. Z. H., 1980, Physiochemical studies of maize starches in relation to extrusion, Ph.D. thesis, Nottingham University.

Blanshard, J. M. V., 1987, Starch granule structure and function: A physicochemical approach, "Starch: Properties and Potential," in: "Critical Reports on Applied Chemistry," (Galliard, T. ed.) 13:16.

French, D., 1984, Organization of starch granules, in: "Starch: Chemistry and Technology," (Whistler, R. L.; BeMiller, J. N.; Paschall, E. J.; eds.), 2nd edn., Academic Press, Inc.

Furo, I., Pocsik, I., Tompa, K., Teeaar, R., and Lippmaa, E., 1987, CP-DD-MAS $^{13}$C-NMR investigations of anhydrous and hydrated cyclomalto-oligosaccharides: The role of water of hydration, Carbohydr. Res., 166:27.

Gagnaire, D., Perez, S., and Tran, V., 1982, Configurational statistics of single chains of α-linked glucans, Carbohydr. Polym., 2:171.

Gidley, M. J., 1988, Conformational studies of α-(1 → 4)-glucans in solid and solution states by NMR spectroscopy, in: "Gums and Stabilizers for the Food Industry," No. 4, IRL Press, p. 71.

Gidley, M. J., Molecular mechanisms underlying amylose aggregation and gelation, Macromolecules, in press.

Gidley, M. J., and Bociek, S. M., 1986, $^{13}$C CP-MAS NMR studies of α- and β-cyclodextrins: resolution of all conformationally-important sites, Chem. Communications, 1223.

Gidley, M. J., and Bociek, S. M., 1988, $^{13}$C CP-MAS studies of amylose inclusion complexes, cyclodextrins, and the amorphous phase of starch granules: relationship between glycosidic linkage conformation and solid-state $^{13}$C chemical shifts, J. Am. Chem. Soc., 110:3820.

Gidley, M. J., and Bociek, S.M., $^{13}$C CP-MAS NMR studies of frozen solutions of (1 → 4)-α-D-glycans as a probe of the range of conformations of glycosidic linkages: the conformations of cyclomaltohexaose and amylopectin in aqueous solution, Carbohyr. Res., in press.

Hewitt, J. M., Linder, M., Perez, S., and Buleon, A., 1986, High resolution CP-MAS $^{13}$C-NMR spectra of solid amylodextrins and amylose polymorphs, Carbohydr. Res., 154:1.

Horii, F., Yamamoto, H., Hirai, A., and Kitamaru, R., 1987, Structural study of amylose polymorphs by cross-polarization magic-angle spinning, $^{13}$C-NMR spectroscopy, Carbohydr. Res., 160:29.

Jaska, E., 1971, Starch gelatinization as detected by proton magnetic resonance, Cereal Chemistry, 48:435.

Katz, J. R., and van Itallie, T. B., 1930, Alle Starkearten haben das gleiche Retrogradation-spectrum, Z. Physik. Chem., A150:90.

Lechert, H. T., 1981, Water binding on starch: NMR studies on native and gelatinized starch, in: "Water Activity: Influences on Food Quality," (Rockland, L. B. and Stewart, G. F. Eds.), Academic Press, New York.

Lelievre, J., and Mitchell, J., 1975, A pulsed NMR study of some aspects of starch gelatinization, Die Starke, 27:437.

Miles, M. J., Morris, V. J., and Ring, S. G., 1985, Gelation of amylose, Carbohydr. Res., 135:257.

Morris, G. A., and Hall, L. D., 1981, Experimental chemical-shift correlation maps from heteronuclear two-dimensional NMR spectroscopy, 1. Carbon-13 and proton chemical shifts of raffinose and its subunits, J. Am. Chem. Soc., 10:4703.

Morris, G. A., and Hall, L. D., 1982, Experimental chemical shift correlation maps from heteronuclear two-dimensional nuclear magnetic resonance spectroscopy, 2. Carbon-13 and proton chemical shifts of $\alpha$-D-glucopyranose oligomers, Can. J. Chem., 60:2431.

Pfeffer, P. E., Hicks, K. B., and Earl, W. L., 1983a, Solid State Structures of keto-disaccharides as probed by $^{13}$C NMR spectroscopy, Carbohydr. Res., 111:181.

Pfeffer, P. E., Hicks, K. B., Frey, M. H., Opella, S. J., and Earl, W. L., 1983b, $^{13}$C-$^{13}$C Dipolar interactions provide a mechanism for obtaining resonance assignments in solid-state $^{13}$C NMR spectra, J. Magn. Reson., 55:344.

Pfeffer, P. E., Hicks, K. B., Frey, M. H., Opella, S. J., and Earl, W. L., 1984, Complete solid state $^{13}$C NMR chemical shift assignment for $\alpha$-D-glucose, $\alpha$-D-glucose $H_2O$ and $\beta$-D-glucose, J. Carbohydrate Chemistry, 3(2):197.

Saito, H., and Tabeta, R., 1981, $^{13}$C chemical shifts of solid (1 $\rightarrow$ 4)-$\alpha$-D-glucans by CP-MAS spectroscopy. Conformation dependent $^{13}$C chemical shift as a reference in determining conformation in aqueous solution, Chem. Letts., 713.

Sarko, A., and Wu, H. -C. H., 1978, The crystal structures of A-, B-, and C-polymorphs of amylose and starch, Staerke, 30:73.

Shashkov, A. S., Lipkind, G. M., and Kochetkov, N. K., 1986, Nuclear Overhauser effects for methyl $\beta$-maltoside and the conformational states of maltose in aqueous solutions, Carbohydr. Res., 147:175.

Veregin, R. P., Fyfe, C. A., Marchessault, R. H., and Taylor, M. G., 1987, Correlation of $^{13}$C chemical shifts with torsional angles from high resolution $^{13}$C CP-MAS NMR studies of crystalline cyclomalto-oligosaccharide complexes, and their relation to the structures of the starch polymorphs, Carbohydr. Res., 160:41.

WillHoft, E. M. A., 1971, Bread Staling 1. - Experimental study, J. Sci. Fd. Agric., 22:176.

Wynne-Jones, S., and Blanshard, J. M. V., 1986, Hydration studies of wheat starch amylopectin, amylose gels and bread by proton magnetic resonance, Carbohydr. Polymers, 6:289.

WATER INTERACTIONS IN BOVINE CASEIN:  $^2$H NMR RELAXATION AND SMALL-ANGLE
X-RAY SCATTERING STUDIES

Helmut Pessen*, Thomas F. Kumosinski and
Harold M. Farrell, Jr.

* ERRC, USDA
  600 E. Mermaid Lane
  Philadelphia, PA   19118

Whole casein occurs in bovine milk as a colloidal calcium-containing
protein complex, commonly called the casein micelle.  Removal of calcium is
thought to result in the dissociation of this micellar structure into non-
colloidal protein complexes called submicelles (Farrell, 1988).  These
submicelles consist of four proteins, $\alpha_{s1}$-, $\alpha_{s2}$-, $\beta$-, and $\kappa$-casein, in the
approximate ratios of 4:1:4:1 (Davies and Law, 1980).  All are phosphorylated
to various extents, have an average monomer molecular weight of 23,300, and
are considered to have little or no secondary structure (Farrell and Thompson,
1988).  The major casein fraction, $\alpha_{s1}$, a single-chain polypeptide of 199
amino acid residues, contains 8 phosphoserine residues (Eigel et al., 1984)
and a large number of hydrophobic residues; the weight-average number of
phosphate groups is 6.6 per monomer.

The isolated fractions exhibit varying degrees of self-association,
mostly hydrophobically driven (Schmidt, 1982).  Little work has been done
on the tertiary and quaternary structure of the native proteins in mixed
association.  However, there is hydrodynamic evidence that, in the
absence of calcium, casein associates to form aggregates, the submicelles, with an apparent upper limit of 94 Å for the Stokes radius
(Pepper and Farrell, 1982); this mainly hydrophobic self-association
increases with increasing temperature and ionic strength (Schmidt, 1982).
One proposed structure of this limiting polymer consists of a hydrophobic
core, composed mostly of the hydrophobic portions of $\alpha_s$- and $\beta$-caseins,
with $\kappa$-casein residing chiefly at the surface because of its ability to
keep $\alpha_s$- and $\beta$-caseins from precipitating at 37°C in the presence of
calcium.  This characteristic of $\kappa$-casein to reside predominantly on the
micelle surface has been shown by electron microscopy on gold-labelled
$\kappa$-casein (Schmidt, 1982) and on ferritin conjugate and anti-$\kappa$-casein
(Carroll and Farrell, 1983).  All charged groups, including the serine

phosphates, are located on the surface of the submicellar structure. In this model the κ-casein content of the submicelles is variable.

It has long been hypothesized that, upon the addition of calcium, these hydrophobically stabilized, self-associated casein submicelles further self-associate via calcium-protein side-chain salt bridges to the colloidal micelles, with average radii of 650 Å (Farrell and Thompson, 1988). However, the exact supramolecular structure of the casein micelle has remained controversial. Models presented have ranged from an assembly of discrete submicelles to the structure of a loose porous gel (Walstra, 1979; Farrell and Thompson, 1988), and to a newer concept of a homogenous sphere with a "hairy" outer layer (Holt and Dalgleish, 1986). According to the nature of the bound or trapped water, its correlation time could thus vary from picoseconds, for a continuum, to nanoseconds, for discrete submicelles.

Small-angle X-ray scattering (SAXS) was undertaken on whole bovine casein both in the absence of calcium, to ascertain whether limiting polymers (submicellar structures) exist, and in the presence of calcium to determine if the colloidal micelle consists of discrete submicellar par-ticles with a particular packing structure or of a nonspecific, unordered, gel-like structure. Another objective was to find an explanation for the ready penetration into the micelle of the enzymes responsible for pro-teolysis, as well as of salts and lactose (Farrell, 1988). From the SAXS data, information on casein hydration was obtained also. For more specific information on the nature of this water binding, given the success in detecting trapped water in earlier β-lactoglobulin work (Pessen and Kumosinski, 1985), NMR relaxation was studied under comparable conditions, as well as at several temperatures. The method was one developed earlier (Kumosinski and Pessen, 1982; Pessen et al., 1985) for determining hydration from the protein concentration dependence of resonance relaxation rates, which permits the calculation of hydrations and correlation times and, from the latter, Stokes radii and apparent molecular weights.

MATERIAL AND METHODS

Sample Preparation

Two liters of warm milk were obtained from the whole milk of an individual Jersey cow. The animal, part of a commercial herd, was in mid-lactation and in good health. Phenylmethylsulfonyl fluoride (0.1 g/L) was added immediately to retard proteolysis. The milk was transported to the laboratory and skimmed twice by centrifugation at 4000g for 10 min. at room temperature.

For the SAXS measurements, 500 mL of this skim milk were diluted with an equal volume of distilled water and warmed to 37°C. Casein was precipitated by careful addition of 1 N HCl to pH 4.6. The precipitate was homogenized with a Polytron ST-10 homogenizer* at low speed and dissolved by addition of NaOH to yield a solution of pH 7.2. The casein was reprecipitated, washed, resuspended, cooled to 4°C, centrifuged at 10,000g for 30 min. to remove residual fat, and then freeze dried. Alkaline urea gel electrophoresis with standard caseins of known structure (Thompson, 1964) showed the following genotype of the animal: $\alpha_{s1}$-BB, $\beta$-AA, and $\kappa$-BB. The lyophilized sodium caseinate was dissolved in PIPES-KCl buffer (25 mM piperazine-N-N'-bis (2-ethane-sulfonic acid), pH 6.75, made to be 80 mM in KCl). Dithiothreitol (0.1 mM) was added to promote self-association of $\kappa$-casein (Pepper and Farrell, 1982). The samples were stirred for 3-5 min., then passed through a 0.44 μm filter; blanks were treated in the same fashion. To produce casein micelles, $CaCl_2$ was added to a final concentration of 10 mM. Alternatively, to produce submicelles, KCl was added to a final concentration of 30 mM (ionic strength comparable to the added $CaCl_2$).

For the NMR measurements, following the initial centrifugation 400 ml of this skim milk was centrifuged for 1 h at 88,000 x g (37°C). The pellets were washed twice in $D_2O$ containing 25 mM PIPES (pH 6.75), 20 mM $CaCl_2$, and 80 mM KCl. The final protein concentration was fixed at about 100 mg/ml (total volume of 5 ml). Subsequent dilutions were made with the same buffer. To produce submicelles, sodium caseinate prepared from the same skim milk was dialyzed and lyophilized; the lyophilized protein was dissolved in $D_2O$, in the same PIPES-KCl buffer without $CaCl_2$, but with added dithiothreitol to promote self-association of $\kappa$-casein (Pepper and Farrell, 1982). These procedures were designed to minimize the concentration of $H_2O$ in the $D_2O$ solutions and thus to eliminate any significant contribution to the relaxation rates from deuterium exchange.

For all samples, casein concentrations were determined spectrophoto-metrically on aliquots diluted 1/50 to 1/100 in 0.1 N NaOH, based on an absorptivity of 0.850 mL $mg^{-1}$ $cm^{-1}$ at 280 nm for whole casein (Pepper and Farrell, 1982). Protein concentrations for the X-ray work were 19.38, 10.77, and 9.19 mg/mL for the submicellar form, 16.4, 12.2, and 4.68 mg/mL for the micellar form. For the NMR work, concentrations obtained by serial dilution with buffer ranged from 10 to 80 mg/mL.

SAXS Measurements and Data Evaluation

All SAXS experiments were performed, as previously described (Kumosinski et al., 1988), on a Kratky camera (Anton Paar, Graz, Austria) with Ni-filtered

---

* Reference to brand or firm name here and in the following does not constitute endorsement by the U.S. Department of Agriculture over others of a similar nature not mentioned.

Cu-K$_\alpha$ radiation (Pilz et al., 1979), equipped with a one-dimensional position-sensitive detection system with pulse-height discrimination (Technology for Energy Corporation); the wavelength, $\lambda$, of the radiation was that of the K$_\alpha$ doublet, 1.542 Å. Absolute intensities were obtained by means of a calibrated Lupolen (polyethylene) platelet sample of known scattering power as a secondary standard. A Paar sample cell with mica windows and a one-millimeter path length was used for the measurements. All SAXS measurements were made at room temperature. Comparison of long and short collection times (2 to 10 h) showed no significant differences, indicating the absence of protein denaturation as a result of sample irradiation or thermal instability.

Data were evaluated by means of equations previously described (Luzzati et al., 1961a; Pessen et al., 1973); the notation used is essentially that of Luzzati. Here, $\rho_1$ is the electron density of the solvent, calculated as 0.355 e$^-$/Å$^3$; $\psi_2$, the electron partial specific volume of the solute, equals $\bar{v}/q = 2.329$ Å$^3$/e$^-$, where $\bar{v}$ is the partial specific volume of the protein, 0.736 mL/g (Kumosinski et al., 1987), and q is the number of electrons per gram of particle, $0.316 \times 10^{24}$, both calculated from the average amino acid composition (Eigel et al., 1984). Slit-smeared curves were deconvoluted by the use of a computer program due to Lake (1967). Tabulated parameters were derived from concentration-dependent parameters by extrapolation to zero concentration.

Customary Guinier analysis (Guinier and Fournet, 1955) was not used for actual calculation of the SAXS results because of the nonlinearity of the plot for the casein micelle. To obviate personal judgment and possible bias regarding the linear portions of a plot, all data were fitted by multiple Gaussian functions by the use of Gauss-Newton nonlinear regression. The rationale for the use of nonlinear regressions and for their analysis by the criterion of randomness of residual plots has been extensively discussed, e.g., by Meites (1979).

NMR Relaxation Measurements and Data Evaluation

Deuteron NMR spectra were obtained by Fourier Transform spectroscopy as previously described (Kumosinski et al., 1987), with a JEOL FX60Q spectrometer operating at a nominal proton frequency of 60 MHz. The frequency of observation was 9.17 MHz. Longitudinal relaxation rates, $R_1$, were measured by the inversion-recovery method (Vold et al., 1968) and transverse relaxation rates, $R_2$, by spin-locking measurement (Farrar and Becker, 1971) of $R_{1\rho}$, the longitudinal relaxation rate in the rotating frame, with appropriate precautions (Pessen and Kumosinski, 1985).

For each sample, $R_2$ was determined with the same number of replications as $R_1$; relative standard errors amounted to 2-3%. Measurements of one mode of relaxation were made on the identical sample, and either immediately following the completion of measurements of the other mode or at latest the next day. Measurements for deuteron relaxation at 9.17 MHz were made at pH 7.0 and at 2°, 15°, and 30°C $\pm$ 1°C, respectively. These rates were measured in $D_2O$ to eliminate cross-relaxation effects between water and protein protons, such as observed by Edzes and Samulski (1978), and by Koenig et al. (1978).

For a two-state model (bound and free water), Pessen and Kumosinski (1985) have shown that for $R_{obs}$, the observed longitudinal or transverse relaxation rate of water in the presence of varying protein concentration,

$$R_{obs} = R_f + (R_b - R_f) \, \bar{\nu}_w \, a_p/W, \tag{1}$$

where $R_f$ is the appropriate relaxation rate of free water ($R_1$ or $R_2$), $R_b$ is the corresponding relaxation rate of bound water, W is the total concentration of water, $a_p$ is the activity of the protein, and $\bar{\nu}_w$ is the degree of hydration (i.e., basically, the average number of molecules of water bound per molecule of dry protein or, in units consistent with the concentration units employed, the number of grams of bound water per gram of dry protein). Although qualifications are necessary, in the following we will, for simplicity and convenience, use the expression "hydration" for short to indicate the quantity $\bar{\nu}_w$ in units of g/g. The protein activity can be expressed in terms of $B_o$, the second virial coefficient of the protein, and concentration, c,

$$a_p = c \exp (2B_oc + \ldots). \tag{2}$$

Data points of the observed relaxation rate (longitudinal or transverse) vs. protein concentration were fitted with a combined function of Eqs. (1) and (2) via iterative Gauss–Newton nonlinear regression, which produced values for $B_o$, $[(R_b-R_f) \, \bar{\nu}_w]$, and $R_f$. Values for $R_{1b}$ or $R_{2b}$, $\bar{\nu}_w$, and $\tau_c$, the correlation time, were obtained by simultaneous solution of the two applicable Kubo–Tomita-Solomon equations (Pessen and Kumosinski, 1985). These relate the foregoing quantities, as well as an order parameter, S, for intermediate asymmetry of the motion of the bound water (Halle et al., 1981), which is applicable to both isotropic (S = 1) and anisotropic (S < 1) motion.


RESULTS

SAXS Data

Figure 1 represents Guinier plots, $\ln i_n(s)$ vs. $s^2$, where $i_n(s)$ is the normalized point-source excess scattered intensity (i.e., scattered intensity of sample, normalized with respect to incident-beam intensity and corrected for scattering of blank) at scattering vector s; s = 2 sin $\theta/\lambda$, where $\lambda$ is the wavelength and $\theta$ is one-half the scattering angle. An immediate indication of differences between the submicellar and the micellar

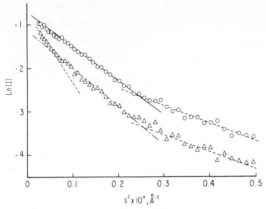

Fig. 1.   Guinier Plots of Submicellar and Micellar Casein.   O,
submicelle at 19.38 mg/mL; Δ, micelle at 16.4 mg/mL.
(Lower intensities for micelles are incidental, since data
are not normalized with respect to concentration.) Solid
line, linear regression for submicelle from s = 0.00173 to
0.00447 $\overset{\circ}{A}^{-1}$; dashed line, linear regression for micelle
from s = 0.00173 to 0.00264 $\overset{\circ}{A}^{-1}$; dot-dashed line, linear
regression for micelle from s = 0.00264 to 0.00447 $\overset{\circ}{A}^{-1}$;
dotted lines, linear regressions for both submicelles and
micelles beyond s = 0.00447 $\overset{\circ}{A}^{-1}$, whose nearly identical
slopes indicate radii of gyration of the compact regions in
agreement with values in Table 2.

forms can be seen:   the curve for the submicelles shows two linear regions,
that for the micelles three.   For the submicellar data, the initial linear
portion (solid), which can be fitted up to a value of s = 0.00447 $\overset{\circ}{A}^{-1}$,
yields a radius of gyration, $R_a$, of 76.0 ± 0.6 $\overset{\circ}{A}$ by linear regression.   For
the micellar data two initial linear regions (dashed and dot-dashed) yield
$R_a$ values of 103.0 ± 1.2 and 82.4 ± 1.4 $\overset{\circ}{A}$ for the corresponding regions.

Analysis of the same data by nonlinear regression with multiple
Gaussian functions was more clear-cut.   The submicellar data were best
fitted by a sum of two Gaussians (Fig. 2A, which also shows the contribution
of each of the two Gaussians), while the micellar data (Fig. 2B) required
three Gaussians.   The quality of the fit is indicated by the standard error
of each derived parameter in Table 1, as well as by the relative root-mean-
square error for the fit and the randomness of the residuals (not shown),
whereas residual plots for single Gaussian fits in all cases were nonrandom.
The smallest Gaussian (labelled "A" in each instance) is so broad as to
resemble a straight line, but is not dispensable as it serves to eliminate
the need for a baseline, which would be incompatible with the requirement
that scattered intensities must approach zero for large angles.

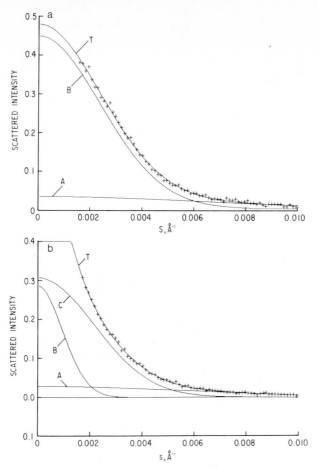

Fig. 2. (A) SAXS of Submicelles. +, Smeared SAXS at 19.38 mg/mL;
curve T, best fit for sum of two Gaussians by nonlinear
regression; curves A and B, two separate Gaussians.
(B) SAXS of Micelles. +, Smeared SAXS at 16.4 mg/mL; curve
T, best fit for sum of three Gaussians by nonlinear
regression; curves A, B, and C, three separate Gaussians.

The deconvoluted intensity at zero angle divided by its corresponding
protein concentration, which yields a value proportional to the molecular
weight of the particle, showed no concentration dependence under any of the
conditions studied. Hence, the interpretation that extreme particle size
polydispersity might have caused the multiple Gaussian character of the two
forms of casein is unlikely (Kratky, 1963; Pessen et al., 1973). Other
models accounting for the double-Gaussian character of the submicellar
scattering might be based on extreme particle asymmetry (e.g., rods), or on
a spherically symmetrical but inhomogeneous particle having regions of
differing electron density. The former would not be in agreement with
accumulating hydrodynamic, light scattering, and electron-microscopic
evidence, which indicates that submicellar casein exists in the form of
spherical particles (Schmidt and Payens, 1976; Schmidt, 1982). Since the

Table 1. Molecular Parameter[*] (Desmeared SAXS)

| Parameter | Submicelle | Micelle |
|---|---|---|
| M | --------- | $882,000 \pm 28,000$ |
| $k_2$ | --------- | $0.308 \pm 0.005$ |
| $M_2$ | $285,000 \pm 14,600$ | $276,000 \pm 18,000$ |
| k | $0.212 \pm 0.028$ | $0.216 \pm 0.003$ |
| $M_C$ | $60,000 \pm 5,650$ | $56,400 \pm 3,700$ |
| $M_L$ | $225,000 \pm 18,500$ | $220,000 \pm 18,700$ |
| $\Delta\rho$, $e^-/\overset{\circ}{A}{}^3$ | --------- | $0.0081 \pm 0.0004$ |
| $\Delta\rho_2$, $e^-/\overset{\circ}{A}{}^3$ | $0.0099 \pm 0.0004$ | $0.0073 \pm 0.0005$ |
| $\Delta\rho_C$, $e^-/\overset{\circ}{A}{}^3$ | $0.0148 \pm 0.0014$ | $0.0128 \pm 0.0007$ |
| $\Delta\rho_L$, $e^-/\overset{\circ}{A}{}^3$ | $0.0091 \pm 0.0003$ | $0.0065 \pm 0.0003$ |
| $H$, $g_{water}/g_{protein}$ | --------- | $7.92 \pm 0.42$ |
| $H_2$, $g_{water}/g_{protein}$ | $6.31 \pm 0.30$ | $8.98 \pm 0.44$ |
| $H_C$, $g_{water}/g_{protein}$ | $3.97 \pm 0.48$ | $4.70 \pm 0.31$ |
| $H_L$, $g_{water}/g_{protein}$ | $6.90 \pm 0.64$ | $9.95 \pm 0.58$ |
| $V$, $\overset{\circ}{A}{}^3$ | --------- | $12,720,000 \pm 250,000$ |
| $V_2$, $\overset{\circ}{A}{}^3$ | $3,330,000 \pm 260,000$ | $4,440,000 \pm 160,000$ |
| $V_C$, $\overset{\circ}{A}{}^3$ | $467,000 \pm 1,520$ | $529,000 \pm 2,600$ |
| $V_L$, $\overset{\circ}{A}{}^3$ | $2,860,000 \pm 400,000$ | $3,910,000 \pm 30,000$ |
| $R_2$, $\overset{\circ}{A}{}^3$ | $80.24 \pm 0.39$ | $90.57 \pm 0.03$ |
| $R_C$, $\overset{\circ}{A}{}^3$ | $37.98 \pm 0.01$ | $39.62 \pm 0.01$ |
| $R_L$, $\overset{\circ}{A}{}^3$ | $88.22 \pm 0.82$ | $100.19 \pm 0.09$ |

[*] Average of three concentrations (see under Materials and Methods).

particles result from a hydrophobically driven self-association of monomer units, it has been considered most likely that they contain a hydrophobic inner core surrounded by a hydrophilic outer layer (Schmidt, 1982). Such an arrangement would theoretically be the thermodynamically most stable and would be in agreement with predictions from primary structure. It would also entail different packing densities in regions of predominantly hydrophobic and hydrophilic side chains and thus would give rise to two concentric regions of differing electron density (Richards, 1974; Lumry and Rosenberg, 1975).

Submicelles. Accordingly, the submicellar data were analyzed by means of a model in which the particle has two regions of different electron densities with the same scattering center. In this model, the scattered amplitudes rather than the intensities of the two regions are additive because of interference effects of the scattered radiation. Luzzati et al. (1961b) have developed expressions for calculating the molecular and structural parameters for such a particle. In the following, subscripts C and L (following Luzzati et al., 1961b) refer to the higher and lower electron density regions, respectively, while subscript 2 designates the particle composed of these two regions. With these expressions, parameters for casein under submicellar conditions were evaluated and are listed in Table 1.

Micelles. For the SAXS results of the casein micelle solutions, i.e. after the addition of 10 mM $CaCl_2$, the same procedure was used for analyzing the two Gaussians having the lower radii of gyration; these constitute the contribution of the submicellar structure to the SAXS results. The third Gaussian, which has the highest radius of gyration, reflects the total number of submicellar particles within the cross-sectional scattering profile. Here, at zero angle, the intensity of the larger Gaussian contribution can be simply added to the intensity of submicellar contribution. A new parameter, $k_2$, the ratio of the mass of the submicelles to the total observed mass ascribable to a cross section, can be expressed in terms of the radii of gyration and the zero-angle intercepts for these Gaussians. The inverse of $k_2$, the packing number, is the number of submicellar particles found within the observed micellar cross section. The meaning of these cross-sectional parameters will be discussed below. The resulting parameters for the micelle are also listed in Table 1, where subscript 2 now designates the corresponding parameters for a submicellar particle when incorporated in the micelle, and unsubscripted parameters refer to the total cross section of the colloidal particle.

NMR Data

$^2$H NMR relaxation measurements, both longitudinal, $R_1$, and transverse, $R_2$, of $D_2O$ with varying concentrations of casein were performed with and without calcium, at 2°, 15°, and 30°C. Figure 3 shows $R_1$ and $R_2$ measurements at 15°C under submicellar and micellar conditions. All data were fitted by Eqs. (3) and (4) and evaluated as described above. The experimental data and the data calculated from the model employed are in excellent agreement, as shown by the solid lines in Fig. 3. Under these and all other conditions the nonlinear portion of the curves yielded a virial coefficient of 0.0032 ± 0.0003 ml/mg, indicating the consistency of the experimental results. The linear portions of the curves were evaluated with a propagated standard error of about 4%; they contain the product of the relaxation rate of the bound water, the hydration, and finally the asymmetry parameter, S. We return to these later.

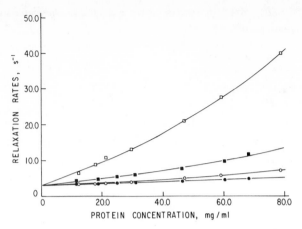

Fig. 3.  Dependence of deuteron relaxation rates of water on casein concentrations in $D_2O$ at pH 6.75 in 0.2 M PIPES/KCl buffer at 15°C. -●-●-, $R_1$ measurements, submicellar form; -■-■-, $R_2$ measurements, submicellar form; -o-o-, $R_1$ measurements, micellar form; -□-□-, $R_2$ measurements, micellar form.

From the linear portion of the concentration dependences of the longitudinal and transverse relaxation rates, together with the Kubo-Tomita-Solomon equations, the following parameters were calculated for the caseins at the various environmental conditions:  (i) correlation times, $\tau_c$, which have been shown to be related to the rotational speed of water bound to the protein; (ii) hydration values, $\bar{\nu}_w$, for an isotropic tumbling model (S = 1), which implies that water is held at the protein surface and therefore is not moving independently; and (iii) the longitudinal and transverse relaxation rates of the bound water, $R_{1b}$ and $R_{2b}$.  The results are shown in columns 3–6 of Table 2.  For submicellar casein, values of $\bar{\nu}_w$ increased from 0.00652 to 0.01201 g water/g protein, and $\tau_c$ values decreased from 38.9 to 29.8 ns, as the temperature decreased from 30° to 2°C; propagated standard errors were about 8% for $\tau_c$ and 6% for $\bar{\nu}_w$.  The same temperature dependence of $\tau_c$ and $\bar{\nu}_w$ was exhibited for micellar casein, although at all temperatures the absolute values were larger for the micellar form than for the submicellar form.

It is appropriate to note that, although the caseins are self-associating, one needs to consider here only the aggregated forms. The concentrations used were high enough so that the association equilibrium at 30° favors polymer formation (Pepper and Farrell, 1982); at lower temperatures, this must be qualified as discussed below.  Also, these $\bar{\nu}_w$ values will in all probability show only a fraction of the total hydration, since at 9.17 MHz any bound water with $\tau_c > 6$ ns would have a $R_2/R_1$ ratio of unity and would not be observable by this procedure.

Table 2.  Hydration and Dynamics of Bound Water, and Derived Molecular Parameters

| Temperature, °C | $\tau_c$, ns | $\bar{\nu}_w$, g $H_2O$/g protein | $R_{1b}$, $s^{-1}$ | $R_{2b}$, $s^{-1}$ | $r$, $\overset{\circ}{A}$ | $M_p$ | $(\bar{\nu}_w)_r$ | $S$ | $(\bar{\nu}_w)_S = 0.237$ |
|---|---|---|---|---|---|---|---|---|---|
| Sub-micelle | | | | | | | | | |
| 30 | 38.9 | 0.00652 | 1904 | 10,510 | 36.4 | 165,000 | | | 0.116 |
| 15 | 34.7 | 0.00824 | 2080 | 9,840 | 30.5 | 97,200 | | | 0.147 |
| 2 | 29.8 | 0.01201 | 2323 | 9,070 | 25.5 | 56,800 | | | 0.214 |
| Micelle | | | | | | | | | |
| 30 | 63.6 | 0.0165 | 1249 | 14,790 | 42.9 | 270,700 | 0.469 | 0.188 | 0.294 |
| 15 | 51.1 | 0.0225 | 1515 | 12,570 | 34.8 | 144,500 | 0.357 | 0.251 | 0.400 |
| 2 | 45.1 | 0.0282 | 1689 | 11,530 | 29.3 | 86,200 | 0.380 | 0.272 | 0.502 |

DISCUSSION

SAXS Data

Submicelles.  Both $M_2$, the molecular weight of the submicellar particle, and k, the mass fraction of the denser or "core" region, were independent of protein concentration, ruling out extreme size polydispersity as a source of the multiple Gaussian character of the data, as mentioned under Results.  The molecular parameters in Table 1, therefore, are measures of the limiting aggregate resulting from the hydrophobically driven self-association of the mixed caseins in the absence of calcium.  Its molecular weight is consistent with those found in other investigations, i.e. 200,000 to 300,000, by a variety of techniques (Schmidt, 1982).  In fact, the value of 285,000 is in excellent agreement with the value of 300,000 observed by small-angle neutron scattering (Stothart and Cebula, 1982).

In the latter work, the data were analyzed on the basis of a model consisting of a homogeneous limiting aggregate.  Here, we find evidence of a heterogeneous particle consisting of two regions of differing electron density, with the mass fraction of the denser region equal to $0.212 \pm 0.028$. This denser region has an electron density difference, $\Delta\rho_C$, of $0.0148 \pm 0.0014$ $e^-/\overset{\circ}{A}{}^3$, a hydration, $H_C$, of $3.97 \pm 0.48$ g water/g protein and a molecular weight, $M_C$, of $60,100 \pm 5,650$ (see Table 1, where the corresponding values of $M_2$, $\Delta\rho_L$, and $H_L$ for the less dense region are also given).  The region of

higher electron density most likely results from the intermolecular hydrophobically driven self-association of the casein monomer units (Farrell,
1988). This hydrophobic inner core would be shielded from interactions with
water (Kuntz and Kauzmann, 1974; Tanford, 1961) by a less dense outer region
containing a preponderance of hydrophilic groups. The low value of hydration
formally ascribed to the core is probably the result of the packing density
(Richards, 1974; Lumry and Rosenberg, 1975) of the hydrophobic side chains
rather than the absolute amount of water within the region. The electron
density of the denser region is still exceedingly low ($0.0148$ $e^-/\overset{\circ}{A}^3$) compared
to compact globular proteins (Pessen et al., 1988), such as lysozyme ($0.078$
$e^-/\overset{\circ}{A}^3$), $\alpha$-lactalbumin ($0.067$ $e^-/\overset{\circ}{A}^3$), ribonuclease ($0.071$ $e^-/\overset{\circ}{A}^3$), or even the
phosphoglycoprotein, riboflavin-binding protein ($0.056$ $e^-/\overset{\circ}{A}^3$ in its acid-
denatured form); this underscores the consequences of the random conformation
of the polypeptide chains in casein. Axial ratios calculated from the appropriate values of V and R (Luzzati et al., 1961b) give values of 1.06 for the
denser region and 1.32 for the total submicelle, using as model a prolate el-
lipsoid of revolution; these ratios confirm the approximate spherical symmetry
of the casein submicelle predicted from electron microscopy (Schmidt, 1982).

To test the assumption of a spherically shaped concentric two-electron
density model for the submicelle, the scattered intensities were normalized
to zero angle and deconvoluted. Fig. 4A compares the results for the
highest concentration of submicellar casein (19.38 g/L) with various
theoretical models calculated from the Debye equation (Debye, 1915), the
fundamental expression for all scattering phenomena, according to which

$$I(h) = \sum_{i=1}^{n} g_i^2 \phi_i^2(h) + 2 \sum_{i=1}^{n-1} \sum_{k=i+1}^{n} g_i g_k \phi_i(h) \phi_k(h) (\sin d_{ik} h)/(d_{ik}h), \quad (3)$$

where $h = 2\pi s$, s is as defined before, $g_i$ is the weighting factor and $\phi_i(h)$
the shape factor of the i-th sphere, and $d_{ik}$ is the center-to-center
distance between the i-th and the k-th spheres. Here $g_i = \rho_i (4\pi/3)$
$R_i^3$, where $\rho_i$ is the electron density and $R_i$ the radius of the i-th sphere,
and, for a sphere, $\phi_i(h) = 3[\sin (R_i h) - R_i h \cos (R_i h)]/(R_i h)^3$.

The radius of gyration of the inner region, 48 $\overset{\circ}{A}$ (calculated from $V_C$,
Table 1), and the total radius of gyration, 92.6 $\overset{\circ}{A}$ (calculated from $V_2$),
were used for all models in Fig. 4A. The models tested were: (i) a homogeneous sphere of radius 92.6 $\overset{\circ}{A}$ (dashed line), a value consistent with that
of 94 $\overset{\circ}{A}$ from gel chromatography and 50 to 150 $\overset{\circ}{A}$, depending on the method of
fixation, from electron microscopy; (ii) a combination of two concentric
spheres of radii 48.1 and 92.6 $\overset{\circ}{A}$ (ticked line); and (iii) two nonconcentric
spheres with the same radii, with centers 20 $\overset{\circ}{A}$ apart (solid line). It may
be noted that the latter two models require an interaction term, since here

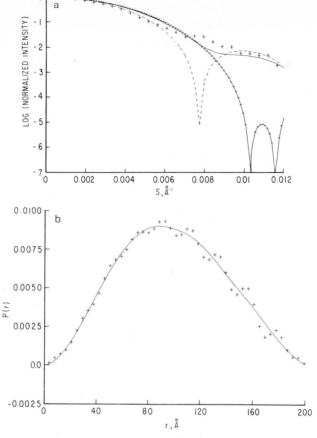

Fig. 4. (A) Normalized SAXS for Submicelles. +, normalized SAXS
for submicelles at 19.38 mg/mL; dashed line, theoretical
for homogeneous sphere with 92.6 Å outer radius; ticked
line, inhomogeneous sphere, i.e., two concentric spheres
with different radii (48.1 and 92.6 Å) for different
electron densities; solid line, two nonconcentric spheres
with different electron densities and radii of 48.1 and
92.6 Å; (B) Distance Distribution of Submicelles. +, p(r)
vs. r for submicellar casein at 19.38 mg/mL; solid line,
best fit for sum-of-three-spheres model (see Discussion).

the amplitudes, not the intensities, are additive. As seen in the figure,
the nonconcentric two-sphere model compares more favorably with the experimental data than do the other models. However, this model does not appear
to fit the experimental results as well as might be expected. To ascertain
if this discrepancy is due merely to experimental error or to a small ordering phenomenon, and to show that the particle is inhomogeneous (whether the
regions are concentric or not), the distance distribution function, p(r),
was calculated (Pilz et al., 1979) from the desmeared, unsmoothed raw data.

To examine the experimental distance distributions (Fig. 4B), a theoretical $p(r)$ should actually be taken as the sum of three $p(r)$ functions for spheres: one for the core region, a second for the possibly nonconcentric region of low electron density, and a third for the interaction term. The experimental $p(r)$ data were therefore fitted by a function calculated from the intensity for three spheres with different fractional contributions, using nonlinear regression (solid line). The analysis of $p(r)$ by this method was quite satisfactory, yielding a relative standard error of 3.7%; the results are shown in Table 3. Here D represents the diameter of a particular sphere and c represents the fraction contributed by that sphere to the total distribution. As seen in the table, $D_1$ could represent the total diameter and $D_3$ the diameter of the compact region of the inhomogeneous particle.

$D_2$, which represents the interaction term, has a value of $150.0 \pm 10.2$ Å, in agreement with the theoretical 148 Å calculated from the expression $(D_1 D_3)^{1/2}$ predicted by this model. Furthermore, the $D_1$ and $D_3$ values lead to radii of gyration of 79.9 Å and 41.0 Å, respectively, in reasonable agreement with the values of 80.2 and 38.0 Å for $R_2$ and $R_C$ in Table 1. Calculation of $R_2$ from the $p(r)$ data in Fig. 4B up to a $D_{max}$ of 198.9 Å yielded a value of 78 Å, again in reasonable agreement with the value of 80.24 (Table 1) obtained directly from desmeared SAXS. Thus the earlier supposition that a limiting polymer, the submicelle, results from predominantly hydrophobically driven self-associations of the caseins is supported. Furthermore, the structure of this submicelle has been shown to have spherical symmetry and to consist of two different spherically shaped electron density regions, the inner, higher density region probably resulting from hydrophobic intermolecular interactions. The question of whether these two regions are exactly concentric has not been completely clarified. However, in view of Fig. 4A, the assumption that the centers are no more than approximately 20–30 Å apart is not unreasonable. The interaction term in Eq. (3) contains the function $(\sin d_{ik}h)/(d_{ik}h)$, which for small values of $d_{ik}$ tends to unity, so that the contribution of this term is virtually constant for small center-to-center distances. The effect of the center separation on the molecular parameters can in any case not be large. Since there is no destructive interference at zero angle, the parameters obtained from $i_n(0)$ (i.e., k, m, and V) remain unaffected; the effect on the radius of gyration is on the order of 5%.

Micelles. As previously reported (Schmidt, 1982; Farrell and Thompson, 1988), addition of 10 mM calcium chloride to casein submicelles causes aggregation of the protein to the colloidal micelles. Whether the integrity of the submicellar structure is maintained within the colloidal micelle is still a subject of much controversy (Walstra, 1979). To address this problem, the scattering of whole casein solutions with 10 mM $CaCl_2$, and no phosphate buffer to compete with the protein calcium binding sites, was studied.

Table 3.  Sum-of-Three-Spheres Model for p(r) (Submicelles)

| Fitting Parameter | Value |
|---|---|
| $c_1$ | 0.773 ± 0.044 |
| $D_1$, Å | 206.3 ± 2.5 |
| $c_2$ | 0.171 ± 0.086 |
| $D_2$, Å | 150.0 ± 10.9 |
| $c_3$ | 0.0600 ± 0.05 |
| $D_3$, Å | 106.0 ± 11.0 |
| RMS | 0.000337 |

As seen in Table 1, $k_2$ for casein micelles was 0.308 ± 0.005, and the packing number, its reciprocal, was 3.2.  The average radius of the micelles, 650 Å as determined by electron microscopy (Farrell and Thompson, 1988), is so large that the corresponding scattering angles are too small to be experimentally accessible.  In place of the total colloidal particle, one can therefore observe only a cross-sectional portion, for which one finds a corresponding molecular weight, M, of 882,000 ± 28,000, an electron density difference, $\Delta\rho$, of 0.0081 ± 0.0004 $e^-/Å^3$, a hydration, H, of 7.92 ± 0.42 g water/g protein and a volume, V, of (12.72 ± 0.25) x $10^6$ $Å^3$ (Table 1).  Since these results refer to cross sections only, whereas, for example the molecular weights of whole casein micelles have been reported to range from 0.5 to 1 x $10^9$ (from porous gel chromatography, Schmidt and Payens, 1976), it appears that only the hydration parameter can be compared with those from other studies.  Our value of 7.92 is somewhat larger than the largest values reported by small-angle neutron scattering (Stothart and Cebula, 1982) of 4.0 to 5.5, in a somewhat less detailed investigation.  Other reported values have ranged from 2 to 7, depending on the method employed (Walstra, 1979).  The micellar parameters, other than hydration, need to be treated with caution, inasmuch as they refer only to a sample of the particle which is restricted in size by a window of scattered intensities bounded by the lowest angle observable by the instrument.  These parameters bear no readily defined relationship to the corresponding inaccessible ones for the entire micellar particle, and hence cannot serve to evaluate the latter.  Nonetheless, they are of value in affording an internal view of the micellar structure.

Most significant is the comparison between parameters of the casein submicellar structure by itself (column 2, Table 1) and within the casein micelle (column 3).  Within experimental error, $M_2$, k, $M_C$, $M_L$, and $\Delta\rho_C$ are the same in both cases.  However, a substantial decrease is observed in $\Delta\rho_2$ and $\Delta\rho_L$.  $H_2$,

$H_L$, $V_2$, $V_L$, $R_2$, and $R_L$ increase substantially, $H_C$, $R_C$, and $V_C$ slightly. From the changes in these parameters it appears likely that the large swelling and hydration in the loose region is due to $Ca^{2+}$ binding to protein electrostatic groups within this region. The relatively small changes in the density and radius of gyration of the internal core region upon the addition of calcium support the conclusion that this region is rich in hydrophobic side chains. Thus, the binding of $Ca^{2+}$ by submicelles and their subsequent incorporation into micelles do not lead to more compact structures; the low electron density of the independent submicelle carries over into the micelle.

To ascertain the spatial arrangement of the three spheres within the observed cross-sectional scattering volume, the distance distribution function p(r) was calculated from the SAXS data for casein micelles as shown in Fig. 5. Calculation of the radius of gyration from the second moment of the p(r) data (Pilz et al., 1979; Glatter, 1980) in Fig. 6, to a $D_{max}$ of 512 Å, yielded a value of 175.2 Å.

The experimental p(r) curve was then compared with theoretical curves calculated by the method of Glatter (1980), using various spatial models. For these, the radii of the outer and inner spheres, calculated from $V_2$ and $V_C$ values of column 3 of Table 1, were 102 and 50 Å, respectively. The equilateral or symmetrical triangular arrangement gave the poorest fit to the experimental data (Fig. 5, dashed line). The Cartesian coordinates for the

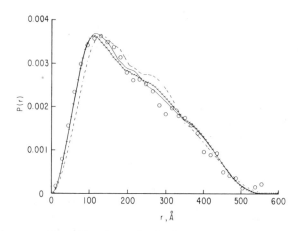

Fig. 5. Distance Distribution of Micelles. O, p(r) vs. r for
micellar casein at 16.4 mg/mL and 10 mM $CaCl_2$; ticked line,
theoretical for three inhomogeneous spheres at coordinates
(0,0), (350,0) and (180,100) with same outer radii (see
text); dashed line, theoretical for three inhomogeneous
spheres with same outer radii in a symmetrical triangular
arrangement (see text); solid line, theoretical for three
inhomogeneous spheres with two different outer radii at
coordinates (0,0), (350,0) and (180,0) (see Discussion).
Theoretical curves were calculated by the method of Glatter
(1980).

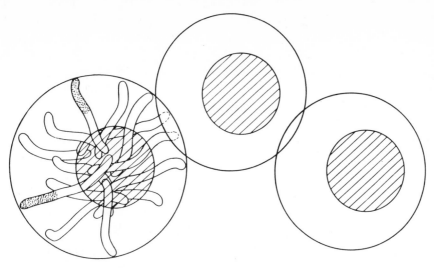

Fig. 6.  Schematic Representation of Submicelles within Micellar
Structure (corresponding to solid line in Fig. 8).  Cross-
hatched area, approximate core region of higher electron density
and higher concentration of hydrophobic side chains.  In submicelle in lower left, several representative monomer chains are
indicated; stippled areas, macropeptide portions of κ-casein.

centers of the three inhomogeneous spheres most compatible with the experimental data were found at nonsymmetrical values of (0,0), (350,0), and
(180,100) (ticked line), but a better fit resulted from an irregular triangular array in which the radius of the (0,0) sphere had been changed from
102 to 125 $\overset{\circ}{A}$ (solid line, and Fig. 6).  In fact, a value of 174.5 $\overset{\circ}{A}$ is calculated for the radius of gyration from the theoretical p(r) curve for this
inhomogeneous, irregular, triangular structure, in excellent agreement with
the value of 175.2 $\overset{\circ}{A}$ found from the experimental p(r) data.  It is notable
that these best-fit coordinates imply interdigitation of the "loose" regions
of the three submicelles (Fig. 6).

It may be concluded from the SAXS results that a discrete hydrophobically
stabilized submicellar structure exists within the colloidal casein micelle,
arguing against models predicting a continuous, porous gel structure, or an
impenetrable homogeneous sphere with a "hairy surface" (Holt and Dalgleish,
1986; Walstra, 1979).  In summary, the submicellar particles consist of an
inner, spherically symmetrical, hydrophobic, relatively electron-dense core,
surrounded by a hydrophilic and less electron-dense region, both much less
dense than globular proteins.  On addition of calcium, the loose region swells
with increased hydration and significantly lower electron density, which may
be caused by calcium binding to hydrophilic groups within this region.
Calculations of the cross-sectional scattering volume support a packing number
of 3 to 1 for the submicelles within the micelle and indicate some interaction
between the loose regions of adjacent submicelles.

## NMR Data

Certain ramifications of the picture painted above for the casein micelle invite further examination, especially with regard to its hydration. While it is widely accepted that trapped water exists within the micelle, the nature of this water is uncertain (Kuntz and Kauzmann, 1974; Walstra, 1979). Hydrodynamic methods generally yield values of 0.25 to 0.75 g/g for hydration of proteins, and calculation of water potentially bound to charges on casein would yield similar values; yet the SAXS data yield hydrations even for submicelles at least an order of magnitude higher, consistent with those for casein micelles from a variety of physical techniques from 2 to 7 g/g (Walstra, 1979). It is clear that not much of this water could be tightly bound, because in that case a tremendous energy barrier would be set up and the well-documented penetration by enzymes, lactose, and salts would be precluded (Farrell, 1988). An interesting observation was made by Kuntz and Kauzmann, who noted that in a drop of water enclosed in a hollow sphere the size of a casein micelle most water molecules might be far enough removed from interaction with the protein atoms to be bulk-like in nature. Although this notion must be modified because we know that the micelle does not contain any smooth-walled spherical cavity, most of the trapped water could be highly mobile. As regards the portion that is bound, the question then is whether it is isotropically or anisotropically bound, that is, whether it does not or does exhibit some motion of its own. To address this question, casein-water interactions and dynamics were studied by means of NMR relaxation techniques.

Most noticeable about the hydration values from NMR (Table 2) are their magnitudes. The values for micelles (column 4) range from 0.017 to 0.028 in this model where the water is considered in fast exchange but isotropically bound to the casein ($S = 1$). Thus only a small fraction of the total water associated with the casein could be expected to be of this nature. Before considering other possibilities we use this information to derive some parameters related to such water and then use their variation with temperature to disclose further information regarding the thermodynamics of casein interactions.

## Hydration and Dynamics

Isotropic Model. Since it has been shown that the $\tau_c$ values derived from NMR relaxation results are those for the unhydrated rather than the hydrated form of the protein (Kumosinski and Pessen, 1982; Pessen and Kumosinski, 1985), the Stokes radius, r, calculated from $\tau_c$ values using the Stokes-Einstein relationship, would indeed be a representation of the quaternary structure for the unhydrated protein. Such r values calculated from the $\tau_c$ results are listed in Table 2.

The Stokes radius of 36.4 Å found at 30°C is near the lower limit of radii reported for submicelles, which range from 40 to 90 Å depending on the method of measurement (Pepper and Farrell, 1982; Schmidt, 1982; Capalja, 1968; Schultz and Bloomfield, 1976). Direct comparison between Stokes radii derived by this NMR method and those calculated from hydrodynamic or small-angle scattering data (although somewhat inappropriate because these latter include the water of hydration, whereas the NMR values, as noted above, pertain to the anhydrous protein) shows a core region of about 38 Å. A reasonable interpretation here is that the most tightly bound water exists within the outer, looser region, but near the surface of the inner, hydro-phobic core. In any case, changes in this apparent radius for protein-associated water which accompanies protein structural changes are of interest in helping to establish structural information. Results for the submicelles show a decrease in hydration and an increase in the Stokes radius with increasing temperature. This suggests that hydrophobic inter-actions are involved in submicelle formation, since, as the temperature is raised, more water would be excluded from the hydrophobic core interface during an association process.

Although the absolute value of the Stokes radius calculated for the micelle is on the same order of magnitude as that of the submicelle, it was not as large as would be expected, because of instrumental limitations. These limitations are due to the large size of the casein micelle ($r \simeq 650$ Å), which would result in a $\tau_c$ value of nearly 200 μs. Such a slow motion would yield a transverse relaxation rate too large to be seen by NMR relaxation at 9.17 MHz. What these data may show is the average hydration of the caseins within the micelle, since the fastest motions dominate relaxation data. Thus, like the SAXS data, they provide a window of information and describe an average feature of the micelle interior. The protein-associated water within the micelle exhibits the same temperature dependence as that of the submicelle, showing hydrophobic interactions in agreement with earlier hypotheses that micelles are formed by aggregation of submicelles via $Ca^{2+}$ salt bridges (Farrell, 1988). The slight increase in r from submicelle to micelle is probably due to a gradual increase in internal hydration (trapped water) as the submicelle is incorporated into the micelle. This is also consistent with the extraordinary hydration (2 to 7 g $H_2O$/g protein) found by hydrodynamic measurements (Walstra, 1979). It is likely that in micelle formation the ultimate micelle size represents a balance between the electrostatic forces involving $Ca^{2+}$ and phosphate or carboxyl groups on the exterior of the submicelle, and the hydrophobic effects within the submicelle.

Derived Molecular Parameters of the Protein. Because the Stokes radius attributable to the bound water as derived from NMR relaxation data can be related to the anhydrous volume (Pessen and Kumosinski, 1985), a molecular weight, listed in Table 2, can be calculated as

$$M_p = 4/3 \pi r^3 N/\bar{v}_p, \tag{4}$$

where r is the Stokes radius, N is Avogadro's number, and $\bar{v}_p$ is the average partial specific volume of the caseins (0.736, as above). Since this $M_p$ is derived from an anhydrous radius, it should be somewhat smaller than $M_2$ calculated from SAXS data. Apart from this, $M_p$ appears to be in rough agreement with $M_2$, as well as with some values ascertained from electron microscopy where dehydration may occur. It may be supposed here that the water is bound to the submicelle and that little change occurs in protein structure on addition of $Ca^{2+}$. The latter supposition has been borne out by Raman spectroscopy of caseins with and without $Ca^{2+}$ (Byler et al., 1988). While each method of measurement views a particular aspect of water of hydration so that precise agreement is not to be expected, changes due to environment should be parallel, and the swelling observed in going from submicelle to the micelle in SAXS is indeed paralleled in the NMR data. The purpose of using Eq. (4) is to obtain information on temperature dependence rather than on molecular weights, $M_p$, whose absolute values have little significance because the equation was derived for globular monomeric proteins, in which, unlike in casein, the packing relationships are less complicated and little or no trapped water exists. Here the increase in $M_p$ for both the submicelle and the micelle as the temperature increases is an indication of hydrophobic self-association, both for the submicelle and within the micelle structure itself.

To quantitate this temperature-dependent variation of $M_p$, apparent equilibrium constants $K_A$ were calculated from $K_A = M_p/23,300$, bearing in mind that the measurements were performed at high casein concentrations, where the equilibrium is driven nearly completely toward the aggregated form. The van't Hoff expression (ln $K_A$ vs. inverse temperature, 1/T) was then used to calculate the apparent enthalpy of self-association, ΔH, for submicellar and micellar casein (Kumosinski et al., 1987). The van't Hoff plots for the two forms of casein (not shown) were linear and essentially parallel. In fact, ΔH was calculated to be 6.34 ± 0.11 kcal for submicelle formation and, only slightly higher, 6.81 ± 0.28 kcal, for self-association to the micelle; both are in good agreement with the value of 4.67 found for the association of purified $\alpha_{s1}$- and κ-caseins (Clarke and Nakai, 1971). It must be remembered that the protein-associated water has been utilized as a surface reporter group, and these parallel changes in water mobility are responsible for similar thermodynamic changes in both submicelles and micelles. This quantitation of the temperature variation of the self-association is strong indication that the integrity of the submicelle is at least partly preserved upon incorporation into the micelle by Ca-phosphate salt bridges.

194

Hydration: Anisotropic Tumbling Model. Up to this point all hydration values have been calculated on the basis of an isotropic motion mechanism of the bound water ($S = 1$). However, the motion of the bound water may, in fact, be anisotropic (not identical with that of the protein, $S < 1$) if the correlation times are long with respect to the Larmor frequency used. This may be the case for the casein micelles, where water may be trapped at the surfaces of submicelles as they self-associate via calcium-phosphate salt bridges into micelles. An attempt may be made to estimate the asymmetry factor S for casein on the following basis.

If it is assumed that the consistently increased Stokes radii of micelles relative to submicelles at the same temperatures (Table 2) are due to increased hydrations accompanying assembly to the micelle, one can use the partial specific volumes of the caseins (Kumosinski et al., 1987) to calculate new hydrations corresponding to the increased Stokes radii. From these, and the isotropic hydration, $(\bar{v}_w)_r$, of Table 2, S values were cal-culated at the three temperatures, as listed in column 8. They average $0.237 \pm 0.033$, in good agreement with the value of 0.23 predicted for anistropic motion by Walmsley and Shporer (1978). New hydrations, $(\bar{v}_w)_{S=0.237}$, can now be calculated for this mechanism, using the $\bar{v}_w$ values of Table 1 and the average S of 0.237. These are listed in the last column. Their absolute values, ranging from 0.116 to 0.502 g water/g protein for submicellar and micellar casein, are closer to hydration values derived by other methods (Pessen and Kumosinski, 1985) and calculated from compositional data (Kuntz and Kauzmann, 1974).

The foregoing calculation does not prove the existence of water with anisotropic motion bound to casein, but it does furnish significant informa-tion. It would not be unexpected that in a large, porous structure, such as the casein micelle, water could be bound to submicelles while retaining some partial motion of its own. More important than the absolute values of hydra-tion is their change with protein quaternary structural changes, which indicates a dynamic state of water within the interior of the casein micelle. The consistent increase in r and $\bar{v}_w$ for the micellar form over the submicellar form clearly demonstrates that trapped water exists in the micellar form of casein, as had been conjectured by Kuntz and Kauzmann (1974). It shows, further, that this trapped water is to a significant extent not a continuum with picosecond motion but is associated in part with hydrophobically stabilized submicelles which exist within the micellar structure.

On comparing the submicellar value of r, from NMR, with those of $R_C$ and $R_1$, from SAXS, it is seen that r and $R_C$ are in fair agreement. In view of the corresponding values of $H_C$ and $H_L$, this makes it apparent that, while some water is bound strongly to the surface of the core, the remainder is loosely bound or trapped. Such water would be highly mobile, unlike water evident in NMR data, whose rotational and translational diffusion coefficients have been profoundly altered by association with the protein. On inclusion into the micelle, the protein-associated water was found to largely retain its physical characteristics and accompanying thermodynamic properties, an indication that the interactions and dynamics observed for submicelles are very little modified on formation of the colloidal micelle.

Regardless of the model assumed, the hydrations measured by NMR are much lower than those measured by SAXS. In contrast to NMR, SAXS observes trapped and bound water, as well as the dynamic motions of the unhydrated protein; it may thus be concluded that much of the SAXS hydration represents trapped water. Subject to little direct interaction with the protein, this water cannot offer any great impediment to, e.g., the penetration of small proteases and clotting enzymes, such as chymosin. At the same time, there is the real possibility that the large hydration values are caused less by water per se than by packing effects of the protein chains. The loose packing or rapid and extensive motions due to molecular dynamics of the relatively unstructured casein may leave a large portion of the interchain volume effectively empty. Either of these possibilities can account for the virtually unique susceptibility of casein to the action of the aforementioned agents as well as for the dynamic equilibria which exist between the various component parts of the colloidal micelle and the serum phase of milk (Farrell, 1988).

## ACKNOWLEDGMENT

We thank Dr. H. Brumberger (Syracuse) for the use of his X-ray scattering facilities, Dr. O. Kratky (Graz) for furnishing a calibrated sample of a Lupolen standard scatterer, Dr. W. E. Damert (ERRC) for special algorithmic developments and computer programs, and Dr. S. J. Prestrelski (ERRC) for computer programs and computations.

## REFERENCES

Byler, D. M., Farrell, H. M., Jr., and Susi, H., 1988, J. Dairy Sci., 71:2622-2629.
Capalja, G. G., 1968, J. Dairy Sci., 35:1-10.
Carroll, R. J., and Farrell, H. M., Jr., 1983, J. Dairy Sci., 66:679-687.
Clarke, R., and Nakai, S., 1971, Biochemistry, 10:3353-3360.
Davies, D. T., and Law, A. J. R., 1980, J. Dairy Res., 47:83-90.
Debye, P., 1915, Ann. Physik, 46:809-823.
Edzes, H. T., and Samulski, E. T., 1978, J. Magn. Reson., 31:207-229.
Eigel, W. N., Butler, J. E., Ernstrom, C. A., Farrell, H. M., Jr., Harwalkar, V. R., Jenness, R., and Whitney, R. McL., 1984, J. Dairy Sci., 67:1599-1631.
Farrar, T. C., and Becker, E. D., 1971, Pulse and Fourier Transform NMR; Academic Press, New York, pp. 92, 105.
Farrell, H. M., Jr., 1988, In Fundamentals of Dairy Chemistry; Van Nostrand Reinhold, New York, pp. 461-510.

Farrell, H. M., Jr., and Thompson, M. P., 1988, In Calcium Binding Proteins; Vol. II, Thompson, M. P., Ed.; CRC Press, Inc., Boca Raton, Florida, pp. 117–137.

Glatter, O., 1980, Acta Phys. Austriaca, 52:243–256.

Guinier, A., and Fournet, G., 1955, In Small-Angle Scattering of X-Rays; Wiley and Sons, Inc., New York, p. 128.

Halle, B., Anderson, T., Forsen, S., and Lindman, B., 1981, J. Am. Chem. Soc., 103:500–508.

Holt, C., and Dalgleish, D. G., 1986, J. Colloid Interface Sci., 114:513–524.

Koenig, S. H., Bryant, R. G., Hallenga, K., and Jacob, G. S., 1978, Biochemistry, 17:4348–4358.

Kratky, O., 1963, In Progr. Biophys. Mol. Biol.; Butler, J. A. V., Huxley, H. E. and Zirkle, R. E., Eds.; Pergamon-MacMillan, New York, pp. 123ff.

Kumosinski, T. F., and Pessen, H., 1982, Arch. Biochem. Biophys., 218:286–302.

Kumosinski, T. F., Pessen, H., Farrell, H. M., Jr., and Brumberger, H., 1988, Arch. Biochem. Biophys., 266:548–561.

Kumosinski, T. F., Pessen, H., Prestrelski, S. J., and Farrell, H. M., Jr., 1987, Arch. Biochem. Biophys., 257:259–268.

Kuntz, I. D., and Kauzmann, W., 1974, Adv. Protein Chem., 28:239–345.

Lake, J. A., 1967, Acta Cryst., 23:191–194.

Lumry, R., and Rosenberg, A., 1975, Colloq. Int. CNRS, 246:53–62.

Luzzati, V., Witz, J., and Nicolaieff, A., 1961a, J. Mol. Biol., 3:367–378.

Luzzati, V., Witz, J., and Nicolaieff, A., 1961b, J. Mol. Biol., 3:379–392.

Meites, L., 1979, CRC Crit. Rev. Anal. Chem., 8:1–53.

Pepper, L., and Farrell, H. M., Jr., 1982, J. Dairy Sci., 65:2259–2266.

Pessen, H., Kumosinski, T. F., and Timasheff, S. N., 1973, Methods Enzymol., 27:151–209.

Pessen, H., and Kumosinski, T. F., 1985, Methods Enzymol., 117:219–255.

Pessen, H., Kumosinski, T. F., and Farrell, H. M., Jr., 1988, J. Ind. Microbiol., 3:89–103.

Pessen, H., Purcell, J. M., and Farrell, H. M., Jr., 1985, Biochem. Biophys. Acta, 828:1–12.

Pilz, I., Glatter, O., and Kratky, O., 1979, Methods Enzymol., 61:148–249.

Richards, F. M., 1974, J. Mol. Biol., 82:1–14.

Schmidt, D. G., 1982, In Developments in Dairy Chemistry; Vol. I, Fox, P. F., Ed.; Applied Sci. Pubs. Ltd., Essex, England, Ch. 2.

Schmidt, D. G., and Payens, T. A. J., 1976, In Surface and Colloid Sci., Matijevic, E., Ed.; Wiley & Sons, Inc., New York, p. 165–229.

Schultz, B. C., and Bloomfield, V. A., 1976, Arch. Biochem. Biophys., 173:18–26.

Stothart, P. H., and Cebula, D. J., 1982, J. Mol. Biol., 160:391–395.

Tanford, C., 1961, Physical Chemistry of Macromolecules, Wiley, New York, p. 236.

Thompson, M. P., 1964, J. Dairy Sci., 47:1261–1262.

Vold, R. L., Waugh, J. S., Klein, M. P., and Phelps, D. E., 1968, J. Chem. Phys., 48:3831–3832.

Walmsley, R. H., and Shporer, M., 1978, J. Chem. Phys., 68:2584–2590.

Walstra, P., 1979, J. Dairy Res., 46:317–323.

CARBON-13 NMR STUDIES OF NATIVE, GELLED, HEAT- AND CHEMICALLY-DENATURED SOY
GLYCININ AND β-CONGLYCININ AT NEUTRAL pH

M. S. Fisher, W. E. Marshall* and H. F. Marshall, Jr.

* USDA, ARS, Southern Regional Research Center
  1100 Robert E. Lee Boulevard
  P. O. Box 19687
  New Orleans, LA   70179

INTRODUCTION

Nuclear Magnetic Resonance Spectroscopy (NMR) is a powerful tool for
obtaining both gross structural and microenvironmental information about
proteins.  NMR studies of food proteins from seeds have emphasized the
alcohol-extractable proteins (prolamines) from cereal grains.  The NMR
characteristics of corn zein (Augustine and Baianu 1986, 1987), wheat
glutenins and gliadins (Baianu, 1981; Baianu et al., 1982), wheat gluten
(Belton et al., 1987) and C hordein from barley (Tatham et al., 1985) have
been studied.  None of these studies has investigated the aqueous
salt-extractable proteins of cereal grains or oilseeds by NMR.

NMR studies of soy proteins have emphasized the non-storage proteins
such as soy leghemoglobin (Trewhella et al., 1986; Mabbutt and Wright,
1983), lipoxygenase (Slappendel et al., 1982) and trypsin inhibitor
(Baillargeon et al., 1980) or metabolism of specific amino acids during
soybean germination (Coker et al., 1987).  While purified soy proteins have
been examined previously by NMR (Kakalis and Baianu, 1985; Baianu, 1989),
and some peaks tentatively assigned, to our knowledge no effort to fully
examine glycinin and β-conglycinin by $^{13}$C-NMR at neutral pH has been made.

To date, the structural studies of soy storage proteins have focused on
gross structural changes observable by either calorimetric (Hermansson,
1978) or spectrophotometric (Dev et al., 1988) methods, particularly CD and
ORD (Ishino and Kudo, 1980; Yamauchi et al., 1979; Koshiyama and Fukushima,
1973).  Recently, the conformational conclusions made with CD and ORD have
been qualitatively confirmed with FT-IR (Dev et al., 1988).

Although $^{13}$C NMR has been used to study a number of food proteins
(Baianu, 1989), some at elevated temperature (Belton et al., 1987; Baianu et

al., 1982), the effects of heat and chemical denaturation and gel formation of purified soy proteins have not been studied by NMR. Hermansson (1978) and Koshiyama et al. (1981), have reported the denaturation temperature for soy 7S and 11S proteins and studied turbidity changes during the denaturation-gelation process. Suresh Chandra et al. (1984) used circular dichroism and optical rotatory dispersion to determine the effects of temperature on the secondary and tertiary structure of glycinin. They found no change in the native structure over the temperature range 15 to 60°C. However, in the presence of urea or guanidine-HCl denaturant, a disordered structure was observed at 15°C which appeared to become more ordered at higher temperatures. The highest temperature used by Suresh Chandra et al. (1984) was 60°C, below the denaturation temperature for both glycinin and β-conglycinin (Hermansson, 1978). Dev et al. (1988) have studied changes in glycinin
conformation upon urea and heat denaturation with FT-IR.

Both glycinin and β-conglycinin can form self-supporting gels upon heating. When glycinin is heated to 95° in the presence of sodium chloride, the gel structure is dominated by ordered strands rather than disordered aggregates; β-conglycinin produces a somewhat less ordered gel matrix after being heated to 85° (Hermansson, 1986). Gel formation occurs at concentrations above 2.5% and 7.5% for glycinin and β-conglycinin, respectively (Nakamura et al., 1986a). To date, structural studies of soy storage protein gels have used electron microscopy (Hermansson, 1985, 1986; Mori et al., 1986), electrophoresis, gel immunodiffusion and other methods (Utsumi and Kinsella, 1985; Nakamura et al., 1986a, 1986b). The methods used by these workers precluded actual observation of protein behavior at particular temperatures. Their methods supply only limited information on specific types of amino acids which may be involved in maintaining gel structure, and are limited in their inability to observe protein conformation at elevated temperatures.

Both a previous NMR study of soy proteins (Kakalis and Baianu, 1985) and a recent study of ours (Fisher et al., 1989a), focused on the [13]C NMR spectra of undenatured materials. A third study (Baianu, 1989) on soy protein isolates was done at elevated pH's and on hydrolysates at very low pH. Using a phosphate-NaCl medium at room temperature and pH 7.0, Fisher et al. (1989a) made peak assignments for most of the amino acids. Also, similarities and differences in protein structure between the two soy storage proteins were discussed. Subsequent work by Fisher et al. (1986b) extended their initial study, determining the effects of gelation, heat and chemical denaturation on the [13]C NMR characteristics of glycinin and β-conglycinin structure and directly observing glycinin and β-conglycinin conformation at elevated temperatures using carbon-13 NMR spectroscopy.

200

MATERIALS AND METHODS

## Protein Isolation and Purification

Crude glycinin was prepared from Nutrasoy 7B flakes (Archer Daniels Midland Co., Decatur, IL) using a modification of the procedure of Thanh, et al. (1975), which is based on the differential solubility of β-conglycinin and glycinin in 0.063 M Tris buffer at pH 6.6, containing 0.010 M 2-mercaptoethanol (2-ME). Soy flakes were dispersed in 0.063 M Tris buffer, pH 7.8, containing 0.010 M 2-ME (1:15 ratio). The suspension was extracted for 2 hr. at ambient temperature and centrifuged for 30 min. at 10000 x g, and the precipitate discarded. The resulting supernatant was filtered through a 0.22μM Millipore filter, then exhaustively dialyzed against water at 4°C. The retentate was centrifuged for 30 min. at 10,000 x g. The precipitate was suspended in deionized water and lyophilized. The crude glycinin (750 mg portions) was dispersed in pH 7.6, 0.035 M phosphate buffer containing 0.4 M NaCl and 0.010 M 2-ME, then fractionated on a 2.5 x 80 cm column of Sepharose 6B using a flow rate of 25 mL/min. Purity of the preparation was ascertained by lithium dodecyl sulfate-polyacrylamide gel electrophoresis using the procedures outlined by Laemmli (1970). The remaining column eluate was dialyzed against water and lyophilized. Glycinin prepared in this manner was approximately 93% pure, and was used without further purification.

β-Conglycinin was prepared by Walter J. Wolf (USDA-ARS, Northern Regional Research Center, Peoria, IL) using a modification of the procedure of Thanh and Shibasaki (1976). In it, the initial extraction was made with 0.03 M Tris-HCl containing 0.005 M ethylenediaminetetraacetic acid (EDTA). The crude β-conglycinin was purified by dissolving 2 g in 25 mL of pH 7.6, 0.5 ionic strength phosphate-NaCl buffer containing 0.01 M 2-ME and 0.01% Thimerosal and pumping it through a 2.6 x 13 cm column of Concanavalin A-Sepharose 4B. The β-conglycinin was eluted with a gradient from 0 to 0.5 M of methyl α-D-glucoside. Samples, assayed for purity by Dr. Wolf by ultracentrifugation, were combined to yield a spectrometric sample containing approximately 92% β-conglycinin and 8% glycinin.

Amino acid analyses were done in duplicate for glycinin and β-conglycinin by Medallion Laboratories, Minneapolis, MN.

## Protein Denaturation

Protein samples were dissolved in an aqueous buffer containing 30% $D_2O$, 0.035 M potassium phosphate, 0.4 M NaCl and 6 M urea, and adjusted to an apparent pH of 7.0. The final protein concentrations were 100 mg/mL. Spectra were obtained at 25°C.

## Formation of Protein Gels

Protein solutions, in the absence of urea, were heated in the NMR
probe to the lowest desired temperature, and data acquisition initiated.
At the conclusion of data acquisition, the temperature was raised to the
next higher temperature and acquisition restarted.  This process was repeated to the highest obtained temperature for each sample.  The glycinin
sample heated to 95°C and the β-conglycinin sample heated to 85°C were
cooled to 4°C immediately after removal from the NMR probe in order to
promote gel formation.  NMR spectra of the gels were subsequently
obtained at 25°C.

## NMR Methods

All samples were measured as 100 mg/mL solutions in an aqueous
buffer containing 30% $D_2O$, 0.035 M potassium phosphate and 0.40 M sodium
chloride, adjusted to an apparent pH of 7.0.  Solvent was filtered
through piggyback 0.45μ and 0.22μ filters to remove fungal and bacterial
contaminants.  Tubes were washed with 95% ethanol, and acetone and dried
at 110°C.  In samples treated this way, the obvious results of decomposition (e.g., foul odor) did not appear until after 10 days storage in
solution at room temperature.  Acetonitrile (0.25%) was used as an
internal reference (chemical shift at pH 7.0, 25°C = 1.30 and
119.61 ppm).

All spectra were obtained with a Varian VXR-200 spectrometer using a
10-mm sample tube containing 3 mL of solution and 0.30 g of protein.
Spectra were acquired at 50.309 MHz over a spectral width of 11148.3 Hz,
proton decoupled with the WALTZ sequence (Shaka et al., 1983) using 256-
pulse blocks, each block preceded by a four-pulse steady-state sequence.
Recycle time was 0.6 sec., tip angle was 62.5°.  Fourier transforms were
done on 13376 points, zero-filled to 16384, processed with 5-Hz exponential line-broadening and 0.05-sec Gaussian apodization [$\exp(-t^2/0.0025)$].
To distinguish between close peaks, lines were converted from Lorentzian
line-shape to Gaussian lineshape by combined exponential [$\exp(t/sec.)$]
and Gaussian functions.  To emphasize narrow peaks, FIDs were convolution
difference weighted with the function [$1-\{0.9 \exp(-t/0.03)\}$].  To
determine peak multiplicity, the Varian-supplied APT sequence (Patt and
Shoolery, 1982), a J-modulated double spin-echo sequence, was used.
Spectra used to estimate spin-lattice relaxation times ($T_1$'s) for the
dominant peaks in the glycinin spectrum were obtained with the fast
inversion-recovery sequence (T-180°-tau-90°; T~2*$T_1$, tau varied in twelve
increments from 8 ms to 4 sec.) (Levy et al., 1980) and the $T_1$'s
calculated with the Varian-supplied software.

202

RESULTS AND DISCUSSION

## General Considerations

Amino acid analysis data are presented in Table 1; chemical shift, peak assignment and bandwidth data are presented as Tables 2, 4, 5 and 6; spin-lattice relaxation times ($T_1$) are in Table 3. Peak assignments were made based on literature values (Howarth and Lilley, 1978; Wüthrich, 1976; Kakalis and Baianu, 1989), aided by convolution difference and resolution enhancement weighting of the free induction decays (FIDs) as stated in NMR Methods above. In several instances, most notably the 10 ppm to 22 ppm region of β–conglycinin (Fig. 1B, Table 2), assignment of peaks that could not be observed in the normal NMR spectrum were made using the APT sequence (Patt and Shoolery, 1983). Bandwidths were determined with software available on the NMR instrument. Unreliable bandwidths resulted when accurate baseline or half-height could not be determined by the software due to peak overlap. All bandwidths reported by the instrument to be unreliable were rejected.

Table 1. Amino Acid Analysis Data (in %).

| | Glycinin Lit[a] | Glycinin Exptl[b] | β–Conglycinin Lit[c] | β–Conglycinin Exptl[b] |
|---|---|---|---|---|
| Lys | 4.4±0.4[d] | 4.7±0.1 | 6.4±0.6[d] | 6.1±0.3 |
| His | 2.0±0.2 | 1.9±0.2 | 2.0±0.3 | 2.3±0.2 |
| Arg | 6.6±1.1 | 7.2±0.2 | 7.9±0.8 | 9.4±0.0 |
| Asn[e] | 7.9±0.5 | 7.7±0.1 | 8.2±0.8 | 7.2±0.0 |
| Gln[e] | 10.2±0.8 | 11.5±0.4 | 9.2±0.3 | 10.4±0.5 |
| Asp[e] | 4.3±0.3 | 4.1±0.1 | 4.7±0.3 | 4.2±0.0 |
| Glu[e] | 8.9±0.6 | 10.1±0.4 | 10.7±0.3 | 12.2±0.5 |
| Thr | 3.9±0.4 | 3.6±0.1 | 2.6±0.2 | 2.2±0.0 |
| Ser | 5.6±1.3 | 5.0±0.0 | 7.0±0.4 | 4.9±0.0 |
| Pro | 6.2±0.7 | 7.8±0.8 | 5.0±1.2 | 4.3±0.7 |
| Gly | 6.4±2.1 | 4.4±0.2 | 4.2±1.2 | 2.6±0.0 |
| Ala | 5.0±1.7 | 3.7±0.2 | 4.5±0.7 | 3.2±0.1 |
| Val | 5.2±0.3 | 4.4±0.3 | 4.7±0.5 | 3.7±0.1 |
| Ile | 4.5±0.3 | 4.1±0.0 | 5.6±0.8 | 4.6±0.0 |
| Leu | 7.0±0.3 | 7.2±0.1 | 9.4±0.7 | 8.6±0.3 |
| Tyr | 2.8±1.0 | 4.1±0.6 | 2.5±1.1 | 3.2±0.3 |
| Phe | 4.4±1.2 | 5.0±0.5 | 6.1±1.1 | 6.3±0.3 |
| Trp | 0.8±0.2 | 0.8±0.0 | 0.2±0.2 | f |
| Met | 1.2±0.5 | 1.4±0.2 | 0.3±0.2 | 0.4±0.1 |
| Cys | 1.3±0.2 | 1.8±0.1 | 0.3±0.0 | 0.7±0.0 |

[a]Unweighted average of values reported in Moreira, *et.al.*, 1979, Kitamura & Shibasaki, 1975 and Badley *et.al.*, 1975; [b]Average of two determinations by Medallion Laboratories (see Methods); [c]Unweighted average of values reported in Coates, *et.al.*, 1985, Thanh & Shibasaki, 1977 and Nielsen, 1985; [d]Standard deviation; [e]Proportions of Asn, Asp, Gln & Glu calculated from Wright, 1987; [f]Not determined.

Table 2.   Glycinin and β-Conglycinin Peak Assignments, Peak Positions and Bandwidths.

| Peak Assignment | Glycinin ppm[a] | bw[b] | β-Conglycinin ppm | bw |
|---|---|---|---|---|
| Glu Cδ | 181.6 | 26.2 | 181.5 | 363.1 |
| Gln Cδ | 178.0 | 38.6 | 178.0 | c |
| C=O + Asn Cγ | 173.6 | 97.1 | 174.0 | 89.0 |
| C=O | 172.1 | 215.0 | 172.4 | 221.2 |
| C=O | d | | 172.0 | 220.8 |
| Arg Cζ | 157.2 | 20.8 | 157.1 | 22.9 |
| Tyr Cζ | 154.8 | 50.6 | e | |
| Phe Cγ | 136.2 | c | 136.5 | 34.4 |
| His Cδ + Cε | 135.6 | 40.5 | 135.6 | c |
| Tyr 2Cδ | 131.1 | 225.5 | 131.1 | 257.9 |
| Phe 2Cδ | 129.5 | 62.4 | 129.6 | 111.6 |
| Phe 2Cε | 129.1 | 68.4 | 129.0 | 113.0 |
| His[+] Cε | 128.5 | 227.3 | f | |
| Phe Cζ + Tyr Cγ + Trp Cδ$_2$[h] | 127.5 | 247.4 | 127.3 | 245.7 |
| Trp Cη[h] | 122.3 | 94.4 | e | |
| His[+] Cδ + Trp Cε[h] | 118.9 | 23.9 | e | |
| His Cδ + Trp Cζ$_2$[h] | 117.6 | 40.4 | e | |
| Tyr 2Cε | 115.8 | 39.0 | 115.8 | 23.0 |
| α-Man C1 | g | | 102.6 | 26.0 |
| GlcNAc(Asn) C1 | g | | 101.2 | 44.1 |
| GlcNAc(Asn) C2 | g | | 78.7 | 50.4 |
| α-Man C5 | g | | 73.5 | c |
| α-Man C2, C3 & C4 | g | | 70.5 | 94.2 |
| Thr Cβ | 68.2 | 45.9 | | |
| + β-Man C4 & C6[g] | | | 67.2 | c |
| Ser Cβ | 61.4 | 171.6 | | |
| + α-Man C6[g] | | | 61.5 | 99.5 |
| Pro Cα | 60.9 | c | | |
| + GlcNAc(Asn) C2[g] | | | 60.7 | 383.4 |
| Ser Cα | 55.9 | c | | |
| + GlcNAc(β-Man) C2[g] | | | 56.2 | 118.1 |
| Cα Env | 55.0 | 296.6 | f | |
| Cα Env | 53.9 | 105.0 | 54.1 | 85.3 |
| Asn Cα | 52.0 | 307.8 | 52.1 | c |
| Pro Cδ | 48.0 | c | 48.3 | c |
| Gly Cα | 42.9 | c | 42.9 | c |
| Lys Cε + Arg Cδ | 41.0 | 144.6 | 41.0 | 115.4 |
| Asn Cβ + Asp Cβ | 39.6 | 40.2 | 39.7 | 36.3 |
| Leu Cβ + Cys2 Cβ[h] | 38.9 | 294.6 | d | |
| Tyr Cβ | 37.9 | c | d | |
| Ile Cβ | 36.9[i] | c | 36.7 | c |
| Phe Cβ + Ile Cβ | 36.5 | c | d | |
| Glu Cγ | 33.9 | 28.0 | 33.9 | 91.2 |
| Gln Cγ | 31.4 | 41.3 | 31.4 | c |
| Lys Cβ + Val Cβ + Met Cβ[h] | 30.7 | c | 30.7 | c |
| Pro Cβ + Met Cγ[h] | 29.7 | c | 29.7 | c |
| Val Cβ | 29.3[i] | c | | |
| Arg Cβ + Glu Cβ | 28.0 | 115.5 | 27.9 | c |
| Lys Cδ | 27.1 | 129.6 | 27.2 | c |
| Gln Cβ | 26.7 | 129.6 | 26.7 | c |
| Arg Cγ + Pro Cγ + Ile Cγ$_1$ | 24.8 | 242.4 | 24.7 | c |
| Leu Cγ | 24.3[i] | c | 25.0[i] | |
| Lys Cγ | 22.3 | 247.0 | 22.3 | c |
| Leu 2Cδ | 21.1 | c | 21.5[i] | c |
| Thr Cγ | 19.2 | c | 19.2[i] | c |
| Val Cγ$_1$ | 18.7 | c | 18.8[i] | c |
| Val Cγ$_2$ | 18.1 | c | d | |
| Ala Cβ | 17.2[i] | c | 17.2[i] | c |
| Ile Cγ$_2$ + Met Cε[h] | 15.2 | c | f | |
| Ile Cδ | 11.0 | c | e | |

[a]Chemical shift in parts per million, acetonitrile as reference at 1.300 ppm; [b]Bandwidth at half height in Hz; [c]Bandwidth calculation unreliable; [d]Merged with preceding peak; [e]Peak intensity too low to permit assignment; [f]Cannot be assigned separately from surrounding peak envelope; [g]β-Conglycinin only; [h]Glycinin only; [i]Peak assigned via the APT pulse sequence (Patt & Shoolery, 1983).

Table 3.  Glycinin Carbon-13 Spin Lattice Relaxation Times.

| ASSIGNMENT[b] | SHIFT(ppm)[a] | $T_1$(sec) |
|---|---|---|
| Glu Cδ | 181.6 | 2.26 |
| Gln Cδ + Asp Cγ | 178.0 | 1.85 |
| Backbone C=O | 173.9 | 2.18 |
| Backbone C=O | 172.4 | 2.24 |
| Arg Cζ | 157.2 | 2.24 |
| Tyr 2Cδ | 131.0 | 1.21 |
| Phe 2Cδ + Phe 2Cε | 129.5 | 0.74 |
| Tyr 2Cγ | 128.1 | 0.74 |
| His Cδ | 117.7 | 0.74 |
| Tyr 2Cε | 116.2 | 0.75 |
| Thr Cβ | 66.2 | 0.93 |
| Ser Cβ | 61.3 | 0.53 |
| Pro Cα | 59.8 | 0.39 |
| Backbone Cα | 53.9 | 0.31 |
| Asn Cα | 52.0 | 0.31 |
| Pro Cδ | 48.2 | 0.35 |
| Gly Cα | 42.9 | 0.33 |
| Lys Cε + Arg Cδ | 40.9 | 0.33 |
| Asn Cβ + Asp Cβ | 39.7 | 0.33 |
| Phe Cα + Ile Cβ | 36.6 | 0.27 |
| Glu Cγ | 33.8 | 0.22 |
| Gln Cγ | 31.4 | 0.17 |
| Pro Cβ + Val Cβ | 29.7 | 0.17 |
| Arg Cβ + Glu Cβ | 27.9 | 0.17 |
| Gln Cβ + Lys Cδ | 26.9 | 0.17 |
| Leu Cγ + Arg Cγ + Pro Cγ + Ile Cγ$_1$ | 24.8 | 0.17 |
| Lys Cγ | 22.4 | 0.20 |
| Leu 2Cδ | 21.3 | 0.31 |
| Thr Cγ + Val Cγ$_{1\ \&\ 2}$ | 19.0 | 0.54 |
| Ile Cγ + Met Cε | 14.5 | 0.51 |
| Ile Cδ | 10.5 | 0.76 |
| Acetonitrile | 1.3 | 10.30 |

[a]From the spectra used in the $T_1$ calculations.  Does not necessarily agree with Table 1.
[b]From Table 1.

Table 4. Native and Denatured Glycinin and β-Conglycinin Peak Positions, Assignments and Bandwidths.

| Peak Assignment | Glycinin −6 M Urea ppm[a] | bw[b] | Glycinin +6 M Urea ppm | bw | β-Conglycinin −6 M Urea ppm | bw | β-Conglycinin +6 M Urea ppm | bw |
|---|---|---|---|---|---|---|---|---|
| Glu C$\delta$ | 181.6 | 26.2 | 181.5 | 24.2 | 181.5 | 363.1 | 181.5 | 80.4 |
| Gln C$\delta$ | 178.0 | 38.6 | 178.0 | 33.1 | 178.0 | c | 177.9 | 33.4 |
| C=O + Asn C$\gamma$ | 173.6 | 97.1 | 174.6 | 170.7 | 174.0 | 89.0 | 173.8 | 95.0 |
| C=O | d | | 173.9 | 111.2 | 172.4 | 221.2 | 172.3 | 191.3 |
| C=O | 172.1 | 215.0 | 172.3 | 205.6 | 172.0 | 220.8 | e | |
| Arg C$\zeta$ | 157.2 | 20.8 | 157.2 | 20.0 | 157.1 | 22.9 | 157.2 | 22.0 |
| Tyr C$\zeta$ | 154.8 | 50.6 | 155.0 | 40.5 | d | | 155.1 | 31.3 |
| Phe C$\gamma$ | 136.5 | c | 136.6 | 44.4 | 136.5 | 34.4 | 136.6 | 26.3 |
| His C$\delta$ + C$\epsilon$ | 135.2 | 40.5 | 136.6 | 143.8 | 135.6 | c | 134.7 | 65.3 |
| Tyr 2C$\delta$ | 131.1 | 225.5 | 131.0 | 34.0 | 131.1 | 257.9 | 130.9 | 35.9 |
| Phe 2C$\delta$ | 129.5 | 62.4 | 129.6 | 49.2 | 129.6 | 111.6 | 129.5 | 47.9 |
| Phe 2C$\epsilon$ | 129.1 | 68.4 | 129.2 | 52.9 | 129.0 | 113.0 | 129.1 | 48.8 |
| His$^+$ C$\epsilon$ | 128.5 | 227.3 | 128.0 | 68.2 | f | | d | |
| Phe C$\zeta$ + Tyr C$\gamma$ (+ Trp C$\delta_2$[h]) | 127.5 | 247.4 | 127.6 | 219.2 | 127.3 | 245.7 | 127.5 | 152.8 |
| Trp C$\eta$[h] | 122.3 | 94.4 | 122.4 | 31.0 | e | | e | |
| His$^+$ C$\delta$ (+ Trp C$\epsilon$[h]) | 118.9 | 23.9 | 118.9 | 177.4 | f | | 118.9 | 88.0 |
| His C$\delta$ (+ Trp C$\zeta_2$[h]) | 117.6 | 40.4 | 117.8 | 49.0 | 118.3 | 69.7 | 117.5 | 59.0 |
| Tyr 2C$\epsilon$ | 115.8 | 39.0 | 115.9 | 38.7 | 115.8 | 23.0 | 115.9 | 23.2 |
| Trp C$\zeta_1$[h] | d | | 112.3 | 14.5 | e | | e | |
| $\alpha$-Man C1[g] | | | | | 102.6 | 26.0 | 102.6 | 15.4 |
| GlcNAc(Asn) C1[g] | | | | | 101.2 | 44.1 | 100.9 | 34.0 |
| GlcNAc(Asn) C2[g] | | | | | 78.7 | 50.4 | 78.8 | 26.2 |
| $\alpha$-Man C5[g] | | | | | 73.5 | c | 73.8 | 23.1 |
| $\alpha$-Man C2, C3 & C4[g] | | | | | 70.5 | 94.2 | 70.5 | 35.6 |
| Thr C$\beta$ (+ β-Man C4 & C6)[g] | 68.2 | 45.9 | 67.6 | 27.4 | 67.2 | c | 67.4 | 36.5 |
| Ser C$\beta$ + (+ $\alpha$-Man C6)[g] | 61.4 | 171.6 | 61.7 | 67.6 | 61.5 | 99.5 | 61.6 | 62.6 |
| trans-Pro C$\alpha$ (+ GlcNAc(Asn) C2)[g] | 60.9 | c | 61.0 | 173.0 | 60.7 | 383.4 | 60.9 | 171.6 |
| cis-Pro C$\alpha$ | f | | 60.1 | 179.9 | f | | 60.0 | 187.2 |
| C$\alpha$ Env | 59.6 | c | 59.3 | 187.1 | f | | f | |
| C$\alpha$ Env | 57.6 | c | 57.6 | 408.3 | f | | f | |
| Ser C$\alpha$ (+ GlcNAc($\alpha$-Man) C2)[g] | 55.9 | c | 56.0 | 166.5 | 56.2 | 118.1 | 56.0 | 244.3 |

Table 4.  Native and Denatured Glycinin and β–Conglycinin Peak
Positions, Assignments and Bandwidths. (Cont'd)

| Peak Assignment | Glycinin −6 M Urea ppm[a] | Glycinin −6 M Urea bw[b] | Glycinin +6 M Urea ppm | Glycinin +6 M Urea bw | β–Conglycinin −6 M Urea ppm | β–Conglycinin −6 M Urea bw | β–Conglycinin +6 M Urea ppm | β–Conglycinin +6 M Urea bw |
|---|---|---|---|---|---|---|---|---|
| Cα Env | f | | 55.2 | 263.0 | 55.9 | c | 55.9 | 266.1 |
| Cα Env | 53.9 | 105.0 | 54.1 | 72.9 | 54.1 | 85.3 | 54.2 | 68.7 |
| Asn Cα | 52.0 | 307.8 | 52.1 | 265.1 | 52.1 | c | 54.2 | 68.7 |
| Pro Cδ | 48.0 | c | 48.3 | 42.5 | 48.3 | c | 48.3 | 42.5 |
| Gly Cα | 42.9 | c | 43.0 | 37.2 | 42.9 | c | 43.0 | 47.0 |
| Lys Cε + Arg Cδ | 41.0 | 144.6 | 41.2 | 31.2 | 41.0 | 115.4 | 41.1 | 28.6 |
| Asp Cβ + Asn Cβ | 39.6 | 40.2 | 39.8 | 55.7 | 39.7 | 36.3 | 39.8 | 31.4 |
| Leu Cβ + Cys2 Cβ[h] | 38.9 | 294.6 | 39.1 | 142.7 | e | | e | |
| Tyr Cβ | 37.9 | c | 38.0 | c | f | | f | |
| Ile Cβ | 36.9[i] | | f | | f | | f | |
| Phe Cβ (+Ile Cβ)[g] | 36.5 | c | 36.6 | 67.3 | 36.7 | c | 36.6 | 68.7 |
| Glu Cγ | 33.9 | 28.02 | 34.1 | 28.6 | 33.9 | 91.2 | 34.0 | 38.2 |
| Gln Cγ | 31.4 | 41.3 | 31.6 | 36.7 | 31.4 | c | 31.5 | 72.2 |
| Lys Cβ + Val Cβ + Met Cβ[h] | 30.7 | c | 30.6 | 286.1 | 30.7 | c | 30.9 | 127.7 |
| Pro Cβ + Met Cγ[h] | 29.7 | c | 29.8 | 274.4 | 29.7 | c | 29.8 | 294.8 |
| Val Cβ | 29.3[i] | | e | | e | | e | |
| Arg Cβ + Glu Cβ | 28.0 | 115.5 | 28.2 | 124.1 | 27.9 | c | 28.2 | 114.4 |
| Lys Cβ | 27.1 | 129.6 | 27.1 | 119.5 | 27.2 | c | 26.9 | 114.4 |
| Lys Cδ + Gln Cβ | 26.7 | 129.6 | e | | 26.7 | c | 26.9 | 114.4 |
| Arg Cγ + Ile Cγ₁ | 24.8 | 242.4 | 25.0 | 55.4 | | | | |
| Pro Cγ (+ GlcNAc methyl)[g] | 24.7 | c | 25.0 | 42.3 | | | | |
| Leu Cγ | 24.4[i] | | e | | 25.0[i] | | e | |
| Lys Cγ | 22.3 | 247.0 | 22.6 | 36.0 | 22.3 | c | 22.6 | 28.8 |
| Leu 2Cδ | 21.1 | c | 21.3 | 130.1 | 21.5[i] | | 21.1 | 35.6 |
| Thr Cγ | 19.2 | c | 19.3 | 92.3 | 19.2[i] | | 19.3 | 79.9 |
| Val Cγ₁ | 18.7 | c | 18.9 | 88.8 | 18.8[i] | | 18.9 | 71.0 |
| Val Cγ₂ | 18.1 | c | 18.4 | 110.7 | d | | 18.4 | 78.8 |
| Ala Cβ | 17.2[i] | | 17.1 | c | 17.2[i] | | 17.0 | 33.1 |
| Ile Cγ₂ + Met Cε[h] | 15.2 | c | 15.2 | 26.5 | f | | 15.2 | 23.8 |
| Ile Cδ | 11.0 | c | 10.7 | 37.0 | d | | 10.7 | 31.3 |

[a]Chemical shift in parts per million, acetonitrile as reference at 1.300 ppm;
[b]Bandwidth at half height in Hz; [c]Bandwidth calculation unreliable; [d]Merged
with preceding peak; [e] Peak intensity too low to permit assignment; [f]Cannot be
assigned separately from surrounding peak envelope; [g]β–Conglycinin only;
[h]Glycinin only; [i]Peak assigned (Fisher, et.al., 1989a) via the APT pulse
sequence (Patt & Shoolery, 1983).

Table 5.  Glycinin Peak Assignments, Positions and Bandwidths at
Different Temperatures.

| Peak Assignment | 25 C ppm[a] | bw[b] | 55 C ppm | bw | 70 C ppm | bw | 95 C ppm | bw | GEL ppm | bw |
|---|---|---|---|---|---|---|---|---|---|---|
| Glu Cδ | 181.6 | 26.2 | 181.5 | 24.9 | 181.4 | 26.4 | 181.4 | 33.0 | 181.5 | 28.0 |
| Gln Cδ | 178.0 | 38.6 | 178.0 | 45.4 | 178.0 | 48.7 | 178.2 | 55.7 | 178.0 | 36.1 |
| C=O + Asn Cγ | 173.6 | 97.1 | 173.7 | 153.9 | 173.7 | 112.4 | 173.8 | 113.5 | 173.8 | 143.9 |
| C=O | 172.1 | 215.0 | 172.2 | 198.6 | 172.2 | 208.2 | 172.5 | 196.7 | 172.2 | 200.8 |
| Arg Cζ | 157.2 | 20.8 | 157.4 | 20.5 | 157.5 | 21.0 | 157.8 | 19.1 | 157.1 | 19.7 |
| Tyr Cζ | 154.8 | 50.6 | 155.2 | 23.0 | 155.4 | 16.2 | 155.4 | 20.0 | 154.9 | c |
| Phe Cγ | 136.5 | c | 136.7 | 124.1 | 136.7 | 96.2 | 136.6 | 64.8 | 136.4 | 37.1 |
| His Cδ | 135.2 | 40.5 | 135.8 | 40.9 | 136.0 | 84.2 | 136.3 | 89.9 | e | |
| His C | e | | 135.1 | 126.2 | e | | e | | 134.8 | 50.7 |
| Tyr 2Cδ | 131.1 | 225.5 | 131.1 | 40.9 | 131.1 | 39.2 | 131.1 | 37.7 | 130.8 | 208.6 |
| Phe 2Cδ | 129.5 | 62.4 | 129.7 | 55.8 | 129.7 | 56.9 | 129.8 | 50.1 | 129.5 | 53.9 |
| Phe 2C | 129.1 | 68.4 | 129.5 | 64.2 | 129.5 | 56.9 | 129.3 | 54.5 | 129.0 | 57.5 |
| His[+] C | 128.5 | 227.3 | 128.1 | 212.1 | 128.0 | 201.1 | e | | 128.0 | 224.9 |
| Phe Cζ + Tyr Cγ + Trp Cδ[2] | 127.5 | 247.4 | 127.5 | 267.1 | 127.6 | 224.4 | 127.7 | 40.0 | 127.5 | 215.9 |
| Trp Cδ[1] | 125.5 | c | e | | 125.6 | e | 125.0 | e | 124.9 | e |
| Trp Cη | 122.3 | 94.4 | 122.3 | 52.3 | 122.6 | 25.3 | 122.2 | 23.5 | e | |
| His[+] Cδ + Trp C[3] | 118.9 | 88.0 | 118.9 | 23.9 | 118.9 | 177.4 | 118.7 | 77.5 | e | |
| His Cδ + Trp Cζ[2] | 117.6 | 40.4 | 117.9 | 82.2 | 118.1 | 49.0 | 118.3 | 76.9 | 117.7 | 53.5 |
| Tyr 2C | 115.8 | 39.0 | 116.3 | 41.2 | 116.4 | 38.0 | 116.3 | 31.1 | 115.8 | 31.4 |
| Trp Cζ[1] | e | | 112.4 | 20.5 | 112.3 | 16.9 | 112.4 | 17.3 | 112.3 | e |
| Thr Cβ | 68.2 | 45.9 | 67.4 | 32.9 | 67.5 | 31.8 | 67.7 | 29.0 | 67.4 | 35.8 |
| Ser Cβ | 61.4 | 171.6 | 61.6 | 178.0 | 61.6 | 180.3 | 61.9 | 175.8 | 61.4 | 162.3 |
| trans-Pro Cα | 60.9 | c | 61.0 | 180.1 | 61.1 | 175.9 | 61.3 | 181.3 | 60.9 | 169.4 |
| Cα Env + cis-Pro Cα | 59.6 | c | 60.0 | c | 60.0 | c | 59.4 | 183.6 | 59.7 | 193.2 |
| Cα Env | 57.6 | c | 57.3 | c | 57.4 | 414.0 | 57.5 | 418.9 | f | |
| Ser Cα | 55.9 | c | 55.8 | 353.5 | 55.9 | 348.6 | 56.1 | 284.7 | 55.9 | 324.4 |
| Cα Env | 55.0 | 296.6 | 55.1 | 306.1 | 55.2 | 301.3 | 55.4 | 234.1 | f | |
| Cα Env | 53.9 | 105.0 | 54.2 | 104.9 | 54.3 | 104.1 | 54.5 | 106.8 | 54.0 | 94.6 |
| Asn Cα | 52.0 | 307.8 | 52.1 | 303.5 | 52.1 | 296.7 | 52.9 | 230.6 | 52.4 | 268.8 |
| Pro Cδ | 48.0 | c | 48.3 | 75.6 | 48.3 | 84.4 | 48.5 | 46.2 | 48.2 | 49.2 |
| Gly Cα | 42.9 | c | 43.1 | 406.1 | 43.1 | c | 43.5 | 32.1 | 42.9 | 46.7 |
| Lys C + Arg Cδ | 41.0 | 144.6 | 41.1 | 151.6 | 41.2 | 143.4 | 41.5 | 31.2 | 40.9 | 35.2 |
| Asp Cβ | f | | 40.2 | 88.5 | 40.3 | 90.5 | 40.6 | 96.7 | 39.7 | 33.2 |
| Asn Cβ | 39.6 | 40.2 | 39.9 | 87.5 | 39.9 | 87.0 | 40.1 | 96.7 | f | |
| Leu Cβ + Cys2 Cβ | 38.9 | 294.6 | 39.3 | 187.2 | 39.4 | 173.0 | 39.4 | 105.0 | 38.7 | 285.7 |
| Tyr Cβ | 37.9 | c | e | | e | | 37.3 | 78.3 | | |
| Ile Cβ | 36.9[g] | c | f | | f | | | | | |
| Phe Cβ | 36.5 | c | 36.8 | 304.4 | 36.9 | 329.6 | 36.9 | 72.8 | 36.3 | 292.4 |
| Glu Cγ | 33.9 | 28.0 | 34.0 | 27.8 | 34.1 | 27.7 | 34.4 | 24.5 | 33.8 | 26.9 |
| Gln Cγ | 31.4 | 41.3 | 31.6 | 37.6 | 31.6 | 35.5 | 31.9 | 26.8 | 31.5 | 33.1 |
| Lys Cβ + Val Cβ + Met Cβ | 30.7 | c | 30.4 | c | 30.4 | c | 30.7 | 275.7 | 30.6 | 292.6 |
| Pro Cβ + Met Cγ | 29.7 | c | 29.7 | 287.8 | 29.7 | 286.8 | 29.9 | 274.5 | 29.7 | 295.2 |
| Val Cβ | 29.3[g] | c | d | | d | | d | | d | |
| Arg Cβ + Glu Cβ | 28.0 | 115.5 | 28.3 | 111.2 | 28.3 | 113.4 | 28.7 | 55.6 | 28.0 | 111.6 |
| Lys Cδ | 27.1 | 129.6 | 27.2 | 287.3 | 27.3 | 285.7 | 27.5 | 125.3 | 27.0 | 116.5 |
| Gln Cβ | 26.7 | 129.6 | 26.8 | 289.0 | 26.8 | 292.0 | 27.1 | 273.7 | 26.7 | 116.5 |
| Arg Cγ + Pro Cγ + Ile Cγ[1] | 24.8 | 242.4 | 24.9 | 49.8 | 24.9 | 46.1 | 25.1 | 30.3 | 24.8 | 47.2 |
| Leu Cγ | 24.4[g] | | | | | | | | | |
| Lys Cγ | 22.3 | 247.0 | 22.4 | 45.4 | 22.4 | 43.9 | 22.7 | 37.2 | 22.5 | 42.5 |
| Leu 2Cδ | 21.1 | c | 21.5 | 408.2 | 21.5 | c | 21.8 | 90.8 | 21.1 | 129.0 |
| Thr Cγ | 19.2 | c | 19.2 | c | 19.0 | c | 19.2 | 44.7 | 19.1 | 80.6 |
| Val Cγ[1] | 18.7 | c | 18.9 | c | 18.9 | c | 18.4 | 99.1 | 18.8 | 80.6 |
| Val Cγ[2] | 18.1 | c | 18.3 | c | 18.4 | c | 18.4 | 114.8 | 18.1 | 102.2 |
| Ala Cβ | 17.2[g] | c | 17.1 | c | 17.2 | c | 17.4 | 39.4 | 17.0 | c |
| Ile Cγ[2] + Met C | 15.2 | 458.2 | 15.3 | 37.1 | 15.3 | 41.2 | 15.6 | 27.2 | 15.1 | 26.6 |
| Ile Cδ | 11.0 | c | 10.6 | 29.6 | 10.6 | 38.2 | 10.9 | 32.3 | 10.5 | 36.6 |

[a]Chemical shift in parts per million, acetonitrile as reference at 1.300 ppm;
[b]Bandwidth at half height in Hz; [c]Bandwidth calculation unreliable; [d]Merged
with preceding peak; [e]Peak intensity too low to permit assignment; [f]Cannot be
assigned separately from surrounding peak envelope; [g]Peak assigned (Fisher,
et.al., 1989a) via the APT pulse sequence (Patt & Shoolery, 1983).

# Table 6. β–Conglycinin Peak Assignments, Positions and Bandwidths at Different Temperatures.

| Peak Assignment | 25 C ppm[a] | bw[b] | 55 C ppm | bw | 70 C ppm | bw | 85 C ppm | bw | GEL ppm | bw |
|---|---|---|---|---|---|---|---|---|---|---|
| Glu Cδ | 181.5 | 363.1 | 181.4 | c | 181.3 | 160.1 | 181.3 | 91.0 | 181.6 | 380.6 |
| Gln Cδ | 178.0 | c | 178.1 | 418.4 | 178.1 | 439.5 | 178.2 | 63.8 | 178.1 | 391.4 |
| C=O + Asn Cγ | 174.0 | 89.0 | 174.0 | 141.3 | 174.0 | 148.6 | 174.0 | 115.7 | 174.0 | 82.7 |
| C=O | 172.4 | 221.2 | 172.2 | 208.9 | 172.1 | 211.4 | 172.0 | 210.2 | 172.4 | 240.9 |
| C=O | 172.0 | 220.8 | d | | d | | d | | 172.0 | 240.9 |
| Arg Cζ | 157.1 | 22.9 | 157.4 | 20.7 | 157.5 | 21.5 | 157.6 | 28.0 | 157.1 | 21.7 |
| Phe Cγ | 136.5 | 34.4 | 136.6 | 128.5 | 136.5 | 83.4 | 137.0 | 60.8 | 136.5 | 68.9 |
| His Cδ + C | 135.6 | c | 135.7 | 131.1 | 135.8 | 172.2 | 135.6 | 127.6 | 135.0 | 38.5 |
| Tyr 2Cδ | 131.1 | 257.9 | f | | 130.9 | 291.0 | 130.8 | 225.8 | 130.8 | 219.8 |
| Phe 2Cδ | 129.6 | 111.6 | 129.7 | 129.8 | 129.6 | 142.0 | 129.6 | 68.3 | 129.6 | 84.0 |
| Phe 2C | 129.0 | 113.0 | 129.2 | 130.3 | 129.1 | 138.6 | 129.2 | 65.0 | 129.0 | 112.9 |
| Phe Cζ + Tyr Cγ | 127.3 | 245.7 | 127.5 | 237.0 | 127.6 | 257.3 | 127.6 | 222.6 | 127.4 | 291.2 |
| His⁺ Cδ | d | | d | | 120.1 | 24.4 | 120.2 | 35.5 | d | |
| His Cδ | 118.3 | 69.7 | 118.6 | 34.6 | 118.6 | 107.0 | 118.4 | 62.8 | f | |
| Tyr 2C | 115.8 | 23.0 | 116.0 | 40.5 | 116.1 | 55.6 | 116.3 | 30.6 | 115.87 | 22.7 |
| α–Man C1 | 102.6 | 26.0 | 102.6 | 17.2 | 102.5 | 23.5 | 102.7 | 27.4 | 102.5 | 17.1 |
| GlcNAc(Asn) C1 | 101.2 | 44.1 | 101.1 | 13.9 | 101.1 | 23.8 | d | | | |
| GlcNAc(Asn) C2 | 78.7 | 50.4 | 78.9 | 39.3 | 78.8 | 27.4 | 78.9 | 54.2 | d | |
| α–Man C5 | 73.5 | c | 73.7 | 51.3 | 73.9 | 42.4 | 73.7 | c | 73.5 | 41.7 |
| α–Man C2, C3 & C4 | 70.5 | 94.2 | 70.5 | 52.4 | 70.7 | 45.7 | 70.8 | 86.7 | 70.7 | 60.2 |
| Thr Cβ + β–Man C4 & C6 | 67.2 | c | 67.4 | 41.9 | 67.5 | 34.9 | 67.6 | 123.8 | 67.3 | 39.9 |
| Ser Cβ + α–Man C6 | 61.5 | 99.5 | 61.6 | 70.3 | 61.6 | 82.5 | 61.8 | 74.7 | 61.4 | 96.4 |
| Pro Cα + GlcNAc(Asn) C2 | 60.7 | 383.4 | 60.9 | 118.6 | 60.9 | 144.7 | 61.1 | 132.6 | 60.7 | 102.9 |
| Cα Env (+ GlcNAc(α–Man) C2[d]) | 56.2 | 118.1 | 55.8 | c | 56.1 | 347.4 | 56.1 | 316.1 | 56.1 | 365.1 |
| Ser Cα | 55.9 | c | f | | f | | f | | f | |
| Cα Env | 54.1 | 85.3 | 54.2 | 79.4 | 54.4 | 74.9 | 54.5 | 73.3 | 54.1 | 73.3 |
| Asn Cα | 52.1 | c | 52.3 | 404.7 | 52.4 | 349.2 | 51.6 | c | 52.1 | 336.0 |
| Pro Cδ | 48.3 | c | 48.3 | c | 48.3 | c | 48.4 | 66.0 | 48.3 | 47.3 |
| Gly Cα | 42.9 | c | 43.1 | c | 43.2 | c | 43.3 | c | 42.9 | c |
| Lys C + Arg Cδ | 41.0 | 115.4 | 41.1 | 39.7 | 41.2 | 33.1 | 41.3 | 32.2 | 40.9 | 30.4 |
| Asp Cβ + Asn Cβ + Leu Cβ | 39.7 | 36.3 | 39.9 | 33.9 | 40.0 | 32.8 | 40.0 | 46.3 | 39.7 | 29.2 |
| Phe Cβ + Ile Cβ | 36.7 | c | 36.6 | c | 36.8 | c | 36.8 | 317.6 | 36.2 | |
| Glu Cγ | 33.9 | 91.2 | 34.0 | 57.1 | 34.1 | 35.5 | 34.2 | 29.9 | 33.9 | 91.2 |
| Gln Cγ | 31.4 | c | 31.6 | c | 31.6 | 91.3 | 31.7 | 92.5 | 31.4 | 126.0 |
| Lys Cβ + Val Cβ | 30.7 | c | 30.8 | c | 30.8 | c | 30.9 | 297.8 | 30.7 | 299.5 |
| Pro Cβ | 29.7 | c | 29.7 | c | 29.7 | c | 29.7 | 308.8 | 29.7 | 307.2 |
| Arg Cβ + Glu Cβ | 27.9 | 116.8 | 28.4 | 50.2 | 28.3 | 111.2 | 28.0 | 123.8 | 28.0 | 107.5 |
| Lys Cβ | 27.2 | c | f | | f | | 27.4 | 287.5 | 27.2 | 133.1 |
| Lys Cδ + Gln Cβ | 26.7 | 135.0 | 26.6 | 133.3 | 26.8 | 133.6 | 26.8 | 177.9 | 26.6 | 121.2 |
| Arg Cγ + Leu Cγ + Pro Cγ + Ile Cγ₁ + GlcNAc methyl | 25.0 | c | 24.9 | 72.8 | 24.9 | 53.6 | 25.0 | 37.3 | 24.9 | 60.5 |
| Lys Cγ | 22.3 | c | 22.4 | c | 22.4 | 508.0 | 22.5 | 44.3 | 22.3 | 35.3 |
| Leu 2Cδ | 21.5[g] | c | f | | 21.4 | c | 21.6 | c | 21.2 | c |
| Thr Cγ | 19.2[g] | c | 19.3 | c | 19.3 | c | 19.0 | 92.4 | 18.8 | c |
| Val Cγ | 18.8[g] | 18.9 | f | | 18.9 | c | 18.4 | 91.5 | 18.2 | c |
| Ala Cβ | 17.2[g] | c | 17.1 | c | 17.3 | c | 17.3 | 48.4 | 17.0 | c |
| Ile Cγ₂ | 15.6[g] | c | 15.6 | c | 15.7 | c | 15.5 | 39.0 | 15.1 | c |
| Ile Cδ | 10.7[g] | c | 10.8 | 36.2 | 10.7 | c | 10.7 | 56.8 | 10.5 | 31.0 |

[a]Chemical shift in parts per million, acetonitrile as reference at 1.300 ppm; [b]Bandwidth at half height in Hz; [c]Bandwidth calculation unreliable; [d]Merged with preceding peak; [e]Peak intensity too low to permit assignment; [f]Cannot be assigned separately from surrounding peak envelope; [g]Peak assigned (Fisher, et.al., 1989a) via the APT pulse sequence (Patt & Shoolery, 1983).

Glycinin is a hexameric protein with an approximate molecular weight of 360,000. Each of its six subunits consists of a pair of polypeptide chains, made up of an "acidic" and a "basic" polypeptide, connected by one or more disulfide linkages (Nielsen, 1985a, 1985b). Glycinin's secondary structure is roughly 20% $\alpha$-helix, 17% $\beta$-structure and 63% random coil (Ishino and Kudo, 1980). $\beta$-Conglycinin is a trimeric glycoprotein of molecular weight 180,000. It is reported to be 20% $\alpha$-helix, 23% $\beta$-structure and 57% random coil (Ishino and Kudo, 1980), and to contain about 5% carbohydrate (Koshiyama, 1968), composed of mannose and N-acetylglucosamine (Yamauchi and Yamagishi, 1979).

## Structure of Native Proteins at 25°C

Although glycinin and $\beta$-conglycinin have large regions of random coil structure (Ishino and Kudo, 1980), neither the one previous [13]C-NMR study (Kakalis and Baianu, 1985) of glycinin nor a recent NMR study of soy protein isolates at high pH (Baianu, 1989) led us to expect that their [13]C-NMR spectra should have as many well-defined, narrow lines at pH 7 as can be seen in Fig. 1. Convolution difference (Fig. 2) and J-modulated double-spin-echo (APT, Fig. 3) spectra confirmed that these peak were in positions very similar to those found for free or "peptide shifted" amino acids (Howarth and Lilley, 1978; Wüthrich, 1976). Consequently, most of the peaks for both the major and minor components of glycinin and $\beta$-conglycinin have been assigned (Table 2), and spin-lattice relaxation times ($T_1$'s) have been determined for many of these peaks in glycinin (Table 3). It was found that previous assignments (Kakalis and Baianu, 1985) of peaks were in error by as much as 5 ppm. Due to low concentrations of tryptophan in glycinin (Table 1), only a few peaks from it could be assigned (Table 2); the even lower concentration in $\beta$-conglycinin precluded assignment of any tryptophan peaks. Similar results were observed for methionine and cysteine/cystine. The peak Kakalis and Baianu (1985; Baianu, 1989) report at 160-63 ppm does not appear in the spectra in this study (Figs. 1, 2 and 3). However, peaks at 157 pm and 155 ppm have been identified as arising from the guanidino carbon (C$\zeta$) of arginine and $\zeta$-carbon of tyrosine, respectively. APT experiments (Fig. 3) were instrumental in differentiating between methyl, methylene and methine carbons in the 10 to 100 ppm region. In particular, the C$\beta$ carbons of valine and isoleucine were assigned as the only CH peaks between 30 and 40 ppm (see Table 2).

This predominance of sharp peaks from a large number of amino acids in glycinin and $\beta$-conglycinin (Fig. 1) contrasts with spectra obtained for proteins from corn (Augustine and Baianu, 1986), wheat (Baianu et al., 1982) and barley (Tatham et al., 1985) whose primary structures are dominated by only a few amino acids. The broader distribution of different amino acids

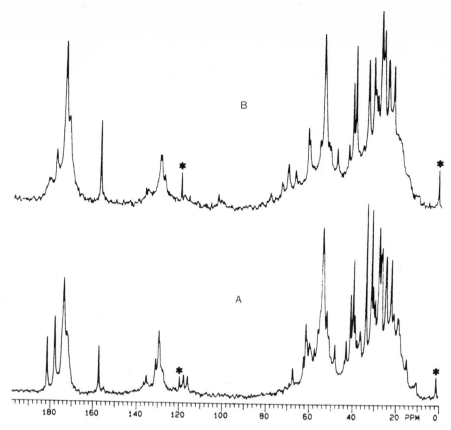

Fig. 1. Normal $^{13}$C-NMR spectra of soy globulins obtained at 25°C in 0.035 M phosphate buffer, pH 7.0, containing 0.4 M NaCl. Recycle time 0.6 sec., 192,000 transients. A: Glycinin; B: β-Conglycinin. Arrows denote peaks from the internal standard.

in glycinin and β-conglycinin (Table 2; Argos et al., 1985; Nielsen, 1985) necessitated long data acquisitions in order to detect and identify minor components. Long data accumulation times also made the APT sequence the method of choice to determine peak multiplicity. APT is not dependent on polarization transfer, which can cause baseline distortions and which requires $T_1$-related delays between pulses which lengthen recycle and data acquisition times. Also unlike polarization transfer techniques, APT does not require an initial 90° pulse, which causes emphasis of regions with short $T_1$'s and deemphasis of regions with long $T_1$'s. Convolution difference weighting, as in Fig. 2, accomplishes this task, emphasizing narrow peaks and suppressing broad ones with minimal loss of signal-to-noise. It can be applied to APT spectra to help in differentiating between close peaks. APT is not limited by decoupler strength or frequency, and $T_2$ line-broadening is reduced with composite 180° pulses.

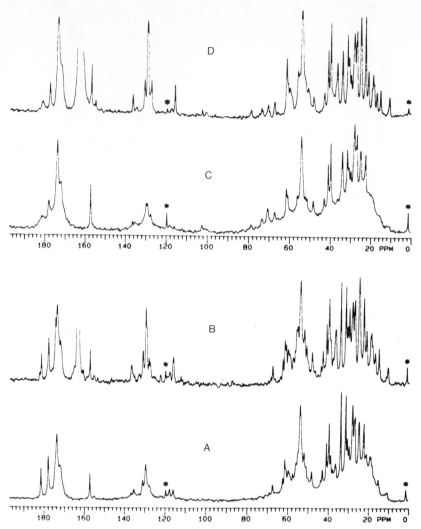

Fig. 2. Convolution difference spectra of soy globulins from Fig. 1, weighted with the function [1-{0.8 exp(-t/0.08)}]. Obtained at 25°C in 0.035 M phosphate buffer, pH 7.0, containing 0.4 M NaCl). Recycle time 0.6 sec., 192,000 transients. A: Glycinin; B: β–Conglycinin. Arrows denote peaks from the internal standard.

The $^{13}$C–NMR spectra of glycinin and β–conglycinin (Fig. 1) show areas (envelopes) where the apparent baseline deviates significantly from the true spectral baseline. These areas contain large numbers of overlapping peaks whose slight differences in chemical shift are due to their location in structured regions (whether helical, β–structure or random/disordered) within the proteins. This effect is most easily seen in the aliphatic region (10–40 ppm) of β–conglycinin (Fig. 1). Structured regions also cause lines to be broadened by spin–spin ($T_2$)–dominated relaxation effects. The $T_1$ and $T_2$ effects are the result of restrictions on sidechain and backbone

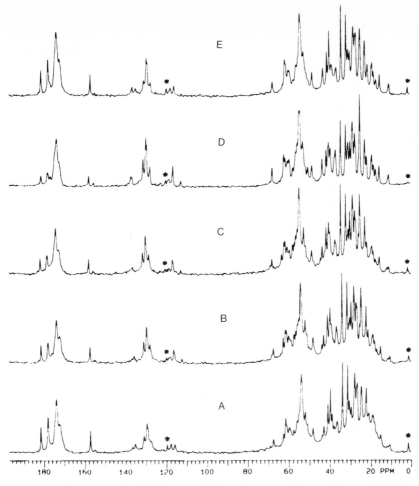

Fig. 3. J-modulated double spin-echo (APT; Patt & Shoolery, 1982)
spectra of soy globulins obtained at 25°C in 0.035 M phosphate
buffer, pH 7.0, containing 0.4 M NaCl. Recycle time 0.6 sec.,
192,000 transients. A: Glycinin; B: β-Conglycinin. Arrows
denote peaks from the internal standard. Upward peaks are
from methylene and quaternary carbons; downward peaks denote
methyl and methine carbons.

motion imposed by the internal and external interactions referred to col-
lectively as "structure." Under conditions of large amounts of structure,
bandwidths become directly related to protein tumbling times in solution
($\tau_r$) rather than sidechain/segmental motions ($\tau_c$). For large proteins such
as glycinin and β-conglycinin, the result is broad peaks. Thus, in general,
the regions within the protein which are highly structured yield spectra
that are difficult to assign, even using pulse sequences such as APT,
leaving qualitative gross structural information available from the apparent
envelope areas. Conversely, the relaxation of nuclei in poorly structured
and/or mobile regions in proteins is dominated by spin-lattice interactions. This plus an accompanying increased nuclear Overhauser enhancement

(nOe) results in a peak that is both narrower and more intense in the unstructured and mobile regions of a protein than the peak for the same carbon in non-random or structured regions (Levy et al., 1980; Jardetzky and Roberts, 1981; Wüthrich, 1976), and therefore relatively easy to assign. Such peaks provide information on the differences and similarities in microstructure and composition between glycinin and β-conglycinin, but only limited amounts of conformational information. Unfortunately, the inter-dependence of frequency, correlation time, bandwidth, spin-lattice relaxation time ($T_1$) and spin-spin relaxation time ($T_2$) limit the usefulness of carbon relaxation measurements when applied to protein dynamics.

In Fig. 1, the aliphatic (10-45 ppm), α-carbon (50-60 ppm) and backbone carbonyl (165-175 ppm) envelopes are all visibly larger for β-conglycinin than for glycinin, denoting regions involving these sidechains that are more structured and motion-restricted in β-conglycinin than in glycinin. This can also be seen in the large number of β-conglycinin peak bandwidths (Table 2) that are unreliable due to the broadness of the peak and inability of the instrument to find an accurate baseline. Convolution difference weighting (Fig. 2) provided only incomplete emphasis of these peaks, especially in the aliphatic region of β-conglycinin, emphasizing how wide these lines are. In these regions, peak assignments were only possible through use of the APT pulse sequence (Fig. 3).

The APT sequence also gave data sufficient to discriminate between close peaks with differing numbers of hydrogens (Fig. 3). Thus the Ser Cβ $CH_2$ at 61.4 ppm (Table 2) could be differentiated from the Pro Cα CH at 60.9 ppm. Similarly, the two CH's from Ile Cβ and Val Cβ (36.9 ppm and 30.7 ppm, respectively, Table 2) can be separated from the surrounding $CH_2$'s, particularly in glycinin (Fig. 3A). In addition, the Trp Cε CH at 118 ppm can be seen clearly in glycinin, and not in β-conglycinin, reflecting amino acid composition differences between the two proteins (Table 1).

The most noticeable similarity between glycinin and β-conglycinin spectra (Fig. 1) is the narrowness (Table 2) of the dominant upfield (21-62 ppm) peaks and the backbone carbonyl envelope (174 ppm), emphasized in the convolution difference spectra in Fig. 2. The sharpness of these peaks and the similarity of their chemical shifts to those of free amino acids are reflections of the large regions of disordered structure in the polypeptide backbone and evidence of its flexibility. These dominant peaks have been identified as arising from the sidechain carbons of glutamine, glutamate, arginine and lysine (Table 2). The flexibility of the protein chain is underscored by the strong, comparatively narrow peaks encompassing the backbone α-carbons and carbonyls around 54 and 174 ppm, respectively (Figs. 1 and 2), and by $T_1$ values (Table 3) intermediate between those observed for free amino acids (Wüthrich, 1976) and highly motion-constrained amino acid residues (Komoroski et al., 1976; Jardetzky and Roberts, 1981). This can also be seen in the amount of fine structure around 54 and 174 ppm that emerges in convolution difference weighted spectra (Fig. 2). The similarities in these regions of the two proteins are such that the general conclusions drawn from the spin-lattice relaxation ($T_1$) measurements (Table 3) for these peaks in glycinin are applicable to β-conglycinin.

There are fewer emphasized peaks around the primary Cα peak (54 ppm) in the convolution difference spectrum of β-conglycinin (Fig. 2B) than in the same region of the glycinin spectrum (Fig. 2A). The β-conglycinin peaks between 10 and 22 ppm are poorly resolved even after convolution difference weighting. These two observations point to fewer mobile α-carbons in β-conglycinin, and indicate not only a higher proportion of structured regions in β-conglycinin, but more importantly, a higher proportion of flexible regions in glycinin. This goes beyond the spectroscopic observations (Dev et al., 1988; Ishino and Kudo, 1980; Yamauchi et al., 1979; Koshiyama and Fukushima, 1973) of the amounts of helical, β-structure and random coil in purified soy storage proteins; these methods are unable to distinguish flexible regions.

Of particular note in the spectra of both glycinin and β-conglycinin (Fig. 1) is the narrow, solitary peak at 157 ppm from the ζ-carbon of arginine (Table 2). Its narrowness (~20Hz), solitary nature, and height suggest both exposure to solvent and location in non-rigid regions of both proteins. This is supported by its $T_1$ (Table 3) between the $T_1$'s anticipated for the free amino acid and motion-constrained residue (Levy et al., 1980; Komoroski et al., 1976; Jardetzky and Roberts, 1981). Sidechain flexibility is also seen in the narrow linewidths of the remaining arginine sidechain peaks at 25, 28 and 41 ppm (Fig. 1 and Table 2).

The δ-carbon of glutamate in β-conglycinin (Fig. 1B, 182 ppm) is sharply different from the δ-carbon of glutamate in glycinin (Fig. 1A, 182 ppm). The δ-carboxyl region in β-conglycinin is unintense and broad (Table 2). It remains so after convolution difference weighting (Fig. 2), losing intensity and gaining noise. The most likely explanation is that the carboxyl is involved in inter- or intra-chain hydrogen bonding at several sites. Furthermore, the β- and γ-carbon peaks from the glutamate sidechain (at 28 and 34 ppm) are wider in β-conglycinin than the equivalent peaks in glycinin (see Table 2). This is consistent with both the carboxyl group and the sidechain being partially motion constrained in β-conglycinin. Glutamine (178 ppm) shows similar effects (Table 2). None of these effects are seen for glutamate, aspartate or glutamine in glycinin. This is particularly important structurally, as glutamate and glutamine are understood to prefer α-helical regions (Levitt, 1978), and not to be buried in proteins (Chothia, 1976).

It can be inferred that glutamate residues contribute to helical and other structure in β-conglycinin, probably via hydrogen bonding to the carbohydrate moiety and/or ionic interactions with positively charged side-chains. While peaks from carbons near the positive charge of deprotonated histidine are inseparable from neighboring peaks (see His[+] Cε, 128.5 ppm in Table 2) or too low and/or broad (see His[+] Cδ, 118.9 ppm in Table 2), the Cε of lysine (41.0 ppm in Table 2) appears to be as mobile or more so than its counterpart in glycinin. Similarly, the Cζ and Cδ peaks of arginine (157.1 and 41.0 ppm in Table 2) are narrow and well-defined as is the Cβ peak from serine (61.5 ppm). Thus, the interaction of glutamate with positively charged residues is speculative at best, with histidine being the only apparent possibility.

The presence of a carbohydrate sidechain in β-conglycinin is readily observed from the peaks between 70 and 110 ppm in Fig. 1B (see also Table 2). Both carbohydrate and amino acid carbons contribute to peaks between 60 and 70 ppm, as can be seen (Table 2) by the presence of these peaks in glycinin. The extension of the α-carbon envelope of β-conglycinin (centered at 54 ppm) to include the carbohydrate peaks (Fig. 1, 65-79 ppm; Table 2) does indicated that the carbohydrate sidechain is locally motion-constrained, consistent with involvement in secondary structure, and make it the most likely site for interaction with glutamate. It should be noted that β-conglycinin is 5% carbohydrate, consisting of seven to nine mannose residues and two N-acetyl glucosamines (Yamauchi and Yamagishi, 1979; Koshiyama, 1968), each with two to four hydroxyl groups available for hydrogen bonding. The absence of carbohydrate from glycinin may therefore explain its observed larger proportion of flexible regions and, in part, its larger proportion of disordered structure.

It is obvious from the large aliphatic and aromatic peak envelopes observed for β-conglycinin (Fig. 1B) that glutamate is not the sole contributor to the structurally ordered regions of β-conglycinin. The large shoulder on this envelope (15-~20 ppm), broadness of the peak at 10.5 ppm and comparison with the equivalent region of glycinin (Fig. 1A) make it apparent that alanine, valine, isoleucine and leucine are also involved. In glycinin, the Arg Cζ peak is accompanied by a small, single peak at 155 ppm for the Tyr Cζ, with other Tyr peaks readily assignable in the aromatic region (120-140 ppm). In β-conglycinin these peaks are either difficult to discern or, as in the case of the Tyr Cζ peak, absent (Fig. 1B). Since the content of tyrosine in these two proteins is very similar (Table 1), this points to the presence of Tyr in motion-constrained regions, most likely in the protein interior. The poorly-resolved aromatic peaks around 130 ppm, plus the presence of a baseline-distorting peak envelope and absence of identifiable tyrosine Cζ peak, imply more involvement of tyrosine and phenylalanine in the structure of β–conglycinin than in that of glycinin (Fig. 1A).

From the peaks at 61, 48, 30 and 25 ppm in glycinin and β–conglycinin (Fig. 1; Table 2), it appears that proline (5 to 6% of the total amino acid content) occurs in the trans configuration in both proteins (Howarth and Lilley, 1978; Wüthrich, 1976). This may be important to understanding the high degree of random coil structure within these proteins, as proline tends to occur in reverse turns and break up α–helical regions (Levitt, 1978). It is also not usually buried in the interior of proteins (Chothia, 1976). This point is supported by Dev et al. (1988) who interpreted their FT–IR results in terms of large amounts of β–sheet, β–turn and disordered structure.

## Influence of Chemical Denaturation

In the presence of a strong denaturant such as 6 M urea, two phenomena occur. First, there is a thorough unfolding of the secondary structure of soy storage proteins to a poorly ordered conformation (Lillford, 1978), decreasing the number of dissimilar internal environments versus that observed in the native protein. Second, the proteins dissociate into sub-units; glycinin into six disulfide–linked, acidic/basic polypeptide pairs and β–conglycinin into three monomeric subunits (Nielsen, 1985). These changes produce a large increase in local motion of both the protein back-bone and amino acid sidechains and a decrease in correlation time ($\tau_c$), resulting in an increase in $T_1$, a larger increase in $T_2$, and, therefore, a decrease in the $T_1/T_2$ ratio. The observable results are narrower peaks, smaller peak envelopes, an increase in signal intensity and peak definition, and a spectrum that more closely resembles a mixture of free amino acids than does the native protein. These changes can be seen in Fig. 4 and Table 4, which summarize chemical shift and bandwidth data for glycinin and β–conglycinin prior to and after denaturation with 6 M urea.

The presence of urea had the anticipated effect on the $^{13}$C NMR spectrum of glycinin as seen in Table 4 and by comparing Fig. 4A to Fig. 4B. Several peaks were observed (Fig. 4B) that had previously been unobservable due to their broadness and/or presence under a peak envelope (Fig. 4A). Several such "new" peaks appeared in the aliphatic region (10–50 ppm) and in the 50–65 ppm region which includes resonances from the peptide backbone carbons (α–carbons). This is emphasized by the appearance in denatured glycinin of the peak at 17.2 ppm, assigned to the β–carbon of alanine. This peak appears in native glycinin (Fig. 4A) as a shoulder which could only be accurately observed and assigned with the APT sequence (Fisher et al., 1989a). After denaturation with 6 M urea (Fig. 4B), it can be clearly observed, although the bandwidth cannot be reliably calculated (Table 4).

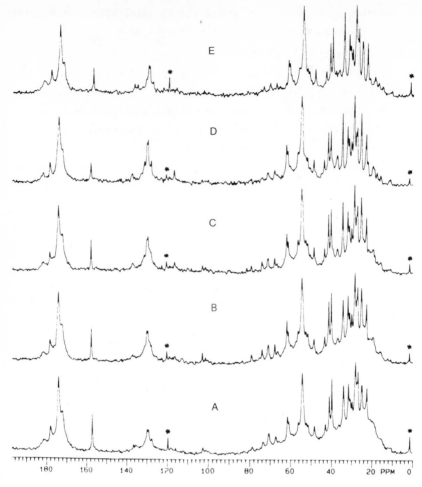

Fig. 4.   Carbon-13 NMR spectrum at 25°C of A:  glycinin in pH 7.0,
0.035 M phosphate buffer containing 0.4 NaCl, 192,000
transients; B:  glycinin denatured in pH 7.0, 0.035 M
phosphate buffer containing 0.4 M NaCl and 6.0 M urea,
48,000 transients; C: β-conglycinin, in pH 7.0, 0.035 M
phosphate buffer containing 0.4 M NaCl, 144,000 transients;
D: β-conglycinin denatured in 6.0 M urea, pH 7.0,
containing 0.035 M phosphate and 0.4 M NaCl, 48,000
transients.   Large peak in (B) and (D) at 160 ppm is urea.
Arrows at 1.3 and 119.6 ppm denote peaks from the
acetonitrile internal reference.

On denaturation, the aromatic region (115-140 ppm) of both glycinin and
β-conglycinin spectra, associated with carbons of phenylalanine, tyrosine and
histidine, show similar changes from the native materials (see Fig. 4).   In
both spectra, the peaks around 130 ppm, assigned to phenylalanine, become
taller and narrower relative to the remainder of the peaks in the spectrum.
This suggests the Phe is far more mobile in the denatured proteins.   This

218

phenomenon was also observed by Dev et al. (1988) using FT-IR. A similar effect is seen in the appearance of the Tyr C$\zeta$ peak at 155 ppm in denatured β-conglycinin, and various tryptophan peaks (112, 118, 119, 122 ppm) in glycinin.

The concentration of tryptophan in glycinin is too low for more than a few protonated carbons in it to be assigned under the experimental conditions used. Also because of low concentrations, methionine and cysteine/ cystine peaks could not be identified as distinct peaks, though they contribute to bands containing other carbons (Table 4). A small number of new peaks were seen in the carbonyl region (165-185 ppm).

Of note in this region is the appearance of a second peak in the area assigned to the glutamate sidechain carboxyl (182 ppm). This new peak can be explained by changes in the involvement of glutamate interactions in the denatured protein, e.g., hydrophilic interactions with hydroxylated side-chains or electrostatic interactions with positively charged sidechains. Many of the existing glycinin peaks (Fig. 4A) became sharper (Fig. 4B) and narrower (Table 4), due to decreased $T_1/T_2$ ratios upon denaturation.

Urea had a similar, though much more pronounced, effect on the β-conglycinin spectrum (compare Fig. 4C with Fig. 4D). The region most affected by chemical denaturation of β-conglycinin was the aliphatic hydro-carbon region (10-45 ppm) which contains the methyl ($-CH_3$) and methylene ($-CH_2-$) groups arising primarily from valine, leucine, isoleucine, alanine and proline. These functional groups readily form stable hydrophobic interactions in the interior of proteins through the exclusion of water, restricting their mobility and resulting in $T_2$-broadened, overlapping peaks that form large envelopes (Fisher et al., 1989a, 1989b) in native β-conglycinin (Fig. 4C). In the denatured protein (Fig. 4D), these interactions are disrupted, producing the effect described above. This effect can also be seen between 60 and 80 ppm, where the apparent extension of the α-carbon peak envelope in Fig. 4C collapses almost to the baseline in Fig. 4D. The peaks in this region are from hydroxyl-containing amino acid sidechains (Serine Cα near 61 ppm, Threonine Cβ near 67 ppm) and the carbo-hydrate moiety (Table 4). These changes in the carbohydrate ring carbon' peaks (61-80 ppm) strongly suggest the involvement of the carbohydrate moiety in β-conglycinin's native secondary structure, supporting observations made on the native proteins themselves (Fisher et al., 1989a).

Very few new peaks were observed in the other regions of the β-conglycinin spectrum (compare Figs. 4C and 4D), although many existing peaks became narrower (Table 4) and more intense (Fig. 4D). This is particularly noticeable for the β-conglycinin tyrosine C$\zeta$ at 155 ppm (Fig. 4D), which cannot be observed in the native (Fig. 4C) protein. The glutamate sidechain carboxyl peak at 182 ppm is of note in that even in the denatured protein it remains broad. This implies that the interactions involving the glutamate carboxyl are strong enough to survive the perturbations of chemical denaturation. The differences in the native and denatured spectra of both glycinin and β-conglycinin make it likely that glutamate is involved in β-conglycinin structure and unlikely that it is involved in glycinin structure.

Major differences between the apparent sizes of the peak envelopes in the spectra of both native and urea-denatured glycinin and β-conglycinin were seen (Fig. 4). Denaturation of glycinin produced a comparatively small reduction in the size of the peak envelopes (Figs. 4A and B), primarily in the aliphatic region (10-45 ppm). By contrast, denaturation of β-conglycinin, caused a large reduction in envelope size in every region of the spectrum (Figs. 4C and D). These differences can also be seen in bandwidth changes (Table 4). Disruption of hydrogen bonds and hydrophobic interactions by urea had some effect on glycinin denaturation, but the presence of an observable peak envelope implies partial retention of structure. A significant role in partial retention of structure by glycinin must be played by disulfide bonds, which are unaffected by urea (Nielsen, 1985). The disulfide bonds in glycinin cause regions of each pair of polypeptide chains to remain in close proximity, even on chemical denaturation, allowing interchain interactions that can be observed as an aliphatic peak envelope that is larger in denatured glycinin than in the equivalent region of the β-conglycinin spectrum (compare Figs. 4B and 4D). The retention of structure can also be seen in Table 4 as bandwidths that in many cases decreased only slightly (e.g., Gln Cγ, 31.4 ppm and Lys Cβ, 27.1 ppm) and in some cases increased (e.g., Glu Cγ, 33.9 ppm and Arg Cβ/Glu Cβ, 28 ppm) on denaturation. There is little or no cysteine found in β-conglycinin (Coates et al., 1985), and no disulfide bonds (Thanh and Shibasaki, 1978). Thus, it should be more susceptible to urea denaturation than should glycinin. That is clearly the case from our results (Fig. 4). The amino acid residues contributing the most to β-conglycinin conformation appear to come from the aliphatic (10-45 ppm) and aromatic (115-140 ppm) regions, though the carbohydrate peak envelope (65-80 ppm) also decreases in area, implying involvement of that moiety in β-conglycinin structure as well. It should also be noted that, for both proteins, the amino acid and (in β-conglycinin) carbohydrate chemical shifts (60-105 ppm) changed very little when the proteins were chemically denatured, confirming that the sharp peaks observed in the native proteins were largely from mobile groups in random regions.

## Influence of Heating

When glycinin and β-conglycinin were heated above room temperature, chemical shift, bandwidth and lineshape of the dominant peaks changed only slightly with increasing temperature (Tables 5 and 6; Figs. 5B and C and 6B and C), supporting the occurrence in both native proteins of large unstructured regions (Ishino and Kudo, 1980) that are not greatly affected by temperatures at or below the gelation point (Suresh Chandra et al., 1984). In both proteins, the dominant peaks in the aliphatic and aromatic regions, which belong to amino acids with charged sidegroups, have bandwidths in the native proteins at 25°C that are very similar to those in the heat-treated, partially unfolded proteins (Tables 5 and 6). Thus, the amino acids associated with the dominant peaks are not greatly motion constrained, consistent with these amino acid sidechains being located on the exterior of the protein.

In both glycinin and β-conglycinin, aliphatic and aromatic interactions appear to be important to the maintenance of structure. Thus, the changes in peak shape and definition during heating is directly related to structural changes. Peaks in the aliphatic (10 to 45 ppm) and aromatic (115 to 140 ppm) regions became sharper and fine structure emerged in the 10-20 ppm envelope of the β-conglycinin spectrum (Figs. 5 and 6) as the temperature was increased. The aliphatic and α-carbon envelopes in both protein spectra (Figs. 5A and 6A) decreased in area with increasing temperature, particularly at 95°C for glycinin (Fig. 5D) and 85°C for β-conglycinin (Fig. 6D). This indicates that the aliphatic sidechains and α-carbons are less dominated by $T_2$ linebroadening (Levy et al., 1980; Jardetzky and Roberts, 1981; Wüthrich, 1976).

The probable cause of these spectral changes is twofold. At the lowest temperatures above 25°C, the polypeptide chains unfold. At higher temperatures, the partly unfolded proteins undergo inter-subunit and/or inter-molecular interactions to form soluble aggregates (Hermansson, 1986; Mori et al., 1986, Nakamura et al., 1986a, 1986b). Glycinin, when heated to 95°C, and β-conglycinin heated to 85°C are referred to as being in a pre-gel state, characterized (Hermansson, 1986; Mori et al., 1986, Nakamura et al., 1986a, 1986b) by the conversion of the protein aggregates formed at temperatures below the gelation temperature to interconnected strands.

These stages are particularly easy to see in β-conglycinin. At room temperature (Fig. 6A), the β-conglycinin methyl peaks (10-21 ppm) appear as blunt shoulders on the aliphatic peak envelope (10-80 ppm). At 55°C (Fig. 6B), these shoulders become flat-topped "shelves," which persist at 70°C (Fig. 6C). At 85°C, the β-conglycinin gelation temperature, these shelves have separated into distinct peaks. A parallel change in the aromatic peaks can also be seen. At the same time, the carbohydrate peaks around 70 ppm have gone from being distinct at 55°C and 70°C to being broad and ill-defined at 85°C.

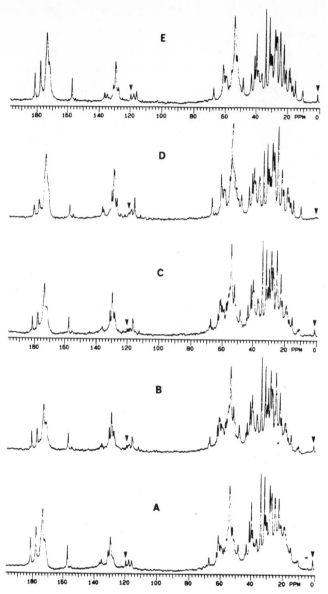

Fig 5.   Carbon-13 NMR spectra in pH 7.0, 0.035 M phosphate buffer containing 0.4 M NaCl of A:   glycinin at 25°C, 192,000 transients;
B:   70°C; C:   85°C; D:   95°C (pre-gel aggregate); E:   glycinin
gel at 25°C.   Spectra (B), (C), (D), and (E) acquired with
144,000 transients.   Arrows at 1.3 and 119.6 ppm denote peaks
from the acetonitrile internal standard.

Equivalent changes in spectral characteristics occur in glycinin as
well, but are much less obvious.   As with β-conglycinin, the glycinin methyl
region undergoes a marked change when heated from 25°C (Fig. 5A) to 55°C
(Fig. 5B).   Little or no change occurs on heating to 70°C (Fig. 6C).   At 95°
(Fig. 5D), methyl peak definition increases, the peaks around 130 ppm become
sharper and those around 60 ppm become fewer.

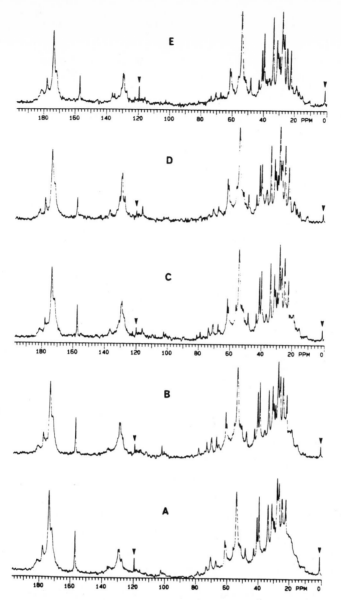

Fig. 6. Carbon-13 NMR spectra in pH 7.0, 0.035 M phosphate buffer
containing 0.4 M NaCl of A: β-conglycinin at 25°C, 192,000
transients; B: 55°C; C: 70°C; D: 85°C (pre-gel
aggregate); E: β-conglycinin gel at 25°C. Spectra (B),
(C), (D), and (E) acquired with 144,000 transients. Arrows
at 1.3 and 119.6 ppm denote peaks from the acetonitrile
internal standard.

The spectra of chemically (Figs. 4B and 4D) and heat-denatured glycinin
(Fig. 5D) and β-conglycinin (Fig. 6D) differ markedly from one another. In
particular, the degree of denaturation, apparent in the number of transients
required to obtain equivalent signal-to-noise ratios (see captions to Figs.
4, 5, and 6), is far larger when 6 M urea is used than when the proteins are

223

heated to their gelation temperatures.  This strongly suggests, as FT-IR results did to Dev et al. (1988), that the configurational changes during chemical and heat denaturation are very different, and probably result from the retention of interactions during the change from native protein to aggregate to strand to gel during heating that are eliminated by chemical denaturation.

## Influence of Gel Formation

Glycinin heated to 95°C and β–conglycinin heated to 85° gave firm gels on cooling to 4°C.  Protein samples heated to lower temperatures did not form gels on cooling.  On gelation, both proteins give spectra that more closely resemble the spectra obtained at temperatures above 55°C than those obtained for either the native or chemically denatured proteins.  Only one possibility can account for these differences:  intermolecular structure having intermolecular regions that are highly random and/or mobile.  Structural regularity has been observed in β–conglycinin gels and glycinin aggregates and gels at room temperature by electron microscopy (Hermansson, 1985, 1986; Mori et al., 1986), though the nature of the technique precludes observations at elevated temperatures.  The work of these investigators and others (Nakamura et al., 1986a, 1986b; Utsumi and Kinsella, 1985), and the sharp differences between the spectra of chemically denatured (Figs. 4B and 4D), and pre-gelstate (Figs. 5D and 6D) and gelstate (Figs. 5E and 6E) glycinin and β–conglycinin support this conclusion.  In the pre-gel state (Figs. 5D and 6D), the spectra of both glycinin and β–conglycinin show peaks throughout the aliphatic and aromatic regions that are more intense and have narrower linewidths than in spectra obtained either before  (Figs. 5B – 5D and 6B – 6D) after gelation (Figs. 5E and 6E).  These peaks decreased in intensity and increased slightly in linewidth upon gel formation (Figs. 5E and 6E), indicating a loss of mobility (but not necessarily randomness) accompanied by shortening of $T_1$ and loss of nOe (Levy et al., 1980; Jardetzky and Roberts, 1981; Wüthrich, 1976).

As in formation of the pre-gel state, this is most obvious for the methyl (10–21 ppm) peaks in both glycinin and β–conglycinin (Figs. 5D and E and 6D and E) and slightly less obvious in the aromatic region around 130 ppm.  Structural changes are also indicated by changes in the carbohydrate region of β–conglycinin (around 70 ppm) and around 60 ppm in glycinin.  These changes result from an increased intermolecular structural order upon cooling, involving aliphatic and aromatic sidechains, evidenced by the decrease in intensity and increase in linewidth of the aromatic and aliphatic sidechain peaks and by the increase in the size of the aliphatic peak envelopes.  These observations, plus those on urea–denatured glycinin and β–conglycinin, indicate that at the gelation temperatures of glycinin and β–conglycinin, loss of native secondary and tertiary structure by both proteins is only partial. On gelation, both proteins increase in their degrees of structure, but, judging from the behavior of the methyl peaks, do not recover native structure. The differences in methyl and aromatic regions between glycinin and β-conglycinin gels indicate that the methyl and aromatic groups in the glycinin gel are more mobile than those in the β-conglycinin gel, strongly implying that the β-conglycinin gel structure is more dense than that of glycinin.

Aggregation and gel formation effects can also be seen in the behavior of the guanidino carbon of arginine peak at 157 ppm in both glycinin and β-conglycinin during heating and cooling. This carbon has no attached protons to provide NOE or $T_1$ dipole-dipole effects and is unaffected by $T_2$ line-broadening. It therefore acts very much like a free amino acid and is relatively insensitive to conformational changes. At temperatures above 25°C in both proteins (Figs. 5 and 6), the peak decreased in both intensity and linewidth, consistent with an increase in $T_1$ resulting from increased freedom of movement in the sidechain and with increased intensity of other peaks resulting from the effects of protein unfolding. Upon gelation (Figs. 5E and 6E), the peak broadens and increases in intensity. This results from loss of mobility of the sidechain itself and loss of peak intensity elsewhere in the spectrum, both due to formation of interstrand structures. Similar effects can be seen (Figs. 5D and E and 6D and E) in the peaks resulting from the carboxyl groups of glutamate (182 ppm) and the carboxamide group of glutamine (178 ppm).

Unlike the aliphatic and aromatic peaks in β-conglycinin, the peaks in the carbohydrate region (60-110 ppm) broaden upon formation of the pre-gel (compare Figs. 6C and 6D), then exhibit an increase in envelope size upon formation of the gel (Fig. 6E). This suggests that at temperatures below that of the pre-gel state (Figs. 6B and C), these groups are in larger numbers of moderately mobile regions, whereas in both the pre-gel and gel these regions become more structured, with some regions sufficiently random and/or mobile to provide small, sharp peaks, much as observed for the chemically denatured protein (Fig. 4D). The most probable structural processes for the carbohydrate chain in the gel are intermolecular interactions involved in pre-gel strand and gel structure formation, most likely interactions with glutamate and other carbohydrate sidechains. This is supported by the shape of β-conglycinin glutamate carboxyl at 182 ppm. At all temperatures, it remains a small, broad peak (Table 6), consistent with restricted motion, probably due to involvement in intra- and/or inter-chain bonding. As pointed out for the urea-denatured protein, the fact that the peak shows little or no change in shape or linewidth during pre-gel or gel formation signifies its importance in stabilizing some regions of β-conglycinin structure in the face of external perturbations to the protein.

The spectra in Figs. 5 and 6 support the theory (Hermansson, 1986; Mori et al., 1986; Nakamura et al., 1986a, 1986b) of a three-step process in soy protein gelation, and augment previous information by supplying direct observations of glycinin and β-conglycinin at specific temperatures. Both proteins undergo partial unfolding between 25°C and 55°C. Small changes around 60 and 130 ppm suggest that glycinin aggregation begins at 70°C and that these aggregates interact to form strands at 95°C. In β-conglycinin, partial unfolding persists through 70°C and both aggregates and strands form at 85°C. Gel formation in both proteins is accompanied by increased restrictions on motion consistent with the coalescence of strands into the gel network. The changes in the methyl region of β-conglycinin imply that its gel is more dense than that of glycinin. In addition, our results support the involvement of the carbohydrate and glutamate sidechains in β-conglycinin structure and gel formation.

SUMMARY

Describing glycinin and β-conglycinin strictly in terms of percentage of structured regions is insufficient. Carbon-13 NMR shows that these two proteins differ sharply in their relative amounts of flexible regions, with glycinin having the higher proportion. The involvement of specific amino acids in both proteins and the carbohydrate moiety of β-conglycinin in protein structure can be seen. It seems likely that these differences contribute to differences in other physicochemical properties of these two proteins.

From the data obtained from carbon-13 NMR spectra of glycinin and β-conglycinin under conditions of gelation, heat and chemical denaturation, we conclude that much of their native structure depends on aliphatic and aromatic interactions. Hydrophilic interactions involving carbohydrate and glutamate are important to β-conglycinin secondary structure. These inter-actions are diminished upon heat and chemical denaturation, but reappear upon gelation, indicating that these interactions are important in formation of glycinin and β-conglycinin gels. Furthermore, from the large number of sharp peaks in all the spectra we conclude that both of the proteins in both native and gelled states have large amounts of relatively mobile structure--random coil and/or flexible β-sheet--consistent with other studies (Dev et al., 1988; Ishino and Kudo, 1980).

The involvement of covalent forces in maintaining gel structure in glycinin and glycinin gels cannot be directly studied by NMR due to the low cysteine content of glycinin. However, the spectra of chemically denatured glycinin lead us to conclude that covalent forces, such as disulfide bonds, also help maintain glycinin structure in part by keeping sections of the acidic and basic polypeptides in close proximity regardless of external perturbations.

ACKNOWLEDGMENTS

We wish to express our gratitude to Dr. Walter J. Wolf, USDA, ARS, Northern Regional Research Center for supplying us with purified β-conglycinin. We also wish to thank Drs. Charles A. Kingsbury of the University of Nebraska-Lincoln and William R. Croasmun of Kraft, Inc. for their critical evaluation of this manuscript. Presented in part at the 196th National Meeting of the American Chemical Society, Los Angeles, CA, September, 1988.

REFERENCES

Augustine, M. E., and Baianu, I. C., 1986, "High-resolution Carbon-13 Nuclear Magnetic Resonance Studies of Maize Proteins," J. Cereal Sci., 4:371-78.

Augustine, M. E., and Baianu, I. C., 1987, "Basic Studies of Corn Proteins for Improved Solubility and Future Utilization: A Physicochemical Approach," J. Food Sci., 52(3):649-52.

Badley, R. A., Atkinson, D., Hauser, H., Oldani, D., Green, J. P., and Stubbs, J. M., 1975, "The structure, physical and chemical properties of the soy bean protein glycinin," Biochim. Biophys. Acta., 412:214.

Baianu, I. C., 1981, "Carbon-13 and Proton Nuclear Magnetic Resonance Studies of Wheat Proteins. Spectral Assignments for Flanders Gliadins in Solution," J. Sci. Food Agric., 32:309-313.

Baianu, I. C., Johnson, L. F., and Waddell, D. K., 1982, "High-Resolution Proton, Carbon-13 and Nitrogen-15 Nuclear Magnetic Resonance Studies of Wheat Proteins at High Magnetic Fields: Spectral Assignments. Changes with Concentration and Heating Treatments of Flinor Gliadins in Solution--Comparison with Gluten Spectra," J. Sci. Food Agric., 33:373-383.

Baianu, I. C., 1989, High-Resolution NMR Studies of Food Proteins, in: "NMR in Agriculture," P. Pfeffer and W. Gerasimowcz, eds.; CRC Press, Cleveland, Ohio.

Baillargeon, M. W., Laskowski, Jr., M., Neves, D. E., Porubcan, M. A., Santini, R. E., and Markley, J. L., 1980, "Soybean trypsin inhibitor (Kunitz) and its complex with trypsin. Carbon-13 nuclear magnetic resonance studies of the reactive site arginine," Biochemistry, 19:5703-10.

Belton, P. S., Duce, S. L., and Tatham, A. S., 1987, "$^{13}$C solution state and solid state n.m.r. of wheat gluten," Int. J. Biol. Macromol., 9:357-63.

Chothia, C., 1976, "The nature of the accessible and buried surfaces in proteins," J. Mol. Biol., 105:1-14.

Coates, J. B., Medeiros, J. S., Thanh, V. H., and Nielsen, N. C., 1985, "Characterization of Subunits of β-Conglycinin," Arch. Biochem. Biophys., 243(1):184-194.

Coker, III, G. T., Garbow, J. R., and Schaefer, J., 1987, "Nitrogen-15 and carbon-13 NMR determination of methionine metabolism in developing soybean cotyledons," Plant Physiol., 83(3):698-702.

Dev, S. B., Keller, J. T., and Rha, C. K., 1988, "Secondary structure of 11 S globulin in aqueous solution investigated by FT-IR derivative spectroscopy," Biochem. Biophys. Acta, 957:272-280.

Fisher, M. S., Marshall, W. E., and Marshall, Jr., H. F., 1989a, "Carbon-13 NMR Studies of Glycinin and β-Conglycinin at Neutral pH," J. Agric. Food Chem., (in press).

Fisher, M. S., Marshall, W. E., and Marshall, Jr., H. F., 1989b, "Carbon-13 NMR Studies of the Effects of Gelation, Heat and Chemical Denaturation at Neutral pH of Glycinin and β-Conglycinin," J. Agric. Food Chem., (in press).

Hermansson, A-M., 1978, "Physico-chemical aspects of soy protein structures formation," J. Texture Stud., 9:33-58.

Hermansson, A-M., 1985, "Structure of Soya Glycinin and Conglycinin Gels," J. Sci Food Agric., 36:822-832.

Hermansson, A-M., 1986, "Soy Protein Gelation," J. Am. Oil Chem. Soc., 63(5):658-666.

Howarth, O. W., and Lilley, D. M. J., 1978, "Carbon-13-NMR of Peptides and Proteins," Prog. Nucl. Magn. Reson. Spectros., 12:1-40.

Ishino, K., and Kudo, S., 1980, "Conformational Transition of Alkali-Denatured Soybean 7S and 11S Globulins by Ethanol," Agric. Biol. Chem., 44(3):537-543.

Jardetzky, O., and Roberts, G. C. K., 1981, Protein Dynamics, in: "NMR in Molecular Biology," O. Jardetzky, and G. C. K. Roberts, eds.; Academic Press, New York.

Kakalis, L. T., and Baianu, I. C., 1985, "Carbon-13 NMR study of soy protein conformations in solution," Federation Proceedings, 44:1855, 1807.

Kakalis, L. T., and Baianu, I. C., 1989, "High resolution carbon-13 NMR studies of soy protein isolates". Referred to as submitted for publication in Baianu, I.C., 1989, High-Resolution NMR Studies of Food Proteins, in: "NMR in Agriculture," P. Pfeffer, and W. Gerasimowicz, eds.; CRC Press, Cleveland, Ohio.

Kitamura, K., and Shibasaki, K., 1975, "Isolation and Some Physico-chemical Properties of the Acidic Subunits of Soybean 11S Globulin," Agric. Biol. Chem., 39:945.

Komoroski, R. A., Peat, I. R. and Levy, G. C., 1976, [13]C NMR Studies of Biopolymers, in: "Topics in Carbon-13 NMR Spectroscopy," v. 2; G. C. Levy, ed.; John Wiley and Sons, New York.

Koshiyama, I., 1968, "Storage Proteins of Soybean," Cereal Chem., 45:394.

Koshiyama, I., and Fukushima, D., 1973, "Comparison of Conformations of 7S and 11S Soybean Globulins by Optical Rotatory Dispersion and Circular Dichroism Studies," Cereal Chem., 50:114-121.

Koshiyama, I., Hamano, M., and Fukushima, D., 1981, "A Heat Denaturation study of the 11S Globulin in Soybean Seeds," Food Chem., 6:309-322.

Laemmli, U. K., 1970, "Cleavage of Structural Proteins during the Assembly of the Head of Bacteriophage T4," Nature (London), 227:680.

Levitt, M., 1978, "Conformational preferences of amino acids in globular proteins," Biochemistry, 17:4277-85.

Levy, G. C., Lichter, R. L. and Nelson, G. L., 1980, Relaxation Studies, in: "Carbon-13 Nuclear Magnetic Resonance Spectroscopy," 2nd ed.; John Wiley and Sons: New York/Chichester/Brisbane/Toronto/Singapore.

Lillford, P. G., 1978, Conformation of Plant Proteins, in: "Plant Proteins," G. Norton, ed.; Butterworths, London.

Mabbutt, B. E., and Wright, P. E., 1983, "Assignment of heme and distal amino acid resonances in the proton NMR spectra of the oxygen and carbon monoxide complexes of soybean leghemoglobin," Biochem. Biophys. Acta, 744(3):281-90.

Moreira, M. A., Hermondsen, M. A., Larkins, B. A., and Nielsen, N. C., 1979, "Partial Characterization of the Acidic and Basic Polypeptides of Glycinin," J. Biol. Chem., 254:9921.

Mori, T., Nakamura, T., and Utsumi, S., 1986, "Behavior of Intermolecular Bond Formation in the Late Stage of Heat-induced Gelation of Glycinin," J. Agric. Food Chem., 34:33-36.

Nakamura, T., Utsumi, S., and Mori, T., 1986a, "Mechanism of Heat-induced Gelation and Gel Properties of Soybean 7S Globulin," Agric. Biol. Chem., 50(5):1287-1293.

Nakamura, T., Utsumi, S., and Mori, T., 1986b, "Interactions During Heat-induced Gelation in a Mixed System of Soybean 7S and 11S Globulins," Agric. Biol. Chem., 50(10):2429-2535.

Nielsen, N. C., 1985a, "Structure of Soy Proteins," New Protein Foods, 5:27-64.

Nielsen, N. C., 1985b, "The Structure and Complexity of the 11S Polypeptides in Sobeans". J. Am. Oil Chem. Soc., 62(12):1680-6.

228

Patt, S. L., and Shoolery, J. N., 1982, "Attached proton test for carbon-13 NMR," J. Magn. Reson., 46:435.

Shaka, A. J., Keeler, J., Frenkiel, T., and Freeman, R., 1983, "An improved sequence for broadband decoupling: WALTZ-16," J. Magn. Reson., 52:335-8.

Slappendel, S., Aasa, R., Falk, K., Malmstrom, B. G., Vaenngaard, T., Veldink, G. A., and Vliegenthart, J. G., 1982, "Proton NMR study on the binding of alcohols to soybean lipoxygenase-1," Biochem. Biophys. Acta, 708(3):266-71.

Suresh Chandra, B. R., Appu Rao, A. G., and Narasinga Rao, M. S., 1984, "Effect of Temperature on the Conformation of Soybean Glycinin in 8 M Urea or 6 M Guanidine Hydrochloride Solution," J. Agric. Food Chem., 32:1402-1405.

Tatham, A. S., Shewry, P. R., and Belton, P. S., 1985, "$^{13}$C-n.m.r. study of C hordein," Biochem J., 232:617-20.

Thanh, V. H., Okubo, K., and Shibasaki, K., 1975, "Isolation and Characterization of the Multiple 7S Globulins of Soybean Proteins," Plant Physiol., 56:19.

Thanh, V. H., and Shibasaki, K., 1976, "Major Proteins of Soybean Seeds. A Straightforward Fractionation and Their Characterization," J. Agric. Food Chem., 24(6):1117-21.

Thanh, V. H., and Shibasaki, K., 1977, "Beta-conglycinin from soybean proteins. Isolation and immunological and physiochemical properties of the monomeric forms," Biochem. Biophys. Acta, 490:370-384.

Trewhella, J., Appleby, C. A., and Wright, P. E., 1986, "Proton NMR studies of high-spin complexes of soy leghemoglobin. Interactions between the distal histidine and acetate, formate and fluoride ligands," Aust. J. Chem., 39(2):317-24.

Utsumi, S., and Kinsella, J. E., 1985, "Forces Involved in Soy Protein Gelation: Effects of Various Reagents on the Formation, Hardness and Solubility of Heat-induced Gels made from 7S, 11S and Soy Isolate," J. Food Sci., 50:1278-1282.

Wright, D. J., 1976, The Seed Gobulins, in: "Developments in Food Proteins - 5," B. J. F. Hudson, ed.; Elsevier Applied Science: London/New York, pp., 81-157.

Wüthrich, K., 1976, Carbon-13 NMR of Amino Acids, Peptides and Proteins, in: "NMR in Biological Research: Peptides and Proteins," North Holland/American Elsevier: Amsterdam/Oxford/New York.

Yamauchi, F., and Yamagishi, Y., 1979, "Carbohydrate Sequence of a Soybean 7S Protein," Agric. Biol. Chem., 43(3):505-510.

Yamauchi, F., Ono, H., Kamata, Y., and Shibasaki, K., 1979, "Acetylation of Amino Groups and Its Effect on the Structure of Soybean Glycinin," Agric. Biol. Chem., 43(6):1309-1315.

# NMR STUDIES OF THE STRUCTURE AND ENVIRONMENT OF THE MILK PROTEIN
## α-LACTALBUMIN

Lawrence J. Berliner(a), Robert Kaptein(b), Keiko Koga(c),
and Giovanni Musci(d)

(a) Department of Chemistry, Ohio State University
    120 West 18th Avenue, Columbus, Ohio  43210
(b) Department of Organic Chemistry, University of Utrecht
    Padualaan 8, Utrecht, The Netherlands NL 3584 CH
(c) Bioenergetics Research Center, Tokushima Research
    Institute, Otsuka Pharmaceutical Co., Ltd.
    463-10 Kagasuno, Kawauchi-cho, Tokushima 771 Japan
(d) Centro di Biologia Molecolare del C.N.R.
    Dipartimento di Scienze Biochimiche
    Universita' "La Sapienza", Piazzale Aldo Moro 5
    00185 Rome Italy

## INTRODUCTION

α-Lactalbumin (α-LA) plays a unique role in milk biochemistry.  That
is, its function appears to be only as a modifier protein in lactose bio-
synthesis by directing the specificity of galactosyl transferase from the
acceptor substrate N-acetylglucosamine (GlcNAc) to glucose (Glc).
Specifically, α-lactalbumin lowers the $K_M$ for Glc from 1.4 M to 5.0 mM,
while the reaction involving GlcNAc as an acceptor becomes inhibited in
the presence of α-lactalbumin (Hill and Brew, 1975).  Actually, it was
not known until very recently that α-LA was a calcium-binding protein.
It had been studied extensively in the past by Kronman and colleagues
(Kronman et al., 1964; Kronman and Andreotti, 1972) yet it was never
realized that the protein existed in several conformations due to
multiple states of cation binding.  In fact, much of the published data
on the physical or structural properties of this protein previous to 1978
are ambiguous since workers were not aware that their α-LA samples
contained calcium and/or other strongly bound cations.

The α-lactalbumins have been fully sequenced from more than eight
species and have been found to be highly homologous in primary structure
(Beg et al., 1985; Bell et al., 1981) and (therefore) tertiary structure.
Several of these structural homologies will be pointed out when comparing
structures by CIDNP NMR spectroscopy.

*NMR Applications in Biopolymers,* Edited by
J. W. Finley *et al.,* Plenum Press, New York, 1990

## NATURE OF THE CALCIUM-BINDING SITE

The first reports of strong cation binding were reported almost simultaneously by three groups (Berliner et al., 1978; Hiraoka et al., 1980; Permyakov et al., 1981). Hiraoka et al. (1980) found by atomic absorption analyses that calcium binds strongly to α–LA with a 1:1 stoichiometry. A study of cation induced structural changes by fluorescence spectroscopy came from the groups of Permyakov (1981) and Berliner (Murakami et al., 1982) who reported evidence for calcium-ion binding with an affinity constant in the nanomolar range. Both groups had worked out detailed methods for total calcium removal.

The structure and nature of the calcium binding site on an α–LA is an intriguing question, especially in light of our knowledge of the calcium-binding sites in parvalbumin, troponin C, intestinal calcium-binding protein, and calmodulin. Most contain a unique α-helical related bend involving carboxyl groups and oxygen donor ligands forming an octahedral calcium coordination site, known as the EF hand (Kretsinger and Nockolds, 1973; Herzberg and James, 1985). Several groups studying calcium binding to α–LA have speculated on the location of the calcium-binding site, based only on inspection of the published amino acid sequence for the bovine species (Hiraoka et al., 1980; Permyakov et al., 1981; Kronman et al., 1981). One group suggested up to four metal-ion sites, although had no experimental evidence at the time for the proposed locations (Kronman and Bratcher, 1984a, 1984b). However, since the sequence of bovine α–LA was recently corrected, there was more confidence in the placement of a putative calcium site, contradicting the original speculations (Kronman and Bratcher, 1981, 1984a, 1984b). Specifically, several new acidic side chains were reported in the corrected sequence, which strongly implicated the region around residues: Asp 82, Asp 83, Asp 84, Asp 87, and Asp 88. The more recent crystallographic evidence has confirmed the involvement of Asp 82, 87, and 88 as well as two carbonyl oxygen atoms from residues 79 and 84 comprising a tight β–turn (Stuart et al., 1986).

In order to learn about the coordination geometry and chemical nature of the calcium site, ESR and NMR were applied with cations especially suited for probing calcium-binding sites. For example, several ligands which bind to calcium sites (e.g., Mn(II) and $Gd^{3+}$) are paramagnetic and therefore exhibit ESR spectra which are representative of the ligand nature and geometry (Berliner et al., 1983; Musci et al., 1986). As another example, NMR experiments with [113]Cd, which also substitutes for Ca(II), yielded the spectra shown in Fig. 1A, which gave a broad single resonance at -80 ppm. The bound [113]Cd could be displaced

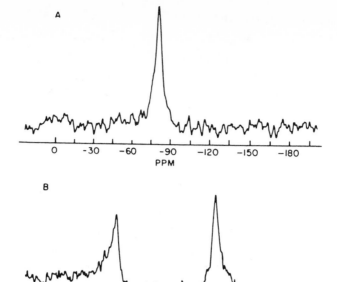

Fig. 1. $^{113}$Cd NMR spectra of α-lactalbumins. (A) 3.4 mM $^{113}$Cd(II)
-bovine α-lactalbumin (pH 6.3, 25 mM Tris-HCl, 20% $D_2O$,
25°C). This spectrum represents 93846 accumulations,
although satisfactory spectra were obtained in 10000-15000
scans where line shape was not critical. (B) 1.5 mM goat
α-lactalbumin [which was initially 37% Ca(II) bound], which
contains ca. 1.1 mM bound $^{113}$Cd(II) and ca. 1.2 mM excess
free $^{113}$CdCl$_2$, pH 6.6. The number of scans was 164000.
Upon addition of equimolar Ca(II), the protein-bound line
at -85 ppm shifted completely to free $^{113}$Cd (II) at 5 ppm
(not shown). Similar results were obtained with bovine
α-lactalbumin in (A) upon Ca(II) addition. All chemical
shifts are relative to Cd(ClO$_4$)$_2$. From Berliner et al.
(1983) with permission.

by addition of Ca$^{2+}$ yielding a line at +5 ppm, consistent with free $^{113}$Cd
in buffer (Fig. 1B). This $^{113}$Cd chemical shift (-80 ppm) compared well
with similar shifts for parvalbumin and troponin C. Perhaps the most
direct experiments were reported by Vogel and colleagues who examined the
$^{43}$Ca NMR of metallobovine-α-LA (Drakenberg and Vogel, 1983). A chemical
shift of -5 ppm was found which was similar, although not exactly that
found for EF-hand calcium-binding proteins. The magnitude and
quadrupolar coupling constant with $^{43}$Ca(II)-α-LA was significantly
smaller than for other EF-hand structures, indicating a more symmetric
environment for the bound metal, i.e., the sequence and geometry of the
EF-hand structure was slightly altered in the α-LA structure vs. other

proteins, as shown later by x-ray crystallography (Smith et al., 1987; Acharya et al., 1989). It was also clear from NMR and ESR that the ligand environment of the calcium loop was essentially all oxygens. In water, proton relaxation-enhancement studies, Murakami et al. (1982) found that for Mn(II) bound at the Ca(II) site, there was direct first sphere contact with at best 0.5 $H_2O$ molecules. That is, the cation-binding site appeared to be almost completely sheltered from the aqueous solution environment since the bound Mn(II) was not a very effective relaxant.

## PROTON NMR STUDIES

High resolution NMR serves as an excellent probe of overall structure of a protein. Where the number of resonances are many, the overlaps between lines frequently create problems in identifying each amino acid residue. This is particularly difficult in the aliphatic region of the proton NMR spectrum (0 to 4 ppm) where literally every $CH_3$, $CH_2$, and CH group on the protein resonate. However, the aromatic region (6 to 9 ppm) and the upfield region (negative ppm) from the DSS standard are usually resolvable due to the smaller number of lines. Figure 2 displays the proton NMR spectrum at 200 MHz for apo- and Ca(II)-bovine-α-LA. After closer examination at 500 MHz, one notes several chemical shift differences between the two species in almost every region of the spectrum, but particularly those affected by aromatic residues, such as the ring current upfield shifted lines from 0 to -2.5 ppm (Fig. 3). The extreme upfield shifted resonance, at -2.45 ppm, is the most diamagnetically shifted line reported to date for a protein (Berliner et al., 1987). Complete assignment of every line in the spectrum requires 2D NMR methods which are obviously quite time consuming and difficult. In order to quickly simplify assignment in the protein spectrum several techniques have been developed recently for examining only specific parts of the protein molecule or resonance spectrum. One such technique, the laser photo CIDNP method, examines only surface exposed His, Trp, and Tyr residues as a result of radical pair formation at the surface induced by a photo-excited flavin dye (Kaptein, 1982). As a result of this phenomenon, the proton NMR spectrum condenses down to just those accessible polarized aromatic amino acids on the surface of the protein.

## LASER PHOTO-CIDNP

The spectacular intensity enhancements of NMR lines observed for the products of free radical reactions is called chemically induced dynamic nuclear polarization (CIDNP). Basically, it arises from a spin-sorting process acting during recombination of radical pairs, which leads to

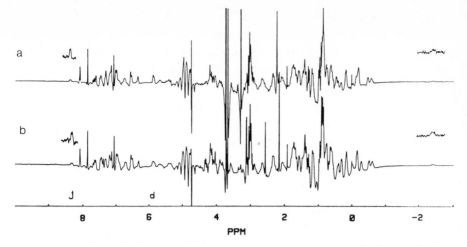

Fig. 2.   High-resolution 200-MHz proton NMR spectra of the aromatic
          and upfield shifted aliphatic region of bovine α-LA (50mM
          Tris-d$_{11}$-DCl, pH 7.2).   The residual HOD signal was

          suppressed by presaturation with a single radio frequency
          for 1 s, which was gated off during acquisition.
          Spectrometer parameters were 8K data points, 1000
          transients, 3000-Hz sweep width, 3.7- μs pulse width (90°
          pulse), and 1.4-s acquisition time.   Spectral resolution
          enhancement was accomplished by the convolution difference
          method (line broadening 0.5 Hz - line broadening 5 Hz).
          (a)1.6 mM apo-α-LA; (b)1.9 mM Ca(II)-α-LA.   From Berliner
          et al., (1987), with permission.

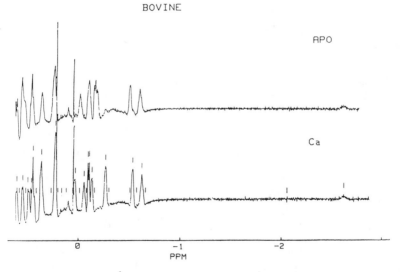

Fig. 3.   The 500-MHz $^{1}$H NMR spectra of α-LA, 50 mM Tris-d$_{11}$, pH 7.3.

          (A) apo-α-LA Ca-(II)-α-LA.   (B) Ca(II)-α-LA.   Water
          presaturation 1 s; pulse repetition time 2.1 s; 2000 scans;
          0.5-Hz line broadening.

greatly perturbed populations of nuclear spin state levels. The application of CIDNP to biochemical problems adds to structural information on biological macromolecules and on their interactions with ligands. The CIDNP effect has been developed to enhance only certain lines in the NMR spectrum of a protein. These intensity enhancements, which can be positive (absorption) or negative (emission), tell us something about the surface structure of the protein. Laser photo-CIDNP is quite simple and involves the following experimental steps:

1) A protein solution containing a dye is irradiated with an argon laser in the NMR probe.
2) The photoexcited dye reacts (reversibly) with certain aromatic amino acid side chains on the protein surface generating dye-protein radical pairs.
3) "Back-reactions" of the radical pairs yield nuclear spin polarization in the reacted residues.
4) "Light" and "dark" NMR spectra are taken in alternating scans in the pulsed Fourier transform mode.
5) The photo-CIDNP difference spectrum contains only lines of the polarized residues.

Thus, the method affords a surface accessibility probe, retains the high intrinsic resolution of NMR, and is capable of identifying individual residues of a protein. Besides characterizing surface residues, the method is also useful for studying protein-ligand interactions. With the flavin-type dyes that have been used almost exclusively thus far, three amino acid residues can be polarized: tyrosine, histidine, and tryptophan. This, of course, limits the "selectivity window" but makes photo-CIDNP spectra of proteins relatively simple and interpretable.

Below we briefly review the photo-CIDNP method as applied to proteins.

THE RADICAL PAIR MECHANISM

When two reactive radicals encounter in solution, two possibilities exist.

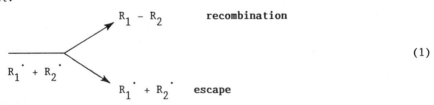

(1)

Here, "recombination" stands for all types of radical pair collapse reactions: coupling, disproportionation, back electron transfer, etc. The rate in reaction (1) depends (to some extent) on the spin states of the nuclei present in the radicals.  This nuclear spin dependent reactivity is a consequence of the radical pair theory:

1) The reaction probability of a radical pair depends on the electronic spin state of the pair (usually recombination occurs only in the singlet state).

2) There are nuclear spin dependent interactions involved in singlet and triplet radical pair states (i.e., reactive and unreactive states).

In the well studied radical recombination reaction where two H atoms react to give a $H_2$ molecule, it is well known that only a singlet pair of H atoms approach on an attractive potential energy curve, while the triplet pair experiences repulsive forces at short distances.  A similar spin state selectivity is generally observed for more complex reactions, as well.  Thus, assumptions 1) and 2) above lead to a sorting mechanism for nuclear spins and therefore to strong nuclear polarizations in radical pair products.  Figure 4 shows a typical photoinduced radical reaction:

1) Compound P is photoexcited to a singlet state $^1P$ which crosses over to a triplet state $^3P$.

2) A radical pair is formed which is in a triplet (T) state initially (a pair of original partners with correlated electron spins is indicated by a bar).

3) Random-walk diffusion sets in and, simultaneously, intersystem crossing (T → S) transitions occur, induced by the coupling of the electron spins with nuclear spins (hyperfine coupling).  The rate of this intersystem crossing depends on the nuclear spin state.  Here it is assumed that it is faster for β nuclei than for α nuclear spins.

4) As a consequence singlet pairs will be formed faster with β nuclear spins than with a spins; i.e., the excess β spins are carried over to the recombination product where only singlet pairs recombine.

5) In the recombination product the β nuclear spin level (which is higher in energy) is overpopulated, leading to an emissive NMR signal for this product.

One might conclude from the previous scheme that no CIDNP should be observed at all in a cyclic reaction without net chemical change.  The escaping radicals would carry the opposite polarization (↑) and (↓) polarizations would cancel exactly.  Fortunately, nuclear spin relaxation causes the escape polarization to leak away and makes it possible to observe recombination polarization in a cyclic reaction, since the

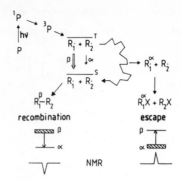

Fig. 4.    Schematic representation of the nuclear spin-sorting
process in a photo induced radical reaction; α and β denote
nuclear spin states.   From Kaptein (1982) with permission.

lifetime of the radicals is not much shorter than the nuclear spin-lattice
relaxation time $T_{1R}$ in the radicals.   $T_{1R}$ is dominated by dipole coupling with
the unpaired electron and is typically of the order of $10^{-4}$ sec. (Fig. 5).

Also, the lifetime of radicals carrying escape polarization is
shortened by degenerate exchange reactions with the parent compound.   In
particular, in cation or anion radicals, degenerate electron transfer may
be very fast:

$$D^{\overline{\cdot}} \uparrow + D \xrightarrow{\text{fast}} D \uparrow + D^{\overline{\cdot}} \tag{2}$$

The equivalent H-atom transfer of the protonated radical is usually
slower:

$$DH \cdot \uparrow + D \longrightarrow D\uparrow + DH \cdot \tag{3}$$

If reactions of this type compete with spin relaxation in the radical,
cancellation of CIDNP effects may again occur.   Degenerate exchange
reactions such as Eq. (2) are slowed down considerably since the large

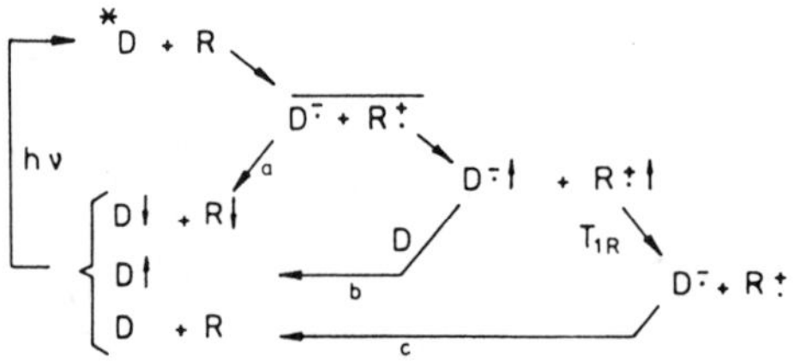

Fig. 5.    Scheme for a cyclic photoreaction of dye D with substrate R
including nuclear spin-lattice relaxation ($T_{1R}$) and
degenerate electron exchange reaction.   Route (a): geminate
recombination; route (b):  degenerate exchange reaction for
dye D, trapping (↑) polarization; route (c):   recombination
of radicals after nuclear spin relaxation.   From Kaptein
(1982) with permission.

macromolecular radicals causes slow translational diffusion. Therefore, it is clear that for the observation of CIDNP in cyclic reactions the role of nuclear spin relaxation in the radicals is quite important. Since relaxation rates differ for different nuclei, CIDNP intensities may be distorted from those expected from radical pair theory. In the case of long radical lifetimes relaxation of escape polarization would be complete and relative CIDNP intensities would be expected to be proportional to the hfc constants $A_i$ in the prevailing situation of very high field where $\Delta g \beta B > A_i$.

CROSS-POLARIZATION

Polarization transfer mechanisms exist, whereby nuclei in close proximity to a primary polarized nucleus can acquire spin polarization. The most important of these is based on dipolar **cross-relaxation**. Cross-relaxation is also responsible for the nuclear Overhauser effect (nOe) and for spin diffusion in proteins. Figure 6 shows the energy level scheme for a two-proton system, with cross-relaxation transitions and level populations for the case where one proton is directly polarized. The dipole-dipole relaxation transition rates are given by

$$w_o = w \tag{4}$$

$$w_1 = 3w/1(+\omega^2\tau_c^2) \tag{5}$$

$$w_2 = 12w/(1 + 4\omega^2\tau_c^2) \tag{6}$$

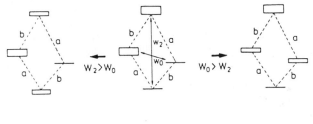

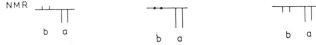

Fig. 6.   The cross-polarization effect.  Four-level scheme for a two-spin-1/2 system where it is assumed that spin is directly polarized.  Energy levels representing nuclear spin states αα, αβ, βα, ββ, respectively (from top to bottom).  Relaxation transitions $w_2$ and $w_0$ are indicated, while single spin-flip transitions $w_1$ are not shown.  Photo-CIDNP NMR spectra are drawn schematically.  For small molecules (left-hand side) $w_2 > w_0$, and cross-relaxation causes transfer of polarization with the opposite sign, while for macromolecules (right-hand side), $w_0 > w_2$, and polarization is transferred with retention of sign. From Kaptein (1982) with permission.

In these expressions $w = \tau_c \gamma^4 h^2/20r^6$, where r is the distance between the nuclei, $\gamma$ the gyromagnetic ratio, $t_c$ the correlation time for molecular tumbling, and $\omega$ the resonance frequency. The sign of the transferred polarization depends on the sign of $w_2 - w_0$ and changes when $\omega\tau_c = 1/2\sqrt{5}$ which, for protons at 360 MHz, occurs at a correlation time $\tau_c = 5 \times 10^{-10}$ sec. As is illustrated in Fig. 6, the cross-polarization effect changes sign when going from a small molecule to a protein, just like the sign of the nOe effect. Even for small proteins correlation times for molecular tumbling are usually longer than $10^{-9}$ sec., so that in macromolecules transfer of polarization by the cross-relaxation effect occurs with retention of sign.

In the slow tumbling limit, $\omega^2\tau_c^2 > 1$, the cross-relaxation rate is proportional to $\tau_c r^{-6}$, so that it is more efficient in larger proteins and has a pronounced distance dependence. Thus, in small proteins transfer of polarization will be limited to a shell of nearest-neighbor nuclei, whereas in large proteins whole regions may become polarized by spin diffusion. Cross-polarization may occur within the same amino acid residue or between different residues. For example, transfer of emission polarization from the 3,5 to the 2,6 protons in the tyrosine ring is commonly observed (Kaptein, 1982).

The experimental distinction between directly polarized and cross-polarized lines in a photo-CIDNP spectrum is possible on the basis of their different time dependence. Figure 7 shows the time evolution of both types of polarization in a photo-CIDNP experiment. The buildup of direct polarization is close to exponential with a time constant of $T_1$. Note how the cross polarization effect shows a sigmoidal behavior which will be relatively small at short light pulses. Conversely, the effect will be relatively enhanced when data acquisition is delayed after switching the light off.

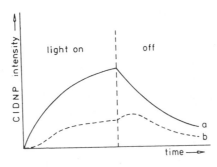

Fig. 7. Time-dependence of the buildup and decay of photo-CIDNP intensity for direct polarization (a), and cross-polarization (b). From Kaptein (1982) with permission.

Cross-relaxation times for internuclear distances of 2.5 - 3 Å in small proteins are of the order of 0.1 to 0.5 sec. Therefore, a light pulse of 50 - 100 msec. at high power is usually sufficient to suppress the cross-polarization effect. On the other hand, including a delay of 0.3 sec after a pulse of 0.6 sec would enhance it. A combination of both short pulse and delay experiments establishes the presence of cross-polarization effects unambiguously.

## THE LASER PHOTO-CIDNP EXPERIMENT

For light irradiation in superconducting magnets a laser is the method of choice. In most laboratories an argon ion laser is employed which has output at suitable wavelengths in the blue-green part of the spectrum (main light power at 488 nm and 514 nm). The most efficient method of sample irradiation is via fiber optics. Most photo-CIDNP spectra have been obtained with 3-N-carboxymethyllumiflavin although other flavins such as riboflavin, FMN, and lumiflavin give very similar results. Flavins have a visible absorption band at $\lambda_{max}$ = 450 nm. They are excited by the 488-nm argon laser line in the shoulder of the absorption band where the extinction coefficient is ca. 1000. Since they are strongly fluorescent compounds, virtually all flavin photochemistry occurs in the triplet state, which is formed after photoexcitation with a quantum yield of about 0.5. For most applications 1 - 5 W output power delivered in pulses of 0.6 sec. is sufficient. In cross-polarization studies, where we have employed short pulses (50-100 msec.), higher powers are needed.

The polarization patterns in Fig. 8 indicate that intermediate radicals are formed in which the unpaired electron spin is delocalized over the imidazole and indole rings, as previously observed by ESR. In the case of tryptophan the spin-density distribution is such that only the C-2 H(singlet), C-4 H(doublet), and C-6 H(triplet) are directly polarized. If delays are included between laser and rf pulse cross-polarization is observed for the C-5 H and C-7 H lines.

With the flavin dyes no polarization has been observed for other unmodified amino acids with the possible exception of a small positive effect for methionine which is not likely to be of practical importance. Free cysteine reacts with photoexcited flavin as evidenced by bleaching. Phenylalanine is the only aromatic residue that is not polarizable.

It should also be noted that flavins form complexes with aromatic compounds such as tyrosine and tryptophan. Stability constants range from 10 to $100M^{-1}$, tending to be somewhat larger for the flavosemiquinone than for oxidized flavin. Formation of tight complexes with proteins,

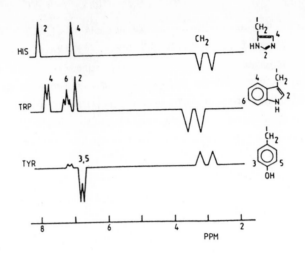

## CIDNP OF AMINO ACIDS

Fig. 8.   Schematic representation of the flavin-induced [1]H photo-CIDNP effects observed for the amino acids histidine, tryptophan, and tyrosine.  From Kaptein (1982) with permission.

however, would be expected to reduce or even eliminate the CIDNP effect, since separation of the radical pair would then not occur.  For example, no CIDNP has been detected for flavodoxins, which contain tightly bound FMN, unless excess flavin is present (Kaptein, 1982).

Other compounds with electron donating properties are known to react with photoexcited flavins and may compete with the amino acid residues. The presence of EDTA during a photo-CIDNP experiment is disastrous.  The same is true for secondary and tertiary amines:  such as "bistris" buffers [2-bis(2-hydroxyethyl)amino-2-(hydroxy-methyl)-1,3-propanediol]. On the other hand, tris buffer does not present the problem, although inert buffers such as phosphate are preferred.  Aromatic amines react readily with triplet flavin and give rise to positive CIDNP effects for the aromatic protons.  This property was used to follow the binding of sulfanilamide to carbonic anhydrase (Kaptein, 1982).

Methyl-ε-lysine (mono-, or dimethylated) has been found to show CIDNP effects in the unprotonated form.  This may render lysine residues in a protein accessible via methylation to the photo-CIDNP method.

## PROTEINS - GENERAL CONSIDERATION

Since the g factor and hfc constants of a radical are not expected to be greatly different in a protein environment, the polarizations of amino acid residues in a protein are qualitatively the same as those of

Fig. 8. Their magnitudes are affected by the larger size of the protein and by the accessibility of the residues. Slower rotational diffusion of a protein causes cross-polarization effects which are often observed from absorption to emission for tyrosine 2,6 protons which change sign. On the other hand the slower translational diffusion diminishes the rate of reaction with the photoexcited dye and, when competing deactivating pathways are available to the dye, the polarization intensities are also reduced. CIDNP effects for <u>very large</u> proteins are often found to be weaker than for smaller ones. However, where electron exchange reactions occur, the slower diffusion of the macromolecular radicals reduces these reactions, thus increasing the observable CIDNP effects. Thus, polarization can be observed in proteins for Trp at pH < 3 and Tyr at pH >1 0, while in the free amino acids it is very weak at this pH due to cancellation effects.

Because radical pair formation requires contact of the photoexcited dye with the amino acid side chain, one should be able to discriminate between surface and internal residues. This is shown schematically for tyrosine residues in Fig. 9. Tyr "a" would be expected to show polarization, but not Tyr "b". Tyr "c" can be polarized depending on the size of the cleft with respect to that of the probing dye molecule. In addition, polarization depends on the mechanism of the primary reaction step, requiring a freely accessible OH group in the case of Tyr and imidazole NH in the case of His residues. Electron transfer reactions (as occurs for Trp) could in principle take place over larger distances by quantum mechanical tunneling. However, the reaction probability for electron tunneling over a distance is considerably smaller than for a contact reaction. Studies on model proteins containing several Trp residues, such as lysozyme suggest that in this case photo-CIDNP also detects surface residues only. Laser photo-CIDNP results are usually in good agreement with other physical surface probes such as solvent perturbation spectroscopy and fluorescence quenching, but sometimes disagree from conclusions based on chemical modification studies.

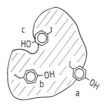

Fig. 9. Globular protein with tyrosine residues with different degrees of exposure. From Kaptein (1982) with permission.

Figure 10 (top) shows the 360 MHz laser photo CIDNP spectrum of bovine α-LA, pH 6.37 (Berliner and Kaptein, 1981). The negative emission lines are due to $Tyr_{3,5}$ ortho protons on the surface of the molecule, accounting for three of the four Tyr in the sequence. By analogy to similar studies with the homologous protein, lysozyme, it was surmised that Tyr 50 was the buried residue. The positive absorption lines are due to His and Trp residues on the surface. Resonances at 7.2 and 8.15 ppm reflect the C-2 and C-4 protons of His 68 on the surface of the protein. Since the human α-LA species is missing the two resonances noted above and lacks a His at position 68 in its sequence, the assignment of this residue was straightforward. The resonances at 6.5, 7.4, and 7.65 ppm are due to Trp 104 which, as we shall see later,

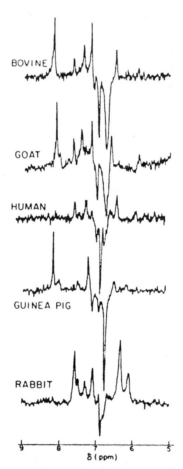

Fig. 10.   Laser photo-CIDNP proton spectra of the aromatic region of several α-lactalbumin species. All proteins were in 0.1M potassium phosphate, $D_2O$, 30°C: bovine, 0.61 mM, pH 6.37; goat, 0.42 mM, pH 6.57; human, 0.93 mM, pH 6.37; guinea pig, 0.99 mM, pH 6.78; rabbit 0.55 mM, pH 6.27. From Berliner and Kaptein (1982) with permission.

resides in a hydrophobic box-like structure with two other aromatic residues observed in the normal NMR spectrum. The small, unusually upfield-shifted resonance at 5.85 ppm results from nuclear spin-cross relaxation from an exposed, directly polarized residue (Trp 104) to a buried Trp 60. This latter 5.85 ppm resonance did not appear at short times in the NMR spectral acquisition, but slowly grew with time as a result of cross relaxation from a polarized (excited) proton on Trp 104. This assignment was corroborated by experiments with guinea pig $\alpha$-LA (which lacks Trp 60), where no resonance appeared at about 5.85 ppm. On the other hand, other $\alpha$-LA species which contain a Trp 60 (i.e., human and goat $\alpha$-LA) all showed a resonance between 5.8 - 5.9 ppm, reflecting the propinquity of Trp 60 to Trp 104 in their structure. A Phe resonance in the 5.8-5.9 ppm has been assigned in bovine $\alpha$-LA by Dobson and coworkers (private communication). Of interest here is that guinea pig $\alpha$-LA has substituted a Phe for Trp 60. It seems, however, that the guinea pig Phe 60 must differ from the "common" Phe residue present in the other species shown earlier. Furthermore, this "common" Phe cannot be the crosspolarized resonance near 5.85 ppm which appeared in the CIDNP spectra for all of the species except guinea pig (Berliner and Kaptein, 1981).

Upon addition of 1:1 $Ca^{2+}$ to apo-$\alpha$-LA, the two Tyr resonances at about 6.8 ppm disappeared (see Fig. 11), indicating almost completely reduced exposure to the (flavin dye) solvent (Berliner et al., 1987). In experiments designed to probe the relationship of these two Tyr residues to the hydrophobic binding region in $\alpha$-LA, we examined the effects of subdenaturing concentrations of SDS or desoxycholate on the CIDNP of bovine $\alpha$-LA. These amphiphilic ligands induced an increased exposure of the two Tyr resonances at 6.8 ppm. This was especially distinct with the apo-conformer, but also, albeit to a smaller degree, with the $Ca^{2+}$ species as well. The addition or removal of $Ca^{2+}$ modulated this interaction. Upon addition of $Zn^{2+}$ to apo- or $Ca^{2+}$-$\alpha$-LA, an increase in solvent exposed Tyr resonances at 6.8 ppm was observed, consistent with the ability of $Zn^{2+}$ to shift the protein to an apo- or apo-like conformation (Musci and Berliner, 1985).

NOE STUDIES

In order to further examine the relationship between Trp 60, Trp 104, and other residues in the protein, the technique of nuclear Overhauser enhancement (nOe) was employed. Here a specific single isolated proton is irradiated with RF power and the effect of this irradiation is monitored on the other protons in the spectrum. Where resonance intensities are altered, as manifested in the difference spectrum between the irradiated spectrum and the off-resonance spectrum (irradiation at some frequency away from the protein resonances), a negative nOe is observed. Figure 12A shows the nOe difference spectrum obtained by irradiation at the unique upfield shifted

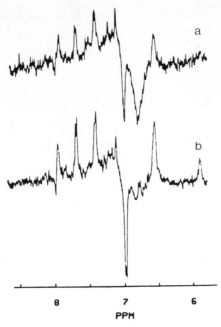

Fig. 11. Laser photo-CIDNP difference spectra of α-LA conformers at 500 MHz. The sample contained 0.5 mM α-LA and 0.4 mM 10-N-(carboxy-ethyl)lumiflavin, pH 7.1. FID's were collected as eight alternating light (laser on) and dark (laser off) scans. Laser power was 7.0 W, and the sample was illuminated for 0.5 s by fiber optic illumination. Spectrometer parameters were 8K data points, 7000-Hz sweep width, and 2-μs pulse width. A 10-s delay was used between each light and dark scan. The HOD signal was presaturated for 1 s preceding each laser pulse (a) Apo-α-LA; (b) Ca(II)-α-LA. From Berliner et al. (1987) with permission.

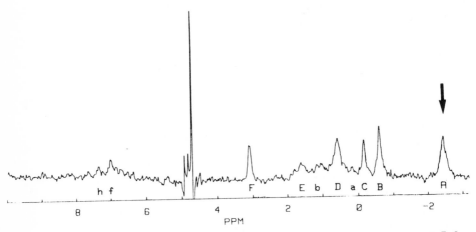

Fig. 12. 200 MHz nOe difference spectrum of 3 mM Ca(II)-α-LA at pH 7.2 (A) after irradiation of peak A (arrow) at -2.45 ppm for 0.4 seconds (5600 scans) (B) after irradiation of peak d (arrow) for 0.45 seconds (8000 scans) sweep width. From Koga and Berliner (1985) with permission.

-2.45 ppm line which corresponds to Ile 95 when compared with the hen egg white lysozyme spectrum where such an extreme upfield shifted aliphatic proton (Ile 98) arose from an unusual hydrophobic box surrounded by two Trp residues (Trp 63, Trp 108). As noted in the table below, the resultant spectrum (Fig. 12A) compares well with that reported for lysozyme (Poulsen et al., 1980).

| Peak | Ile Proton | $\alpha$-LA | Lyozpyme |
|------|-----------|---------|----------|
| A | $H^{\gamma 12}$ | -2.45 ppm | -2.10 ppm |
| B | $CH_3^{\delta}$ | -0.62 | -0.01 |
| C | $CH_3^{\gamma 2}$ | -0.17 | -0.26 |
| D | $H^{\gamma 11}$ | -0.56 | 0.63 |
| E | $H^{\beta}$ | 1.59 | 1.56 |
| F | $H^{\alpha}$ | 3.07 | 2.88 |

Figure 12B shows difference nOe spectra for $Ca^{2+}$-$\alpha$-LA irradiated at Trp 60 (5.85 ppm). The difference nOe spectrum shows peaks at 7.18 (g), 7.64 (i), 6.98 (f), 6.48 (e), 3.63, 3.19, 3.07 (F), 1.12 (b) and 0.43 ppm which are reflective of nuclear cross relaxation to the following residues: Ile 95, Try 103, Trp 104, and Trp 60 (Koga and Berliner, 1985). In order for strong nOe resonances to be observed in a protein spectrum, the respective protons must be less than approximately 4-6 Å apart. Where neighboring residues are within the distance measurements noted above, the nOe dif- ference spectrum will contain their proton lines. Thus, the nOe observed between Trp 60 and Trp 104 in these experiments was in precise agreement with the cross polarization noted between Trp 104 and Trp 60 in the CIDNP results discussed earlier (Berliner and Kaptein, 1981). The nOe irradia- tions were also performed on "isolated" proton resonances at -5.85 and 8.36 ppm in the $Ca^{2+}$-$\alpha$-LA spectrum. The results implicated a close structural relationship between Ile 95, Trp 104, Tyr 103, and Trp 60 A structural sketch of this portion of the $\alpha$-LA molecule is shown in Fig. 13. It should be noted that the Tyr 103 resonance at 6.98 ppm appears to overlap the Tyr resonance in the CIDNP Fig. 10 above. The CIDNP results, in conjunction with the nOe results, suggests that the -OH moiety of Tyr 103 and edge of the Trp 104 ring are accessible to the flavin solvent in the photo CIDNP experiments. Interestingly, the differences between the apo- and $Ca^{2+}$-form of the protein were relatively minor in this region of the structure although differences in structural rigidity were found as reflected in the extent of nOe for these protons. The results were con sistent with a more rigid structure in the $Ca^{2+}$ form.

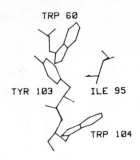

TRP 60

TYR 103      ILE 95

TRP 104

Fig. 13.  Two dimensional projection of the hydrophobic box region of
bovine α-LA.  From Koga and Berliner (1985) with permission.

PARAMAGNETIC METAL ION BINDING

Identification of specific nuclei situated close to the putative
calcium-binding site has been more difficult to assess by NMR due
principally to the slow chemical exchange of cation with this site.
Furthermore, proton relaxation enhancement by bound paramagnetic cations has
not been as informative as desired, since the metal ion is in slow chemical
exchange on the NMR time scale.  In a $Mn^{2+}$ or $Gd^{3+}$ titration of α-LA one
observes a broadening of the majority of the proton resonances to such an
extent where only the narrow line apo-α-LA spectrum remains.  Several
lanthanide NMR shift reagents were examined with α-LA by Berliner et al.
where a significantly shifted multi-line spectrum appeared with increasing
lanthanide concentration while the apo-α-LA spectrum correspondingly
disappears (Berliner et al., 1987).  For example Fig. 14 depicts the effects
of Tb(III) on α-LA.  Since Murakami et al. (1982) showed earlier that the
lanthanides bind to the calcium with stronger affinity constants than for
calcium, the added lanthanide binds essentially stoichiometrically as shown
in Fig. 14b.  Note the profuse number of shifted lines compared with the
Ca(II)-α-LA from (Fig. 14a).  However, all of the new lines grew
proportionally with the stoichiometric addition of Tb(III) shift reagent as
expected for slow exchange.  Several other lanthanides were shown to binding
α-LA, causing either large paramagnetic shifts or extensive paramagnetic
broadening, depending on the nature of the lanthanide.  Although distances
could not be derived from these experiments due to the slow exchange
problem, relative affinities were extractable from the NMR measurements of
calcium displacement by these reagents:  Dy(III), Tb(III), Pr(III)>Ca(II)>
Yb(III).  These relative affinities also agreed with correlations to ionic
radius (Berliner et al., 1987).

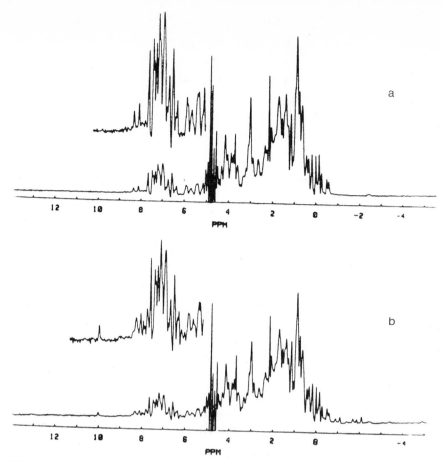

Fig. 14.   200 MHz proton NMR spectra of 2 mM Ca(II)-α-LA in the absence (a) of and presence (b) of 0.5 mM Tb(III).   The insets are high gain plots of the downfield regions. Conditions were pH 7.2 (50 mM Tris-d11-DCl), 16K data points, 4000 Hz sweep width.   From Berliner et al. (1987) with permission.

On the other hand, when a paramagnetic ligand is undergoing fast exchange with a single cation site (to which it binds weakly) amino acid residue assignments and the cation-proton distance calculations are more feasible.   For example, Gerken measured a distance of 7.5Å from the Mn(II) bound to a low-affinity secondary calcium site to the($^{13}$C-dimethylated) amino terminus (Gerken, 1984).   Likewise, Berliner et al. (1987) observed significant fast-exchange paramagnetic broadening of His-68 and 107 by $^1$H NMR Mn(II) bound to secondary site(s) on Ca(II)-α-LA.

INTRAMOLECULAR DISTANCE MEASUREMENTS

While with NMR it is difficult to overcome the slow exchange problems noted above, it is possible with ESR to examine paramagnetic interactions in

$\alpha$-LA between the calcium site (I) and a zinc site (II). Upon binding Cu(II) to site II (zinc site) of Gd(III)-$\alpha$-LA, no change in the ESR spectrum of either bound cation was observed, accounting for no electron-electron dipolar interactions (Musci et al., 1986). When examining the theory behind this interaction, it was clear that the dipolar interaction becomes negligible at an intrasite separation of about 10 to 12Å. This was also reconfirmed in fluorescence distance measurements between Eu(III), or $_T$b(III) and Co(II) (Gerken, 1984).

Spin labels are powerful probes of protein conformation and dynamics. An iodoacetamido nitroxide analog was covalently attached to the single Met-90 thioether moiety of bovine apo-$\alpha$-LA (Musci et al., 1988). The protein was labeled with a 10-fold excess of the nitroxide spin label 4-(2-bromoacetamido) -2,2,6,6-tetramethylpiperidine-N-oxyl at pH 3.6 for several days. The resultant ESR spectrum, which was reflective of a rapidly tumbling nitroxide moiety on the methionine side chain, yielded a correlation time (related to a rotational tumbling rate) of 1.2 nsec. Upon addition of 1:1 Ca(II) to the apo-protein, the rotational correlation time changed to 0.9 nsec consistent with a calcium-induced change in structure. Upon substitution with the paramagnetic lanthanide Gd(III) at the calcium site, an 88% decrease in the spin label ESR intensity was observed, consistent with an 8.0 ± 1.0 Å separation between the two sites (Taylor et al., 1969). As we discussed earlier, several intramolecular distances between the N-O moiety and residues on the protein were also determined from the paramagnetic proton relaxation between the nitroxide and amino acid residues. A topographical map, which incorporates fluorescence measurements, as well, is shown in Fig. 15.

The nitroxide was paramagnetically broadened when Gd(III) was sub-stituted for Ca(II) at the strong site. Utilizing theory developed by Leigh and colleagues (Kaptein, 1982), a distance of 8.0 ± 1.0Å was calculated as the closest distance between the electron on the N-O group of the spin-label moiety and the calcium site. The distance measured was consistent with the x-ray determined calcium binding loop assuming the N-O moiety enjoys the freedom of rotation reflected in its ESR spectra.

Overall, we now have several pieces of evidence in solution which are consistent with a calcium site in the vicinity of residues Asp 82, Asp 83, Asp 84, Asp 87, and Asp 88; however, it is clear that additional experiments are always desirable to further triangulate upon this site.

PROTEIN FOLDING

As noted in the earlier NMR discussions, $\alpha$-LA is known to undergo well-characterized thermal unfolding transitions. In particular, apo-$\alpha$-LA as a mixture of the folded and unfolded state(s) at room temperature is also

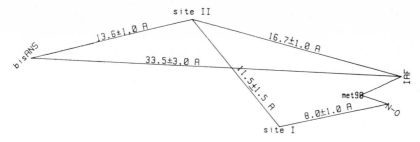

Fig. 15.   A distance of α–lactalbumin derived from ESR (Site I to Met 90) and fluorescence energy transfer distance measurements. From Musci and Berliner (1986) with permission.

evident from optical and NMR spectroscopy thermal unfolding curves as well. While this unfolding transition is sensitive to the concentrations of other monovalent salts to some degree, it is most dramatically affected by the binding of 1:1 calcium.   As noted in Fig. 16 where apo–α–LA at 39° is completely unfolded with no ring current shifted lines, the Ca(II)–α–LA, which is absolutely normal at 45°, does not unfold until extremely high temperatures are reached.

Permyakov, Sugai and others have found spectroscopic evidence for Na(I) and K(I) binding, which partially stabilizes apo–α–LA against thermal denaturation although not at all as effectively as Ca(II) (Permyakov et al., 1985; Kuwajima et al., 1986).   The results also suggested multiple binding sites for these monovalent cations.   For example, the effects of KCl on thermal unfolding, as monitored by high–resolution proton NMR, were shown to be more local structural phenomena. While it is clear that the strong

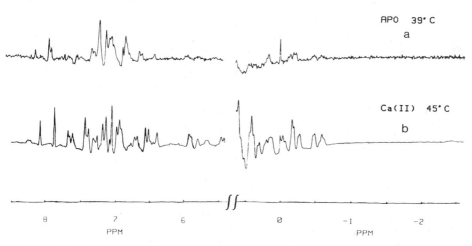

Fig. 16.   200 MHz H NMR spectra of (a) 1.3 mM apo–α–LA (50 mM ($d_{11}$) tris–DCl, pH 7.5 at 39°; (b) 1.9 mM Ca(II)–α–LA at 45°.

binding of Ca(II) and probably Zn(II) are significant both physiologically and in vitro, the additional known sites are more difficult to evaluate for physiological relevance. Since cytosolic concentrations of monovalent cations and Ca(II) are rather high, the phenomena stated above, whether they be specific or nonspecific, must contribute to the physical state of the protein in-vivo.

## CONCLUSIONS

The examples shown here of NMR applications to α-lactalbumin represent several approaches to elucidating conformational details about this protein. Several chapters in this volume address 2D-NMR techniques which may be applied to α-lactalbumin and yield even more information more rapidly. Of course the tedious and laborious process of specific assignment precedes 2D methods as well. It is hoped that these details will be available within the next year.

## ACKNOWLEGEMENTS

The work described herein was generously supported by the National Science Foundation (DMB 8703794) and the U.S.P.H.S. (HD17270). The NMR equipment was supported by a core research grant NIH GM 27431 and partial support to the OSU Campus Chemical Instrumentation Center (S10 RR01458).

## REFERENCES

Acharya, K. R., Stuart, D. I., Walker, N. P. C., Lewis, M., and Phillips, D. C., 1989, J. Mol. Biol., 208:99.
Beg, O. U., von Bahr-Lindstrom, H., Zaidi, Z. H., and Jornvall, H., 1985, Eur. J. Biochem., 147:233.
Bell, K., McKenzie, H. A., and Shaw, D. C., 1981, Mol. Cell. Biochem., 35:113.
Berliner, L. J., Andree, P. J., and Kaptein, R., 1978, Proc. 7th Int. Conf. Magnetic Resonance Biological Systems, Nara, Japan, 115.
Berliner, L. J., Ellis, P. D., and Murakami, K., 1983, Biochemistry, 22:5061.
Berliner, L. J., and Kaptein, R., 1981, Biochemistry, 20:799.
Berliner, L. J., Koga, K., Nishikawa, H., and Scheffler, J. E., 1987, Biochemistry, 26:5769.
Drakenberg, T., and Vogel, H. J., 1983, Calcium-Binding Proteins (B. de Bernhard, G. L. Sottocasa, G. Sandria, E. Carafoli, A. N. Taylor, T. C. Vanaman, and R. J. P. Williams, eds.), Elsevier Science, Amsterdam, 25:73.
Gerken, T. A., 1984, Biochemistry, 23:4088.
Herzberg, O., and James, M. N. G., 1985, Biochemistry, 24:5298.
Hill R. L., and Brew, K., 1975, Adv. Enzymol. Relat. Areas Mol. Biol., 43:411.
Hiraoka, Y., Segawa, T., Kuwajima, K., Sugai, S., and Moroi, N., 1980, Biochem. Biophys. Res. Commun., 95:1098.
Kaptein, R., 1982, Biol. Magn. Reson., 4:145.
Koga, K., and Berliner, L. J., 1985, Biochemistry, 24:7257.
Kretsinger, R. H., and Nockolds, C. E., 1973, J. Biol. Chem., 248:3313.
Kronman, M. J., and Andreotti, R. E., 1964, Biochemistry, 3:1145.
Kronman, M. J., and Bratcher, S. C., 1984a, J. Biol. Chem., 259:0887.

Kronman, M. J., and Bratcher, S. C., 1984b, J. Biol. Chem., 259:10875.
Kronman, M. J., Hoffman, W. B., Jeroszko, J., and Sage, G. W., 1972,
    Biochim. Biophys. Acta, 285:124.
Kronman, M. J., Sinha, S. K., and Brew, K., 1981, J. Biol. Chem.,
    256:8582.
Kuwajima, K., Harushima, Y., and Sugai, S., 1986, Int. J. Peptide Protein
    Res., 27:18.
Murakami, K., Andree, P. J., and Berliner, L. J., 1982, Biochemistry,
    21:5488.
Musci, G., and Berliner, L. J., 1985, Biochemistry, 24:3852.
Musci, G., and Berliner, L. J., 1986, Biochemistry, 25:4887.
Musci, G., Koga, K., and Berliner, L. J., 1988, Biochemistry, 27:1260.
Musci, G., Reed, G. H., and Berliner, L. J., 1986, J. Inorg. Biochem.,
    26:229.
Permyakov, E. A., Morozova, L. A., and Burstein, E. A., 1985, Biophys.
    Chem., 21:21.
Permyakov, E. A., Yarmolenko, U. V., Kalinchenko, L. P., Morozova, L. A.,
    and Burstein, E. A., 1981, Biochem. Biophys. Res. Commun., 100:191.
Poulsen, F. M., Hock, J. C., and Dobson, C. M., 1980, Biochemistry,
    19:2597.
Smith, S. G., Lewis, M., Aschaffenburg, R., Fenna, R. E., Wilson, I. A,
    Sundaralingam, M., Stuart, D. I., and Phillips, D. C., 1987,
    Biochem. J., 242:353.
Stuart, D., Acharya, K., Walker, N., Smith, S., Lewis, M., and Phillips,
    D., 1986, Nature, 324:84.
Taylor, J. S., Leigh, J. S., and Cohn, M., 1969, Proc. Natl. Acad. Sci.
    U.S.A., 64:219.

APPLICATIONS OF MULTINUCLEAR NMR IN THE SOLID STATE TO STRUCTURAL AND
DYNAMICAL PROBLEMS IN MACROMOLECULAR CHEMISTRY

Robert G. Bryant[*+], S. D. Kennedy[+], C. L. Jackson[*],
Thomas M. Eads[**], William R. Croasmun[**] and
Allen E. Blaurock[**]

Biophysics Department
University of Rochester
Medical Center
601 Elmwood Avenue
Rochester, NY  14642

INTRODUCTION

Nuclear magnetic resonance is a long established molecular spectroscopy
that provides both dynamical and structural information about the molecules
bearing the nuclear spin observed.  Currently, data acquisition is accomplished in the time domain by the application of radio frequency excitation
followed by acquisition of the transient nuclear induction signal using
rapid analog to digital conversion for storage of the data set in a fast and
usually dedicated digital computer.  A Fourier transform of the time domain
signal provides the usual frequency domain spectrum that is most often
interpreted to yield structural information about the molecules observed.

The dominant application of these methods has been to liquids, even
liquid components in heterogeneous systems such as whole tissues (Ernst et
al., 1987).  Using acquisition methods appropriate to liquid samples, the
acquisition trigger is set so that the signals from solids, which decay very
rapidly following the excitation pulse, are not captured.  In this mode the
signals from a solid component of a heterogeneous system are not observed
while the rapidly moving molecules are.  This scheme is the basis for most
in vivo NMR spectroscopy practiced today, and provides the basis, for
example, for studying the high energy metabolites in tissues such as
adenosine triphosphate or creatine phosphate as a function of what is

---

[*]  Chemistry Department, University of Rochester, Rochester, NY  14627
[**] Kraft, Inc. – Technology Center, Glenview, IL  60025
[+]  Biophysics Department, University of Rochester, Rochester, NY  14642

done physiologically to the host (Gorenstein, 1984). Clearly similar methods may be applied to agricultural problems in plants and this symposium provides ample evidence of the rich opportunities available from these experimental approaches.

The crucial components of a complex material, however, may not always be liquid. In the case of a solid, the magnetic dipole-dipole coupling between nuclear spins that are close together cause a significant broadening of the observable nuclear magnetic resonance signals that obscures the high resolution chemical and dynamical information inherently available in the spectrum. However, there are now many methods that may be used to recover this information and characterize both the structural and dynamical nature of the molecules in a solid or heterogeneous system. These methods provide new and important opportunities for understanding such systems at a molecular level (Mehring, 1976).

Though solid state line narrowing methods may be applied to observe both dominant spins like protons and rare spins like carbon-13, only the rare spin experiments will be discussed in the present context. It is perhaps useful to note in passing that the homonuclear line narrowing methods are consider- ably more demanding technically than the heteronuclear experiments that are used to observe the relatively rare spins like carbon-13, nitrogen-15 or phosphorus-31. Using the carbon case as an example, the essence of the problem is to eliminate the proton-carbon dipole-dipole broadening sensed at a carbon spin. The solution is the application of a strong radio frequency field at the proton resonance frequency during the acquisition of the carbon spectrum, i.e., the protons are decoupled. To affect the decoupling, the rf field must be strong compared with the strength of the dipole-dipole coupl- ing, roughly 50 kHz, which is considerably larger than the scalar couplings of about 125 Hz that are routinely suppressed in liquid spectroscopy. One consequence of the higher rf field is that sample heating may be a problem, especially in samples, which includes most biological materials. The problem of proton-carbon rf channel isolation, at one time a limiting experimental factor, is now routinely solved in commercial instrumentation. The carbon spectrum that results from such an experiment is rather different from a liquid spectrum in that the lines are still rather broad and not of the usual Lorentzian shape. The line shape in static samples generally results from the anisotropy inherent in the chemical shift at the observed nucleus. In liquids this anisotropy, which is on the order of 10-100 ppm, is averaged completely by the rapid rotation of the molecules; however, in a polycrystal- line solid no averaging occurs, and the carbon resonance observed reports the superposition of spectra for the random distribution of crystals and is usually called the powder spectrum. A simple representative spectrum is shown in Fig. 1 for calcium acetate. The downfield resonance is the more

256

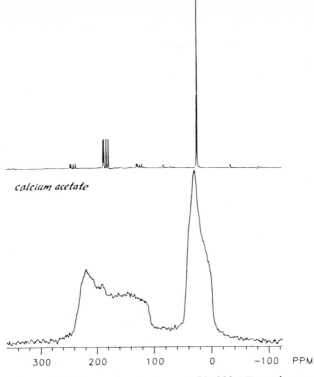

*calcium acetate*

Fig. 1. Carbon-13 NMR spectra taken at 50.289 MHz using cross-polarization methods. At the bottom is the static spectrum acquired in 3260 transients with a contact time of 1 ms into a 1 k data file which was exponentially weighted with a line broadening of 40 Hz. At top is the magic angle spinning spectrum acquired in 460 transients with a contact time of 1 ms into an 8 k data file which was exponentially weighted with a line broadening of 5 Hz. The small satellites are spinning side bands. The broad carbonyl tensor splits into four resolved lines indicating four distinct environments while the methyl resonance remains an single line. The anisotropy of both resonances is readily apparent in the static spectrum.

anisotropic and is due to the carboxyl carbon atoms. The high field peak is much less anisotropic, and is due to the methyl carbons which are averaged in part by rapid methyl rotation about the carbon-carbon bond at room temperature. In the case where motion in the solid may reorient the chemical shift tensor on a time scale short compared with the width of the pattern, the motionally induced averaging causes NMR line shape changes that may be used to study intramolecular motions in the solid. Alternatively, mechanical motion may be imparted to the sample; this motion may also cause the line shape to change as well. In fact if the mechanical motion is a rotation about an axis tilted at an angle of 54.7° from the direction of the static magnetic field, the chemical shift anisotropy may be made to collapse completely and a sharp liquid-like spectrum results as shown in Fig. 1B. The increased resolution available from a spectrum taken on a spinning sample is impressive, and permits the same sorts of structural information to be obtained from such samples as would normally be obtained from a standard one dimensional carbon spectrum of a liquid. In addition, longitudinal relaxation rates or other relaxation rates may be measured on spinning samples at high resolution that permit dynamical characterization of the solid based on the relaxation rates of specific resonance lines.

A difficult feature of carbon spectroscopy both in liquids and solids is the long spin-lattice relaxation time generally associated with either very non-viscous liquids or crystalline solids. This problem may be overcome in part, with considerable additional advantage in signal-to-noise ratio, if in addition to decoupling, the spectrum is acquired by deriving the carbon magnetization from a thermal contact with the proton spins. If the carbon and proton spins may be made to have the same resonance frequency, energy or magnetization will flow between the two spin systems and they will come to a common spin temperature. Since the proton population is very large, the proton system has a large heat capacity and may easily pump the carbon spin population. In addition, the proton spins usually relax more rapidly than the carbons; thus, signal averaging may be executed much more rapidly using a recycle time appropriate to the proton spins rather than the carbons. The two spin systems are brought to the same resonance frequency using a pair of radio frequency fields, the amplitudes of which are adjusted so that in these fields the proton spins and the carbon spins precess at the same rate. Since they are at different radio frequencies, these fields may be adjusted independently while the static or dc magnetic field may not. The result when the fields are properly matched is a magnetization transfer from the proton system to the carbon system and approximately a factor of 4 improvement in signal-to-noise ratio. For the transfer to be effective, the carbon-proton dipole-dipole interaction must be strong, i.e., not averaged by rapid internal molecular motion, otherwise the transfer rate will be prohibitively slow. The overall experiment is often called a cross-polarization experiment and it may be done on spinning samples as well as static ones. The failure of a cross-polarization experiment is one indication of internal motions in the sample that average the dipole-dipole coupling. Of course, once the carbon magnetization is prepared, one may manipulate the magnetization in the usual variety of ways so that the whole cadre of relaxation and multiple pulse experiments may be executed once the cross-polarization step is completed. The cross-polarization experiment may present some difficulties because the spectral intensities depend on the proton relaxation time, the carbon and proton relaxation times in the radio frequency fields, and often different rates for the magnetization transfer. Usually these times are sufficiently long that reliable intensities may be obtained; however, in some relatively proton poor samples such as coal or large polynuclear aromatic systems, the intensity problem may be major (Dudley et al., 1982). There may also be intensity shifts in the case where part of a large molecule moves, such as protein side chains, and another part does not, such as the protein backbone. Nevertheless, such problems are usually obvious and a source of information.

Perception of food texture is a sensory experience of the mechanical responses of food to manipulation and shear. Mechanical responses originate largely in the structure and dynamics of networks of food macromolecules. For example, for a polysaccharide, there is a direct relationship between polymer conformational fluctuations and solution viscoelastic properties. Similarly, observation of an elastic response in semisolid foods suggests the existence of macromolecular elements with significant conformational freedom involved in a network that has a sufficient number of cross-links to provide reversibility. Since rheological response strongly depends on intensive properties of the system, such as temperature or the symmetry constraints provided by the network, an examination of the structural and dynamical states of the macromolecular constituents under native or near-native conditions must provide useful information for control of the sensory properties.

The application of solid state NMR methods to heterogeneous systems in the agricultural and food industries is very attractive because a number of materials that have been resistant to study may be characterized at a molecular level (Jacob et al., 1985). We will examine two situations: 1) the dynamical and structural response of proteins and a polysaccharide to hydration, and 2) the effect of crystal lattice constraints on motions of deuterated triglycerides.

HYDRATION RESPONSE

The response of molecules and organisms to hydration is central to a number of issues including draught survival, cold hardiness, materials storage, product texture, product development as in doughs or baking or fermentation, etc., (Kuntz and Kauzman, 1974). The dynamical and structural response of molecules may be rather different, but solids NMR methods provide a means for addressing aspects of both.

Casein is an abundant milk protein that is present in a large number of food products that may have very different textures and water contents. The NMR experiment provides a step towards relating the internal dynamics of the protein to the influence on food textures. The magic-angle-spinning spectrum taken under cross-polarization conditions of sodium alpha-S$_1$

caseinate is shown in Fig. 2A for dry protein.  It is typical of the spectra associated with proteins.  The resolution is not impressive because of the very large number of overlapping resonances from both main-chain or backbone carbon atoms and side-chain carbon atoms.  That the spectrum was accumulated efficiently using cross-polarization methods indicates that the proton-carbon dipolar coupling is unaveraged by motion in this dry material.  By contrast, the spectrum of alpha-$S_1$ hydrated at $a_w$ = 0.92 shown in Fig. 2B was taken without rapid sample spinning to narrow the lines, and without the benefit of cross-polarization because the attempts to acquire such spectra yielded very low signal-to-noise ratios.  The reason for the failure of the cross-polarization experiment is apparent in the highly narrowed spectrum of Fig. 2B.  Note that there are now three resonances for the carbonyl carbon atoms near 180 ppm while only one is apparent in spectrum A.  Therefore, there is sufficient local motion of the macromolecule to very largely collapse the chemical shift anisotropy of the backbone carbon atoms as well as the side chains.  By this criterion, the protein is essentially dis-solved, though the resulting solution is so highly viscous that the material has the consistency of neoprene rubber.  Somewhat higher resolution is possible when the same highly hydrated sample is rotated rapidly at the magic angle as shown in Fig. 2 spectrum C.

A similar hydration response may be seen in the carbon-13 spectra of glycogen, which is a highly branched polymer of glucose linked alpha-1,4 with alpha-1,6 branches approximately every 12 residues.  This very open structure makes the molecule difficult to pack (Jackson and Bryant, 1989), and the carbon spectrum of the dry material shown in Fig. 3B obtained with magic-angle spinning is considerably broadened relative to that of the crystalline glucose, spectrum A, because of the combination of the effects of local motions and a distribution of local environments in the non-uniformly packed material.  Nevertheless, the cross-polarization static spectrum shows only a single broad line indicating that the proton-carbon dipolar interaction is largely unaveraged and the chemical shift anisotropy of the carbon resonances is not eliminated.  The hydration response of this molecule is dramatic.  At a hydration level of 1 g water per gram of gly-cogen, the pasty sample fails to provide a cross-polarization spectrum, and the resolution of the static spectrum implies very considerable local motion of the glucose units in this polymer.  Relaxation rate measurements indicate that changes in the motion occur at lower water contents, with the C-6 carbon more mobile than the rest (Jackson and Bryant, 1989).  However, by the high water contents such as 50% the carbon resonances relax with nearly maximal efficiency, similar to that expected for a viscous isotropic liquid.

The hydration response of all macromolecules is not like the first two examples given above, i.e., that they begin dissolving with the onset of

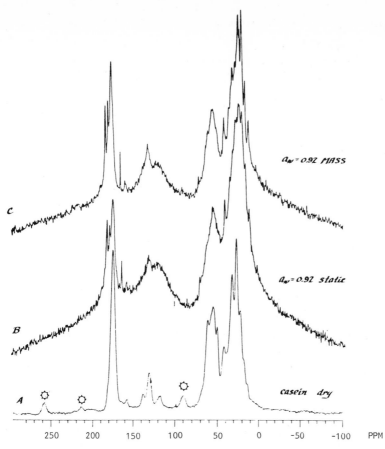

Fig. 2. The carbon-13 NMR spectra of casein taken at 50.3 MHz at ambient temperature. Bottom: the spectrum of a dry powder using a cross-polarization contact time of 1 ms and magic angle spinning at 4.2 kHz. 37800 transients were acquired into a 4 k data file exponentially weighted with 10 Hz line broadening. The circular symbols mark the spinning side bands at 4.2 kHz. Middle: casein hydrated against 92% relative humidity using a carbon 90° pulse with a 50 ms spin-locking pulse to improve phase cycling cancellation of instrument noise. 14,804 transients were acquired into 2 k data file and exponentially weighted with a line broadening of 10 Hz. The broad rolling base line is due in large measure to the Kel-F stator assembly for the high speed rotor that now shows up because the carbon magnetization is no longer derived from contact with the protons. Bottom: a spectrum of the same material as in the middle acquired with the sample spinning at the magic angle at 2.9 kHz using 11016 transients acquired into a 2 k data file exponentially weighted with a line broadening of 10 Hz.

considerable local motion in both the backbone and the side chains. The CP-MAS carbon-13 spectrum of lysozyme in the lyophilized state is shown in Fig. 4A. The resolution is not unlike that for alpha-$S_1$ casein in a similar state shown in Fig. 2. However, in this case the addition of water preserves the cross-polarization spectrum, yet the spectrum sharpens, as

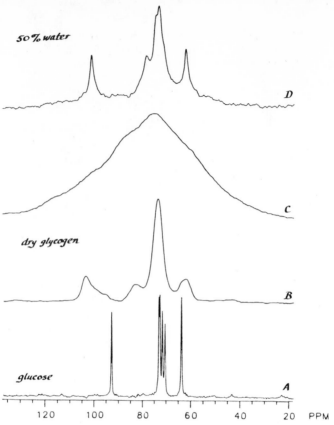

*50% water*

*D*

*C*

*dry glycogen*

*B*

*glucose*

*A*

120    100    80    60    40    20    PPM

Fig. 3. Carbon-13 NMR spectra of glycogen and glucose obtained at 50.3 MHz. A) The cross-polarization magic-angle spinning spectrum obtained using a 1 ms contact time and a rotor frequency of 3.0 kHz. 1008 transients were accumulated into a 4 k data set and exponentially weighted with a line broadening of 5 Hz. B) The cross-polarization magic-angle spinning spectrum of dry glycogen obtained with a 0.75 ms contact time at a rotor frequency of 3.0 kHz. 1188 transients were accumulated into a 2 k data set and exponentially weighted with a line broadening of 30 Hz or 0.6 ppm. C) The cross-polarization of dry glycogen obtained with a contact time of 0.75 ms under static conditions. 13704 transients were accumulated into a 2 k data set and exponentially weighted with a line broadening of 30 Hz. D) The spectrum of a static non-spinning glycogen sample at 50% water content obtained using a 90° carbon pulse only. 2676 transients were accumulated into a 2 k data set and exponentially weighted with a line broadening of 30 Hz.

shown clearly in the aliphatic region, spectrum B of Fig. 4. The preservation of the cross-polarization spectrum indicates that the dipole-dipole coupling between carbon and protons is unaveraged by any motion in the system. The spectrum remains heterogeneously broadened in the lyophilized state, i.e., the line breadth is dominated by the distribution of chemical

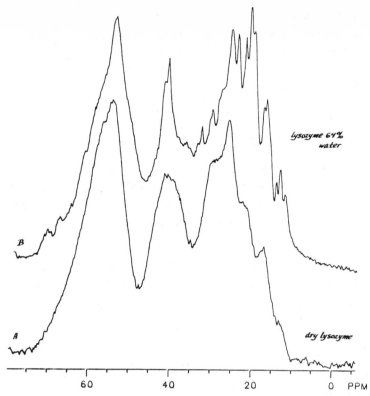

Fig. 4.  The cross-polarization magic-angle-sample spinning
carbon-13 NMR spectra of lysozyme acquired at 50.3 MHz and
ambient temperature.  Bottom:  the aliphatic region of the
lyophilized dry material acquired with a cross-polarization
contact time of 1.0 ms and magic-angle spinning at 4.5 kHz.
24,528 transients were accumulated into a 4 k data table
exponentially weighted with 10 Hz line broadening.  Top:
the aliphatic region of the lyophilized lysozyme rehydrated
to a level of 0.64 g water per gram of protein acquired
using cross-polarization contact time of 0.9 ms.  107,468
transients were accumulated into a 4 k data table
exponentially weighted with 5 Hz line broadening.

shifts presumably created by a distribution of local environments for the
carbon atoms in proteins that are folded slightly differently in the
lyophilized state (Kennedy and Bryant, 1990).  The significant sharpening of
the spectrum on hydration indicates that the protein relaxes to a much
narrower distribution of local conformational states.  In a sense, the water
in this case permits the protein to crystallize rather than dissolve.
Further, what we now know about the comparison between crystalline enzymes
and enzyme solution structures suggests that the structures are not very
different except for side chains in the regions of protein-protein contact
(Wüthrich et al., 1989).  Thus, for this enzymatic protein, the hydration
response is very different from that for alpha-S$_1$ casein or glycogen,
namely, the structural heterogeneity of the lyophilized lysozyme is

significantly reduced by the addition of water, and the increase in local motions often associated with partial solvation of a polymer are no where near as dramatic as those observed in the first two examples.

FAT CRYSTALS

In foods containing semicrystalline fats, rheological properties critical to perception may be conceptualized in models involving crystalline and liquid domains. The mechanisms connecting unit cell symmetry, crystallite size and shape, and super-crystalline morphology of triglycerides to thermal-mechanical properties, including mouthfeel, are not well understood. However, in semicrystalline polymers, molecular motion in the solid state, determined primarily by NMR, can be related to functional properties (Komoroski and Mandelkern, 1986).

Proton NMR applications to solid fats have shown that, at any particular temperature, polymorphic form determines molecular mobility (Norton et al., 1985; Gibon et al., 1986), and that proton relaxation rates can be used to detect motional changes caused by disruption of packing by emulsifiers (Azuory et al., 1988). However, because protons are ubiquitous, and their interactions may normally produce broad and poorly resolved NMR spectra, molecular sites cannot usually be resolved. Thus, localized motions cannot be individually studied. New solids NMR methods can overcome such problems, as shown in high resolution studies of polymorphism in n-alkanes (Derbyshire et al., 1969), membranes and lipid bilayers (Griffin, 1981), and deuterium NMR studies of membranes and bilayers (Davis, 1983; Smith and Oldfield, 1984). The chemical detail obtainable by NMR sometimes complements the structural detail of X-ray diffraction methods. NMR work of this kind was reported for tripalmitin and tristearin by the Unilever group (Norton et al., 1985; Bociek et al., 1985) who proposed molecular packing origins of carbon-13 chemical shifts of different polymorphs. Molecular packing is only known in detail for the beta polymorph, from single crystal studies (Larsson, 1964). Bociek et al. (1985) also showed qualitatively by carbon-13 relaxation experiments that the glycerol backbone was relatively rigid, that methylene motions increase from beta to beta prime to alpha, and that methyl groups retain their mobility in all forms at near ambient temperatures. We report here preliminary experiments utilizing deuterium NMR measurements that are sensitive to the motions in the solid at different and shorter time scales.

Glycerol positions 1 and 3 of commercial tripalmitin (Aldrich) were substituted with perdeuterated palmitic acid by enzymatic interesterification using Mucor miehei lipase (Lipase 3A, Novo Enzyme Co.). The tripalmitin, in which 75% of the fatty acid chains in the 1 and 3 positions were perdeute rated, was purified by preparative thin layer chromatography and recovered by solvent evaporation. Differential scanning calorimetry (DSC) experiments showed that the deuterated sample could be prepared in alpha, beta prime, and beta forms, which melted at the same temperatures to within 2°K and with the same specific enthalpy to within 10 J/g as unsubstituted tripalmitin employing the same protocols. The polymorphic forms were separately prepared from the same sample. Their X-ray powder diffraction patterns, which are similar to those determined on analogous samples of unsubstituted tripalmitin, and which correspond to typical hexagonal (alpha), orthorhombic (beta prime), and triclinic (beta) patterns (Small, 1986), were stable for hours at 293°K. The alpha and beta preparations showed virtually no contamination by other forms. The beta prime sample was found to contain between 83 to 89% beta prime after the NMR experiment with a maximum beta contamination of 0.5%, the rest being alpha as judged by the X-ray diffraction pattern. This sample was found by DSC to convert more rapidly to beta with increasing temperature.

The deuterium NMR spectrum is dominated by the interaction between the nuclear electric quadrupole moment of this spin-1 nucleus and the electric field gradient sensed at the nucleus (Abragam, 1961). The quadrupole contribution to the Hamiltonian generally dominates the deuterium spectrum; though the chemical shift contribution exists, it is essentially the same as that for protons on a ppm scale. The electric field-gradient tensor is usually well approximated by axial symmetry in a carbon-deuterium bond, however, for a powder sample, there are all orientations of this gradient represented in the sample leading to a superposition of each orientational contribution to the spectrum. Unlike the carbon chemical shift tensor, the deuterium NMR spectrum arises from two transitions each with an axially symmetric powder pattern that is related by symmetry to the other. As in the carbon case, motions that mix orientations on a time scale short compared with the reciprocal of the line separations cause significant powder-pattern line-shape changes. Fast motions about a single axis, such as the rapid rotation of methyl groups, preserves the axial nature of the powder pattern, but reduces its width. Fast motions of other types may cause both a line-shape change and a decrease in width. Detailed line shape analysis generally requires comparison of experimental line shapes with those computed from various motional models; however, often considerable information may be obtained without an attempt to make a detailed fit of the line shape.

Deuterium NMR spectra were recorded on a Bruker MSL400 NMR spectrometer at the Kraft Technology Center. A quadrupole echo pulse sequence (Spiess, 1985) was used with a 90° pulse width of 2.2 ms, and a typical recycle delay of 20 s. Sample size was generally about 50 mg, and temperature was maintained at 298°K using a nitrogen flow over the sample.

Deuterium spectra obtained by Fourier transformation of the quadrupole echo using an echo delay of 17 ms are shown in Fig. 5 for these three

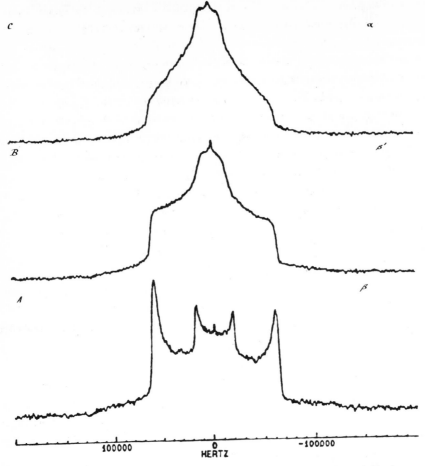

Fig. 5. Deuterium NMR spectra taken using a solid echo with the echo delay of 15 μs at 61.40 MHz at 298°K on tripalmitin that is perdeuterated at chains 1 and 3 prepared in different polymorphic forms. 56 transients were accumulated for each spectrum with a recycle delay of 20 s, exponential weighting was applied with a 400 Hz line broadening. A) beta form (triclinic); B) beta prime form (orthorhombic); C) alpha form (hexagonal).

polymorphic crystalline forms. The beta form, spectrum A, shows a superposition of axial powder patterns that are common in deuterium spectroscopy. The narrow component has a horn-to-horn separation of 35 kHz while that expected for a rapidly rotating methyl group experiencing no other motion is 43 kHz (Spiess, 1985). This difference suggests that the terminal methyl group in this form experiences additional motion that retains the essentially axial symmetry of the spectrum. The broader components of the spectrum contain contributions from all the remaining methylene deuterons. Unlike phospholipid membranes that may display unique effective quadrupole splittings for each position down the chain from the headgroup, this spectrum shows only one with a horn-to-horn separation of approximately 120 kHz while that expected for a rigid methylene group is 127 kHz (Jelinski, 1986; Burnett and Muller, 1971). This spectrum requires that, though there may be some high frequency motions present that reduce the effective quadrupole coupling constant somewhat in the methylene region of the structure, the motions are sufficiently restricted that the methylenes do not display distinctly quadrupole coupling constants for different positions down the chain from the glycerol headgroup. Further, these motions are such that the axial character of the spectrum is retained.

By contrast, the spectra of both alpha and beta prime forms are drastically different from the axial pattern of the beta form. Though the width of the spectrum is preserved, the clear distinction between methyl positions and methylene positions is lost. It is clear that there are motions present on the time scale of microseconds or faster that alter the inherently axial character of the deuterium spectrum. A detailed model of the motion requires considerably more data; however, we note that similar spectral changes are observed in polymeric systems when two site tetrahedral jumps are present in alkane chains (Spiess, 1985; Jelinski, 1986).

The dramatic dynamical differences between the beta and the other two forms are shown clearly in the inversion recovery data summarized in Figs. 6, 7, and 8. The data for the beta form (Fig. 6) indicate that even at a recycle time of 20 s, the spectrum is partially saturated. As a result the methyl to methylene intensity ratio is slightly distorted. The methylene deuterons relax much more slowly than the methyl deuterons in the beta form, which may be expected from the rapid methyl rotation. Though there are no major pattern changes in the partially relaxed spectra, $T_1$ is not uniform across the powder pattern, which is expected (Torchia and Szabo, 1982). The data summarized for the beta prime and alpha forms in Figs. 7 and 8 demonstrate much more rapid spin-lattice relaxation for all the deuterons in the spectrum, which is consistent with the line-shape changes requiring considerable and rapid motion of the chains.

The major differences among these polymorphs as displayed in the deuteron spectra are clearly the much more restrictive requirements on the motions in the beta form compared with the other two. Nevertheless, the spectra for beta prime and alpha are obviously different in detail. The beta prime spectrum has sharper edges at the extreme and more apparent line shape changes in the inversion recovery experiment. Presently, there is insufficient data to attempt a detailed model of these differences; however, we note that the beta prime spectrum, which at first glance might appear to be in some sense intermediate between the alpha and beta spectrum, cannot be constructed as a linear combination of the alpha and beta spectra.

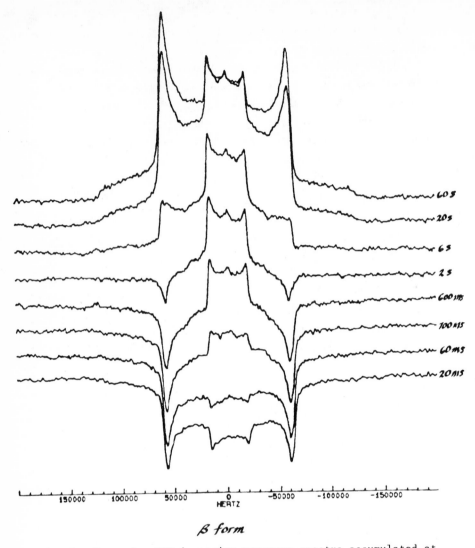

150000    100000    50000    0    −50000    −100000    −150000
HERTZ

*β form*

60 s
20 s
6 s
2 s
600 ms
100 ms
60 ms
20 ms

Fig. 6. Deuterium NMR inversion recovery spectra accumulated at
61.04 MHz and 298°K using solid echo methods with an echo
delay of 15 μs on the beta form of tripalmitin
perdeuterated at chains 1 and 3. Delay times are
indicated; other parameters are summarized in Fig. 5.


Indirect physical evidence has led to the supposition that the hexagonal
lattice of the alpha form affords the opportunity for rotational freedom of
the aliphatic chains. The previous carbon–13 spectra (Norton et al., 1985;
Bociek et al., 1985) and the proton spectra (Gibon et al. 1986) imply an
increased motion present in the alpha compared with beta structures. The
present deuterium spectra clearly support this result which appears to be
consistent with the increased volume per methylene unit available in the
hexagonal and orthorhombic packing of the alpha and beta prime forms.

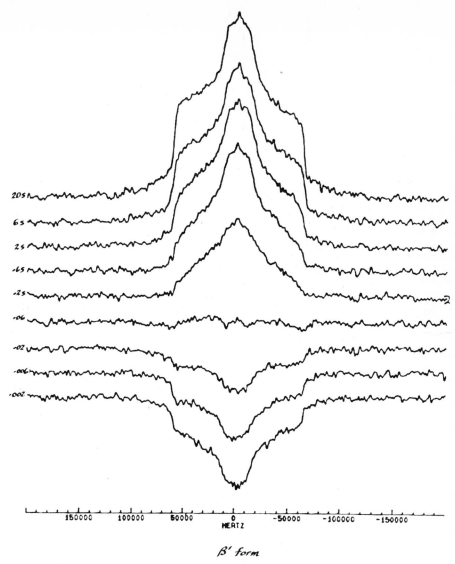

205s

6s

2s

·6s

·2s

·06

·02

·006

·002

150000    100000    50000    0    -50000    -100000    -150000

HERTZ

β' form

Fig. 7. Deuterium NMR inversion recovery spectra accumulated at 61.04 MHz and 298°K using solid echo methods with an echo delay of 15 μs on the beta prime form of tripalmitin perdeuterated at chains 1 and 3. Delay times are indicated; other parameters are summarized in Fig. 5.

ACKNOWLEDGMENTS

The deuterated tripalmitin was prepared by David Hayashi, Robert Dinwoodie, and Michael Kern. Frank Sasevich performed DSC measurements. The assistance of these Kraft scientists is gratefully acknowledged. Portions of this work were supported by the National Institutes of Health, (GM34541), The University of Rochester, and an extramural grant in aid from Kraft, Inc.

269

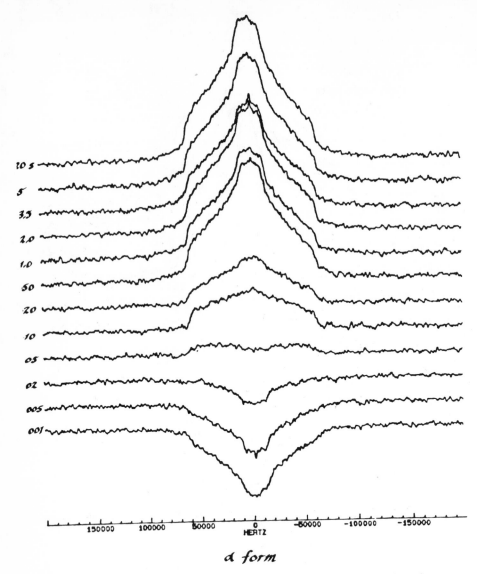

*d form*

Fig 8.   Deuterium NMR inversion recovery spectra accumulated at
61.04 MHz and 298°K using solid echo methods with an echo
delay of 15 μs on the alpha form of tripalmitin
perdeuterated at chains 1 and 3.   Delay times are
indicated; other parameters are summarized in Fig. 5.


REFERENCES

Abragam, A., 1961, Principles of Nuclear Magnetism, Clarendon Press,
    Oxford, Ch. 7.
Azuory, R., Aronhime, J. S., Sarig, S., Abrashkin, S., Mayer, I., and
    Garti, N., 1988, J. Am. Oil Chem. Soc., a6:964.
Bociek, S. M., Ablatt, S., and Norton, I. T., 1985, J. Am. Oil Chem.
    Soc., 62:1261.
Burnett, L. J., and Muller, B. H., 1971, J. Chem. Phys., 55:5829.

Davis, J. H., 1983, Biochem. Biophys. Acta., 737:117.

Derbyshire, W., Gorvin, T. C., and Warner, D., 1969, Mol. Phys., 17:401.

Dudley, R. L., and Fyfe, C. A., 1982, Fuel, 61:651.

Ernst, R. R., Bodenhausen, G., and Wokaun, A., 1987, "Principles of Nuclear Magnetic Resonance In One and Two Dimensions," Clarendon Press, Oxford.

Fyfe, C. A., 1983, Solid State NMR for Chemists, C.F.C. Press, (P.O. Box 1720, Guelph, Ontario, Canada).

Gerstein, B. C., and Dybowski, C. R., 1985, Transient Techniques in NMR of Solids, Academic Press, New York.

Gibon, V., Durant, F., and Deroanne, C., 1986, J. Am. Oil Chem. Soc., 63:1047.

Gorenstein, D. G., ed., 1984, Phosphorus-31 NMR, Academic Press, New York.

Griffin, R. G., 1981, Methods in Enzymology, 72:108.

Jackson, C. L., and Bryant, R. G., 1989, Biochemistry, 28:5024.

Jacob, G. S., Schaefer, J., and Wilson, Jr., G. E., 1985, J. Biol. Chem., 260:2777.

Jelinski, L. W., 1986, "High Resolution NMR Spectroscopy of Synthetic Polymers in Bulk," R. A. Komoroski, ed., VCH Publishers, Deerfield Beach, Florida, p. 335.

Kennedy, S. D., and Bryant, R. G., 1990, Biopolymers, in press.

Komoroski, R. A., and Mandelkern, L., 1986, High Resolution NMR Spectroscopy of Synthetic Polymers in Bulk, R.A. Komoroski, ed., VCH Publishers, Deerfield Beach, Florida, pp. 1-17.

Kunts, I. D., and Kauzman, W., 1974, Adv. Protein Chem., 28:239.

Larsson, K., 1964, Ark. Kemi., 23:1.

Mehring, M., 1976, High Resolution NMR Spectroscopy in Solids, Springer-Verlag Series: NMR Basic Principles and Progress, Volume 11.

Norton, I. T., Lee-Tuffnell, C. D., Ablett, S., and Bociek, S. M., 1985, J. Am. Oil Chem. Soc., 62:1237.

Small, D. M., 1986, "The Physical Chemistry of Lipids, from Alkanes to Phospholipids," V. 4 in Handbook of Lipid Research, Plenum Press, New York, pps. 26-7 for hydrocarbon chain packing volumes.

Smith, R. L., and Oldfield, E., 1984, Science, 225:280.

Spiess, H. W., 1985, Adv. Polymer Sci., 66:24.

Torchia, D. A., and Szabo, A., 1982, J. Magn. Reson., 107.

Wuthrich, K., et al., 1989, J. Am. Chem. Soc., 111:1871.

HETEROGENEITY OF INTACT COLLAGEN BY SOLID STATE MULTINUCLEAR MAGNETIC
RESONANCE

Yukio Hiyama[*] and Dennis A. Torchia

Bone Research Branch
National Institute of Dental Research
National Institutes of Health, Bethesda, MD   20892

INTRODUCTION

Molecular structure of collagen by X-ray diffraction and electron microscopy suggests heterogeneity of the molecule in the gap and overlap regions (Fig. 1).  Previous $^{13}$C NMR data of collagen showed no heterogeneity (Sarkar et al., 1983; Torchia et al., 1985).  Therefore we have designed the following two experiments.

1)   To study ultra-slow motions of the backbone region via hydrogen exchange rate measurements at GLY residues through the monitoring of proton cross-polarized $^{15}$N NMR signals at 25.38 MHz.

2)   To investigate the dynamical properties of the surface of the molecule via lineshape and relaxation time measurements of fluorinated phenylalanyl residues by $^{19}$F NMR at 470 MHz (Sarkar et al., 1987).

EXPERIMENTAL

($^{15}$N GLY) Labeled Collagen

Newborn rats were labeled with ($^{15}$N)glycine by feeding a nursing mother rat this amino acid for 21 days at birth.  Calvaria and tail tendon were removed from the 21 day old animals.  The calvaria were defatted after the periosteum was scraped.  Then the calvaria were washed, cut into small pieces, equilibrated with 0.15 M NaCl and packed into a 5mm NMR tube.  Tendons were pulled out of the tail, washed with Triton solution in 0.45 M NaCl, defatted, washed and equilibrated with 0.15 M NaCl.  GC-MS showed that the level of $^{15}$N incorporation was 18%.

[*] Present address:  Control Research and Development, Upjohn Pharm. Ltd., Tsukuba 300-42, Japan

*NMR Applications in Biopolymers*, Edited by
J. W. Finley *et al.*, Plenum Press, New York, 1990

Typical collagen sequence:

Gly – X – Y – Gly – Pro – Y – Gly – Pro – Hyp

Minor helix:

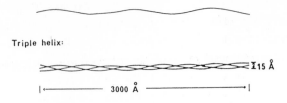

Triple helix:

$\lesssim$ 15 Å

|←――――― 3000 Å ―――――→|

Molecular assembly into fibrils:

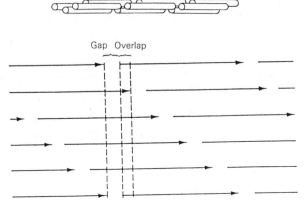

Gap  Overlap

Fig. 1.   Molecular Structure of Intact Collagen.


## Hydrogen Exchange Experiments

Hydrogen exchange was initiated by replacing the 0.15M NaCl $H_2O$ with a 0.15 M NaCl $D_2O$ solution.  The sample was suspended in a fresh deuterated solution every 12 hours during the first 5 days, and then once every 5 days up to 60 days to keep the environment 100% deuterated.  When NMR spectra were recorded, dry nitrogen gas was blown over the sealed sample. At other times the samples were stored in a sealed vial.

## Solid State $^{15}N$ NMR

All solid state $^{15}N$ NMR spectra were obtained at 25.38 MHz with proton cross polarization (with a contact time $t_1$) and proton decoupling using a home built 250 MHz spectrometer described elsewhere (Hiyama et al., 1988). The 90 degree pulse was 5 μs corresponding to a decoupling field strength of 50 kHz.  The proton spin lattice relaxation time ($T_1$) of the intact collagen was on the order of one second in the two tissues.  $T_1$ in the rotating frame ($T_1\rho$) for the tendon collagen was ca 2 ms while $T_1\rho$ of calvaria collagen was longer than 5 ms which were determined by $^{15}N$ NMR signal detection.

274

## (4-Fluorophenylalanyl) Labeled Collagen

L-(4-Fluorophenyl)alanine was incorporated into rabbit Achilles tendon collagen fibers by feeding DL-(4-fluorophenyl)alanine starting at 2 weeks of age for six months. After the animals were sacrificed with sodium phenobarbital, all tissues were washed with 0.15 M NaCl, defatted, and equilibrated against 0.15 M NaCl before packing the samples into NMR tubes. The level of enrichment of L-(4-fluorophenyl)alanine was determined to be 2% by comparing the $^{19}$F NMR signal intensity of a known weight of a hydrolyzed collagen sample with the $^{19}$F signal intensity of a standard solution containing a known weight of the amino acid.

## Solid State $^{19}$F NMR

The 470 MHz fluorine 19 spectra were taken on a Nicolet NIC-500 spectrometer with a 2090 Explorer digitizer and a homemade probe. The probe contained a 4- or 5-mm id solenoid coil, two Johansen capacitors, and a glass Dewar. The 90 degree pulse was 8 μs for the tissue samples while the pulse width was only 3 μs for the crystalline samples. To minimize distortion of the chemical shift powder line shape resulting from finite pulse power, the EXORCYCLE pulse sequence was employed (Bodenhausen et al., 1977; Rance and Byrd, 1983). Free induction decay signals were acquired in quadrature with a 1 MHz sampling rate and 2048 points per channel. Theoretical spectra were calculated on a DEC-10 computer and were corrected for effects of finite pulse power (Hiyama et al., 1986).

## RESULTS AND DISCUSSION

## Hydrogen Exchange at Glycyl Residues-Solid State $^{15}$N NMR

$^{15}$N NMR of glycyl residues of collagen rat calvaria and rat tendon have revealed essentially solid state characteristics (Fig. 2) (Torchia et al., 1985). Table 1 lists principal components of the three powder spectra. Most of the $^{15}$N NMR spectra of glycyl residues in other peptides and proteins have shown nearly axially symmetric chemical shift anisotropy (Hiyama et al., 1988). This is also the case for the glycyl residues in collagen. Determination of the orientation of the $^{15}$N chemical shift tensor will be presented in the next section. The $^{15}$N NMR spectra of collagen have shown moderate mobility of the collagen backbone at 22°C (Torchia et al., 1985) and did not show any heterogeneity.

As hydrogen exchange occurs at glycyl residues, $^{15}$N-$^{1}$H pairs become $^{15}$N-$^{2}$H pairs. Solid state $^{15}$N NMR spectra of such pairs obtained by the conventional proton cross polarized and decoupled method show a triplet $^{15}$N-$^{2}$H dipolar structure in addition to the anisotropic nitrogen chemical

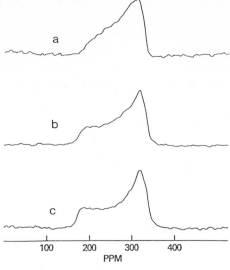

Fig. 2.   Comparison of 25.4 MHz proton-enhanced/decoupled $^{15}$N
          spectra of [$^{15}$N]glycine labeled rat tail tendon and rat
          calvaria (a) tendon, at 22°C, (b) calvaria, at 22°C, (c)
          tendon, at –35°C.  Spectra were obtained using matched
          Hartmann–Hahn contacts of 0.4 ms.

Table 1.   Chemical Shift Anisotropy of Nitrogen – 15 in $^{15}$N–Gly
           Collagen

|  | $\sigma_{xx}$(ppm) | $\sigma_{yy}$(ppm) | $\sigma_{zz}$(ppm) | $\eta$ | Total CSA(ppm) |
|---|---|---|---|---|---|
| Tendon –35°C | 65.7 (–1667, | 43.7 –1109, | –109.5 2779) | 0.20 | 175.2 |
| Calvaria 22°C | 57.8 (–1466, | 42.0 –1066, | –99.8 2533) | 0.16 | 157.6 |
| Tendon 22°C | 55.2 (–1401, | 36.7 –931, | –91.9 2332) | 0.19 | 147.1 |

$|\sigma_{zz}| \geq |\sigma_{xx}| \geq |\sigma_{yy}|$, $\sigma_{xx} + \sigma_{yy} + \sigma_{zz} = 0$, and $\eta = (\sigma_{yy} - \sigma_{xx})/\sigma_{zz}$.
Values in parentheses are $\sigma_{xx}$, $\sigma_{yy}$ and $\sigma_{zz}$ in Hz at 25.38 MHz.

shift.  Figure 3 illustrates the various lineshape components that result
when both the dipolar and chemical shift (CS) interactions are present.
Since the unique axis of the CS tensor is nearly along the nitrogen-
hydrogen bond vector, the anisotropy of the $S_z=1$ multiplet (Sz is the
deuterium magnetic quantum number) shows a larger anisotropy while
anisotropy of the Sz=–1 component decreases.  The detailed procedure for
the analysis of the lineshape has been described for Boc–glycyl–glycyl–
glycine benzyl ester (hereafter TGLY) (Hiyama et al., 1988).

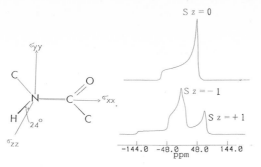

Fig. 3.  Triplet structure of deuterium coupled $^{15}$N NMR spectrum in a
peptide bond with orientation of the chemical shift tensor.

We consider two ways to follow the hydrogen exchange: one is to monitor
the $^{15}$N NMR signal amplitude of the protonated (unexchanged, singlets under
the proton decoupling) glycyl sites.  The other is to measure the $^{15}$N NMR
spectrum of deuterated, exchanged $^{15}$N signals (triplets because of the $^2$H-$^{15}$N
dipolar interaction).

A series of $^{15}$N NMR spectra of TGLY obtained as a function of the contact
time, $t_1$, (Fig. 4) showed that a $t_1$ of 80 μs is sufficient to generate most of
the signal if the Hartman-Hahn condition is carefully matched.  On the other
hand, the NMR signal of 100% deuterated TGLY was hardly observed with such a
short contact time because of long nitrogen-proton interatomic distance
(Fig. 5).  However the deuterium exchanged amide $^{15}$N will contribute to the
cross-polarized spectrum when the contact time exceeds 400 μs, and will attain
full intensity when the contact time is 1500 μs, provided that $T_1$ of the
protons is larger than ca 2 ms.

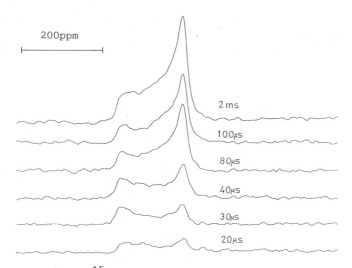

Fig. 4.  25.4 MHz $^{15}$N NMR spectra of a tripeptide, TGLY, with a
series of contact times, $t_1$.

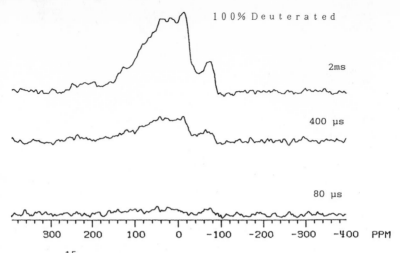

100% Deuterated

2ms

400 µs

80 µs

| 300 | 200 | 100 | 0 | -100 | -200 | -300 | -400 | PPM |

Fig. 5.   $^{15}$N NMR spectra of 100% deuterated TGLY with several contact times.

When amides are partially deuterated, a short contact time eg. 80 µs gives only the protonated portion and a long contact time e.g. 2 ms provides the triplet structure as shown in Fig. 6.  Since the protonated amide $^{15}$N signal is very sensitive to mismatch of the Hartmann Hahn condition when the contact time is short, we make the contact time larger than 1500 µs and obtain both the $^{15}$N-$^{1}$H and the $^{15}$N-$^{2}$H signals in the hydrogen exchange study.

Figure 7 shows proton cross polarized and decoupled $^{15}$N NMR spectra of rat calvaria collagen after heavy water introduction.  After a few days the Sz=-1 component of the $^{15}$N-$^{2}$H dipolar interaction was clearly evident in the spectra.  The amplitude of the dipolar signal component grew with a time constant of ca 10 days.  After 30 days, the deuterium component (Fig. 8) attained a constant amplitude, and the sample gave a

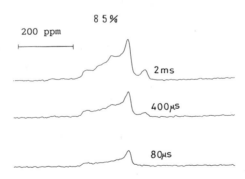

85%

200 ppm

2ms

400µs

80µs

Fig. 6.   $^{15}$N NMR spectra of 85% deuterated TGLY with several contact times.

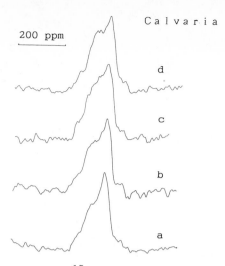

Fig. 7. Proton decoupled $^{15}$N NMR spectra of [$^{15}$N]Gly labeled rat calvaria collagen at (a) 3 days, (b) 9 days, (c) 15 days, (d) 60 days after deuterium oxide introduction.

protonated $^{15}$N NMR signal with 80us contact time (Fig. 9) whose amplitude suggested that about 20% of the glycyl residues remained unexchanged. A spectrum taken after one year is essentially the same as the spectrum taken at 30 days. Therefore it is concluded that about 20% of the glycyl amide sites are essentially unexchangeable. In tendon collagen, there are also two time constants in the hydrogen exchange experiment. The faster one is on the order of 50 hours. The slower one is undetectably long (Fig. 10).

The two tissues have significantly different shorter time constants. In tendon, the soft tissue, which has a more flexible molecular backbones than the hard tissue collagen (Sarkar et al., 1983), provides the glycyl amide sites more chance to expose to outside of the molecule. We think that the nonexchangeable (up to a year) sites correspond to the overlap region that is supposed to be rigid (Sarkar et al., 1983).

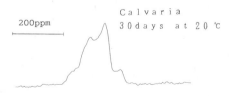

Fig. 8. Proton decoupled $^{15}$N NMR spectra of [$^{15}$N]Gly labeled rat calvaria collagen at 30 days after deuterium oxide introduction.

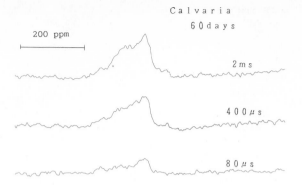

Calvaria
60days

200 ppm

2 m s

400 μ s

80 μ s

Fig. 9.   Proton decoupled $^{15}$N NMR spectra of [$^{15}$N]Gly labeled rat
calvaria collagen at 30 days after deuterium oxide
introduction with several contact times.

## $^{15}$N–$^2$H Dipolar Interaction and Orientation of $^{15}$N Chemical Shift Tensor at Glycyl Residues in Collagen

The $^{15}$N–$^2$H dipolar coupled $^{15}$N NMR powder spectrum provides the
orientation of the dipolar tensor (D=γ($^{15}$N). γ($^2$H)r$^3$) in relative to the
orientation of the chemical shift tensor (Hiyama et al., 1988).  First we
analyse the spectrum (Fig. 11) at –35°C where the backbone motion of collagen
is essentially frozen (Sarkar et al., 1983; Torchia et al., 1985).  Table 2
lists principal frequencies ($\nu_{ii}$) of the powder spectrum for the triplet and a
computer calculation that fit best for the observed principal frequencies.
From the calculation, the orientation of the nitrogen chemical shift tensor
was determined as shown in Fig. 12.  The orientation is esssentially identical
to that found in TGLY (Fig. 3).  The polar angles, (Θ, Φ) in Table 2 define
the orientation of the $^{15}$N–$^2$H dipolar interaction in the principal axis system
of the nitrogen chemical shift tensor.  The calculation for the spectra at
22°C showed that the dipolar interaction is also averaged.

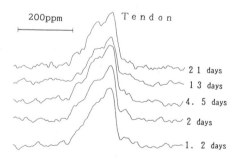

200ppm

Tendon

2 1 days

1 3 days

4. 5 days

2 days

1. 2 days

Fig. 10. Proton decoupled $^{15}$N NMR spectra of [$^{15}$N]Gly labeled rat
tendon collagen at (a)28 hrs., (b)48 hrs., (c)4.5 days.,
(d)13 days and (e)21 days after deuterium oxide
introduction. The spectra were obtained at 20°C.

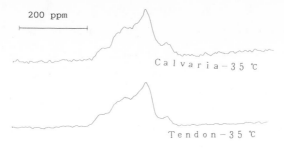

200 ppm

Calvaria-35 ℃

Tendon-35 ℃

Fig. 11. Low temperature(-35°C) proton decoupled $^{15}$N NMR spectra of
[$^{15}$N]Gly labeled rat (a)tendon collagen and (b)calvaria
collagen.

Table 2.  Principal Frequencies(a) of $^{2}$H Coupled $^{15}$N
Powder Pattern in ($^{2}$H-$^{15}$N)(Gly) Collagen

| | $m_s$=+1 | | | | | $m_s$=-1 |
|---|---|---|---|---|---|---|
| | ν11 | ν22 | ν33 | ν11 | ν22 | ν33 |
| Tendon, Calvaria -35°C | | | | | | |
| Experimental(b) | 5482 | -2614 | -2867 | 1464 | 542 | -2006 |
| Calculated(c) | 5574 | -2692 | -2883 | 1566 | 468 | -2034 |
| Calvaria  22°C | | | | | | |
| Experimental(b) | 5533 | -2600 | -2800 | 1333 | 500 | -1800 |
| Calculated(c) | 5374 | -2616 | -2757 | 1269 | 484 | -1752 |
| Tendon  22°C | | | | | | |
| Experimental(b) | ---- | ≤ -2000 | -2664 | 1302 | 402 | (-1704) |
| Calculated(c) | | -2230 | -2520 | 1380 | 406 | -1786 |

(a)   in Hz

(b)   Exerimental values were obtained by fitting the observed spectra.

(c)   Calculated values were obtained with (θ, Φ) =(23°, 0°), D=1.56 kHz
for the spectra at -35°C, with (θ, Φ) =(20°,0°), D=1.55 kHz for the
Calvaria at 22°C and with (θ, Φ) = (24°, 0°), D= 1.35 kHz for the
tendon spectrum at 22°C.

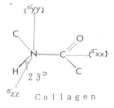

Fig. 12. Orientation of [$^{15}$N]Gly chemical shift tensor in collagen.

The [19]F spectra of rabbit Achilles tendon collagen, labeled with DL-4-fluorophenylalanine, exhibit major changes in lineshape as the temperature decreases from 26°C to -37°C (Fig. 13a-d). At -37°C a broad chemical shift powder pattern is observed (Fig. 13d and Table 3) whose lineshape and width are similar to that observed for the model amino acid (Fig. 14) (Hiyama et al., 1986). In contrast to the similar lineshapes observed for collagen and the model compound, the fluorine $T_1$ value in collagen at -37°C, ca. 2s, is an order of magnitude smaller than the $T_1$ measured for the model amino acid at room temperature. The smaller $T_1$ value observed for the protein can be accounted for by small amplitude angular fluctuations that modulate the [19]F chemical shift anisotropy (e.g. ca. 10°rms rolling motions about the $C^\beta$-$C^\gamma$ bond axis), provided that such motions have correlation times in the 0.1-1 ns range. An alternative explanation of the $T_1$ difference is that the phenyl rings in collagen undergo 180° ring flips at -37°C. Such ring flipping is observed at elevated temperatures in the crystalline amino acid, and while this type of motion does not modulate the [19]F chemical shift tensor (and therefore does not significantly affect the lineshape) it does modulate the [19]F-[1]H dipolar interaction, thereby causing spin-lattice relaxation. Ring flips having correlation times in either the 0.1-1 ns or in the 10-100 ns range can account for the [19]F $T_1$ values observed in collagen at -37°C.

In contrast with the -37°C spectrum which contains a single broad component, the rabbit Achilles tendon spectra observed at -4°C, and above, are composed of several components. Computer simulation shows that the -4°C spectrum is a superposition of four different lineshapes;

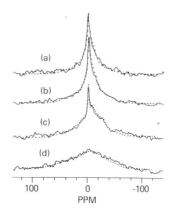

Fig. 13. Comparison of experimental and calculated (.....) 470.5 MHz [19]F NMR spectra of rabbit Achilles tendon collagen at four temperatures.

Table 3. Chemical shift tensor principal elements[a] and relative areas[b] of lineshape components used to obtain the calculated $^{19}$F spectra of rabbit Achilles tendon collagen in Fig. 13.

| T(C) | $\sigma_1$(ppm) | $\sigma_2$(ppm) | $\sigma_3$(ppm) | Relative Area | |
|------|------|------|------|------|------|
| -37 | -69 | 6 | 56 | 1.0 | |
|  | -69 | 6 | 56 | 0.63 | |
| -4 | -32 | 6 | 36 | 0.21 | |
|  | -18 | 6 | 29 | 0.11 | |
|  | isotropic component | | | 0.05 | |
|  | -53 | 3 | 53 | 0.63[c] | 0.56[d] |
|  | -25 | 3 | 28 | 0.11 | 0.13 |
| 10,26 | -12 | 3 | 19 | 0.11 | 0.13 |
|  | -6 | 3 | 16 | 0.11 | 0.13 |
|  | isotropic component | | | 0.05 | 0.05 |

[a] Uncertainty: 20% for broad component; 10% for narrow components.
[b] Uncertainty 20%.
[c] At 10°C.
[d] At 12°C.

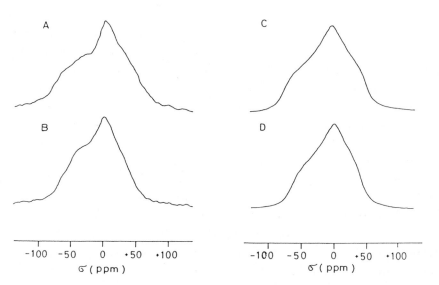

Fig. 14. Comparison of 470 MHz experimental $^{19}$F NMR powder spectra and calculated spectra of 4-fluoro-D,L-phenylalanine at two temperatures: A, 22°C; B, 143°C; C, simulation of 22°C spectrum; D, simulation of 143°C spectrum.

283

(a) a broad component having a width of ca. 125 ppm, approximately equal
to the width observed at −37°C, (b) a narrow component, at the isotropic
shift, having a width of less than 2 ppm and (c) two overlapping motionally averaged components having widths of 50–65 ppm. The relative
intensities and chemical shift tensor principal elements ($\sigma_1, \sigma_2, \sigma_3$) for
each of the lineshape components are listed in Table 3. We have corre-
lated these lineshape components to particular kinds of 4-fluorophenyl-
alanyl motion in collagen in the following way. We assign the broad
spectral component to 4-fluorophenylalanyl residues whose sidechain
motions are confined to either ring flips or small amplitude fluctuations. Because this component contains ca. 63% of the total signal
intensity, we estimate that 25–30 of the 44 Phe residues in the collagen
molecule (Fraser and Trus, 1986) undergo restricted motions of these
types. The narrow isotropic component contains ca. 5% of the signal
intensity, 2–3 Phe residues per molecule, and is assigned to residues
whose sidechains undergo essentially isotropic molecular motion. We
think that the motion is isotropic, because both the chemical shift and
dipolar interactions are completely averaged. It is difficult to see how
sidechain motion alone, which is limited to rotation about only two C–C
bonds, can completely average both types of NMR interactions, and we
therefore conclude that the Phe residues corresponding to the isotropic
spectral component are located in regions of the collagen molecule that
have substantial backbone flexibility.

The remaining signal intensity in the −4°C powder pattern,
corresponding to 12–16 Phe residues per molecule, consists of two powder
patterns having widths of 50–65 ppm. In order for rotation about the Phe
$C^{\beta}$–$C^{\gamma}$ bond axis to reduce the 125 ppm "static" [19]F linewidth to 60 ppm,
the phenyl ring must undergo nearly free rotation about the β–γ bond
axis. We think that such a motion is unlikely in view of the large
activation energies reported for phenyl ring flips in solids, 5–17
kcal/mol (Hiyama et al., 1986), and therefore ascribe the observed 50–65
ppm powder linewidths to motion about the Phe α–β bond axis. In a survey
of Phe sidechain conformations in protein crystal structures (Janin et
al., 1978), it was found that the relative populations of the trans,
gauche-plus and gauche-minus α–β rotamers were 0.31, 0.58 and 0.12,
respectively. If we assume that these rotamers have these equilibrium
populations in collagen, and that there is rapid interconversion among
the three rotamers, one calculates the following motionally averaged
principal frequencies for the Phe [19]F chemical shift tensor: $\sigma_1 = -40$
ppm, $\sigma_2 = 15$ ppm, $\sigma_3 = 26$ ppm. The linewidth, $\sigma_3 - \sigma_1$, of the calculated
powder pattern, 65 ppm, is in qualitative agreement with the 50–65 ppm
value observed at −4°C.

Upon increasing the temperature from $-4°C$ to $26°C$, the widths ($\sigma_3$ – $\sigma_1$) of the broad and intermediate components of the spectrum are observed to decrease by 30 to 50% (Table 3) and at $26°C$ the intensity of the broad component decreases by 25%, without a concomitant increase in signal intensity of any of the narrower spectral components. The first observation is ascribed to the onset of backbone motions, above $-4°C$, and to an increase in the amplitude rolling motions about the sidechain C-C bond axes, for the Phe residues corresponding to the intermediate width spectral components. The second observation is explained by noting that the onset of a slow motion (correlation time ca. 10 μs) about the $\alpha$–$\beta$ bond axis shortens the homogeneous $T_2$ values of the fluorine nuclei which contribute signal intensity to the broad component at low temperature. The reduction in $T_2$ will cause a substantial loss of signal intensity, because the EXORCYCLE pulse sequence does not refocus magnetization that dephases as a consequence of a homogeneous $T_2$ process.

In addition to spectra of rabbit Achilles tendon, spectra of rabbit calvaria collagen were also obtained in order to study the effect of mineral upon Phe sidechain dynamics. Unfortunately the fluorine signal from collagen was completely masked by the one-hundred fold larger fluorine signal from the mineral component, fluoroapatite, of the tissue.

## REFERENCES

Bodenhausen, G., Freeman, R., and Turner, D. L., 1977, J. Magn. Reson., 27:511.
Fraser, R. D. B., and Trus, B. L., 1986, Biosci. Rep., 6:221-226.
Hiyama, Y., Niu, C. H., Silverton, J. V., Barvoso, A., and Torchia, D. A., 1988, J. Amer. Chem. Soc., 110:2378-2383.
Hiyama, Y., Silverton, J. V., Torchia, D. A., Gerig, J. T., and Hammond, S. J., 1986, J. Amer. Chem. Soc., 108:2715.
Janin, J., Wodak, S., Levitt, M., and Maigret, B., 1978, J. Mol. Biol., 125:357-386.
Rance, M., and Byrd, R. A., 1983, J. Magn. Reson., 52:221.
Sarkar, S. K., Hiyama, Y., Niu, C. H., Young, P. E., Gerig, J. T., and Torchia, D. A., 1987, Biochemistry, 26:6793-6800.
Sarkar, S. K., Sullivan, C. E., and Torchia, D. A., 1983, J. Biol. Chem., 258:9762-9767.
Torchia, D. A., Hiyama, Y., Sarkar, S. K., Sullivan, C. E., and Young, P. E., 1985, Biopolymers, 24:65-75.

FORAGE DIGESTIBILITY AND CARBON-13 SOLID STATE NMR

Natsuko Cyr(a), R. M. Elofson(b) and G. W. Mathison(b)

(a) Alberta Research Council, P.O. Box 8330
    Postal Station F, Edmonton, Alberta  T6H 5X2
(b) University of Alberta, Animal Science Department
    Edmonton, Alberta  T6G 2P5

INTRODUCTION

There are many factors that influence the nutritive value of forages for ruminant animals such as cattle and sheep.  Economically, however, the most important of these is the provision of energy and protein to the animal.

Cellulose and hemicellulose are the main sources of chemical energy in forages.  Monogastric animals such as humans and pigs cannot utilize significant quantities of cellulose or hemicellulose because these animals do not produce the cellulase enzymes and microbial digestion is limited.  The ruminant animal has a system of four stomachs, however, with the first two of these (rumen and reticulum) acting as large fermentation vats where extensive microbial degradation occurs.  A relatively small portion of hydrolyzed carbohydrate is incorporated with nitrogen into microbial material which eventually passes from the ruminoreticulum and is digested by the animal.  A much larger portion of the cellulose is metabolized by the microorganisms to volatile fatty acids (primarily acetic, propionic and butyric) which are absorbed by the animal and used as energy sources and substrates for synthetic reactions.  Carbon dioxide and methane are also produced as waste products of rumen fermentation.

Not all of the carbohydrate is available to the microorganisms since the lignin component in forages is not broken down in the anaerobic environment of the rumen.  Cellulose and hemicellulose which are chemically bonded to, or surrounded by, lignin are thus not digested within the animal and pass out in the feces.  A very important attribute of forage quality, then, is the digestibility of the forage organic matter which is primarily determined by the digestibility of the lignocellulose component of the forage.  Digestibility of forages decreases as the plant matures since the lignocellulose components of the plant increase.

*NMR Applications in Biopolymers*, Edited by
J. W. Finley *et al.*, Plenum Press, New York, 1990

The rate of digestion of cellulose in the ruminoreticulum is relatively slow, thus a portion of potentially degradable fiber is not fermented by the time it is moved to the lower digestive tract. The extent of digestion of potentially degradable fiber is thus dependent upon the rate of digestion. In addition, a lowered rate of digestion and breakdown of fiber in the rumen results in a reduced voluntary intake since the animal cannot put more into its digestive system than is removed in the processes of digestion and passage. The rate of digestion of cellulose is decreased by factors such as increased lignin content of the plant, the presence of protective waxes, and increased crystallinity of cellulose (Van Soest, 1982a).

From the above discussion it is obvious that the proportion of plant dry-matter which is digested within the ruminant animal is variable. The percentage digestibility is approximately 30% for aspen wood (the most digestible of common woods), 45% for cereal straw, 50 to 75% for grasses and legumes, and greater than 75% for most cereal grains that contain starch as the main carbohydrate. Numerous laboratory procedures have been used to predict digestibility and to obtain a better understanding of the factors influencing the nutritive value of forages. The complexity of the lignocel-lulose component of plants has slowed progress in the area, however, and solid state NMR provides a real opportunity to advance our understanding of the forage chemistry and its influence upon the nutritive value of feeds for animals.

CARBON-13 SOLID STATE NMR APPLICATIONS FOR FORAGE DIGESTIBILITY EVALUATION

The C-13 NMR technique known as cross-polarization and magic angle spinning with proton decoupling (CP/MAS) offers an opportunity to examine the forage components which are important to its nutritive value for ruminant animals. Relative quantities of these components in various forages can be obtained simultaneously from the spectra without doing separate wet chemical analyses. Closer spectral examination reveals that the polymerization state of the lignin and the crystallinity of cellulose influence the cellulose digestibility of forages by ruminant animals. Some of the studies we have done in our laboratories will be discussed in the following three sections.

Quantitative Composition Analysis by NMR

Forage organic components which are important to animal nutrition are protein, fat, lignin and carbohydrates. They are commonly analyzed in the laboratory by wet chemical methods. For example, crude protein is calculat-ed as total nitrogen multiplied by 6.25, based on the assumption that the average protein contains 16% nitrogen. Fiber contents are often measured as "neutral detergent fiber (NDF)" which represents cell-wall components (cellulose, hemicellulose and lignin) or "acid detergent fiber" (ADF, cellulose and lignin). It is a useful measure of the total slowly digestible plant carbohydrates. However, lignin is generally indigestible and its presence interferes with the utilization of other cell-wall components. The lignin content is usually reported as "Klason lignin".

Different analyses must be done for different measurements in currently used laboratory methods. On the other hand, NMR spectra, of the solid forage, can be easily related to the total chemical structure and it is possible to obtain multiple pieces of information from one spectrum. A carbon-13 solid-state NMR spectrum of mature Timothy Hay is shown in Fig. 1. Spectra of isolated components are shown in Fig. 2 for comparison. Since different components in forages are probably chemically bonded to each other, isolation procedures inevitably alter some of the original structures and hence the spectra of the components. Nevertheless, the spectra of the isolated components will provide information on identification of most of the peaks. Assignments of major peaks are listed in Table 1.

It would be simple to obtain quantitative information of individual components from NMR spectra of forages if they show unique (non-overlapping) peaks. However, as Himmelsback et al. (1983) pointed out, even some dominant peaks are overlapping with minor peaks of other components. For analysis of forage components, they suggested the use of the peak at 105 ppm for carbohydrate, the oxygenated aromatic peak at 140-160 ppm for lignin, and the carbonyl peak at 165-190 ppm for protein for analysis of forage components. The present authors took a slightly different approach and tried to solve simultaneous linear equations using signal intensities of designated spectral regions (Cyr et al., 1987). In Fig. 2 those spectral regions used for calculations are also shown. Region 2 (62-115 ppm) is dominated by carbohydrate peaks. Region 3 is mainly aromatic carbons of lignins. Main peaks in regions 1 and 4 are from proteins. The combination of regions 3 and 4 yielded the best agreement (within ± 5%) between the actual contents and calculated contents of lignin and protein by NMR. The results are shown in Fig. 3.

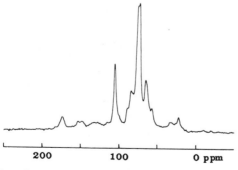

Fig. 1. A Solid State C-13 NMR Spectrum of Timothy Hay.

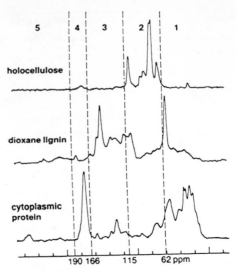

Fig. 2.   C-13 NMR Spectra of Forage Components.

However, there are still some problems to be discussed and solved to make use of this approach to evaluate the method.   They are:

1.   There are difficulties of uniform sampling of inhomogeneous solid particulate samples.   Every time vertical shock is given to the mixture of powdered components, large and light particles tend to rise to the top and small and heavy particles tend to settle at the bottom.   Did the NMR spectra represent the true sample?

2.   Very subjective spectral phase adjustments by the operator influence the spectral integral dramatically.   A large error may be introduced especially for the analysis of minor components.

Table 1.   Functional Group Identification

| Chemical Shift (ppm from TMS) | Description |
| --- | --- |
| ~ 20 | $-\underset{O}{C}-CH_3$ of hemicellulose |
| 20 - 50 | aliphatic carbons in protein and fat |
| ~ 55 | $-OCH_3$ of lignin and $C\alpha$ of peptide |
| 60 - 85 | carbohydrates C2 - C6 |
| ~ 105 | C1 of carbohydrates |
| 105 - 140 | aromatic carbons of lignin |
| 140 - 160 | oxygenated aromatic carbons in lignin |
| ~ 175 | C=O from protein and acetyl groups |

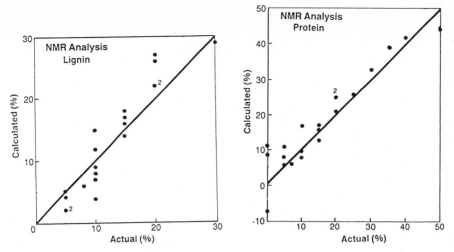

Fig. 3.   Results of Actual vs. Calculated Proportions in Artificial
          Mixtures using NMR.

3.  Relative spectral intensities per unit weight of different
    components are difficult to obtain since a simple reference
    material is not found for intensity calibration of CP/MAS spectra.

4.  Other forage components such as lipid may be significant in
    quantity.  The presence of this component should not be ignored.

Once these experimental difficulties are overcome, the advantage of
using the NMR technique is the convenience of being able to obtain those
quantities simultaneously.  Not only the quantities of components but
chemical and physical information on these components will lead to other
valuable information for evaluating forages.  These approaches will be
discussed in the next two sections.

Polymeric State of Lignin and Its Digestibility

The presence of lignin in forages is deleterious because it decreases
digestibility of other components.  The attachment of lignin prevents access
to the plant fiber by microbial polysaccharidases.  Low grade, poorly
digestible herbage is often associated with a relatively high lignin
content.  In order to improve the digestibility problem caused by the
presence of lignin, alkali treatment of forages are commonly used.  Alkali
swells the fiber and cleaves alkali-labile lignin-hemicellulose bonds.  The
molecular weight or the amount of cross-linking in the lignin may be as
important as its quantity in the determination of the digestibility of a
forage (Harkin, 1973).

Lignin is a cross-linked polymer of p-hydroxyl phenylpropane units
(Fig. 4).  One of the most important bonds holding these phenylpropane units

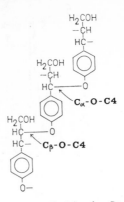

Fig. 4.    General Lignin Structure.

together is the Cα-O-C4 linkage.   These bonds are readily attacked by all the major delignification processes.  A second important and recurring bond is the Cβ-O-C4 and it is cleaved by fairly mild acid treatment.   In both cases, the result is the production of aromatic C4-OH groups substituted with methoxyl groups on the C3- and/or C5- carbons.  It has been accepted that the cleavage of these bonds results in a shift of the NMR resonance from 150-155 ppm to about 145-150 ppm (Cyr et al., 1988).  Polymerization causes the shift in the reverse direction.  In other words, C-13 NMR spectra show some indication of changes in the degree of polymerization in lignin.

## Effect of Digestion

In Fig. 5a there is shown a spectrum of a standard hay that was fed to a steer.  Figure 5b is the sample of fecal fiber after passage through the steer.  Not only is the lignin content of the fecal fiber higher than that in the original hay, the major peak now is ca. 153 ppm instead of 147 ppm indicating that the remaining lignin has a higher molecular weight; hence difficult-to-digest lignin remain undegraded.

## Effect of Maturity

The spectrum in Fig. 6a is of young timothy hay (13% protein) cut in May.  It shows that the peak at ca. 147 ppm is stronger than that at ca. 153 ppm indicating that the lignin is not highly polymerized.  Figure 6b is the spectrum of intermediate timothy (cut in July) whose nutritional value is lower than that of young timothy.  This is not only because of the lower protein content but also because of the low accessibility of the carbohydrate due to the increased lignin content and its further polymerization.  In this spectrum the peak at ca. 153 ppm has become dominant.

## Effect of Chemical Treatments

1.  Acid.  Figure 7 shows some examples of the effect of various treatments of aspen wood on the appearance of the peaks in the region of

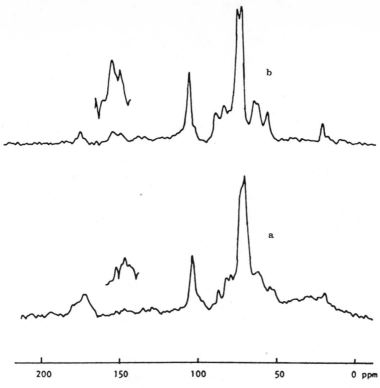

Fig. 5.   Effect of Digestion seen by NMR on Standard Hay
          a. standard hay, b. fecal fiber.

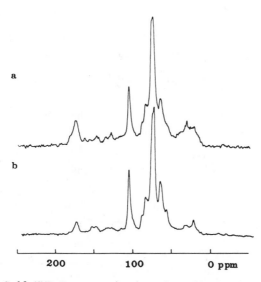

Fig. 6.   C-13 NMR Spectra showing the Effect of Maturity
          a. young timothy hay, b. intermediate timothy hay.

140-160 ppm. Figure 7a is a typical spectrum of aspen wood. The peak is centered at ca. 153 ppm and hardly any shoulder on this peak at ca. 147 ppm is seen. Figure 7b is a spectrum of aspen sawdust cooked in 1% sulfuric acid. Here, the intensity of the 147 ppm peak is approximately the same as that of 153 ppm. $C\alpha,\beta$-O-C4 bonds have been broken to produce phenolic OH groups under this mildly acidic condition. Figure 7c is a spectrum of Klason lignin from aspen wood. Since this preparation involved treatment with 72% sulfuric acid followed by cooking in 3% sulfuric acid under 2 atm. pressure, it is not surprising that the peak at 147 ppm has become dominant. In Fig. 7d is the spectrum of a steam-exploded aspen wood and again those two peaks are approximately the same height. Hydrolysis by acetic acid produced during the steam-explosion process cleaved most of the $C\alpha$-O-C4 bonds.

2. Alkali. Alkali alone can cause some destruction of the ether bonds in lignin. The effect of alkali on straw is shown in Fig. 8. A straw sample (Fig. 8a) was boiled for 1 hour in 5% NaOH solution. Most of the material was dissolved. The spectrum of the insoluble residue is shown in Fig. 8b. Two peaks at ca. 147 and 153 ppm are visible indicating that several phenylpropane units remain chemically attached to each other through C-O-C4 bonds. The solution was acidified and the precipitate was a mixture of hemicellulose and lignin. Some acid soluble material which may have included lignins of very low molecular weight may have been removed in the acidified solution. They would have the resonance at ca. 147 ppm only.

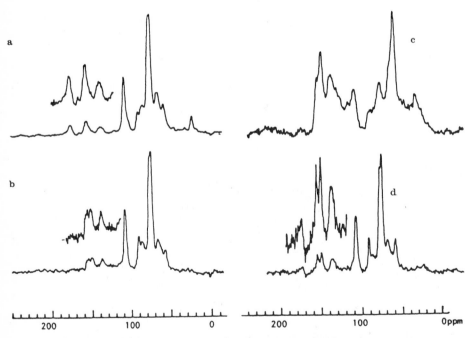

Fig. 7. C-13 NMR Spectra showing the Effect of Treatments on Aspen Wood, a. untreated, b. cooked in 1% $H_2SO_4$, c. Klason lignin, d. steam-exploded.

294

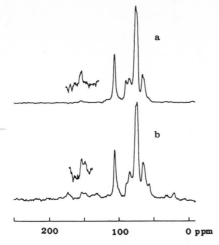

Fig. 8.   Effect of Alkali on Straw seen in C-13 NMR Spectra
a. untreated, b. 5% NaOH residue.


3.   $SO_2$.   It is well-known that treatment of low quality forage with
$SO_2$ enhances the feeding value (Ben-Ghedalia and Miron, 1983/84).   When a
mature timothy sample was treated with $SO_2$ gas the result is shown in
Fig. 9a.   In the spectrum of an over ripe timothy hay, the dominant peak in
the region of interest is at ca. 153 ppm with possibly a shoulder at ca.
147 ppm.   Exposure of this sample to $SO_2$ gas for 1 hour at 100°C produced a
fairly strong resonance at 147 ppm.   When a sealed sample under $SO_2$ atm. was
left for five years at room temperature, all the lignin became water-soluble
together with hydrolyzed hemicellulose and amino acids.   The peak at ca. 147
ppm dominated (Fig. 9b).

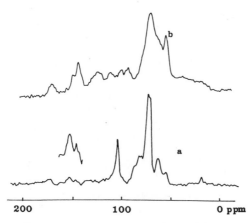

Fig. 9.   Effect of $SO_2$ Gas on Timothy Hay seen in C-13 NMR Spectra
a. mature, b. dissolved lignin.

295

4. Ammonia. Ammonia has been used to improve digestibility of forages by ruminant animals (Waiss et al., 1972) because the swelling of the cellulosic material is known to increase with this treatment (Tarkow and Feist, 1969).

To demonstrate the effect of ammonia on forage components, fecal fiber which contained a higher proportion of lignin and higher cellulose "crystallinity" than the feed was treated with ammonia. The spectrum remained unchanged (Figs. 10a and 10b). As will be discussed later the cellulose "crystallinity" decreased drastically after the treatment but the polymeric state of the lignin did not seem to have been altered as seen from the relative signal strength in the region of interest (140-160 ppm). This suggests that the treatment of forages with ammonia affects cellulose but does not break bonds in lignin as NaOH does.

Digestibility and Crystallinity of Cellulose Measured by NMR

Cellulose is present in all plants and is the structural polysaccharide which contributes most to the rigidity and strength of plant structures. Ruminant animal microorganisms are able to digest virtually all plant structural polysaccharides subject only to the interfering effect of lignin and the time available for digestion. However, the digestibility depends on several important properties. They are molecular size, crystallinity and solubility in various solvents. It has long been known that cellulose consists of crystallites and less ordered (amorphous) interlinking regions (Krassig, 1985). Amorphous regions and the surface areas of crystallites

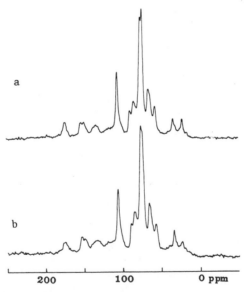

Fig. 10. Effect of Ammonia on Fecal Residue seen in C-13 NMR Spectra. a. fecal fiber, untreated, b. fecal fiber, NH₃-treated.

are more accessible to a variety of reagents and microorganisms than crystal-line regions. Therefore, all these properties are essentially very well correlated to each other. Largely by inference from the above work it has been concluded that this factor (cellulose crystallinity) is of considerable importance in the digestion of forages by animals (Van Soest, 1982b).

Several papers using CP/MAS solid state NMR have studied the different crystal structures of cellulose and the interconversion of some of them (Atalla et al., 1980; Earl and VanderHart, 1980; Teeaar and Lippmaa, 1984). References have been made in particular to the shape of the C4 and C6 resonance of cellulose as a measure of the crystallinity of the cellulose (or as an inverse measure of the nutritive value of forage), but in our laboratory this has proven to be a relatively insensitive measure of the digestibility of a forage. In wood and forages the CP/MAS spectra of cellulose are partly obscured by the presence of other plant components such as hemicellulose, lignin and sugars.

VanderHart and coworkers (1986) have shown that the spectra of crystal-line regions in polyethylene can be preferentially observed by the use of a delayed contact pulse sequence (see diagram below) in solid state C-13 NMR. In this pulse sequence the contact between the spin-locked proton and the carbon reservoir under the Hartman-Hahn condition is established after a

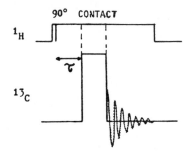

pre-contact delay. During the pre-contact delay, the proton spin-locking is maintained so that those protons with short $T_1\rho$ (spin-lattice relaxation time in the rotating frame) lose magnetization. Protons associated with the amorphous region have shorter $T_1\rho$ than those associated with the crystalline region. Therefore, the signals of carbons in the amorphous region are decreased in intensity or eliminated from the spectra and those in the crystalline regions are relatively enhanced. Here we discuss how the relative crystallinity of cellulose fibers can be measured by solid state C-13 CP/MAS NMR spectroscopy and how we can use this technique to assess the apparent crystallinity of the cellulose in forages and naturally occurring plant materials.

The amount of crystalline cellulose is the greatest in natural cellulose fibers (cellulose I) and decreases drastically in macerized cellulose (cellulose II). There are various physical and chemical methods to assess the crystallinity of pure cellulose samples in bulk (Browning, 1967). This new technique of solid state NMR can be added to the methods to estimate the crystallinity of cellulose in materials where cellulose is only one of many components. It should be pointed out that none of these methods, nor the NMR method described here, can tell if the accessibilities of the crystallites of different types of cellulose are actually different or if the amount of interlinking regions vary.

Pre-Contact Delay and Signal Intensities of Some Cellulose Samples

In order to demonstrate that this technique does discriminate against the amorphous portion of cellulose samples, three types of cellulose samples were used and signal intensities with various delay times were measured. Figure 11 shows plots of total C-13 signal intensities of hydrolyzed cotton (cellulose I), commercial rayon (cellulose II) and cellulose ground in liquid nitrogen (cellulose III) versus the length of the delay. As one can see readily from Fig. 11, after a delay of 20 msec., the total signal of rayon diminished nearly to zero whereas approximately 15% of the signal of a more crystalline hydrolyzed cotton, remained. The reduction in the "crystallinity" in the sample ground in liquid $N_2$ is also easily seen.

When a sample of cellulose is hydrolyzed by acids, the attack is initially confined to amorphous regions. Therefore, the initial rate of hydrolysis should reflect the amounts of accessible, amorphous regions in cellulose. The

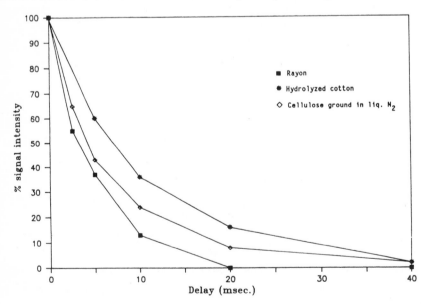

Fig. 11.   Variation of C-13 Signal Intensities.

residues of the hydrolysis represent the relative amount of crystalline cellulose. A good correlation has been observed between the initial rate of hydrolysis and the "crystallinity" measured by the NMR method (Cyr et al., 1990).

## Application of the Spin-Lock Technique to Carbohyrdates in Forages

The NMR "crystallinity" of cellulose samples was measured by comparing the areas under the total signals. However, it is not simple to measure this quantity in forage samples since signals from other plant components overlap with those of cellulose. From Fig. 2, the C1 peak (ca. 105 ppm) of the carbohydrates in the plant does not significantly overlap with other signals (the fraction of lignin in grass is small). It was found that the relative signal intensities of this signal can be used satisfactorily to assess the "crystallinity" instead of the total signal intensity. This method simplifies the measurement procedure.

Figure 12 depicts the influence of maturity on the "crystallinity". Figures 12a and 12b are CP/MAS spectra without and with a 20 msec delay, respectively, on a sample of ripe timothy. Figures 12c and 12d are spectra of new growth harvested in June 1988. If low "crystallinity" of the cellulose is indicative of high nutritive value, these results accord with prediction. The present authors (Elofson et al., 1984) had previously shown that not only the protein content decreased with maturity and weathering but the lignin content increased as well. These results agreed with predictions from NDF and ADF measurements. Clearly, lignin content, protein content and the nature of the cellulose are factors in the nutritional evaluation of timothy hay. Cellulose "crystallinity" value of several forage samples are listed in Table 2. It is easily seen that the fecal fiber contains only highly crystalline cellulose compared with the original hay which was fed. This highly crystalline cellulose in fecal fiber became less crystalline by treatment with ammonia. This process of treating forages by ammonia has been used to improve animal digestibility (Waiss et al., 1972) and the reason for the improvement may be due to the decrease in cellulose "crystallinity".

While the results presented here provide a measure of the cellulose "crystallinity" which agree qualitatively with laboratory experimental and animal nutritional significance, it must be emphasized that a low value in the "crystallinity" scale can arise from other sources. While the values assessed for the same hay and fecal fiber represent differences in ratios of crystalline cellulose and amorphous cellulose, the low "crystallinity" measured after the conversion of cellulose I to cellulose III by grinding under liquid $N_2$ or by simply immersing it in liquid ammonia (Cyr et al., 1989) may not provide the corresponding improvement in the digestive value. Otherwise this method of assessing "crystallinity" appears to accord with our feeding experience.

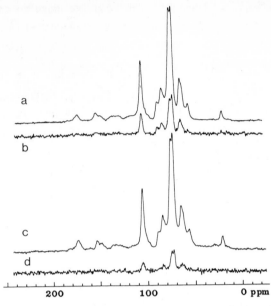

Fig. 12.  CP/MAS and SLCP/MAS C-13 NMR Spectra of Timothy Hay
a. mature, CP/MAS, b. mature, SLCP/MAS, c. young, CP/MAS,
d. young, SLCP/MAS.

Neutral detergent fiber is a measure of the content of lignin, cellulose and hemicellulose.  Acid detergent fiber provides a measure of lignin
and cellulose.  C-13 CP/MAS solid state NMR by examination of the lignin
(Cyr et al., 1988) and now cellulose can provide further insight into the
reasons for the negative relationship which acid detergent fiber has with
the nutritive value of forages for ruminant animals (Elofson et al., 1984).

CONCLUSIONS AND PERSPECTIVES

In this presentation we have tried to illustrate by example some of the
problems and some of the newly achieved accomplishments that we have experienced in our laboratories with the application of solid state C-13 CP/MAS
NMR to the examination of forages.  Our efforts to attain quantitative

Table 2.  Cellulose "Crystallinity" in Forage Samples Measured
by NMR.  ("crystallinity" of cellulose powder was taken
as 100)

| | |
|---|---|
| timothy hay, stem (June) | 7 |
| timothy hay, stem (July) | 16 |
| mature timothy, stem | 20 |
| control hay fed to cows | 10 |
| fecal fiber from above | 49 |
| fecal fiber treated with $NH_3$ | 7 |

determinations of lignin carbohydrate and protein have to date met with limited success. Perhaps, in retrospect, a less ambitious approach wherein spectra were correlated with standard procedures for nitrogen, carbohydrate and lignin would have been more successful. As we have shown repeatedly, attempts to isolate the individual components and thence to reconstitute spectra face the difficulty that the components are modified during isolation and can only approximately be put back together again. This applies to both chemical and physical factors as already discussed. The work on the polymerization and depolymerization of lignin shows how new information, not readily available by conventional studies, can almost inadvertently be obtained by careful examination of a single group of resonances. The last section of this report shows how the standard solid state NMR procedure can be modified to elucidate the physical structure, that is, the crystallinity, of cellulose. We have been able to detect lipid material, presumably solid, which should yield interesting information on such characteristics as heat and cold tolerance of forage species. We have chosen to omit our studies in this area because the results are as yet fragmentary. Additional discoveries on the protein component may be anticipated in relation to its availability to the ruminant by further application of CP/MAS techniques.

NUCLEAR MAGNETIC RESONANCE MEASUREMENTS

Solid state carbon-13 CP/MAS NMR spectra were obtained on a Bruker CXP-200 at 50.3 MHz using a Doty probe. Samples were packed in cylindrical sapphire rotors with Kel-F plugs and turbines. They were spun at ca. 4-5 kHz. Six times more FID's were usually accumulated for spectra with a 20 msec. delay to improve the signal-to-noise ratios. The pulse repetition and the contact times used were 2 sec. and 2 msec., respectively.

ACKNOWLEDGMENTS

Financial support for some of this work was provided by the Agricultural Research Council of Alberta and the Alberta Research Council.

REFERENCES

For example, Atalla, R. H., Gast, J. C., Sindorf, D. W., Bartuska, V. I., and Maciel, G. E., 1980, J. Am. Chem. Soc., 102:3249; Earl, W.L., and VanderHart, D. L., 1980, J. Am. Chem. Soc., 102:3251; Teeaar, R., and Lippmaa, E., 1984, Polymer Bulletin, 12:315.
Ben-Ghedalia, D., and Miron, J., 1983/84, J. Anim. Food Sci. Technol., 10:269.
Browning, B. L., 1967, "Methods of Wood Chemistry," J. Wiley and Sons, NY, ch. 24, p. 499.
Cyr, N., Elofson, R. M., and Mathison, G. W., 1990, Can. J. Anim. Sci., to be printed.
Cyr, N., Elofson, R. M., Mak, A., Mathison, G., and Milligan, L. P., 1987, presented at the 70th Can. Chem. Conference, Quebec, June.

Cyr, N., Elofson, R. M., Ripmeester, J. A., and Mathison, G. W., 1989, J. Agric. Food Chem., 36:1197 and the references therein.

Cyr, N., Elofson, R. M., and Mathison, G. W., 1989, unpublished results.

Elofson, R. M., Ripmeester, J. A., Cyr, N., Milligan, L. P., and Mathison, G., 1984, "Nutritional Evaluation of Forages by High-Resoltion Solid State C-13 NMR," Can. J. Amin. Sci., 64:93-102.

Harkin, J. M., 1973, "Chemistry and Biochemistry of Herbage," Butler, G. W., and Bailey, R. W., (Eds.), Academic Press, NY, Vol. 1, p. 323.

Himmelsback, D. S., Barton, F. E., and Windham, W. R., 1983, J. Agric. Food Chem., 31:401-404.

Krassig, H., 1985, "Cellulose and Its Derivatives: Chemistry, Biochemistry and Applications," Kennedy, J. F., Phillips, O. G., Wedlock, D. J., and Williams, P. A., (Eds.), J. Wiley and Sons, NY, p. 3.

Tarkow, H., and Feist, W. G., 1969, "Cellulases and Their Applications," Hajny, G. J., and Reese, E. T., (Eds.), Advances in Chemistry Series 95, A.C.S., Washington D.C., p. 197.

VanderHart, D. L., and Perez, E., 1986, Macromolecules, 19:1902-1909.

Van Soest, P. J., 1982a, "Nutritional Ecology of the Ruminant," O & B Book Inc., Corvallis, Oregon.

Van Soest, P. J., 1982b, "Nutritional Ecology of the Ruminant," O & B Book Inc., Corvallis, Oregon, p. 110.

Waiss, Jr., A. C., Gugglolz, J., Kohler, G. O., Walker, Jr., H. G., and Garrett, W. N., 1972, J. Anim. Sci., 35:109.

NMR OF CARBOHYDRATES AT THE SURFACE OF CELLS

Harold C. Jarrell* and Ian C. P. Smith

* Division of Biological Sciences
  National Research Council of Canada
  Ottawa, Ontario Canada K1A OR6

Carbohydrates attached to lipid and proteins are a major element of the cell surface, where, anchored in the membrane they modulate interactions with the outside world. Glycolipids may be divided into two major classes distinguished by the nature of the hydrophobic anchor. The glycolipids of bacteria and plants generally consist of a mono- or oligo-saccharide glycosidically linked to the glycerol 3-position of 1,2-diacylglycerol (Curatolo, 1987a). The major glycolipids of animals are in the second class which consist of carbohydrate glycosidically linked to ceramide glycosphingolipids (Curatolo, 1987a). Glycolipids are intimately involved with membrane structure, recognition, immune function, interaction with toxins and biological pathogens, and growth control in both normalcy and disease (Curatolo, 1987a; Thompson and Tillack, 1985; Critchley, 1979). While the important roles that these lipids play is well established, how these roles are fulfilled and influenced by the environment in which these molecules exist is far from being understood. For these reasons, there has been increasing interest in attempting to systematically delineate various physio-chemical properties of these lipids as pure systems and in mixtures (Curatolo, 1987b). The ultimate goal of such studies is to gain insight into how these parameters may influence the various biological roles that the glycolipids can assume.

It is relatively easy to characterize the conformation and dynamics of an isolated glycolipid molecule free in solution, using a variety of techniques of which high resolution nuclear magnetic resonance (NMR) spectroscopy has proven to be particularly useful (Yu et al., 1984; Bloom and Smith, 1985). However, the natural environment of these molecules is not that of an isotropic solution; it is anisotropic and of relatively high viscosity due to the fundamental characteristics of the membrane bilayer. Thus, when the glycolipid is mixed in a sea of other components which together form a large anisotropic bilayer system, the high resolution techniques are not applicable.

*NMR Applications in Biopolymers*, Edited by
J. W. Finley *et al.*, Plenum Press, New York, 1990

Fig. 1 illustrates the dramatic difference between [1]H NMR spectra of biological molecules in solution and in a membrane (Bloom and Smith, 1985).
Bovine pancreatic trypsin inhibitor (BPTI) is a water-soluble protein which
reorients sufficiently rapidly and isotropically to give a high resolution [1]H
NMR spectrum (Fig. 1a).  Such a spectrum contains thousands of resolved lines
distributed over just a few kHz which in combination with additional NMR
techniques provides structural information.  In the case of membrane systems,
the situation is dramatically different as illustrated for rhodopsin, an
integral membrane protein.  In the liquid crystalline phase of the membrane
lipid, rhodopsin reorients too slowly to average [1]H dipolar interactions to
any significant degree.  The resulting [1]H NMR spectrum (Figure 1c) is broad
and featureless.  In such circumstances, structural information is not readily
obtained.  Under these circumstances wideline NMR has proven to be very
powerful.  The quantities measured are quite different from those of high
resolution studies, but they lead to similar desired information: conformation
and mobility.  Cell membranes are dynamic structures whose components may be
moving in some fashion on time scales as short as $10^{-12}$s or as long as 1s.
Thus, what is meant by structure and mobility has an intimate dependence on

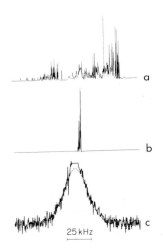

25 kHz

Fig. 1.   Comparison of [1]H NMR spectra associated with a water
soluble protein and a membrane bound protein.  (a) [1]H NMR
spectrum (400 MHz) of BPTI in aqueous solution exhibits an
enormous number of resolved resonances which may be used to
obtain structural information.  (b) Spectrum of (a) on the
same frequency scale as that shown in (c).  (c) [1]H NMR
spectrum of rhodopsin reconstituted in model membranes of
dimyristoylphosphatidylcholine (DMPC-$d_{54}$).  Note that the
spectrum is a broad featureless line which is approximately
25 kHz wide.  As a result of the slow anisotropic reorientation of the protein in the membrane, the [1]H NMR spectrum is
dominated by the large proton-proton dipolar interactions.

the time frame being probed.  It is important to remember that usually one is
not so much interested in the absolute value of these parameters but rather in
characterizing the relative values as a result of changes to the system.  Thus
in order to develop a representative picture of a glycolipid in a bilayer
matrix, as many time frames as possible should be probed.  Solid state NMR
techniques can be sensitive to time scales of $1s-10^{-12}s$ which has made it one
of the techniques of choice for studying membrane systems (Smith and Jarrell,
1983; Davis, 1983).  A variety of nuclei are applicable for this purpose:
$^{14}N$, $^{15}N$, $^{13}C$, $^{31}P$, $^{2}H$ and even $^{1}H$.  $^{2}H$ NMR has proven particularly valuable
because of the ease of obtaining information about specific molecular sites by
isotopic labelling, and the straightforward spectral analysis.

## $^{2}H$ SOLID STATE NMR

$^{2}H$ NMR is dominated by the Zeeman and the nuclear quadrupole interactions.  Details can be found in a number of articles (Smith and Jarrell,
1983; Davis, 1983; Abragam, 1961) and will not be described here.  Only the
essential features of $^{2}H$ NMR of membrane systems will be outlined.  In a
slowly tumbling structure such as a membrane, a single immobile deuteron
would yield a pair of resonances, a quadrupole doublet, whose separation is
dependent on the value of the quadrupole coupling constant, and the angle
between the static applied magnetic field and the principal direction of the
quadrupole interaction (frequently the $C-^{2}H$ bond direction) (Fig. 2).  In a
powder sample the $C-^{2}H$ bonds are randomly distributed in space (Fig. 2) so
that the resulting $^{2}H$ NMR spectrum is a superposition of doublet spectra
arising from each angle.  When rapid molecular motion ($> 10^{5}s^{-1}$) occurs
there is some averaging of the quadrupole interactions.  This leads to a
general narrowing of the overall spectrum.  The actual shape of the $^{2}H$
spectrum depends on the symmetry of the molecular motion.  In membranes in
the liquid crystalline phase, molecular motion has high effective symmetry.
As a result the residual splitting is related to the extent of the
averaging, the so-called order parameter.  The higher this parameter
(ranging from 0 to 1), the less conformational freedom the molecule has.  In
the case of a rigid molecule or molecular fragment, if several positions are
labelled, the location within the molecule of the axis of motional averaging
can be determined.  Finally, the response of the $^{2}H$ spectral characteristics
after different radiofrequency pulse sequences can yield information on the
types and rates of the various motions undergone by the molecule (Jeffrey,
1981; Meier et al., 1986; Wittebort et al., 1987).

## Structure and Ordering of the Carbohydrate Moieties

The objective of these studies is to relate structural and motional
properties of glycolipids in biological membranes to their biological roles.

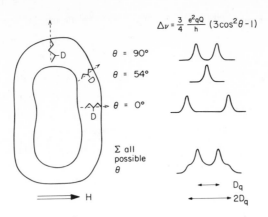

$$\Delta \nu = \frac{3}{4} \frac{e^2 qQ}{h} (3\cos^2\theta - 1)$$

Fig. 2.  Origin of the $^2$H NMR spectrum of a deuteron in a sample in which the C-$^2$H bond is randomly oriented with respect to the magnetic field direction, H, of the spectrometer.

It seemed that a fundamental prerequisite to such studies is to reach some understanding of the properties of these lipids in simpler systems.  This view has motivated an increasing number of physical studies of a wide range of simple glycolipid systems (Curatolo, 1987b).  While such artificial systems appear to be a poor representation of nature, they have been used effectively to shed light on phospholipid organization in natural membranes (Smith and Jarrell, 1983; Davis, 1983).

With this motivation, we have labelled the glucosyl, galactosyl and mannosyl moieties of a series of glycolipids based on ditetradecylglycerol (Jarrell et al., 1986; 1987a; 1987b; Renou et al., 1989; Carrier et al., 1989).  These lipids form multilamellar structures which are good models for membrane bilayers.  Figure 3 shows typical $^2$H NMR spectra of a pure monoglucosyl lipid, β-DTGL, labelled at different sites of the gluco-pyranosyl group, at temperatures corresponding to the fluid bilayer phase.  Segmental order parameters (related to the amplitude of segmental angular fluctuations) for this phase are as follows:  glucose ring - 0.45; glycerol, C-3 - 0.63; chains, C-4 - 0.40.  It is clear that the glycerol moiety is the most ordered of the three lipid segments.  The results for the sugar region suggest that motion about the glucose-glycerol linkage is limited.

Since the sugar rings in glycolipids are rigid on the $^2$H NMR time-scale, their average orientation relative to their motional axis can be determined if enough observables can be measured.  In the case of bilayer systems the average head group orientation relative to the membrane surface may be obtained.  In one study, the effects of minor structural changes in glycolipid head group structure were probed (Jarrell et al., 1987b).  Thus

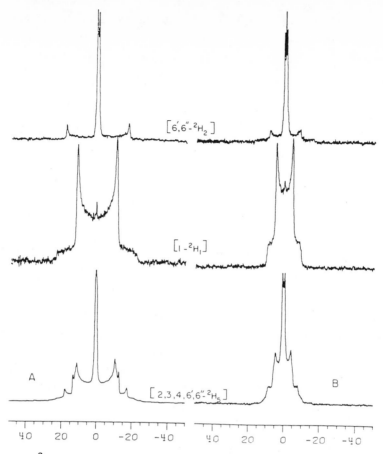

$[6',6''-^2H_2]$

$[1-^2H_1]$

A

$[2,3,4,6',6''-^2H_5]$

B

| 40 | 20 | 0 | -20 | -40 |
| 40 | 20 | 0 | -20 | -40 |

Fig. 3.  $^2$H NMR spectra (30 MHz) of aqueous dispersions of 1,2-di-
0-tetradecyl-3-0-β-glucopyranosyl-rac-glycerol (β-DTGL)
variously labelled in the glycosyl moiety:  (A) fluid
lamellar phase (52°C) and (B) hexagonal phase (58°C).

the glycolipids containing β-glucopyranosyl, α-glucopyranosyl, and
α-mannopyranosyl head groups were compared (Table 1) and the head group
orientations and orientational fluctuations were found to be substantially
different in both the lamellar (L) and hexagonal (H) mesophases.

While glyceroglycolipids are of interest, glycosphingolipids are the
subject of the greatest interest.  In Fig. 4, we compare the glucose head
group orientation of β-DTGL with that of the corresponding glucocerebroside
(Skarjune and Oldfield, 1982).  It is interesting to note that the results
for both systems are very similar.  This suggests that what is learned about
glycoglycerolipids may give some insight into the properties of the corresponding glycosphingolipid systems.  In natural membranes glycolipids, in
particular glycosphingolipids, usually represent a minor component in
natural membranes.  Thus, having studied pure glycolipid systems, it is
important to ask what, if any, effect does diluting the lipid in a matrix

307

**Table 1.  Order Parameter for the Glycosyl Residues of Glycolipids**

| Glycolipid[*] | Residue | Temp. | Phase[+] | Order Par. |
|---|---|---|---|---|
| β–DTGL | Glu | 52° | L | 0.45 |
|  | Glu | 58° | H | 0.38 |
| α–DTGL | Glu | 52° | L | 0.56 |
|  | Glu | 58° | H | 0.60 |
| α–DTML | Man | 52° | H | 0.41 |
| GC | Glu | 90° | L | 0.32 |
| DTLL | Gal | 75° | L | 0.51 |
|  | Glu | 75° | L | 0.53 |

[*] α–DTML and DTLL represent the mannosyl– and lactosyl– analogues of β–DTGL.

[+] L – lamellar phase; H – hexagonal phase.

composed of another lipid have on the structural properties?  When simple glycolipids were dispersed (10 mole %) in phospholipid bilayers (Jarrell et al., 1986; 1987a; Skarjune and Oldfield, 1982), the average head group orientation changed only slightly.  This suggests that for simple glycolipids in a fluid bilayer strong inter–head group interactions either do not exist or do not determine the head group orientations.

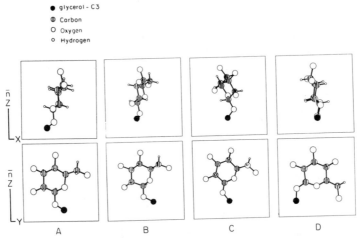

Fig. 4.  Calculated orientations of the glucose moiety of (A) fully extended β–DTGL in the lamellar structure, (B) DTGL in the lamellar structure at 52°C, (C) a glucocerebroside (GC) (18) in the lamellar structure at 92°C, and (D) β–DTGL in the hexagonal phase at 58°C.  (Top) View of the XZ plane of the rotation axis system:  (Bottom) top view rotated about the motional axis, **n**, by 90°.  The carbon at 0–1′ (dark circle) is shown to indicate the point of attachment of the head group to the membrane surface.

308

While monosaccharide glycolipids have received considerable attention, glycolipids having more complex carbohydrate head groups are biologically the most interesting. Such lipids have the added dimension of conformation about intersaccharide bonds. Thus, in addition to probing the membrane surface orientation, it is equally essential to elucidate conformational aspects in the carbohydrate region of the lipids. In order to begin to address this issue, we have extended our studies to disaccharide-containing lipids (Renou et al., 1989; Carrier et al., 1989). Results for the lactosyl residue (DTLL) show segmental ordering (Table 1) which is comparable to the simple monosaccharide lipids, in the region of 0.5. Unlike the situation for the monoglycosyl lipids where axially symmetric ordering was assumed, a complete order parameter analysis for DTLL was performed. Results showed that the head group is extended away from the bilayer surface. This is contrary to what has been proposed based on molecular models. The conformation of the lactosyl residue of DTLL (Fig. 5) was found to differ from that of lactose in solution or in crystals. If this is a general phenomenon for oligosaccharide head groups, it may have significant implications for understanding recognition at the membrane surface.

Molecular Dynamics

To this point, the discussion has focussed on molecular ordering (molecular fluctuation) and on average orientation. In order to obtain as complete a description as possible, it is important to ask what are the nature and rates of molecular motion which characterize angular fluctuations about the average direction and what are the motions which describe the system on longer time scales ($>10^{-4}$s). As part of our investigation of the β-DTGL system, the glycerol backbone was probed (Jarrell et al., 1987a).

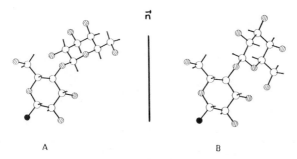

Fig. 5.   Orientation of the lactosyl residue relative to the local motional axis as calculated from the quadrupole splittings. The conformation about the intersaccharide linkage is (A) that calculated from the NMR data in this study; (B) that for crystalline lactose; (o) hydrogen; (⊜) oxygen; (0) carbon; (●) 03 of the glycerol backbone.

309

The glycolipid was labelled at C-3 of glycerol which gave a $^2$H spectrum characterized by two quadrupolar splittings. Using oriented multilamellar films and proton-decoupling, the relatively small dipolar couplings between the hydrogen at C-2 of glycerol and the deuterons at C-3, and the homonuclear dipolar couplings between the geminal deuterons at C-3 were estimated (Jarrell et al., 1987a). The results were interpreted in terms of segmental ordering and a fixed conformation for the glycerol backbone (Fig. 6). However, when the glycolipid is cooled to 25°C the $^2$H spectrum is characteristic of axially asymmetric motion (Auger et al., in press) (Fig. 7, Top). Since the lipid is in the more ordered gel phase, molecular motion is expected to be less complex than that which obtains in the liquid crystalline phase. The longitudinal relaxation rate, $T_{1Z}^{-1}$, reflects the rate of return of Zeeman order to its equilibrium value after a perturbation from equilibrium (such as in an inversion-recovery experiment) and is sensitive to motions of frequency near that of the NMR experiment, in this case $30 \times 10^6 s^{-1}$. Inspection of partially-recovered $^2$H spectra (inversion-recovery experiment) for the C-3 labelled position of glycerol (Fig. 7) reveals that the rate of recovery depends on the orientation of the bilayer normal relative to the direction of the static magnetic field; relaxation is anisotropic. Such behaviour is important since it has a sensitive dependence on the type of molecular reorientation giving rise to relaxation. The experimental $^2$H lineshapes indicate that there is a three site hop about the glycerol C2-C3 bond with rotameric site populations of 0.46, 0.34 and 0.2, and an interconversion rate of $5 \times 10^8 s^{-1}$. This means that the glycerol conformation discussed above, which was deduced from the residual quadrupole and dipolar splittings, must be an average structure resulting from this internal motion.

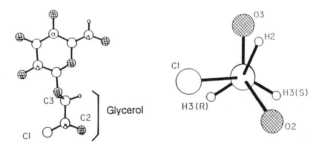

Fig. 6.   Left: Head group orientation of β-DTGL relative to the bilayer normal as calculated from $^2$H NMR data (Jarrell et al., 1986; 1987a). Right:   Glycerol backbone of β-DTGL. View is along the C2-C3 bond from C2 towards C3. The conformation shown is that calculated from the estimated $^1$H2-$^2$H3 (S) and $^1$H2-$^2$H3 (R) dipolar couplings (1) (Jarrell et al., 1987a).

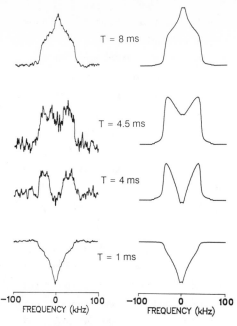

−100    0    100    −100    0    100
FREQUENCY (kHz)      FREQUENCY (kHz)

Fig. 7.   Experimental (left) and simulated (right) partially
recovered ($T_1$) spectra of a multilamellar dispersion of
$[3,3\text{-}^2H_2]\beta$–DTGL at 25°C as a function of the delay T in the
inversion recovery sequence.  Simulations were performed
using a three-site jump model with an exchange rate of 5 x
$10^8$ rad $s^{-1}$ and site populations 0.46, 0.34, 0.20.

At 25°C the internal motion dominates the $^2H$ NMR spectrum for the
glycerol C–3 position.  However, as the temperature is elevated to that of
the phase transition temperature (52°C) there is a dramatic loss in spectral
intensity.  This feature is shared by all the glycolipid systems studied in
this laboratory and by some phospholipid systems (Auger et al., unpublished
results).  Since $^2H$ NMR spectra of lipid dispersions are acquired by the
quadrupole echo technique ($90°_x$ – $\tau$ –$90°_y$ – $\tau$ – acquire), molecular motion
occurring on a timescale comparable to $\nu_Q^{-1}$ ($10^{-4}$ – $10^{-6}$s) will lead to loss
in signal intensity (Jeffrey, 1981; Meier et al., 1986; Wittebort et al.,
1987), the degree being dependent on the type and rate of motion.  By
exploiting this feature of the glycolipid spectra, the second motion was
shown to be reorientation about the long axis of the lipid molecule with a
rate of $<10^3 s^{-1}$ at 25°C and $>10^6 s^{-1}$ above 52°C.  Thus, in the gel phase
$\beta$–DTGL exhibits a whole body motion about its long axis which occurs on a
timescale which is well separated from that of its internal motional modes,
rotameric interconversion about the glycerol C2–C3 bond.  This motional
description is not unique to $\beta$–DTGL but is, we believe, general to a number
of lipid systems (Auger et al., unpublished results, b).

311

At this point one may ask, what relevance do these conclusions on the dynamics in pure lipid systems have to those of the more complex biomembrane systems? At present the answer is not clear. However, there is some evidence which suggests that the above results offer insight into such systems. Some eight years ago, while studying the glycolipid containing membranes of the microorganism Acholeplasma laidlawii enriched with $^2$H-labelled fatty acids, we noted that as the temperature was elevated from 25°C to above the gel-liquid crystalline phase transition temperature, there was a loss in signal intensity. This was also true for the total lipid extracts of the membranes (Fig. 8). These spectral changes may have their origin in motional changes similar to those already demonstrated for the simpler systems.

$^2$H spin-lattice relaxation and lineshape studies have been used to probe glycolipid system for motion in the range of $10^{-12}$ to $10^{-3}$s. If there are molecular motions on longer time scales, is it possible to study these? Using $^2$H two-dimensional chemical exchange spectroscopy, we have been able to probe motion having a timescale of 100 ms (Auger et al., unpublished results, a). In the gel phase β-DTGL exhibits a slow long axis motion, the nature of which cannot be defined unambiguously from $^2$H NMR lineshape studies. At 25°C this motion is too slow to affect the $^2$H spectrum. However, the motion can be detected by 2D chemical exchange spectroscopy (Fig. 9) which correlates changes in frequency (quadrupole splittings), and therefore orientation, during a fixed mixing time period (Auger and Jarrell, in press; Auger et al., unpublished results, a). The nature of the cross peak patterns can be interpreted in terms of the type of reorientation occurring during the mixing time. In the case of β-DTGL the slow long axis motion is most likely a large angle jump (>60°), not small angle diffusion, about the long molecular axis on a timescale of milliseconds at 25°C.

## Future Prospects

We have shown that $^2$H NMR is capable of providing information about conformation and molecular reorientation in membrane systems. By a judicious choice of the possible observables available in $^2$H NMR, this information can be obtained over a broad range of time scales. The challenge in the future is not only to attempt similar studies with the biologically more interesting complex glycolipids, but also to examine how orientational and motional parameters change as a result of membrane surface interactions. These studies will be exceedingly difficult but we are now attempting to label such lipids and to probe recognition.

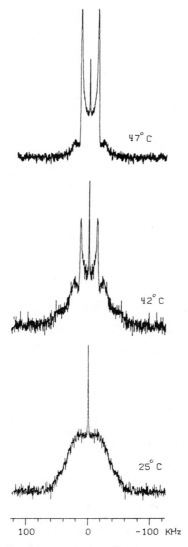

Fig. 8. Temperature dependence of the $^2$H NMR spectrum (46.1 MHz) of the total lipid extract of <u>A. laidlawii</u> membranes enriched with pentadecanoic acid $^2$H-labelled at C5 (Jarrell and Smith, unpublished results).

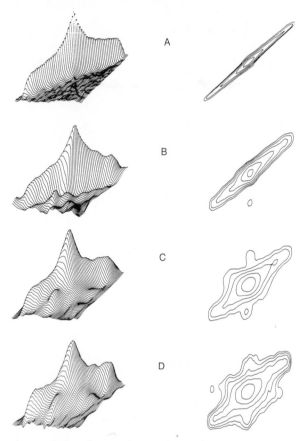

Fig. 9. Experimental 2D absorption mode exchange spectra and corresponding contour plots for a multilamellar dispersion of $[3,3-^{2}H_{2}]\beta$–DTGL. A. 25°C, $t_{mix}$ = 1 ms, B. 35°C, $t_{mix}$ = 0.5 ms, C. 35°C, $t_{mix}$ = 2 ms and D. 40°C, $t_{mix}$ = 2 ms.

REFERENCES

Abragam, A., 1961, The principles of nuclear magnetism, Oxford University Press, London.

Auger, M., Carrier, D., Smith, I. C. P., and Jarrell, H. C., Elucidation of motional modes in glycoglycerolipids - a $^2$H NMR relaxation and lineshape study, J. Amer. Chem. Soc., in press.

Auger, M., and Jarrell, H. C., Elucidation of Slow Motions in Glyco-glycerolipid Bilayers by Two-Dimensional Solid-State Deuteron NMR, Chem. Phys. Letters, in press.

Auger, M., Smith, I. C. P., and Jarrell, H. C., unpublished results, a.

Auger, M., van Calsteren, M. R., Smith, I. C. P., and Jarrell, H. C., unpublished results, b.

Bloom, M., and Smith, I. C. P., 1985, Manifestations of lipid- protein interactions in deuterium NMR, in: Progress in Protein-Lipid Interactions, Watts, A., and de Pont, J., eds. Elsevier, Amsterdam, 61-88.

Carrier, D., Giziewicz, J. B., Moir, D., Smith, I. C. P., and Jarrell, H. C., 1989, Dynamics and orientation of glycolipid headgroups by $^2$H-NMR: gentiobiose, Biochim. Biophys. Acta., 983:100.

Critchley, D. R., 1979, Glycolipids as membrane receptors important in growth regulation, in: Surfaces of Normal and Malignant Cells, R.O., Hynes, Ed., John Wiley and Sons, New York.

Curatolo, W., 1987a, Glycolipid function, Biochim. Biophys. Acta., 906:137.

Curatolo, W., 1987b, The physical properties of glycolipids, Biochim. Biophys. Acta, 906:111.

Davis, J. H., 1983, The description of membrane lipid conformation, order and dynamics by $^2$H-NMR, Biochim. Biophys. Acta., 737:117.

Jarrell, H. C., Giziewicz J. B., and Smith, I. C. P., 1986, Structure and dynamics of a glyceroglycolipid: A $^2$H NMR study of head group orientation, ordering and effect on lipid aggregate structure. Biochemistry, 25:3950.

Jarrell, H. C., Jovall, P. A., Giziewicz, J. B., Turner, L. A., and Smith, I. C. P., 1987a, Determination of conformational properties of glycolipid head groups by $^2$H NMR of oriented multibilayers, Biochemistry, 26:1805.

Jarrell, H. C., Wand, A. J., Giziewicz, J. B., and Smith, I. C. P., 1987b, The dependence of glyceroglycolipid orientation and dynamics on head-group structure, Biochim. Biophys. Acta., 897:69.

Jeffrey, K. R., 1981, Nuclear magnetic relaxation in a spin 1 system. Bull. Magn. Reson., 3:69.

Meier, P., Ohmes E., and Kothe, G., 1986, Multipulse dynamic nuclear magnetic resonance of phospholipid membranes, J. Chem. Phys., 85:3598.

Renou, J. P., Giziewicz, J. B., Smith, I. C. P., and Jarrell, H. C., 1989, Glycolipid membrane surface structure: orientation, conformation, and motion of a disaccharide headgroup, Biochemistry, 28:1804.

Skarjune, R., and Oldfield, E., 1982, Physical studies of cell surface and cell membrane structure. Deuterium nuclear magnetic resonance studies of N-palmitoyl-glucosylceramide (cerebroside) head group structure, Biochemistry, 21:3154.

Smith, I. C. P., and Jarrell, H. C., 1983, Deuterium and phosphorus NMR of microbial membranes, Acc. Chem. Res., 16:266.

Thompson, T. E., and Tillack, T. W., 1985, Organization of glycosphin-golipids in bilayers and plasma membranes of mammalian cells, Ann. Rev. Biophys. Biophys. Chem., 14:361.

Wittebort, R. J., Olejniczak, E. T., and Griffin, R. G., 1987, Analysis of deuterium nuclear magnetic resonance lineshapes in anisotropic media, J. Chem. Phys., 36:5411.

Yu, R. K., Koerner, T. A. W., Demori, P. C., Scarsdale, J. N., and
    Prestegard, J. H., 1984, Recent advances in structural studies of
    ganglioside: primary and secondary structures, in: Ganglioside
    Structure, Function and Biomedical Potential, Leden, R. W., Yu, R.
    K., Rapport, M. M., and Suzuki, K., eds. Plenum Press, New York.

STUDIES OF EVOLVING CARBOHYDRATE METABOLISM IN VIVO BY $^{13}$C SURFACE-COIL NMR
SPECTROSCOPY

Nancy N. Becker* and Joseph J. H. Ackerman

Department of Chemistry
1 Brookings Drive
Washington University
St. Louis, Missouri   63130-4899

The capability of NMR spectroscopy to study biological systems noninvasively was extended to whole animal applications by the introduction of the surface-coil receiver by Ackerman et al. (1980). While this first demonstration utilized a $^{31}$P-tuned receiver for the detection of phosphorus signals, the technique has been extended to other nuclides, including proton ($^{1}$H) and fluorine ($^{19}$F), and, in conjunction with proton-decoupling, carbon ($^{13}$C).

The $^{13}$C NMR spectrum of cellular biological systems displays carbohydrate resonances, especially the C-1 carbons of glycogen and glucose, which are well-resolved from resonances of lipids and other carbon species. Thus, when sequential spectra are obtained with adequate time resolution, $^{13}$C NMR is an excellent tool for the analysis of carbohydrate metabolism as it evolves in vivo. This was demonstrated by high-field, high-resolution studies of carbohydrate metabolism in cells and perfused organs (Cohen et al., 1979).

Reo et al. (1984a) extended the capability of high-field $^{13}$C NMR spectroscopy to the study of surface-accessible tissues of animals in vivo. They described the design of a double-resonance [$^{13}$C-{$^{1}$H}] surface-coil NMR probe for performance of in vivo high-field (8.5 T) proton-decoupled $^{13}$C experiments. With this surface-coil probe, they obtained in vivo $^{13}$C spectra of the leg muscle, brain, and liver of a rat. The probe consisted of two coaxial surface coils individually tuned to observe $^{13}$C at 90.56 MHz and decouple protons at 360.13 MHz. The circuit (see Fig. 1) consisted of a 4-turn, 1-cm diameter coil tuned to the $^{13}$C frequency for transmission and detection, placed inside a 1-turn, 2-cm diameter coil tuned to the proton

---

* Current Address: Department of Chemistry, University of San Diego, Alcala
  Park, San Diego, CA  92110.

frequency for decoupling. Variable capacitors allowed frequency tuning and impedance matching of each coil. Isolation between the two circuits was achieved with an open quarter wavelength (for proton) coaxial transmission line positioned at the low-frequency (carbon) input. This open-ended quarter wavelength line produces an apparent short circuit to ground (at point A, in Fig. 1) for the proton frequency but not for the $^{13}C$ frequency. With this design, ca. 40 dB attenuation of 360 MHz pickup by the $^{13}C$ channel was achieved. Appropriate standard rf filters (e.g., low pass on the $^{13}C$ channel) were used to further enhance isolation prior to signal amplification by the preamplifier.

An efficient high frequency $^{1}H$ channel is required to achieve optimal $^{13}C$ sensitivity and resolution by effective $^{1}H$ decoupling and maintenance of the nuclear Overhauser enhancement. To ensure adequate "broadband" decoupling at moderate power levels (ca. 3-5 W) it is useful to make the $^{1}H$ coil larger than the $^{13}C$ coil so that the $^{1}H$ transverse magnetic field ($B_1$) spatial distribution significantly exceeds the "sensitive volume" of the $^{13}C$ observe coil. Adequate proton-decoupled spectra are obtained with a $^{1}H$ coil which has twice the radius of the $^{13}C$ coil. This design also results in a $^{1}H$ $B_1$ which is reasonably homogeneous over the sensitive volume of the $^{13}C$ observe coil, thus enabling acquisition of well-decoupled $^{13}C$ spectra using the Waltz-16 decoupling mode. The resolution and sensitivity attainable with this probe are apparent in the 8.5 T $^{13}C$ spectrum obtained from a rat liver in situ using Waltz decoupling, displayed in Fig. 2. This spectrum

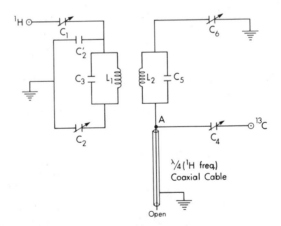

Fig. 1.  Circuit diagram for double-resonance surface-coil receiver. $L_1$ and $L_2$ are separate coils tuned for $^{1}H$ and $^{13}C$ and arranged coaxially. Capacitors $C_1$ and $C_2$ range from 0.8 to 10 pf while $C_4$ and $C_6$ are variable over a range of 1 to 30 pf. (From Reo et al., 1984a.)

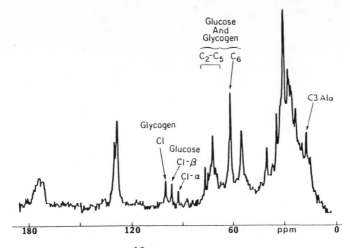

Fig. 2.  Proton-decoupled $^{13}$C NMR spectrum describing gluconeogenesis and glycogen synthesis from alanine, obtained using
the $^{13}$C-{$^{1}$H} surface-coil circuit described in Fig. 1
(N. N. Becker, unpublished results).

was obtained in 5 min from the liver of a fasted rat after administration of
[3-$^{13}$C]alanine and unlabeled glucose, demonstrating the in vivo incorporation of label from alanine into the C-1 position of glycogen.  The C-1
glucose resonance has a linewidth of 16 Hz or ca. 0.2 ppm.

The utility of high-field NMR spectroscopy to study carbohydrate
metabolism as it evolves in vivo was demonstrated by the use of this
proton-decoupled $^{13}$C surface-coil probe to monitor glucose and glycogen
metabolism in a rat liver in situ (Reo et al., 1984b).  A representative
natural abundance spectrum of the liver of a rat fed ad libitum is shown in
Fig. 3a.  The change in the glycogen C-1 resonance over time after i.v.
administration of glucagon is shown in Fig. 4, demonstrating the ability of
proton-decoupled $^{13}$C surface-coil spectroscopy to monitor the time course of
liver glycogenolysis in vivo.  Glycogen formation may also be monitored.  As
shown in Figs. 3b and 3c, introduction of exogenous [1-$^{13}$C]glucose to a
fasted rat resulted in a rapid increase in the hepatic C-1 glucose
resonances, followed by an increase in the hepatic C-1 glycogen resonance
due to incorporation of label into glycogen.

Hormonal effect on incorporation of [1-$^{13}$C]glucose into hepatic
glycogen in vivo was also investigated using the $^{13}$C surface-coil probe
designed by Reo et al.  The role of glucagon as an antiglycogenesis and
glycogenolytic agent is well established.  However, controversy has
surrounded the in vivo role of insulin in the incorporation of glucose into
glycogen and in glycogenolysis.  There is considerable evidence implicating

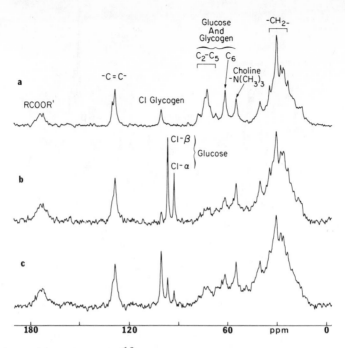

Fig. 3. Proton-decoupled $^{13}$C surface-coil NMR spectra at 90.56 MHz
from rat liver in vivo. A gated bilevel (power: 5/0.5
watt) broad-band decoupling sequence was employed to
minimize sample heating. a) A 10-min accumulation from a
rat fed ad libitum depicting the natural abundance $^{13}$C
resonances from hepatic glycogen (61-100 ppm). The C-1
carbon resonance of the alpha-1,4 glycosidic linkage at
100.4 ppm is clearly resolved. b) A 3-min accumulation
from a 271-g rat fasted for ca. 15 h. This spectrum was
initiated 3 min after an injection of 100 mg of
D-[1-$^{13}$C]glucose (90% enriched) into a femoral vein and
depicts signals at 92.7 and 96.5 ppm arising from the C-1
carbon of the alpha and beta anomers of glucose,
respectively. This glucose is converted into hepatic
glycogen as shown by the spectrum in c) which spans the
time period 30-33 min after glucose administration and
displays signals from the C-1 carbon of both glucose and
glycogen. Chemical shifts are reported relative to
tetramethylsilane using IUPAC recommendations. (From Reo
et al., 1984b.)

insulin as an activator of glycogen synthase which catalyzes the rate-
limiting step of glycogen synthesis, but insulin's ability to significantly
enhance hepatic glycogenesis in vivo has not been clearly demonstrated.
Because of this confusion concerning insulin's role in hepatic glycogen
metabolism, Siegfried et al. (1985) investigated the effect of insulin and
one of its principal counter-regulatory hormones, glucagon, on incorporation
of labeled glucose into liver glycogen using high-field NMR spectroscopy of
rat liver in vivo.

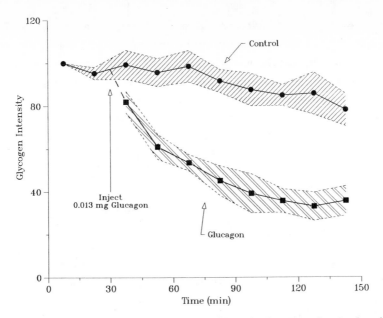

Fig. 4.  Time course for glucagon-stimulated glycogenolysis in the
rat liver <u>in vivo</u> as monitored from C-1 glycogen signal
intensity by natural abundance $^{13}C-\{^1H\}$ NMR.  The signal
intensity in the first spectrum from each animal was
normalized to 100, and subsequent data points represent the
average relative peak heights from six control animals
(circles) and four animals which were administered glucagon
via a femoral vein (squares).  Assuming constant relaxation
times, changes in NMR peak heights are directly comparable
to concentration changes.  The shaded regions indicate
$\pm$ S.E.  (From Reo et al., 1984b.)

Results of this study indicated that, after prior labeling of hepatic
glucose and glycogen, glucagon causes the loss of labeled glucose monomers
from glycogen and inhibits incorporation of labeled glucose into glycogen.
Data were consistent with the model in which, following an intravenous bolus
of glucose in the fasted rat, glycogen synthesis in the liver increases
linearly with hepatic glucose concentration after a threshold hepatic
concentration is exceeded.  Exogenous insulin did not alter the overall rate
constant or the threshold for glycogen synthesis.  Insulin did accelerate
the loss of labeled glucose; glucagon was found to counteract this effect.
Insulin did not perturb the time course of the dose-dependent inhibition of
glycogen formation by glucagon.

The quantitative accuracy of these studies purporting to monitor
glycogen concentrations by NMR may be questioned because of the high
molecular weight of hepatic glycogen.  Glycogen is the major storage form of
glucose in the liver.  It is a highly branched, spherical polymer of
D-glucopyranose residues in $\alpha$-(1→4) glycosidic linkages with branching by
means of $\alpha$-(1→6) glycosidic bonds.  The molecular weight of each glycogen

particle can reach $10^7$-$10^9$ daltons, and total glycogen can be up to 6% of hepatic weight. Considering its structure and molecular weight, one might anticipate that as increasing numbers of glucopyranose residues are attached to the glycogen particle, such as occurs when an animal ingests carbo-hydrates after a fast, there would be a marked increase in the correlation time of the inner core of residues due to decreasing mobility. An increase in the correlation time of the carbons in these core residues would lead to a decrease in the transverse relaxation time, $T_2$, and an increase in the observed linewidth of the resonance if $T_2^*$ were significantly reduced. If the correlation time of the core residues became greater than $\omega_0^{-1}$, these residues would experience an increase in the longitudinal relaxation time, $T_1$, relative to the residues attached to the outer, presumably more mobile, portions of the glycogen particle. However, as our laboratory and others have shown, the carbons of hepatic glycogen give rise to high-resolution $^{13}C$ NMR signals in vivo that appear to increase appropriately under conditions of synthesis. Validation of uniform visibility is critical for quantifying results from liver studies of hepatic glycogen synthesis or degradation performed under conditions of rapid pulse repetition for optimization of signal-to-noise ratio and time resolution.

This issue has been addressed quantitatively in an in vitro analysis by Sillerud and Shulman (1983) in which they found hepatic glycogen to be uniformly visible under conditions of active, glucagon-stimulated glyco-genolysis in a nonperfused, excised rat liver. However, since many experiments are done in vivo or in perfused livers under conditions of glycogen synthesis, NMR visibility under these conditions also must be examined. In addition, the possibility of subtle motional differences between glycogen undergoing synthesis vs. degradation due to changes in enzyme binding and activity may not allow results obtained under conditions of glycogen breakdown to be generalized to conditions of synthesis.

A "pulse-chase" study was performed to determine whether the visibility of the signal from the inner portion of the glycogen molecule is affected as the particle grows in size during active glycogen synthesis in the rat liver in vivo (Shalwitz et al., 1987). A "pulse" of [1-$^{13}$C]glucose (99% enrichment) was administered to a fasted rat, followed by a one hour "chase" of unlabeled glucose. This protocol enabled synthesis of glycogen in which an inner core of [1-$^{13}$C]glucopyranose residues was laid down, followed by an outer layer of unlabeled residues. Because the natural abundance level of $^{13}$C is 1.1%, the outer, unlabeled residues of the glycogen particle were virtually undetected by the NMR experiment. Thus, the peak height and linewidth of the inner residues could be monitored as they were progressively "buried" during the chase. The first pulse-chase was then immediately followed by an identical pulse-chase. Rapid pulse-repetition acquisition conditions were used.

Under in vivo conditions of active glycogen synthesis, we have determined the $T_1$ of the C-1 resonance of glycogen to be approximately 0.4 sec. Therefore, the C-1 magnetization is partially saturated under the rapid pulse-repetition rates widely used to maximize signal-to-noise and time resolution. Substantial changes in $T_1$ are expected to result in a change in signal intensity due to a change in the degree of saturation of the resonance. The results of the pulse-chase experiment, shown in Fig. 5, indicate that there was not a marked increase in $T_1$ for the labeled glucopyranose residues laid down prior to and during the initial portions of the chase. Linewidth changes of the resonances were analyzed, revealing no significant change in linewidth attributable to changes in glycogen relaxation.

Two complementary studies were also performed. First, to demonstrate that glycogen continued to be formed during the chase period, a separate in vivo NMR experiment (n=1) was performed in identical fashion to the one described above, except that only a single pulse-chase was performed and [1-$^{13}$C]glucose was given throughout the experiment, i.e., both the pulse and the chase consisted of labeled glucose. The NMR peak height of C-1 glycogen

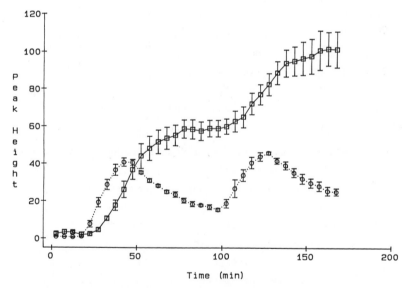

Fig. 5. Mean peak heights ($\pm$ SE) from n=3 during the pulse-chase study. Baseline spectra were acquired for the first 20 min. At time = 20 min, a pulse of 60 mg/100 g body weight of D-[1-$^{13}$C]glucose was given over 20 min. This was followed immediately by a chase with unlabeled glucose at 1 mg/(100g/min) for 60 min. At time = 100 min, an identical pulse-chase was repeated. Each spectrum represents 5 min of data accumulation, with 1083 scans per spectrum. Peak heights were normalized to an external $CS_2$ standard. The circles represent the beta anomer of C-1 glucose, and the squares represent C-1 glycogen. (From Shalwitz et al., 1987.)

was found to rise continuously throughout the chase. The ratio of the peak height for C-1 glycogen at the end of the chase compared to the beginning of the chase was ca. 50% greater than the mean ratio (n=3) for the original pulse-chase study. This indicates that the infusion rate used during the chase was adequate to maintain active glycogen synthesis.

In the second complementary study, a series of benchtop experiments was performed to assess the quantity of glycogen synthesized under the conditions of this experiment. For these, an experimental protocol identical to that described above was used, except that all glucose was unlabeled. At the end of the second chase, the liver was immediately removed and frozen in liquid nitrogen. Glycogen was extracted and quantified by degradation to glucose followed by determination of glucose concentration. Approximately 14 mg glycogen per gram wet weight liver was formed during the pulse-chase study, representing an approximately seven-fold increase in glycogen from the fasted state.

These results demonstrated that the labeled portion of glycogen remained visible throughout the pulse-chase study and indicated that there was no major change in either $T_1$ or $T_2$. It can be concluded that hepatic glycogen remains uniformly visible to the NMR experiment despite its high molecular weight. This is true for conditions of synthesis as well as conditions of breakdown, as demonstrated by Sillerud and Shulman. The demonstration of uniform visibility is important for in vivo studies of glycogen metabolism which require optimal time resolution (i.e., rapid pulse-repetition rate conditions) and a reproducible assessment of $^{13}$C-labeled glycogen under varying physiologic conditions.

A major advantage of the surface-coil NMR technique for monitoring carbohydrate metabolism in the liver is the ability to determine both substrate (e.g., glucose and alanine) and product (e.g., glycogen) concentrations nondestructively. In fact, this is the only method currently available which allows in a single animal the concurrent analysis of hepatic substrate and product over time, thus allowing determination of pseudo-first-order rate constants for glycogen synthesis for each experimental animal. Furthermore, $^{13}$C-labeling of substrates for glycogen synthesis allows elucidation of the pathways involved by NMR detection of randomization of the $^{13}$C label into the various carbon positions of glycogen ("label scrambling").

Our laboratory has used these unique properties of NMR to determine the relative contribution of direct and indirect pathways and the kinetics of hepatic glycogen synthesis from duodenally administered glucose and alanine (Shalwitz et al., 1989). Recently, several studies have concluded that glucose is not the primary precursor for glycogen synthesis following an oral glucose load or an intraduodenal infusion of glucose in a fasted rat or man (Shulman et al., 1985). These studies have also shown that a significant, but variable, portion of the glucose load is converted into 3-carbon precursors and then reincorporated into glycogen via gluconeogenesis. While the mechanisms involved are not yet clear, one explanation may be a limiting supply of glucose-6-phosphate derived directly from orally administered glucose. McGarry (1984) has recently proposed that the liver has a relatively low capacity for phosphorylating glucose. Hepatic glycogen synthesis from gluconeogenic precursors would thus effectively compete with hepatic glycogen synthesis from glucose, if gluconeogenesis continued to provide substantial levels of glucose-6-phosphate following an oral glucose load.

A surface-coil $^{13}C-\{^1H\}$ NMR study was performed to compare the kinetic time course of glycogen synthesis from $[3-^{13}C]$alanine and $[1-^{13}C]$glucose given in combined, equimolar amounts by intraduodenal infusion. The metabolic fate of the duodenally administered $^{13}C$-labeled substrate load, glucose or alanine, can be addressed through analysis of label randomization in glycogen. In this way, glycogen synthesis from $^{13}C$-labeled glucose occurring via either the "direct" pathway (glucose → glucose-6-phosphate → glycogen) or the "indirect" pathway in which glucose is first broken down to lactate (glucose → glucose-6-phosphate → lactate → glucose-6-phosphate → glycogen), or from $^{13}C$-labeled alanine occuring via gluconeogenesis can be monitored. Analysis of pathways of glycogen synthesis from label scrambling pertains only to that glycogen formed from introduced labeled substrate. By administering a substrate mixture in which only one of the substrates is $^{13}C$-labeled, the fate of label from that substrate can be monitored unambiguously. Glycogen synthesis via the direct pathway from glucose labeled at the C-1 carbon ($[1-^{13}C]$glucose) results in glycogen labeled only at the C-1 carbon. Glycolytic breakdown of glucose labeled at the C-1 carbon, followed by gluconeogenesis and glycogenesis (the "indirect" pathway) results in glycogen labeled approximately randomly at the C-1, C-2, C-5 and C-6 positions. Glycogen synthesis (via gluconeogenesis) from alanine labeled at the C-3 carbon ($[3-^{13}C]$alanine) results in glycogen labeled approximately randomly at the C-1, C-2, C-5 and C-6 positions.

In this study, fasted rats were administered duodenally a substrate mixture of glucose and alanine (0.67 mmol/100 g of each substrate) after collecting baseline $^{13}C-\{^1H\}$ spectra from the liver for 20 min. In one group (n=3), rats were given a substrate mixture containing $[1-^{13}C]$glucose and unlabeled alanine, while in another group (n=3), rats were given a substrate mixture containing unlabeled glucose and $[3-^{13}C]$alanine. Glycogen synthesis from the labeled substrate was followed in each group for 150 min (see Fig. 6). Glycogen synthesis from unlabeled substrate was virtually undetected, as the natural abundance concentration of $^{13}C$ is 1.1%. Hepatic

325

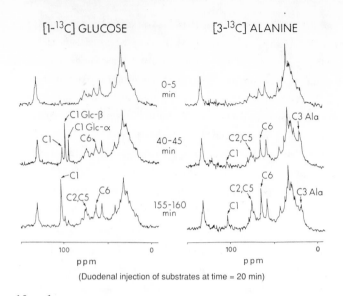

[1-$^{13}$C] GLUCOSE          [3-$^{13}$C] ALANINE

0-5 min

C1 Glc-β
C1 Glc-α
C1    C6    40-45 min    C2,C5    C6    C3 Ala
C1

C1
C2,C5    C6    155-160 min    C6
C2,C5    C3 Ala
C1

100          0          100          0
ppm                     ppm

(Duodenal injection of substrates at time = 20 min)

Fig. 6.    $^{13}$C-{$^{1}$H} surface-coil NMR spectra of rat liver in situ
representing time course of hepatic metabolism of
[1-$^{13}$C]glucose and [3-$^{13}$C]alanine.  Representative spectra
are shown for each protocol where the first spectrum (0-5
min) in each case represents background signal contribution
before duodenal administration of substrate mixture at
time = 20 min.  Headings indicate the labeled substrate
given in that series of spectra ([1-$^{13}$C]GLUCOSE:  substrate
mixture contained labeled glucose and unlabeled alanine;
[3-$^{13}$C]ALANINE: substrate mixture contained labeled alanine
and unlabeled glucose).  Resonance peak labels are as
follows:  C1 Glc-beta, beta anomer of C-1 glucose; C1
Glc-alpha, alpha anomer of C-1 glucose; C1, C-1 glycogen;
C2, C5, combined C-2 and C-5 glucose and glycogen; C6, C-6
glucose and glycogen; C3 Ala, C-3 alanine.  Chemical shift
is expressed in parts per million (ppm) relative to
tetramethylsilane.  (From Shalwitz et al., 1989.)

concentrations of the $^{13}$C label in glucose, alanine, and glycogen were
obtained from the appropriate resonance in each spectrum throughout the time
course.  Experimentally determined correction factors were applied to
resonance areas to account for differences in partial magnetization saturation, nuclear Overhauser effect, and $T_1$-dependent spatial sensitivities of
the carbon species analyzed.  Thus, true label concentrations, in arbitrary
units, were obtained for the C-1, C-6, and C-2 plus C-5 carbon positions of
glycogen, the C-1 position of glucose, and the C-3 position of alanine.

It was found in this study that while [1-$^{13}$C]glucose is predominantly
incorporated into glycogen labeled at the C-1 position, there is substantial
incorporation of label into the C-6, C-2 and C-5 positions of glycogen.  The
percentage of glycogen formed from the administered label via the indirect
pathway may be estimated by comparison of the concentration of label at the

various positions of glycogen. The contribution of the indirect pathway to glycogen synthesis was determined to be at least 30% under the conditions of this study. Figure 7 displays the time course of glycogen synthesis via the direct pathway from [1-$^{13}$C]glucose overlaying the time course of total glycogen synthesis from [3-$^{13}$C]alanine (via gluconeogenesis). Also displayed in Fig. 7 are the time courses of the hepatic concentrations of the labeled substrates. Apparent in this graph is the result that, while the hepatic concentration of labeled alanine never reaches that of labeled glucose, more glycogen (total $^{13}$C label at all positions) is actually synthesized from labeled alanine via the indirect pathway than from labeled glucose via the direct pathway.

Because both the rate of synthesis of hepatic glycogen (i.e., the slope of the glycogen versus time curve) and the hepatic concentration of substrate (labeled glucose or alanine) may be determined simultaneously for each animal studied, in vivo $^{13}$C NMR spectroscopy uniquely allows the

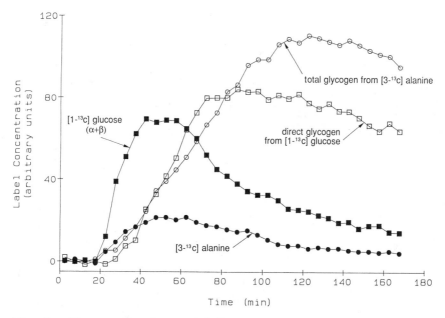

Fig. 7.  Time course of mean label concentrations of glycogen derived from the direct pathway of glycogen synthesis from [1-$^{13}$C]glucose compared to total glycogen synthesis from [3-$^{13}$C]alanine. Baseline spectra were collected for 20 min, at which time a substrate mixture was administered duodenally. The symbols are:  filled squares, [1-$^{13}$C]glucose; open squares, direct glycogen from [1-$^{13}$C]glucose; filled circles, [3-$^{13}$C]alanine; open circles, total glycogen from [3-$^{13}$C]alanine. (From Shalwitz et al., 1989.)

determination of pseudo-first-order rate constants for glycogen synthesis in a single animal. This allows each animal to serve as its own control. Under the conditions used in this study, the mean rate constant for glycogen synthesis from [3-$^{13}$C]alanine, 0.075 min$^{-1}$ ($\pm 0.026$ S.E.), was significantly greater (p<.05) than that from [1-$^{13}$C]glucose, 0.025 min$^{-1}$ ($\pm 0.005$ S.E.). Because these two pathways of glycogen synthesis (the direct from glucose and indirect from alanine) meet at glucose-6-phosphate and then proceed in identical fashion to glycogen, these results support the proposal that there may be a limitation in hepatic phosphorylation of glucose by glucokinase.

The hepatic studies discussed here demonstrate proton-decoupled $^{13}$C surface-coil NMR spectroscopy to be a unique tool for quantitative analysis of the pathways of carbohydrate metabolism. Extension of the technique to carbohydrate metabolism in other tissues, including brain (Kotyk et al., 1989), has recently been reported. Future solutions to metabolic questions await the application of this powerful technique.

REFERENCES

Ackerman, J. J. H., Grove, T. H., Wong, G. C., Gadian, D. G., and Radda, G. K., 1980, Nature (Lond.), 283:167-170.
Cohen, S. M., Ogawa, W., and Shulman, R. G., 1979, Proc. Nat. Acad. Sci. USA, 76:1603; Cohen, S. M., Shulman, R. G., and McLaughlin, A. C., 1979, ibid, 76:4808.
Kotyk, J., Deuel, R. K., and Ackerman, J. J. H., 1989, J. Neurochem, 53:1620-1628.
Kuwajima, M., Newgard, C. B., Foster, D. W., and McGarry, J. D., 1984, J. Biol. Chem., 261:8849-8853.
Reo, N. V., Ewy, C. S., Siegfried, B. A., and Ackerman, J. J. H., 1984a, J. Magn. Reson., 58:76-84.
Reo, N. V., Siegfried, B. A., and Ackerman, J. J. H., 1984b, J. Biol. Chem., 259:13664-13667.
Shalwitz, R. A., Reo, N. V., Becker, N. N., and Ackerman, J. J. H., 1987, Magn. Reson. Med., 5:462-465.
Shalwitz, R. A., Reo, N. V., Becker, N. N., Hill, A. C., Ewy, C. S., and Ackerman, J. J. H., 1989, J. Biol. Chem., 264:3930-3934.
Shulman, G. I., Rothman, D. L., Smith, D., Johnson, C. M., Blair, J. B., Shulman, R. G., and DeFronzo, R. A., 1985, J. Clin. Invest., 76:1229-1236.
Siegfried, B. A., Reo, N. V., Ewy, C. S., Shalwitz, R. A., Ackerman, J. J. H., and McDonald, J. M., 1985, J. Biol. Chem., 260:16137-16142.
Sillerud, L. O., and Shulman, R. G., 1983, Biochemistry, 22:1087.

UPTAKE, METABOLISM, AND STORAGE OF PHOSPHATE AND NITROGEN IN PLANT
CELLS; AN NMR PERSPECTIVE

Hans J. Vogel* and Peter Lundberg

* Department of Biological Sciences
  University of Calgary
  2500 University Drive, N.W.
  Calgary, Alberta, Canada   T2N 1N4

A number of Nuclear Magnetic Resonance (NMR) Spectroscopy techniques
can be used to study pH regulation and various aspects of nutrient
metabolism in plant material.  In this study phosphorus-31 NMR has been
used to determine the energy state (ATP) and the intracellular cytoplasmic
and vacuolar pH of cultured plant cells and algae.  For the algae it was
found that the chemical shift of the terminal polyphosphate resonance
provided a good monitor of the vacuolar pH which was estimated at pH 5.5.
A cytoplasmic pH of 7.2 was determined from the chemical shifts of the Pi
and glucose-6-phosphate resonances.  Phosphate uptake could also be
followed by $^{31}$P NMR and these studies showed that Pi was stored as
polyphosphates in algae, but as vacuolar Pi in certain higher plants such
as Catharanthus roseus and Nicotiana tabacum.

A series of nitrogen-14 and nitrogen-15 NMR studies of white spruce
buds and embryos was also performed.  It was found that $^{14}$N NMR was
particularly useful to study the intracellular levels of $NH_4^+$ and $NO_3^-$.
In contrast, $^{15}$N NMR provided great detail on the existing amino acid
pools.  In addition, in $^{15}$N incorporation studies with perfused plant
material, the flux of the isotopic label through the amino acid pools
could be followed, thus providing information regarding the pathways that
are involved in the assimilation of inorganic nitrogen nutrient.  These
NMR methods are generally applicable to the study of a large variety of
plant material.

INTRODUCTION

In vivo NMR has rapidly emerged as an important experimental tool for the study of metabolism in all kinds of living matter. Presently the main focus of this active research field is to develop and implement methods for clinical applications (Cohen, 1987). However, a large body of work also deals with studies that are of a more basic scientific nature in fields such as biochemistry, physiology and botany. The major appeal of the NMR approach in most of these studies is that they can be performed in a totally noninvasive manner. As such, information can be obtained about intracellular pH and metabolite levels repeatedly from the same sample (Gadian and Radda, 1981; Ingwall, 1982; Gadian, 1983; Vogel et al., 1987). This allows one to combine in vivo NMR studies with other noninvasive studies of function, growth, morphogenesis, hydrodynamics, etc. Another attractive feature of in vivo NMR is that information can sometimes be obtained about intracellular compartmentalization and binding of ligands to macromolecules. This aspect has been particularly useful in studies of plant material and yeast, where the signals for compounds that are sequestered in the large acidic vacuole can generally be separated from those in the cytoplasm and hence they can be individually quantitated (Roberts et al., 1980; Nicolay et al., 1982, 1983; Martin et al., 1982; Kime et al., 1982; Wray et al., 1983; Vogel and Brodelius, 1984; Brodelius and Vogel, 1985; Roberts, 1984; Vogel, 1987). In addition, information about the binding of specific nucleotides to proteins, or the degree of saturation of ATP by $Mg^{2+}$ is also obtained (Ingwall, 1982; Gadian, 1983; Vogel et al., 1987; Morris, 1988; Hellstrand and Vogel, 1985; Gupta et al., 1984). Clearly, these latter types of information are difficult to obtain with classical invasive methods for the study of metabolism.

Although NMR studies dealing with noninvasive determinations of pH and metabolite levels in living tissues were reported as early as 1973 and 1974 (Moon and Richards, 1973; Hoult et al., 1974) the study of plant metabolism has only received major attention over the last decade or so, in spite of some early reports showing its feasibility (Schaefer et al., 1975). One of the main problems was that the average metabolite levels are generally low in highly vacuolated older plant tissues, which renders them rather difficult to detect with the relatively insensitive NMR technique. Moreover, the paramagnetic properties of the $O_2$ which is sequestered in the air spaces of leaves for example, causes substantial broadening of NMR signals which again gives rise to poor detectability (Waterton et al., 1983). Consequently, it is not surprising that most of the successful reports concerning NMR studies of plant metabolism have dealt with the study of actively growing submerged root tips or tissue cultured plant cells (Martin, 1985a). In addition, a large number of NMR studies have been reported which deal with metabolism in a number of microalgae (Sianoudis et al., 1986; Kugel et al., 1987; Sianoudis et al., 1987; Watanabe et al., 1987; Thomae and Gleason, 1987; Gimmler et al., 1988; Bental et al., 1988) and certain macroalgae (Lundberg et al., 1989; Weich et al., 1989).

NMR studies can in principle be performed for all elements in the periodic table, because generally an isotope can be found which has the property of nuclear spin.  However, not all of these are equally useful for use in in vivo NMR studies.  For example, a large number of nuclei have a rather large electrical quadrupole moment (spin > 1/2) that broadens their resonances beyond detection.  This applies for example to the only NMR-sensitive isotope of oxygen ($^{17}O$) and consequently there are few NMR studies dealing directly with oxygen in biological systems.  Other nuclei are difficult to study because they have an intrinsically low resonance frequency and sensitivity.  Moreover, the natural abundance of certain NMR isotopes can be so low that one needs to work with isotopically enriched material to get good sensitivity.  Isotopically labelled compounds can be quite expensive and hence it may not always be economically feasible to pursue NMR studies of a given nucleus although it can technically be done.  Despite these contraints there is still quite a large range of nuclei that can be used for in vivo NMR studies (see Table 1).  Inspection of the numbers in this table already gives considerable insight into the advantages and disadvantages of NMR of the various nuclei.  For example, although the low receptivity of $^{13}C$ and $^{15}N$ severely limits natural abundance NMR studies, good sensitivity can be obtained by using 99% isotopically enriched precursors which are fed to and metabolized by the cell (Martin, 1985a).  The use of $^{1}H$ NMR would be attractive because of its intrinsic sensitivity, but the large amount of $H_2O$ in the samples may cause some problems, although techniques are now available to suppress this (Vogel et al., 1987).  Because solutions of $Na^+$, $K^+$ and $Cl^-$ in $H_2O$ give rise to relatively narrow resonances, quadrupolar $^{23}Na$, $^{39}K$ and $^{35}Cl$ NMR studies can be used to measure the gradients of these monovalent ions across the membrane.  Resonances for intra- and extra-cellular resonances are generally separated in the spectra by adding aqueous shift reagents (Gupta et al., 1984; Springer, 1987).  $^{14}N$ NMR has turned out to be mainly useful to study $NO_3^-$ and $NH_4^+$ uptake, metabolism, and storage in plant cells (Lundberg et al., 1989; Belton et al., 1985; Thorpe et al., 1989).  However, of all nuclei, phosphorus-31 NMR has been studied most because it has a good receptivity and it provides information about the intracellular pH, the energy status of the cells and the mode of phosphate uptake and storage (for reviews see Roberts, 1984; Vogel, 1987; Martin, 1985a).  In the following some

Table 1. NMR properties of isotopes that can be used for the study of plant cell metabolism. [a]

| NMR Isotope | Spin quantum number | NMR frequency (MHz) at 9.4 Tesla | Natural abundance (%) | Relative sensitivity at constant field [b] | Relative receptivity [c] |
|---|---|---|---|---|---|
| $^1$H | 1/2 | 400 | 99.98 | 1.0 | 1.0 |
| $^{13}$C | 1/2 | 100.6 | 1.11 | $1.6 \times 10^{-2}$ | $1.8 \times 10^{-4}$ |
| $^{14}$N | 1 | 28.9 | 99.63 | $1.0 \times 10^{-3}$ | $1.0 \times 10^{-3}$ |
| $^{15}$N | 1/2 | 40.5 | 0.37 | $1.0 \times 10^{-3}$ | $3.7 \times 10^{-6}$ |
| $^{23}$Na | 3/2 | 105.8 | 100 | $9.3 \times 10^{-2}$ | $9.3 \times 10^{-2}$ |
| $^{31}$P | 1/2 | 162 | 100 | $6.6 \times 10^{-2}$ | $6.6 \times 10^{-2}$ |
| $^{35}$Cl | 3/2 | 39.2 | 75.5 | $4.7 \times 10^{-3}$ | $3.6 \times 10^{-3}$ |
| $^{39}$K | 3/2 | 18.7 | 93.1 | $5.1 \times 10^{-4}$ | $4.7 \times 10^{-4}$ |

[a] Adapted from Morris, 1988.
[b] $^1$H is arbitrarily given a sensitivity of 1.0.
[c] Receptivity is defined as sensitivity x natural abundance.

examples of $^{31}$P NMR and $^{14}$N and $^{15}$N NMR will be discussed as they relate to the regulation of pH and the phosphate and nitrogen metabolism in some cultured plant cells and algae. Studies dealing with the use of other NMR nuclei for the study of plant metabolism will not be discussed, but the interested reader can find the necessary information in the following references: $^1$H (Fan et al., 1986, 1988), $^{13}$C (Schaefer et al., 1975; Martin, 1985a; Bental et al., 1988; Stidham et al., 1983; Thomas and Ratcliffe, 1985; Dijkema et al., 1988; Chang and Roberts, 1989), $^{23}$Na (Sillerud and Heyser, 1984; Gerasimowicz et al., 1986; Bental et al., 1988), $^{39}$K (Yazaki et al., 1988; Lundberg, 1989) and $^{35}$Cl (Lundberg, 1989).

PHOSPHORUS-31 NMR

Figure 1 shows a typical $^{31}$P NMR spectrum that is obtained for a suspension of tissue-cultured Catharanthus roseus. The spectrum gives quite some detail and resonances for many compounds can be distinguished as they have a different chemical shift and they can be assigned to various metabolites (Martin et al., 1982; Kime et al., 1982; Wray et al., 1983; Vogel and Brodelius, 1984; Brodelius and Vogel, 1985). The resonance labeled PME in fact comprises contributions mainly from glucose-6-phosphate and fructose-6-phosphate (Lundberg, 1989), while the unlabeled

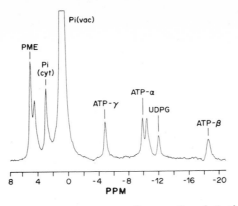

Fig. 1. Phosphorus-31 NMR spectrum of a perfused Catharanthus roseus suspension. The assignment of the resonances is indicated in the figure and discussed in the text (PME = phosphomonoesters).

peak in between those labeled ATPα and UDPG comprises both NAD(H) and one of the two phosphates of UDPG. Interestingly ADP is not detected in this spectrum; in a spectrum of intact tissue its two peaks would normally overlap with those for ATPα and ATPγ but in [31]P NMR spectra of tissue extracts these resonances can be resolved. However, given that the area under the ATPβ signal (which does not contain contributions from any other metabolites) is roughly equal to that of ATPγ we can conclude that the levels of free ADP are in fact very low. It should be pointed out here that in a spectrum of this kind it is impossible to separate the different nucleotides. Hence GTP, CTP, and UTP would all contribute to the ATP resonances. Be that as it may, in most plant material ATP is generally the most abundant nucleotide (Hooks et al., 1989).

Of particular interest in Fig. 1 is the emergence of two peaks, which are assigned to the inorganic phosphate (Pi) that resides in the cytoplasm and the vacuole, respectively. The chemical shifts of Pi and glucose-6-phosphate display a marked pH dependence. Hence Pi residing in the more acidic vacuole (pH 5.7) appears as a resonance which is separate from that for the Pi in the cytoplasm (pH 7.1). Hence by using appropriate calibration curves, we can read out the pH and pH changes in the cytoplasm and vacuole simply by recording the chemical shifts of both resonances. This pattern of separate resonances for cytoplasmic and vacuolar pH is not unique for Catharanthus roseus, but it is in fact observed in virtually all higher plant material studied to date (Roberts, 1984), as well as in yeasts (Nicolay et al., 1982, 1983) and in blue-green-, micro- and macro-algae (Vogel, 1987; Sianoudis et al., 1986; Kugel et al., 1987; Sianoudis et al., 1987; Watanabe et al., 1987; Thomae and Gleason, 1987; Gimmler et al., 1988; Bental et al., 1988; Lundberg et al., 1989; Weich et al., 1989). Despite the fact that plants generally have a series of other intracellular organelles such as chloroplasts, amyloplasts and mitochondria, resonances for Pi residing in these environments have generally not been observed.[*] However, the lack of $^{31}$P NMR detectable pools of Pi in these other organelles, can to some extent be explained by their relatively small size compared to the vacuoles for example. Other factors that could contribute would be a high viscosity in these organelles, which could significantly broaden these resonances. Furthermore, if these organelles would have the same pH as the vacuole or the cytoplasm, their NMR resonance would be obscured by the vacuolar and cytoplasmic Pi resonances.

It is of considerable interest that the vacuolar Pi resonance in Fig. 1 is of much higher intensity than the other resonances. This represents intracellular storage of Pi in the vacuole. In fact, Catharanthus roseus rapidly accumulates all the Pi that is present in the growth medium into the vacuole (Vogel and Brodelius, 1984; Brodelius and Vogel, 1985; Lundberg, 1989) where the concentration can reach as high as 150 mM (Mathieu et al., 1989). The Pi accumulation does not cause a change in the vacuolar pH value (Brodelius and Vogel, 1985; Mathieu et al., 1989). This rapid active Pi uptake can be inhibited by the use of metabolic inhibitors which lower the ATP levels and by proton and cation ionophores which collapse the membranous proton and potassium gradients (Lundberg, 1989). A similar, albeit slower, accumulation of Pi into the vacuole is observed for Nicotiana tabacum, but not for ten other cultured higher plant strains that we have tested (Wray et al., 1983; Brodelius and Vogel, 1985; Lundberg, 1989). Thus although this strategy of accumulating and storing Pi would appear to be a useful one for all plant cells, it seems to occur only in certain higher plant species, while others apparently only take up Pi as they need it for growth and consequently they take it up very slowly from the medium (Brodelius and Vogel, 1985).

Recently, the high vacuolar levels of Pi in Catharanthus roseus have been utilized experimentally, to determine the proton pumping capacity and the regulation of vacuolar pH in isolated C. roseus vacuoles by $^{31}$P NMR (Mathieu et al., 1989; Guern et al., 1989). This elegant study comprises

---

[*] In all fairness, it should be pointed out here that there has been some confusion and disagreement concerning the assignment of Pi resonances in the case of the micro-algae Chlorella (Vogel, 1987; Sianoudis et al., 1987; Mimuro and Kirino, 1984; Mitsumiro and Ito, 1984).

one of the few studies of isolated plant organelles (see also Foyer
et al., 1982). The chemical shift of the vacuolar Pi resonance was used
to study the vacuolar pH regulation under a variety of experimental
conditions; the pH ranged from 6.7 to 5.0 in this study. It is important
to point out here that pH 5.0 is somewhat beyond the limit of pH values
that can be reliably detected by [31]P NMR using the Pi chemical shift. The
pKa of Pi is around 6.7 and hence the Henderson-Hasselbach equation for
acid-base titrations dictates that 90% of the molecules are in either the
acid or base form at a $\Delta$ pH from the pKa of $\pm 1$ pH unit. Consequently, the
titration curve for Pi (for which the chemical shift is the fast exchange
equilibrium of all species) is quite steep between pH 5.7 and 7.7 but very
flat outside this range. As a result the inaccuracy of the pH
determinations rapidly increases the further the pH is away from the
actual pKa of the compound: a small error in the chemical shift
measurement can give rise to a large error in the pH. This situation is of
particular concern, because the Pi chemical shift is also somewhat
dependent upon salt concentration, the presence of divalent cations and
binding to macromolecules (Gadian and Radda, 1981; Vogel et al., 1987;
Roberts, 1984). Hence the use of Pi as a pH indicator below pH 5.7 should
be restricted to situations where the latter parameters are known not to
fluctuate. Because of its lower pKa of 6.2, glucose-6-phosphate would
make a better [31]P NMR pH indicator below pH 5.7, but it is unfortunately
not present in the vacuole. Consequently other compounds such as organic
acids with lower pKa values have become important tools in the determination of low vacuolar pH values (Stidham et al., 1983; Chang and
Roberts, 1989).

Virtually all algae that have been studied to date contain polyphosphates, a linear phosphate polymer which plays a role as a phosphate store
(Vogel, 1987) and possibly as an energy store (Nicolay et al., 1983).
This compound is also found in most yeast strains (Nicolay et al., 1982,
1983). The polyphosphates give rise to some characteristic [31]P NMR
resonances that can be readily assigned and quantitated. In particular,
the resonance for the core of the linear polyphosphate chain at -21 or
-23 ppm can be easily discerned. Figure 2 shows an NMR spectrum of a
solution of a polyphosphate in the presence and absence of $Mg^{2+}$. The
marked dependence of the spectrum on $Mg^{2+}$ can be used to estimate the
degree of saturation of polyphosphates by $Mg^{2+}$ (Lundberg et al., 1989); in
most algae studied to date, the polyphosphates appear to be saturated with
$Mg^{2+}$. Moreover, if the resonance for the terminal phosphate can be
resolved in the in vivo NMR spectra, the average chain length can be
determined by comparing the areas of the various polyphosphate peaks
(Lundberg et al., 1989). Finally, the polyphosphate provides a useful pH

335

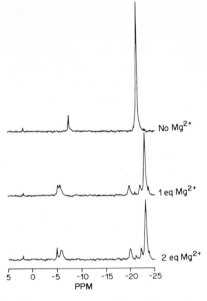

Fig. 2.  Phosphorus–31 NMR spectra of samples of polyphosphate with
an average chain length of 18 (pH = 6.5, 130 mM KCl).  The
resonance for the terminal group (PP1) can be readily
discerned around –5 ppm.  The sample contains some
pyrophosphate which contributes a sharp resonance around
–5 ppm.  The resonances for the central phosphates (PPn)
appear at –21 and –23 ppm.  The resonances for PP2 and PP3
overlap with PPn in the absence of $Mg^{2+}$, but they can be
separated in this sample (but not always in in vivo
situations) upon addition of $Mg^{2+}$.  Please note that
addition of more than one equivalent of $Mg^{2+}$ does not cause
further changes in the spectrum.

indicator to determine the vacuolar pH (Lundberg et al., 1989).  Figure 3
shows the pH titration curves that were measured for the same samples
which were used to obtain Fig. 2.  Panel A shows that the chemical shift
of the terminal polyphosphate resonance is particularly sensitive to the
pH, both in the presence and absence of $Mg^{2+}$.  Most importantly however,
the largest chemical shift changes for $Mg^{2+}$-poly P occur between pH 4.5
and 6.5, which is the pH range expected for the vacuole, thus rendering it
a good probe for registering vacuolar pH changes.

A further problem which may give rise to errors in the determination
of pH by in vivo NMR is inaccuracies in the determinations of chemical
shift.  These are always measured with respect to some reference compound.
If such a reference compound is endogenous to the sample and not sensitive
to changes in pH, cations, or protein binding, then one can simply measure
the difference in chemical shift between the standard and the resonance
with the pH-sensitive shift, thus circumventing any errors in the

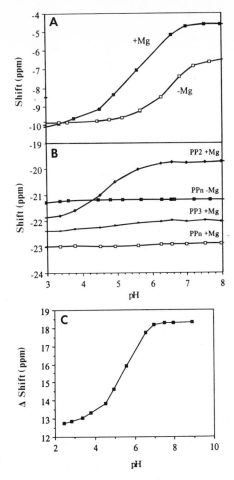

Fig. 3. pH-titration curves for the various resonances of polyphosphates in the presence and absence of $Mg^{2+}$. A) PP1, B) PP2, PP3 and PPn resonances. Panel C shows a titration curve which gives the difference in chemical shift between the PP1 and PPn resonance in the presence of $Mg^{2+}$ (see text for explanation).

calibration of the chemical shifts. Such compounds are not always available, however, and hence external standards are generally used. In the case of the polyphosphates where the PPn resonance of the polyphosphates is not sensitive to the pH (see Fig. 3, Panel B), its resonance can be used as an internal reference compound, and hence the difference in chemical shift between PPn and the terminal polyphosphate resonance (see Panel C) provides a very reliable means of measuring the vacuolar pH in algae and yeast by NMR (Lundberg et al., 1989).

Table 1 shows that in principle two nuclei can be utilized when one wants to study nitrogen metabolism. Both nuclei have about equal sensitivity of detection, but $^{15}$N has the advantage that it is a spin – 1/2 nucleus. $^{14}$N NMR has the advantage that this nucleus has a high natural abundance, however, it has a spin = 1, and hence its quadrupole moment generally renders its signals relatively broad. This is shown in Fig. 4, where the $^{14}$N and $^{15}$N NMR spectra of the two amino acids glutamine and arginine are compared (see also Richards and Thomas, 1974). As expected the $^{15}$N NMR spectra display relatively sharp resonances, whereas the $^{14}$N NMR resonances are quite broad. Be that as it may, the low natural abundance of $^{15}$N (see Table 1) makes it virtually impossible to study $^{15}$N NMR in a reasonable time for concentrations below 100 mM. Since most biologically interesting metabolism does not involve such high amino acid concentrations, it is mandatory to feed and incorporate isotopically labeled precursors such as $^{15}NH_4^+$, or $^{15}NO_3^-$ to obtain sufficient sensitivity. This isotopic-labeling strategy allows one to study the flux of isotopic label through metabolic pathways as we will show below for plant metabolism (Thorpe et al., 1989; Lundberg, 1989; Parry, 1986; Monselise et al., 1987). The same strategy has also been used to study nitrogen metabolism in fungi (Kanamori and Roberts, 1983; Legerton et al., 1983; Martin, 1985b) and bacteria (Kanamori et al., 1988; Choi and Roberts, 1985; Haran et al., 1983).

The linewidth of a quadrupolar resonance, such as for example in $^{14}$N NMR, depends to a large extent on the symmetry of the chemical substitution on the $^{14}$N nucleus. Hence, compounds with a tetrahedral symmetrical substitution such tetramethylammonium give rise to a very sharp $^{14}$N resonance. The same substitution pattern is also found in compounds such as betaine and phosphorylcholine etc. and hence these can be readily detected by $^{14}$N NMR in vivo. In addition, resonances for $NH_4^+$, and $NO_3^-$ (Lundberg et al., 1989; Belton et al., 1985; Thorpe et al., 1989; Lundberg, 1989) are very sharp and thus these resonances can also be quantitated with $^{14}$N NMR. This has led to some studies in which the intracellular levels of $NO_3^-$ and $NH_4^+$ could be followed in vivo (Lundberg et al., 1989; Belton et al., 1985). Moreover by using aqueous shift reagents (Gupta et al., 1984; Springer, 1987) it is also possible to separate resonances for intra- and extracellular $NH_4^+$ and hence its uptake can be studied directly (Lundberg, 1989). By using metabolic inhibitors for $NH_4^+$ metabolizing enzymes it is even possible to study the combined activity of the nitrate and nitrite reductase enzymes by $^{14}$N NMR (R. Joy and H. J. Vogel, unpublished observations); unfortunately the broad $^{14}$N NMR resonance for $NO_2^-$ does not allow one to study these two enzymes

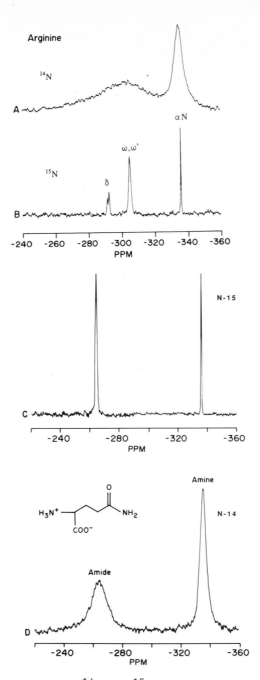

Fig. 4.   Comparison of the $^{14}$N and $^{15}$N NMR spectra of the amino acids
arginine (panels A and B) and glutamine (panels C and D).
These spectra were recorded at pH 7.0.   Note that particularly the $^{14}$N resonances for the side chain nitrogens are
quite broad as a result of their asymmetric substitution.
This renders it difficult to detect these in spectra of
living matter, because they cannot always be distinguished
from other phenomena (such as probe ringing) that may
contribute to baseline distortions.

separately by $^{14}$N NMR. A major advantage of $^{14}$N over $^{15}$N NMR in this type of application is that it has a shorter spin lattice relaxation time $T_1$ and hence $NH_4^+$ and $NO_3^-$ are easier to detect quantitatively by $^{14}$N than by $^{15}$N NMR. Surprisingly, also atmospheric dinitrogen ($N_2$) which dissolves to some extent in $H_2O$ and hence is present in all samples that have not been degassed, has a very narrow $^{14}$N NMR signal and this is often observed as a resonance around –66 ppm (McIntyre et al., 1989). Figure 5 shows an expansion of a representative $^{14}$N NMR spectrum of freshly dissected white spruce buds and of an extract of the same sample. Resonances for Pro (–321 ppm), $\alpha NH_2$ (–336 ppm) and $NH_4^+$ (–355 ppm) can be readily discerned in both spectra. However, the broad resonance which represents the nitrogen atoms in the Arg side-chains only contributes a baseline distortion to the NMR spectrum of the intact buds, but it can be clearly detected as a broad resonance centered around –300 ppm in the extract spectrum (compare with Fig. 4). This spectrum also contains a resonance for $NO_3^-$ (at 0 ppm) which is not shown in this expansion (Thorpe et al., 1989). The detection of the amino acid proline in the freshly obtained bud samples is of particular

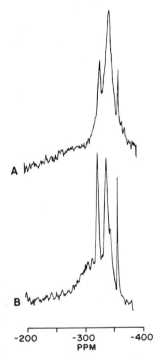

Fig. 5.  Nitrogen-14 NMR spectra of white spruce buds.  Panel A is an intact sample, panel B is a perchloric acid cell extract neutralized to pH 7.0.  Resonances for Pro, $\alpha NH_2$, $-NH_2$, $NH_4^+$ and Arg sidechain nitrogens can be discerned in these spectra (see text).

interest, because this compound is considered to play a major role in osmoprotection in various plants (Pahlich et al., 1983); this Pro pool disappears when the spruce buds start to grow (Thorpe et al., 1989).

Another noteworthy feature is that the resonance for the aliphatic $-NH_2$ groups (at 343 ppm) is not resolved from the $\alpha NH_2$ resonance in the case of the freshly dissected sample. However, it is seen as an upfield flank on the $\alpha NH_2$ resonance in the extract. In the spectra of the buds, we also did not observe a resonance around -260 ppm for the amide in glutamine or asparagine (-265 ppm). This resonance has a reasonably narrow linewidth which is comparable to that of $\alpha NH_2$ (see Fig. 4) and hence it should have been observed in the extract spectrum of Fig. 5. However, its absence indicates that the levels of these two amino acids are relatively low in fresh buds. It should be noted that in this type of $^{14}N$ NMR spectrum, as is also the case for in vivo $^{31}P$ and $^{15}N$ NMR spectra, the resonances for nitrogen or phosphate groups that are a part of a membrane or other macromolecules are not detected because they are too broad. Only the resonances for small metabolites that are characterized by a high rotational mobility are detected in this type of analysis.

The much better resolution that can be obtained in in vivo $^{15}N$ NMR spectra of white spruce embryos is illustrated in Fig. 6. In addition to the major sharp resonances for the Gln amide and the Arg $\delta$ and $\omega$, $\omega'$ side chain nitrogens, two intense resonances are observed which represent the sum of all $\alpha$-amino groups (most amino acids resonate at -336 ppm) and the sum of all aliphatic amino groups (lysine, ornithine, $\gamma$-aminobutyric acid and polyamines all resonate at -343 ppm). Furthermore, a few minor resonances can be detected for amino acids such as Pro, Ala, and Ser. It should be noted that amide protons in Gln and Asn cannot be separated in such spectra, hence additional information from amino acid analyses (Martin, 1985b) or proton NMR (Lundberg, 1989) would be needed to assign the amide resonance with confidence to Gln. It should also be noted that a broad resonance labeled PP is present in the spectra. This represents the backbone of smaller proteins and peptides which have become labeled with $^{15}N$ in the course of the incubation. This peak is in fact of much larger intensity than its appearance in Fig. 6 would suggest. This is due to the different heteronuclear nuclear overhauser enhancements (NOE) that the resonances experience. Whereas the signals for all amino acids have an almost full NOE-enhancement of a factor of -4, which is indicative of their high rotational mobility inside the cells, the resonance for the protein backbone has an NOE of close to -1. Thus, because of cancelation effects, its intensity is close to zero in Fig. 6 and it can only be studied under conditions where the nOe is suppressed (Thorpe et al., 1989; Lundberg, 1989; Kanamori and Roberts, 1983; Legerton et al., 1983).

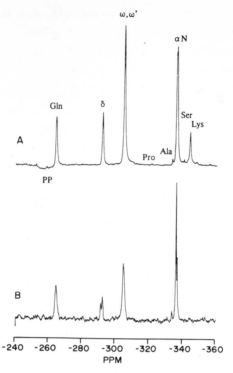

Fig. 6.  A) Nitrogen-15 NMR spectra of white spruce embryos that have been incubated for 2 weeks on a medium containing $^{15}NH_4^+$ (Lundberg, 1989).  B) Nitrogen-15 NMR spectrum of a mixture of amino acids (Gln, Arg, Glu, Ala) which resembles the amino acid composition of white spruce embryos.  The resonance labeled Lys contains contributions from Lys, ornithine, polyamines and γ-aminobutyric acid.

Over the last few years a lot of interest has been generated by the introduction of modern multipulse NMR techniques which allow for the enhanced detection of insensitive nuclei such as $^{15}N$.  Invariably these sequences make use of a more sensitive nucleus (generally $^1H$) which is attached to and spin-coupled to the insensitive nucleus.  There are in principle two variations of this experiment.  In the first one, the insensitive nucleus is detected following a series of radio frequency pulses on the $^{15}N$ as well as the $^1H$ nucleus which results in enhanced sensitivity with a factor of approximately 10 (Derome, 1987).  In the second one, the protons that are coupled to the insensitive nucleus are detected and this experiment may result in a 1000-fold enhancement (Live et al., 1984).  This proton-detected $^{15}N$ experiment has been applied successfully in an in vivo NMR study of amino acid metabolism in yeast (Juretscke, 1984).  Thus, because of the large enhancements that can be obtained, these experiments are potentially very useful as it would allow for much better time-resolution, or much lower levels of isotopic labels in in vivo $^{15}N$ NMR studies of plant

metabolism.  Figure 7 shows an INEPT spectrum (Derome, 1987) which was
obtained for a $^{15}$N labeled white spruce embryo sample.  Clearly, the top
spectrum does not show much of an enhancement in sensitivity over the
standard spectrum (bottom) and in fact the resonances for $\alpha$ NH$_2$ and –NH$_2$ are
not even detected. The major problem that limits the use of sensitivity
enhancement techniques in this type of $^{15}$N NMR application is that the
protons which are directly attached to the nitrogen atom exchange rapidly
with the surrounding solvent, thus drastically reducing the efficiency of
the various nuclear-spin-transfer steps that give rise to the enhanced
sensitivity.  The rate of hydrogen exchange is strongly dependent on the pH,
with the exchange being considerably slower at lower pH values as the data
in Fig. 8 illustrate.  Of all amino acids tested, the amide group in Gln and
Asn has the slowest exchange rate and hence their attached protons can still
be detected at pH values above 7.0.  It is therefore not surprising that the
only successful in vivo proton detected $^{15}$N NMR study reported to date dealt
with Gln metabolism (Juretscke, 1984).  Furthermore, Fig. 8 shows that the
protons on $\alpha$ NH$_2$ and –NH$_2$ exchange faster and can only be reliably detected
below pH 4.0.  This explains the disappearance of the $\alpha$ NH$_2$ and –NH$_2$
resonances from the top spectrum in Fig. 7.  Inspection of the hydrogen
exchange characteristics of the Arg side-chain nitrogens in Fig. 8 suggests
that these should be observable below pH 6.0.  Hence, a tentative
interpretation of the Arg side-chain NH protons in the top spectrum of
Fig. 7 would be that the Arg resides in the acidic vacuole, rather than in

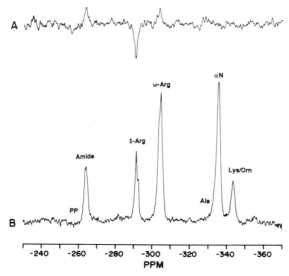

Fig. 7.    An in vivo Nitrogen-15 INEPT (panel A) and normal spectrum
with close to full nOe (panel B) of white spruce embryos,
which were grown as described in Fig. 6.  Only the resonances
for the Gln amide and the Arg sidechain nitrogens are seen in
the INEPT spectrum because of the hydrogen exchange properties
of the nitrogen-bound hydrogens (see text).

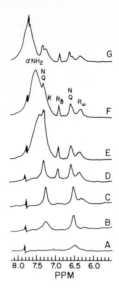

**Fig. 8.** Proton NMR spectra of a mixture of amino acids dissolved in 90% $H_2O$/10% $D_2O$ (150 mM KCl) which were recorded with the jump-and-return pulse sequence to eliminate the water resonance. Only the region of the spectrum is shown in which hydrogens directly bonded to nitrogen appear. The spectra were recorded at the following pH values: A) 8.9, B) 7.9, C) 6.7, D) 5.6, E) 4.0, F) 3.0, G) 2.0. The large downfield shift for the $\alpha$ NH$_2$ resonance at lower pH is caused by titration of the carboxylate group of the amino acids. The sharp resonances around 7.8 ppm are an instrumental artifact.

the neutral cytoplasm. Be that as it may, these hydrogen-exchange characteristics severely limit the use of these techniques in in vivo $^{15}$N NMR studies of metabolism. Similar restrictions apply to the study of surface residues in proteins (Leighton and Lu, 1987). However, backbone amide nitrogens in proteins can generally be detected readily with these techniques, because they have slow intrinsic hydrogen exchange rates and, in addition, they are often shielded from the solvent (Leighton and Lu, 1987; Smith et al., 1987; Torchia et al., 1988).

The in vivo $^{15}$N NMR data discussed to this point gave an example of batch studies in which $^{15}$N label is fed to algae or plant material and the distribution of the label through the amino acid pools was studied after a certain period of incubation has elapsed. However, the incorporation of $^{15}$NH$_4^+$, $^{15}$NO$_3^-$ or $^{15}$NO$_2^-$ can also be followed in a more dynamic sense where isotopic labeling is done in a perfusion experiment.*

---

\* Plant cells can be maintained and perfused in the NMR tube in the spectrometer with a nylon net (Vogel et al., 1987; Vogel and Brodelius, 1984). However yeast and bacterial cells are too small and they would float away. Hence we perform perfusion studies for the latter cells after immobilization in agarose beads (Vogel et al., 1987; Vogel and Brodelius, 1984).

For this type of experiment, the plant cells are transferred under sterile conditions to an NMR tube, which is perfused with sterile oxygenated medium containing the desired $^{15}N$ labeled nitrogen sources, while $^{15}N$ NMR spectra are continuously recorded. Hence, the emergence of the $^{15}N$ label in various metabolites can be recorded as a function of time. An example of such a study is shown in Fig. 9 for the yeast, <u>Candida tropicalis</u>, perfused with $^{15}NH_4^+$ containing medium. In this case, the first amino acid peak to emerge is $\alpha\,NH_2$, which is followed after some time by two other resonances demonstrating incorporation of isotopic label into amides and aliphatic amines. This labeling pattern suggests that $NH_4^+$ is incorporated into the amino acid pools via the enzyme, glutamate dehydrogenase, in the following fashion:

$$NH_4^+ + \alpha\text{-ketoglutarate} + NAD(P)H \rightarrow Glu + NAD(P)^+ + H_2O \;.$$

This labeling pattern is consistent with earlier studies concerning nitrogen metabolism in yeast (Sims and Folkes, 1964; Harder and Dijkhuizen, 1983). When similar studies were done with white spruce buds (Thorpe et al., 1989) and embryos (Lundberg, 1989), the labeling pattern was quite different and the $^{15}NH_4^+$ was first incorporated into the Gln amide resonance. Subsequently it emerged as $\alpha\,NH_2$ and after some time it also contributed to Arg and aliphatic amine resonance (Thorpe et al., 1989; Lundberg, 1989). This labeling pattern is indicative of the operation of the following pathway, which is thought to play a major role in plant nitrogen assimilation (Oaks and Hirel, 1985; Skokut et al., 1978) and involves the sequential action of the two enzymes, glutamine synthetase and glutamate synthase:

$$NH_4^+ + Glu + ATP \rightarrow Gln + ADP + Pi + H^+$$
$$\alpha\text{-ketoglutarate} + Gln + NAD(P)H + H^+ \rightarrow Glu + Glu + NAD(P)^+ + H_2O \;.$$

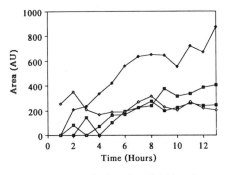

Area (AU)

Time (Hours)

Fig. 9. Perfusion experiment of the immobilized yeast <u>Candida tropicalis</u> with a glucose-based medium containing $^{15}NH_4^+$ as the nitrogen source. The following four resonances could be detected: $NH_4^+$ ( ◇ ), $\alpha\,NH_2$ ( ◆ ), Gln + Asn Amide ( □ ), aliphatic $NH_2$ ( ■ ). For interpretation see text.

The operation of this pathway was further confirmed by pulse-chase labeling studies [$^{15}NH_4^+$ followed by $^{14}NH_4^+$, (Lundberg, 1989)] and by using metabolic inhibitors of specific enzymes in the pathways (unpublished observations).

CONCLUSIONS

The examples discussed throughout this chapter demonstrate that the uptake of two of the most important inorganic nutrients by plant material can be readily studied by $^{31}P$ and $^{14}N$, $^{15}N$ NMR techniques in a totally noninvasive manner. Phosphorus-31 NMR is most useful to obtain information about the energy status of the cell (ATP, UDPG, and phosphorylated sugars), and to study the mode of Pi uptake and storage. For example we have discussed that algae store Pi as polyphosphates, whereas higher plants may use a large vacuolar Pi pool for storage. In addition, information about the intracellular pH in the cytoplasm and the vacuole can be obtained directly from analyzing the chemical shift of specific resonances in the $^{31}P$ NMR spectrum. Compared to other techniques that have been used to study pH, cellular energy and Pi uptake, a wealth of information is obtained in one single experiment by $^{31}P$ NMR, thus rendering it a rather unique tool for the study of these aspects of plant metabolism.

The combination of $^{14}N$ and $^{15}N$ NMR allows for unique opportunities to study the intracellular pools of $NO_3^-$, $NH_4^+$ and total amino acids, while at the same time dynamic information can be obtained about the rates at which specific amino acid pools become labeled. Because the radioactive $^{13}N$ has a very short lifetime and is not always convenient to use (Skokut et al., 1978), the main method for the study of nitrogen metabolism has become the analysis of isotope ratios ($^{15}N/^{14}N$) in extracted and purified amino acid samples by gas chromatography and mass-spectrometer (GCMS) methods (Rhodes et al., 1981, 1989). Compared to this technique NMR has the advantage that it is noninvasive and that it readily shows the labeling of amino acid sidechain nitrogens versus that of the $\alpha NH_2$ group. In addition, information about intracellular $NH_4^+$ and $NO_3^-$ can be obtained simultaneously. However, GCMS is superior when it comes to quantitation and when one wishes to distinguish between the labeling of the $\alpha NH_2$ groups of all the different amino acids. Hence the two methods are complimentary to a large extent and the maximum amount of information can be obtained by applying both methods simultaneously.

ACKNOWLEDGEMENTS

The research work on plant cell metabolism is sponsored by operating grants from the Natural Sciences and Engineering Research Council. Funds for the purchase of the NMR spectrometer were provided by the Alberta

Heritage Foundation for Medical Research (AHFMR). HJV and PL are the recipients of an AHFMR scholarship and studentship, respectively. We gratefully acknowledge many stimulating discussions regarding the topic of this review article with our colleagues: K. Bagh, P. Brodelius, E. Lohmeier, D. McIntyre, R. Joy, and T. Thorpe. We are indebted to Susan Stauffer for expert secretarial assistance.

REFERENCES

Belton, P. S., Lee, R. B., and Ratcliffe, R. G., 1985, J. Exp. Bot., 36:190.
Bental, M., Degani, H., and Avron, M., 1988, Plant Physiol., 87:813.
Bental, M., Oren-Shamir, M., Avron, M., and Degani, H., 1988, Plant Physiol., 87:320.
Brodelius, P., and Vogel, H. J., 1985, J. Biol. Chem., 260:3556.
Chang, K., and Roberts, J. K. M., 1989, Plant Physiol., 89:197.
Choi, B. S., and Roberts, M. F., 1985, Biochem. Biophys. Acta, 928:259-265.
Cohen, S. M., Ed., 1987, "Physiological NMR Spectroscopy: From isolated Cells to Man," Ann. N.Y. Acad. Sci., 508:1.
Derome, A. E., 1987, "Modern NMR Techniques for Chemistry Research," Pergamon Press, Oxford, 129-153.
Dijkema, C., DeVries, S. C., Booij, H., Schaafsma, T. J., and vanKammen, A., 1988, Plant Physiol., 88:1332.
Fan, T. W. -M., Higashi, R. M., and Lane, A. N., 1986, Arch. Biochem. Biophys., 251:674.
Fan, T. W. -M., Higashi, R. M., and Lane, A. N., 1988, Arch. Biochem. Biophys., 266:592.
Foyer, C., Walker, D., Spencer, C., and Mann, B., 1982, Biochem. J., 202:429.
Gadian, D. G., 1983, Annu. Rev. Biochem. Biophys. Bioeng., 12:69.
Gadian, D. G., and Radda, G. K., 1981, Annu. Rev. Biochem., 50:69.
Gerasimowicz, W. V., Tu, S. -I., and Pfeffer, P. E., 1986, Plant Physiol., 81:925.
Gimmler, H., Kugel, H., Leibfritz, D., and Mayer, A., 1988, Physiol. Plant, 74:521.
Guern, J., Mathieu, Y., Kurkdjian, A., Manigault, P., Manigault, J., Gillet, B., Beloeil, J. -C., and Lallemand, J. -Y., 1989, Plant Physiol., 89:27.
Gupta, R. K., Gupta, P., and Moore, R.D., 1984, Annu. Rev. Biophys. Bioeng., 13:221.
Haran, N., Kahana, Z. E., and Lapidot, A., 1983, J. Biol. Chem., 258:12929.
Harder, W., and Dijkhuizen, L., 1983, Annu. Rev. Microbiol., 37:1.
Hellstrand, P., and Vogel, H. J., 1985, Am. J. Physiol., 248:C320.
Hooks, M. A., Clark, R. A., Nieman, R. H., and Roberts, J. K. M., 1989, Plant Physiol., 89:963.
Hoult, D. I., Busby, S. J. W., Gadan, D. G., Radda, G. K., Richards, R. E., and Seeley, P. J., 1974, Nature, 252:285.
Ingwall, J. S., 1982, Am. J. Physiol., 242:H729.
Juretscke, H. P., 1984, FEBS Lett., 178:123.
Kanamori, K., and Roberts, J. D., 1983, Acc. Chem. Res., 16:35.
Kanamori, K., Weiss, R. L., and Roberts, J. D., 1988, J. Biol. Chem., 263:2817.
Kime, M. J., Ratcliffe, R. G., Williams, R. J. P., and Loughman, B. C., 1982, J. Exp. Bot., 33:656.
Kugel, H., Mayer, A., Kirst, G. O., and Leibfritz, D., 1987, Eur. J. Biophys., 14:461.
Legerton, T. L., Kanamori, K., Weiss, R. L., and Roberts, J. D., 1983, Biochemistry, 22:899.

Leighton, P., and Lu, P., 1987, Biochemistry, 26:7262.
Live, D. H., Davis, D. G., Agosta, W. C., and Cowburn, D., 1984, J. Am. Chem. Soc., 106:6104.
Lundberg, P., 1989, Ph.D. Thesis, Univ. of Calgary.
Lundberg, P., Weich, R. G., Jensen, P., and Vogel, H. J., 1989, Plant Physiol., 89:1380.
Martin, F., 1985a, Physiol. Veg., 23:463.
Martin, F., 1985b, FEBS. Lett., 182:350.
Martin, J. B., Bligny, R., Rebeille, F., Douce, R., Lequay, J., Mathieu, Y., and Guern, J., 1982, Plant Physiol., 70:1156.
Mathieu, Y., Guern, J., Kurkdjian, A., Manigault, P., Manigault, J., Zielinska, T., Gillet, B., Beloeil, J. -C., and Lallemand, J. -Y., 1989, Plant Physiol., 89:19.
McIntyre, D. D., Applett, A., Lundberg, P., Schmidt, K., and Vogel, H. J., 1989, J. Magn. Reson., 83:377.
Mimuro, T., and Kirino, Y., 1984, Plant Cell Physiol., 25:813.
Mitsumiro, F., and Ito, O., 1984, FEBS Lett., 174:248.
Monselise, E. B. I., Kost, D., Porath, D., and Tal, M., 1987, New Phytol., 107:341.
Moon, R. D., and Richards, J. M., 1973, J. Biol. Chem., 248:7276.
Morris, P. G., 1988, Annu. Rep. NMR Spectr., 20:1.
Nicolay, K., Scheffers, W. A., Bruinenberg, P. M., and Kaptein, R., 1982, Arch. Microbiol., 133:83.
Nicolay, K., Scheffers, W. A., Bruinenberg, P. M., and Kaptein, R., 1983, Arch. Microbiol., 134:270.
Oaks, A., and Hirel, B., 1985, Annu. Rev. Plant Physiol., 36:345.
Pahlich, E., Kerres, R., and Jager, H. J., 1983, Plant Physiol., 72:590.
Parry, D. L., 1986, Mar. Biol., 87:219.
Rhodes, D., Myers, A. C., and Jamieson, G., 1981, Plant Physiol., 68:1197.
Rhodes, D., Rich, P. J., and Brunk, D. G., 1989, Plant Physiol., 89:1161.
Richards, R. E., and Thomas, N. E., 1974, J. Chem. Soc., Perkins II, 368.
Roberts, J. K. M., 1984, Annu. Rev. Plant Physiol., 35:375.
Roberts, J. K. M., Ray, P. M., Wade-Jardetzky, N., and Jardetzky, O., 1980, Nature, 283:870.
Schaefer, J. O., Stetsjkal, O., and Beard, F., 1975, Plant Physiol., 155:1048.
Sianoudis, J., Kusel, A. C., Mayer, A., Grimme, L. H., and Leibfritz, D., 1986, Arch. Microbiol., 144:48.
Sianoudis, J., Kusel, A. C., Mayer, A., Grimme, L. H., and Leibfritz, D., 1987, Arch. Microbiol., 147:25.
Sillerud, L. O., and Heyser, J. W., 1984, Plant Physiol., 75:269.
Sims, A. P., and Folkes, B. F., 1964, Proc. Roy. Soc. B., 159:479.
Skokut, T., Wolk, C. P., Thomas, J., Meeks, P. W., and Shaffer, P. W., 1978, Plant Physiol., 62:299.
Smith, G. M., Yu, L. P., and Domingues, D. J., 1987, Biochemistry, 26:2702.
Springer, C. S., 1987, Annu. Rev. Biophys. Chem., 16:375.
Stidham, M. A., Moreland, D. E., and Siedow, J. N., 1983, Plant Physiol., 73:517.
Thomae, W. J., and Gleason, F. K., 1987, Biochemistry, 26:2510.
Thomas, T. H., and Ratcliffe, R. G., 1985, Physiol. Plant, 63:284.
Thorpe, T., Bagh, K., Cutler, A. J., Dunstan, D. I., McIntyre, D. D., and Vogel, H. J., 1989, Plant Physiol., 91:193.
Torchia, D. A., Sparks, S. W., and Bax, A., 1988, Biochemistry, 27:5135.
Vogel, H. J., 1987, Ann. N.Y. Acad. Sci., 508:164.
Vogel, H. J., and Brodelius, P., 1984, J. Biotechn., 1:159.
Vogel, H. J., Brodelius, P., Lilja, H., and Lohmeier-Vogel, E. M., 1987, Meth. in Enzymol., 135:512.
Watanabe, M., Kohata, K., and Kunugi, M., 1987, J. Phycol., 23:54.
Waterton, J. C., Bridges, I. G., and Irving, M. P., 1983, Biochem. Biophys. Acta., 763:315.
Weich, R., Lundberg, P., Vogel, H.J., and Jensen, P., 1989, Plant Physiol., 90:320.
Wray, V., Schiel, O., and Berlin, J., 1983, Z. Pflanzenphysiol, 112:215.
Yazaki, Y., Mahi, K., Sato, T., Ohta, E., and Sakata, M., 1988, Plant Cell Physiol., 29:1417.

**IN VIVO** PHOSPHORUS NMR STUDIES OF THE HEPATIC METABOLISM OF AMINO SUGARS

Robert E. London, Scott A. Gabel, and Michael E. Perlman*

\* Laboratory of Molecular Biophysics
National Institute of Environmental Health Sciences
NIH
Box 12233
111 T. W. Alexander Drive
Research Triangle Park, NC  27709

INTRODUCTION

Although the amino sugars glucosamine and galactosamine occur naturally and, as N-acetyl derivatives, are constituents of various cellular glycoconjugates, the administration of large doses can result in significant toxicity.  This aspect of their metabolism has led to evaluations of these compounds as potential chemotherapeutic agents (Quastel and Cantero, 1953; Bekesi et al., 1969; Keppler et al., 1985), and to studies aimed at elucidating the mechanisms involved in the production of hepatocellular injury (Decker and Keppler, 1974, 1972).  As with a number of other cellular toxins, the toxicity results from biochemical activation or "lethal synthesis" (Decker and Keppler, 1972; Shull et al., 1966), since the metabolic transformations involved begin with phosphorylation by the corresponding enzymes of glucose and galactose metabolism (Ballard, 1966; Walker and Khan, 1968; Oguchi et al., 1977).  This metabolism leads ultimately to uridylate trapping effects and, in the case of galactosamine, to the production of metabolites which are not observed in untreated tissue.

The role of phosphorylation and the involvement of phosphorus-containing metabolites in the toxic action of these sugars offers the potential of studying amino sugar metabolism by phosphorus-31 NMR, both **in vivo** and **in vitro.**  A particular advantage of the former approach is the ability to follow the time course of the metabolic response in individual experimental animals (Ackerman et al., 1980; Iles et al., 1982).  Since the animal acts as its own control, changes which would otherwise rest on inferences from statistical comparisons of larger numbers of animals can be directly observed.  Furthermore, determination of the time course of the metabolic response to various chemical and physical perturbations is central to the development of an understanding of the cause and effect relationships which characterize the production of irreversible cell injury (Decker and Keppler, 1974, 1972; Murphy and London, 1988). We report here on the results of both **in vivo** and **in vitro** studies of the hepatic metabolism of glucosamine and galactosamine in the anesthetized rat.

METHODOLOGY

The study of hepatic metabolism by NMR methods requires spectral localization strategies which can involve either surgical or technological approaches. For the present studies, we have utilized the previously developed protocol (London et al., 1985) in which a portion of muscle directly in front of the liver is surgically removed and the skin sutured, allowing the rat to recover and observations to be carried out over an extended period of time (Fig. 1). Rats (275–325 gms) were anesthetized with the long acting anesthetic Inactin (Lockwood Assoc. Imports, East Lansing, MI) as described previously (Smith et al., 1987). Concentric surface coils 1.1 and 1.5 cm in diameter, and tuned to the $^{31}$P and $^{1}$H resonance frequencies of 146.15 and 361.04 MHz, respectively, are then placed directly over the skin above the liver (Ackerman et al., 1980). The $^{1}$H tuned coil was used for shimming on the tissue water. The resulting $^{31}$P NMR spectrum of the liver obtained over a five minute accumulation time is shown in Fig. 2. The effectiveness of the surgical technique is monitored by the degree of elimination of the phosphocreatine resonance which arises from muscle but not liver tissue. Various lines of evidence suggest that mitochondrial ADP (Iles et al., 1985; Stubbs et al., 1984) and ATP (Murphy et al., 1988) do not contribute to the observed NMR spectra.

**IN VIVO** SPECTROSCOPIC STUDIES

Injection of 100 mg of glucosamine (1.85 mmol/kg) into the rat leads to the appearance of a phosphate monoester resonance at 7.8 ppm which is readily apparent in the second five minute period post injection (Fig. 3). This resonance continues to increase in intensity over the first hour of the study, and subsequently increases slightly over the remaining four hours during which **in vivo** observations were carried out. Based on the observed shift as well as the known biochemistry, this resonance may correspond to glucosamine 6–phosphate or to N–acetylglucosamine 6–phosphate. NMR studies of extracts obtained five hours after the administration of the glucosamine indicate that the observed resonance corresponds almost exclusively to the N–acetyl derivative (Perlman et al., in press). A well resolved resonance at −9.8 ppm corresponding to the β–phosphate (hexose phosphate) group of the UDP–sugars (Ugurbil et al., 1978) is also observed in these spectra (Figs. 2 and 3), the intensity of which increases more slowly during the study to a

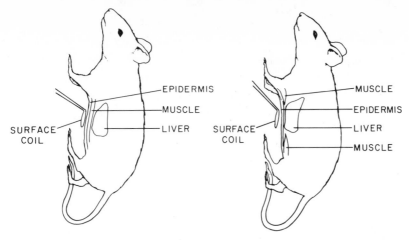

Fig. 1.   Surgical modification of the rat for selective **in vivo** NMR
observation of liver involves removal of the layer of
muscle directly over the liver, followed by resuturing the
skin.   The surface coil is then positioned directly over
the skin lying above the organ.

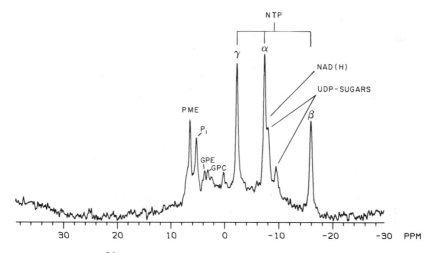

Fig. 2.   **In vivo** $^{31}$P NMR spectrum obtained from the liver of a control
rat as illustrated in Fig. 1.   The spectrum was obtained on a
Nicolet NT-360 NMR spectrometer operating at a frequency of
146.15 MHz using 8 K data points and a sweep width of +/-
5 kHz and an interpulse delay of 1.2 sec.   Spectrum corresponds to 252 accumulations obtained over a period of 5
minutes.   An exponential multiplier of 20 Hz was used to
improve signal/noise.   Resonance assignments are:   PME,
phosphate monoesters; $P_i$ inorganic phosphate; GPE,

glycerophosphoryl ethanolamine; GPC, glycerophosphoryl
choline; NTP, $\alpha$, $\beta$, and $\gamma$ phosphate resonances of nucleoside
triphosphates;   NAD(H), both of the phosphodiester resonances
of the pyridine nucleotides; UDP-sugars, the $\alpha$ (downfield,
nucleotide) and $\beta$ (upfield, hexose) phosphates.   Chemical
shifts are referenced to the $\alpha$-ATP resonance at −7.5 ppm.

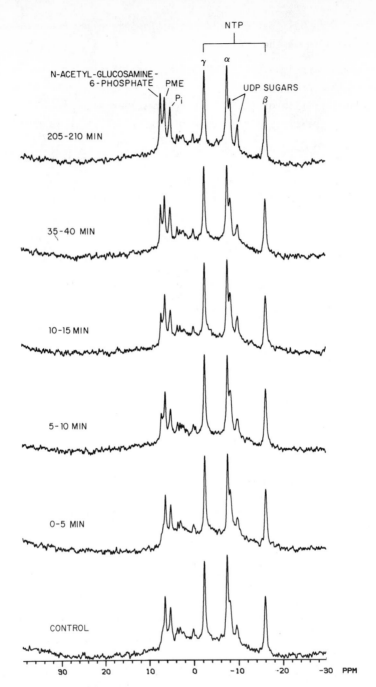

Fig. 3.   **In vivo** $^{31}$P NMR spectra obtained prior to (control) and
subsequent to injection of 1.85 mmol/kg glucosamine.   Each
spectrum corresponds to a five minute accumulation time
with spectral parameters as given in Fig. 2.

final value approximately twice the initial intensity.   Thus, although
changes in the composition of the UDP-sugar pool cannot be monitored, the **in**
**vivo** studies indicate a net increase in total UDP-sugar content which would

.be associated with utilization of uridine base and hence consumption of the uridine nucleotides in the cell.

A corresponding **in vivo** $^{31}$P NMR study utilizing galactosamine rather than glucosamine was carried out, and several time points are shown in Fig. 4. Administration of the galactosamine is associated with a large increase in the intensity of a resonance at 5.2 ppm which is essentially

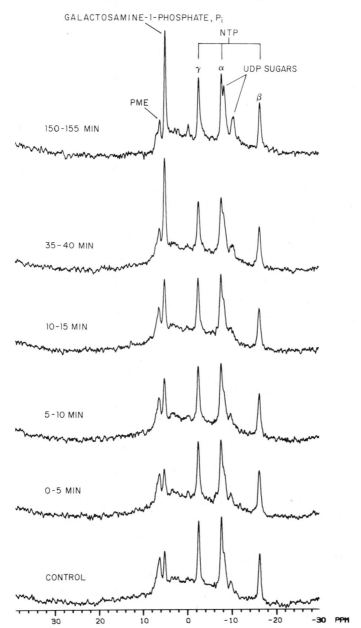

Fig. 4. In vivo $^{31}$P NMR spectra obtained prior to (control) and subsequent to the injection of galactosamine (1.85 mmol/kg) i.v. (at t = 0) at the times indicated.

coincident with that of inorganic phosphate. However, [31]P NMR studies carried out on extracts obtained five hours after galactosamine administration reveal that this resonance corresponds to galactosamine 1-phosphate (Perlman et al., in press). As in the glucosamine study, the UDP-sugar β-phosphate (hexose phosphate) resonance intensity also increases, but at a significantly greater rate, ultimately approaching a level approximately four times the initial value.

Changes in the levels of the phosphate monoesters, the UDP-sugars, and nucleoside triphosphate pools can be more clearly revealed in difference spectra (Fig. 5). In the second (5-10 minute) block studied, an increase in galactosamine 1-phosphate as well as a loss of nucleoside triphosphate content is readily apparent (Fig. 5). The slower increase in the UDP-sugar content is evident from the later difference spectra. Despite the progressive consumption of UTP in the synthesis of these UDP-sugars, there is no corresponding progressive deficit in the [31]P resonance intensity of the nucleoside triphosphates. This presumably reflects the fact that the UTP represents only a small fraction of the nucleoside triphosphate pool. The initial rapid phosphorylation of the administered galactosamine, which is reflected in the intensity of the galactosamine 1-phosphate resonance, may be sufficient to perturb the ATP pool, so that the observed negative resonances in the difference spectra correspond to decreased cellular ATP.

Quantitative estimates of the levels of phosphomonoesters and UDP-sugars present in the cells were made based on measurements of peak heights (Fig. 6). From this figure, differences in the quantitative response to the two amino sugars, and the extent of uridine trapping, are readily apparent. Thus, after a period of 200 minutes, the total concentration of UDP-sugars increases by a factor of 2 in the glucosamine study, and by a factor of 4 in the galactosamine study. For the phosphomonoesters, the increase observed over 200 min. in the glucosamine study, which on the basis of analysis of extracts corresponds primarily to N-acetylglucosamine 6-phosphate, is significantly less than the level of galactosamine 1-phosphate observed in the galactosamine study. Since these compounds are not present in significant amounts prior to treatment, intensity evaluations must be based on comparisons with other resonances. In the first case, the phosphomonoester resonance rises to approximately 2.5 times the initial intensity of the UDP-sugar β-phosphate peak, while in the galactosamine study the corresponding phosphomonoester resonance for galactosamine 1-phosphate is 6.5 times the initial intensity of the UDP-sugar β-phosphate peak. We note, however, that although changes in the UDP-sugar content can be quantitatively estimated from these spectra, comparisons between the levels of the hexose phosphates and the UDP-sugars require that account be taken of differences in the spin-lattice relaxation rates of the corresponding phosphorus resonances. Since the sugar phosphates generally

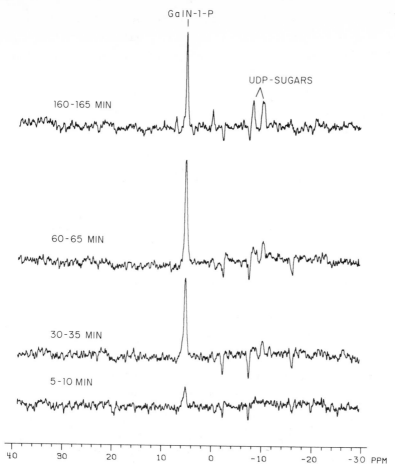

Fig. 5. NMR difference spectra relative to control, for the galactosamine treated rat. The decreased level of nucleoside triphosphates and increased levels of UDP-sugars are readily apparent. It is noted that a fall in the former is observed prior to the increase in the latter, supporting the conclusion that this initial decrease reflects primarily a loss of ATP which is consumed in the initial phosphorylation step. Additionally, the total increase in intensity of the phosphorylated metabolites, which reflects primarily the formation of galactosamine 1-phosphate, greatly exceeds the loss in intensity which is observed primarily as a loss of nucleoside triphosphate.

have a longer $T_1$ than the UDP-sugar phosphates (Schleich, et al., 1984), such a comparison underestimates the increase in the sugar phosphate levels. Further quantitation based on analyses of extracts will be presented elsewhere (Perlman et al., in press).

NMR ANALYSIS OF EXTRACTS

Changes in the content of the UDP-sugar pool can be monitored in detail by [31]P NMR analyses of perchloric acid extracts. The portions of the [31]P

355

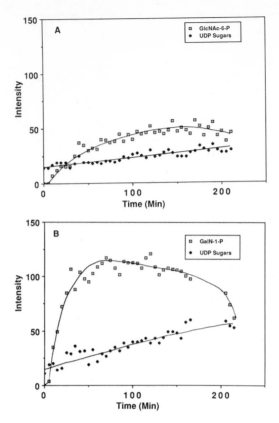

Fig. 6. Time dependence of peak heights measured in **in vivo**
studies. (A) glucosamine treated rat; (B) galactosamine
treated rat. Data obtained from analysis of extracts
suggests that the phosphomonoester resonance increasing
with time in the glucosamine study corresponds primarily to
N-acetylglucosamine 6-phosphate, while that observed in the
second study arises from galactosamine 1-phosphate.
UDP-sugar intensity reflects the height of the β-phosphate
peak; individual UDP-sugars are not resolvable **in vivo**.

spectrum containing the β-phosphate resonances of the UDP-sugars corresponding
to extracts obtained from the livers of normal fed rats, and from rats five
hours after receiving 1.85 mmol/kg glucosamine are shown in Fig. 7. Resonances corresponding to the UDP derivatives of glucose, galactose, glucuronic
acid, N-acetylglucosamine, and N-acetylgalactosamine are readily observed in
the spectra derived from the untreated rat. The changes in intensity of the
corresponding UDP-sugars resulting from treatment with glucosamine are readily
quantitated from these spectra. Several additional resonances are apparent in
the extract from the galactosamine treated rat, corresponding to
UDP-glucosamine and UDP-galactosamine.

Additionally, it is clear from the extract data that the overlap between
the chemical shifts of galactosamine 1-phosphate and inorganic phosphate is
sufficient to limit quantitation of the **in vivo** spectra in the galactosamine

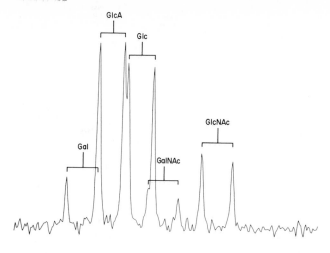

A) CONTROL

GlcA

Glc

GlcNAc

Gal

GalNAc

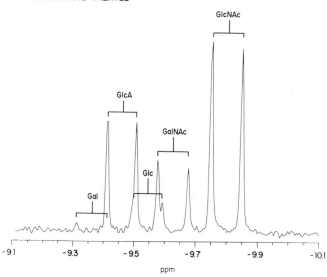

B) GLUCOSAMINE TREATED

GlcNAc

GlcA

GalNAc

Glc

Gal

-9.1    -9.3    -9.5    -9.7    -9.9    -10.1

ppm

Fig. 7.   Proton-decoupled $^{31}$P NMR spectra of the region containing
the β-phosphate resonances of the nucleotide sugars derived
from liver extracts (note, however, that the much less
prominent UDP-mannose resonances which are sometimes
observed are further upfield).   (A) control spectrum from a
fed rat; (B) spectrum corresponding to an extract obtained
five hours subsequent to treatment with glucosamine.
Excised livers were immediately frozen in liquid nitrogen,
extracted with 1.8 N perchloric acid, and neutralized to
pH 8.2.   Spectra were run at 202.447 MHz on a GE GN-500 NMR
spectrometer.   Abbreviations correspond to the UDP-sugar
derivatives:   Gal: galactose; GalNAc: N-acetylgalactosamine; GlcA: glucuronic acid; Glc: glucose; GlcNAc:
N-acetylglucosamine.

study.  Analysis of the glucosamine data indicates that the phosphomonoester resonance observed **in vivo** corresponds primarily to α- and β-N-acetyl-glucosamine 6-phosphate.  Extract spectra also reveal a small amount of N-acetylglucosamine 1-phosphate, the resonance of which would probably not be resolved from that of inorganic phosphate **in vivo**.  This may arise at least in part due to degradation of UDP-N-acetylglucosamine in the extraction proce-dure, and in any case appears to be sufficiently small to allow quantitation of the phosphate resonance in the spectrum.  A more complete analysis of the extracts based on 2-dimensional NMR studies is given elsewhere (Perlman et al., in press).

## DISCUSSION

The toxicity associated with large doses of amino sugars has been of interest in connection with the use of these compounds as chemotherapeutic agents (Quastel and Cantero, 1953; Bekesi, et al., 1969; Keppler, et al., 1985), as well as with their use in studies of the mechanisms of cell injury (Decker and Keppler, 1974, 1972; Murphy and London, 1988).  The importance of following the time course of the metabolic response to the administration of the amino sugars for understanding the coordinated changes in cell metabolism which occur has been emphasized by Decker and Keppler (1974).  From this standpoint, **in vivo** NMR techniques provide an ideal probe since the time dependent changes of various metabolite levels can be monitored continuously in individual experimental animals.  Combining such studies with analysis of tissue extracts provides further insight into the specific composition of various metabolite pools, such as the uridine diphosphosugars, the resonances of which cannot be resolved **in vivo**.

**In vivo** $^{31}$P NMR data are in good agreement with conclusions based on the analysis of hepatic extracts which have established the biochemical conver-sions of these amino sugars (Ballard, 1966; Walker and Khan, 1968; Oguchi, et al., 1977; Weckbecker and Keppler, 1982).  These are summarized in Fig. 8.  As shown in the figure, administration of glucosamine is likely to result in the accumulation of either glucosamine 6-phosphate or of the N-acetyl derivative. The chemical shift difference between these species is too small to permit resolution **in vivo**.  Extract data obtained five hours after glucosamine administration indicate only the N-acetyl form.

Although changes in the levels of individual uridine diphosphosugars cannot be determined **in vivo**, the change in the intensity of the total level of nucleotides as reflected by the β-phosphate resonance increases nearly four fold after three hours in the galactosamine study.  These data are in good quantitative agreement with studies of extracts by Weckbecker and Keppler (1982).  The inability to separately monitor the levels of

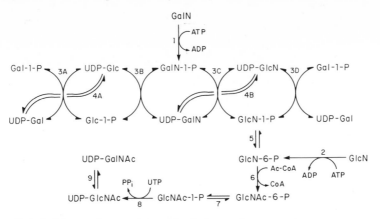

Fig. 8.  Metabolic pathways determined for galactosamine and glucosamine.
The numbered transformations correspond to: (1) galactokinase;
(2) glucokinase; (3) hexose-1-phosphate uridylyl transferase;
(4) UDP-glucose-4'-epimerase; (5) phosphoglucomutase;
(6) glucosamine-phosphate acetyltransferase; (7) acetylglucosamine phosphomutase; (8) UDP-acetylglucosamine
pyrophosphorylase; (9) UDP-acetylglucosamine-4'-epimerase.
Abbreviations are as given for Fig. 7.  Pathways as determined
by Weckbecker and Keppler (1982) and by Oguchi et al. (1977).

individual UDP-sugars constitutes a significant limitation of the **in vivo**
NMR studies.  However, since it has been proposed that the trapping of the
uridine base in the various UDP-sugars represents the primary basis for
amino sugar toxicity (Decker and Keppler, 1972), the measurements of changes
in the levels of total UDP-sugars represent a meaningful parameter
in the development of cell injury.

From the difference spectra obtained subsequent to treatment with galactosamine, it is evident that the total level of phosphorylated metabolites
in the liver is not constant.  This can reflect either additional entry of
phosphate into the cell from the blood, or the recruitment of intracellular
phosphate from some non-observed pool.  The latter could include, for example,
phosphorylated proteins exhibiting relatively broad resonances.  It has been
reported that in isolated rat hepatocytes, the level of intracellular phosphate could be increased by addition of phosphate to the medium and decreased
by exogenous D-galactosamine (Stermann and Decker, 1978).  Hence, it is likely
that the observed net change in the level of phosphorylated species reflects
some uptake of phosphate from the blood.  It cannot be determined whether this
uptake is sufficient to overcome the drop in $P_i$ expected to result from the
net utilization of phosphate in the synthesis of galactosamine 1-phosphate
since the two resonances overlap almost identically.

The analysis of hepatic extracts by $^{31}$P NMR provides a powerful method
for the determination of the composition of the complex UDP-sugar mixtures
found in control or treated liver.  As a consequence of the significant

dependence of the chemical shifts on concentration and salt levels in the extracts, $^{31}P-^{1}H$ heteronuclear shift-correlated spectroscopic methods are extremely useful for the analysis of these compounds (Perlman et al., in press). Resonance assignments can be made in this way without the adulteration of the sample with added standards. In this way, quantitative analysis of complex mixtures can be carried out without the need for physical separation of the components.

REFERENCES

Ackerman, J. J. H., Grove, T. H., Wong, G. G., Gadian, D., and Radda, G. K., 1980, Nature, 283:167-170.
Ballard, F. J., 1966, Biochem. J., 98:347-352.
Bekesi, G. J., Molnar, Z., and Winkler, R. J, 1969, Cancer Res., 29:353-359.
Decker, K., and Keppler, D., 1974, Rev. Physiol. Biochem. Pharmacol., 71:77-106.
Decker, K., and Keppler, D., 1972, Galactosamine Induced Liver Injury, in: "Progress in Liver Diseases IV" (Eds. H. Popper and F. Schaffner), Grune and Stratton, NY, pp. 183-199.
Iles, R. A., Stevens, A. N., and Griffiths, J. R., 1982, Progr. in NMR Spectrosc., 15:49-200.
Iles, R. A., Stevens, A. N., Griffiths, J. R., and Morris, P. G., 1985, Biochem. J., 229:141-151.
Keppler, D., Holstege, A., Weckbecker, G., Fauler, J., and Gasser, T., 1985, "Advances in Enzyme Regulation," 24:417-427.
London, R. E., Galvin, M. J., Thompson, M., Jeffreys, L., and Mester, T., 1985, J. Biochem. Biophys. Methods, 11:21-29.
Murphy, E., and London, R. E., 1988, In Vivo NMR Spectroscopy and Cell Injury, Revs. Biochem. Toxicol., 9:131-184.
Murphy, E., Gabel, S. A., Funk, A., and London, R. E., 1988, Biochemistry, 27:526-528.
Oguchi, M., Sato, M., Miyatake, Y., and Akamatsu, N., 1977, J. Biochem., 82:559-567.
Perlman, M. E., Davis, D. G., Gabel, S. A., and London, R. E., 1990, Biochemistry, in press.
Quastel, J. H., and Cantero, A., 1953, Nature, 171:252-254.
Schleich, T., Willis, J. A., and Matson, G. B., 1984, Exp. Eye Res., 39:455-468.
Shull, K. H., McConomy, J., Vogt, M., Castillo, A., and Farber, E., 1966, J. Biol. Chem., 241:5060-5070.
Smith, L. J., Murphy, E., Gabel, S. A., and London, R. E., 1987, Toxicol. Appl. Pharmacol., 88:346-353.
Stermann, R., and Decker, K., 1978, FEBS Lett., 95:214-216.
Stubbs, M., Freeman, D., and Ross, B. D., 1984, Biochem. J., 224:241-246.
Ugurbil, K., Rottenberg, H., Glynn, P., and Shulman, R. G., 1978, Proc. Natl. Acad. Sci. U.S.A., 75:2244-2248.
Walker, D. G., and Khan, H. H., 1968, Biochem. J., 108:169-175.
Weckbecker, G., and Keppler, D. O. R., 1982, Eur. J. Biochem., 128:163-168.

MULTINUCLEAR SPIN RELAXATION AND HIGH-RESOLUTION NUCLEAR MAGNETIC
RESONANCE STUDIES OF FOOD PROTEINS, AGRICULTURALLY IMPORTANT MATERIALS
AND RELATED SYSTEMS

I. C. Baianu(a), T. F. Kumosinski(c), P. J. Bechtel(b),
P. A. Myers-Betts(a), P. Yakubu(a), and A. Mora(a)

(a) University of Illinois at Urbana, Department of Food
    Science, Agricultural and Food Chemistry NMR Facility,
    905 S. Goodwin Avenue, 580 Bevier Hall, Urbana, IL
(b) University of Illinois at Urbana, Department of Animal
    Science, Urbana, IL
(c) United States Department of Agriculture, ARS, NAA,
    ERRC, Philadelphia, PA

ABSTRACT

An overview of the applications of Nuclear Magnetic Resonance (NMR)
techniques in Agriculture and Food Chemistry, covering high-resolution,
solid-state, pulsed gradient and two-dimensional techniques is presented.
The systems investigated by such techniques range from purified proteins
to mixtures, starch granules and wheat grains.

Both hydration and structural/composition approaches that employ NMR
techniques are discussed from the point of view of applications rather
than technique development.

Hydration models derived from multinuclear studies are discussed for
food proteins and enzymes. The role of protein-protein interactions in
the analysis of the NMR results on hydration is also discussed. Amongst
the food proteins considered are: wheat gliadins, glutenins, corn zeins,
soy glycinins and conglycinins, as well as muscle proteins.

Several new applications and new directions of development of NMR in
Agriculture and Food Chemistry are suggested, and potential practical
applications are pointed out.

INTRODUCTION

Over the last ten years the applications of Nuclear Magnetic
Resonance (NMR) techniques in Agriculture and Food Chemistry have
sharply increased in their number and importance. Since most of the
agricultural and food materials are hydrated solids the interest in the

applications of solid-state and pulsed gradient NMR techniques to such systems is growing at an accelerated pace. At the same time, analytical applications of high-resolution NMR of food components in solution have become established and are widespread in the food industry.

Proton NMR investigations of the hydration of foods have begun more than twenty years ago but they were much less common and less sophisticated than they are today.

The first report of high-resolution $^{13}$C NMR, cross-polarization magic angle spinning (CP/MAS) of a very important group of food materials (solid gluten, wheat grains and wheat starch powders) was published in 1980 (Baianu and Förster, 1980). In the following years several, detailed high-resolution $^{13}$C NMR reports were published for such food materials both in the solid-state (Schofield and Baianu, 1982; Maciel et al., 1981; Baianu, 1985; Schaefer et al., 1979; Kricheldorf and Müller, 1984) and in solution (Baianu, 1981; Baianu et al., 1982; Augustine and Baianu, 1984, 1986, 1987; Baianu, 1989).

Following an early report published in 1979 on the water diffusion in wheat grains measured by $^{1}$H NMR (Callaghan et al., 1979), several NMR papers were published on the hydration of proteins by NMR spin-echo techniques applied to deuteriated solutions (Kumosinski and Pessen, 1982; Baianu et al., 1986), fully hydrated wheat glutens (Lioutas et al., 1987) and wheat grains (Baianu et al., 1983). Subsequently, NMR images of hydrated wheat grains were reported with 50 μm resolution (Baianu, 1983), and only recently with 25 μm resolution (Callaghan, 1988). Wheat flours and wheat doughs have also been studied by $^{13}$C CP/MAS (Baianu and Förster, 1980; Baianu, 1983; Callaghan, 1988) and $^{1}$H/$^{2}$H spin-echo NMR techniques (Richardson et al., 1985, 1986). During the last few years, the emphasis of such hydration studies of food systems has shifted rapidly from proton and deuterium to oxygen-17 ($^{17}$O) NMR (Baianu et al., 1985; Lioutas et al., 1986, 1988; Kakalis and Baianu, 1988).

This review is primarily concerned with the agricultural and food chemistry applications of the NMR techniques rather than the instrumental and technique developments in NMR; such technical developments outnumber by far the applications and have had a tremendous impact on the NMR field over the last ten years, with several dramatic improvements in sophistication, resolution and sensitivity. Several, excellent (Bax, 1984; Hull, 1987; Kumosinski and Pessen, 1985; Wüthrich, 1976), or almost comprehensive (Bodenhausen, 1981; Levy, 1979; Breitmeier and Bauer, 1984), reviews of such technical improvements and new techniques were recently published and the interested reader may want to refer to these publications before embarking on high-resolution NMR applications.

MULTINUCLEAR SPIN RELAXATION OF FOOD PROTEIN HYDRATION AND ACTIVITY. THE
EFFECTS OF CHARGE-CHARGE INTERACTIONS AND CHARGE FLUCTUATIONS IN RELATION
TO MOLECULAR DYNAMICS OF HYDRATED PROTEINS

Much of the research on protein hydration has been, and is, carried
out on globular proteins in solution; furthermore, animal proteins were
more extensively studied than vegetable ones. Amongst these, hen egg
white (HEW) lysozyme is perhaps the most intensely studied because of its
relatively low molecular weight, good solubility in aqueous solutions,
known sequence and three-dimensional structure, as well as its enzymatic
activity. Lysozyme has become, in fact, a model system on which new
experimental and theoretical approaches are being tested.

NMR Relaxation Studies of Lysozyme in Solution

There have been relatively numerous [1]H NMR relaxation studies of
lysozyme in solution, single crystals or powders; much of the earlier
work was reviewed by R. G. Bryant (1978).

Recently, in addition to [1]H NMR relaxation studies (Lioutas et al.,
1987; Baianu et al., 1985; Fullerton et al., 1986), deuterium ([2]H) and
oxygen-17([17]O) NMR measurements were also reported (Lioutas et al., 1986,
1987; Kakalis and Baianu, 1988; Halle et al., 1981; Picullel and Halle,
1986). Most of the [1]H and [2]H NMR relaxation studies on lysozyme
solutions involved longitudinal (spin-lattice) relaxation time ($T_1$)
measurements, although the transverse ("spin-spin", or $T_2$) relaxation
times are often found to vary more rapidly than $T_1$ with protein
concentration, or pH of the solution. Important reasons for the
selection of the $T_1$ measurements are: (1) the availability of field-
cycling instruments that can readily determine proton or deuterium $T_1$'s
(but not $T_2$) as a function of magnetic field strength, and (2) the
potential for estimating the water correlation time(s) from the presence
of a $T_1$ minimum (if found) with varying temperature. The current
consensus is that both $T_1$ and $T_2$ measurements are needed to test
critically the proposed models of lysozyme hydration.

The simplest model of protein hydration consistent with NMR
relaxation data in very dilute solutions involves a fast exchange between
"bound" and "free" (or "bulk") water populations (Zimmerman and Brittin,
1957; Derbyshire, 1982), in this "two-state" model, the exchange rate of
water molecules is fast compared with the nuclear spin relaxation rates
of water in protein solutions. The observed relaxation rates are:

$$R_{i,obs} = R_{i,B} \cdot P_B + R_{i,F} \cdot P_F \tag{1}$$

where i = 1, 2, and $R_1 = 1/T_1$; the subscripts B and F stand, respectively, for "bound" and "free" water molecules, and $P_B$ is the fraction of
"bound" water molecules.

Such a model predicts single exponential relaxation for the water nuclei (or single Lorentzian water peaks in the frequency domain) in protein solutions, as well as the equality of $T_1$ and $T_2$ values at NMR frequencies in the "extreme narrowing" limit, (following a monotonic decrease of $T_1$ with increasing frequency), (Derbyshire, 1982).

Marked deviations from this simple model are however, found for lysozyme when one determines the NMR relaxation rates of water as a function of lysozyme concentration (Figs. 1A, 1B, 1C), or ionic strength (Figs. 1D, 1E and Fig. 2), respectively. First of all, the concentration dependences of the nuclear spin relaxation rates of water in lysozyme solutions are nonlinear, in the absence (Figs. 1A to 1C), or presence of added NaCl (Figs. 1D, 1E and Fig. 2), at various pH values (data not shown). Secondly, the lysozyme concentration dependence is different in the absence of salt (Figs. 1A to 1C) from that in the presence of added NaCl (Figs. 1D, 1E and Fig. 2). Last but not least, the $T_1$ and $T_2$ values are not equal for lysozyme solutions in the high frequency region (Kakalis and Baianu, 1988). One has to consider, therefore, appropriate modifications and additions to the "two-state" model in order to be able to explain the observations discussed above. At temperatures close to 20°C the assumption of fast exchange is borne out by the Lorentzian NMR lineshapes of both $^2H$ and $^{17}O$ for hydrated lysozyme up to reasonably high concentrations (> 80% w/w), either in the presence or absence of salt (Fig. 3).

Furthermore, the $T_1$ relaxation for both $^2H$ and $^{17}O$ nuclei is also observed to be a single exponential for lysozyme solutions, in agreement with a fast exchange mechanism. Deuterium NMR relaxation behavior of lysozyme solutions as a function of magnetic field strength, pH and lysozyme concentration is, however, significantly different from the $^{17}O$ NMR relaxation (Kakalis and Baianu, 1988, and references cited therein).

In spite of the fact that the exchange is fast between the bound and free water populations, the lysozyme concentration dependence is non-linear, with or without NaCl added. Such observations indicate the need for a more complex model to explain the hydration behavior of lysozyme. Non-idealities of protein solutions are often related to the presence of significant protein-protein interactions in solution, or to protein activity (Kumosinski and Pessen, 1982). By taking into account the protein activity one can regain the linearity required by Eq. (1), if activities are employed instead of concentrations. In this case, protein activities are determined by non-linear regression analysis of the NMR relaxation data as previously described (Kumosinski and Pessen, 1982;

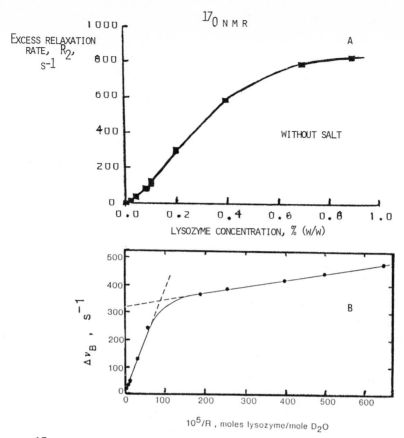

Fig. 1. $^{17}O$ NMR linewidth dependence on lysozyme concentration in $D_2O$, (linewidths are corrected for inhomogeneity broadening and are measured at half-height).
A. Lysozyme solutions at pD 7.4 without salt;
B. Lysozyme solutions at pD 7.4 with 1 M NaCl added;

Myers-Betts and Baianu, 1987, 1990). This involves replacing Eq. (1) with Eq. (2):

$$R_{i,obs} = n_H \cdot c_p \cdot (R_{i,B} - R_{i,F}) \times \exp(2B_o c_p + 1.5 B_2 \cdot c_p^2 + \cdots) \qquad (2)$$

as proposed by Kumosinski and Pessen (1982). If the salt concentration is sufficient to surround each protein with ions, all virial coefficients except $B_o$ become negligible, as it is the case for lysozyme at pD 5.1, with 0.1 M NaCl added.

A second modification of the hydration model in Eq. (1) involves the introduction of two-correlation times for bound water: a slow (ns) correlation time related to the tumbling rate of the protein in solution, and a short, (~30ps) correlation time related to the rapid reorientation of the water molecules near the protein surface. Furthermore, the bound

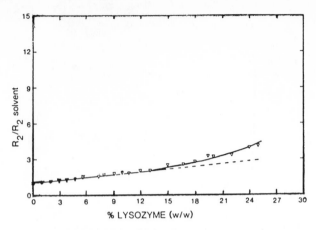

Fig. 1.  C.  Lysozyme solutions at pD 5.1 with 0.1 M NaCl added.  1,000
scans were accumulated at 20°C and with 10 mm diam. sample tubes
and 0.3 s recycling time, using 90° pulses (15μs), on a Bruker,
model CXP 200, NMR spectrometer operating at 4.7 Tesla (Fig. 1C;
27.109 MHz [17]O resonance); 16k data points were collected with a
sweepwidth of 8 kHz and 32k Fourier transforms were carried out
with 16k zero-filling.  Data in Figs. 1A, 1B were recorded at
5.4 T a with a laboratory-assembled NMR instrument operating
with an Oxford Instruments magnet. Conditions were the same as
for Fig. 1C but for the FT size which was limited to 16k.   For
the data in Fig. 1C a digitally-controlled, Bruker broadband NMR
probe was employed; its signal-to-noise ratio was ~150:1 (100
scans) for distilled water.

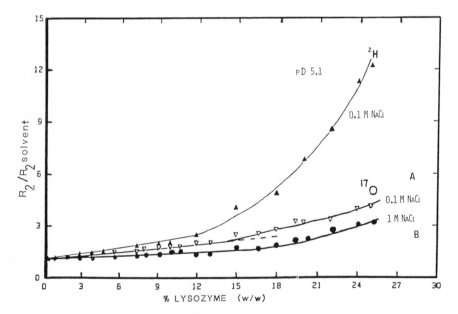

Fig. 2.  Variation of the [17]O NMR linewidth with ionic strength for
lysozyme solutions in $D_2O$ at pD 5.1, with NaCl added.

A.  With 0.1 M NaCl added. B. With 1 M NaCl added.  Other
conditions were as in Fig. 1C.

366

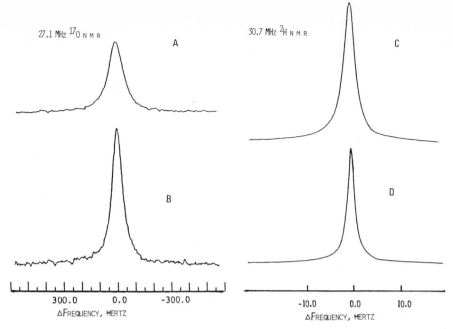

Fig. 3. $^{17}O$ and $^{2}H$ NMR lineshapes for lysozyme solutions in $D_2O$ at 20°C. Lorentzian fits were carried out by employing DISCXP programs (Bruker Co.) and GEM Software (General Electric, Co., Freemont, CA); lineshapes were found to be Lorentzian, up to 98-99% goodness-of-fit. Other conditions were as in Fig. 1C. $^{17}O$ NMR spectra of $D_2O$ in lysozyme solutions: A. with salt; B. without salt. $^{2}H$ NMR spectra of $D_2O$ in lysozyme solutions: C. with salt; D. without salt.

water motions are assumed to be <u>anisotropic</u>. The modified model was previously described as the <u>dual-motion</u> model of hydration (Fig. 4), and the
pertinent equations were discussed by Halle et al. (1981), Picullel and
Halle (1986); and Kakalis and Baianu (1988). With Eq. (2) and the dual-
motion model, all available nuclear spin relaxation data can be utilized to
derive the hydratio'n characteristics of lysozyme, including its activity/
virial coefficients in solutions. Table 1 summarizes the relevant parameters of lysozyme hydration derived from NMR data as a function of pD with
Eq. (2) and the dual-motion model. The anisotropy parameter employed in
such calculations has the value derived from theory. Notably, the slow
correlation time (~5ns) of bound water in lysozyme agrees remarkably well
with the value expected from the Stokes-Einstein equation and with the value
recently reported from frequency-domain fluorescence measurements on lysozyme solutions (Gratton et al., 1986). The fast correlation time (~30ps) of
bound water in lysozyme is consistent with previous $^{17}O$ NMR reports (Lioutas
et al., 1986; Halle et al., 1981; Picullel and Halle, 1986). The hydration
number $n_H$ is, however, in agreement only with X-ray diffraction data (about

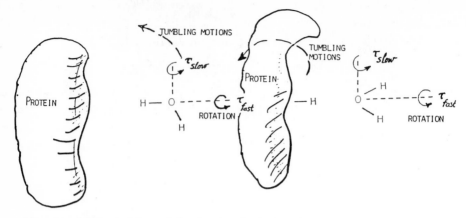

Fig. 4. Dual-motion model of protein hydration; motions of bound
water molecules occur simultaneously as indicated by
arrows, and water of hydration exchanges with the bulk
water at rates which are faster than bound water transverse
relaxation rate.

Table 1. Changes in Lysozyme Hydration[c] with pD in the presence of
0.1 M NaCl at 21°C.

| pD | $R_{2,obs}(s^{-1})$ | Hydration[a] | $\tau_{Bs}$[b] (ns) |
|---|---|---|---|
| | (27.13 MHz) | g "bound" $D_2O$/g protein | |
| 5.33 | 400.9 | 0.53 | 12.1 |
| 5.76 | 422.9 | 0.61 | 15.8 |
| 6.33 | 448.0 | 0.70 | 19.8 |
| 6.94 | 438.6 | 0.67 | 18.4 |
| 7.27 | 485.7 | 0.84 | 25.6 |
| 7.40 | 482.5 | 0.83 | 25.1 |
| 8.33 | 476.3 | 0.81 | 24.2 |
| 9.14 | 463.7 | 0.76 | 22.3 |

[a] Assuming $\tau_{Bs}$ = 4.7 ns and $\tau_{Bf}$ = 15.5 ps for all pD values.

[b] Assuming $P_B$ = 0.040 (0.36 g bound $D_2O$/g protein) and $\tau_{Bf}$ =
15.5 ps for all pD values.

[c] Assumptions a and b are mutually exclusive; assumption b appears much
more reasonable by comparison with sorption isotherm data at pH 7.0
(Lioutas et al., 1987).

180 moles $H_2O$/mole lysozyme) and with our previous reports (Lioutas et al., 1986, 1987). The most probable source of disagreement with other reports is the omission of lysozyme activity from the analysis of [17]O NMR data in previous reports (Halle et al., 1981; Picullel and Halle, 1986).

## NMR Relaxation Studies of Corn Zeins in Alkaline Solutions

Corn zeins are a major fraction (> 60%) of the corn storage proteins that are utilized either as feed, coating, or glue in several industrial applications. Because of their high hydrophobicity corn zeins rapidly precipitate in water at pH 7.0; they become, however, somewhat soluble upon raising the pH of an aqueous solution to ~11.4, or if treated with mixtures of certain organic solvents (Augustine and Baianu, 1986, 1987). At such high pH values the deamidation reaction needs also be considered. The protein activity model [(Eq. 2)] was employed to analyze the NMR relaxation data by non-linear regression analysis using Simplex and quasi-Newton algorithms (Myers-Betts and Baianu, 1987, 1990). Significant improvements of corn zein functionality are found either as a result of heat processing of corn zeins or through the addition of more than 20% soy storage proteins (Figs. 5 and 6, respectively).

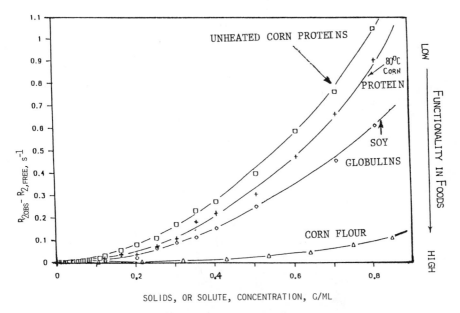

SOLIDS, OR SOLUTE, CONCENTRATION, G/ML

Fig. 5. Comparison of food protein concentration dependences of the 10 MHz [1]H NMR transverse relaxation rates (measured with the CPMG multipulse sequence) for corn zeins, soy globulins and corn flour. A. Corn zeins (data from Myers-Betts and Baianu, 1987, 1990); B. Soy globulins; C. Corn flour; D. Heated corn zeins; Bruker/IBM Instruments, Multispec PC-10 and 20 pulsed NMR spectrometers were employed to obtain the data using 10mm NMR tubes. (Note that food protein functionalities are in reverse order to food protein activities.) Qualitatively similar dependences were obtained by [17]O and [2]H NMR for corn zeins using a Bruker, CXP 200 NMR Spectrometer.

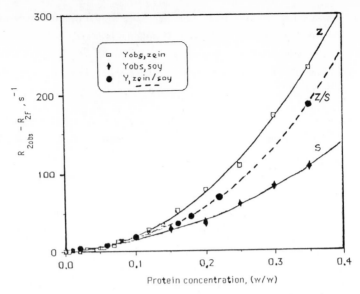

Fig. 6. Comparison of food protein concentration dependences of 10 MHz $^1$H NMR transverse relaxation rates for corn and soy protein mixtures. Other conditions as in Fig. 5.

The hydration and activity characteristics of corn zeins and zein/soy protein mixtures derived from our NMR relaxation data are summarized in Table 2. It must be stressed that these are number-average values since corn zeins are a fairly heterogeneous fraction of storage proteins with more than 20 fractions that can be separated by isoelectric focusing (Fig. 7); their molecular weights range from ~19kD to ~44kD). The virial coefficients, $B_o$, listed in Table 2 are, therefore, only average numbers for the set of more than 20 corn protein fractions present, which have a significant spread of net charges (Fig. 7).

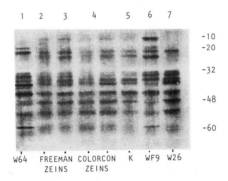

Fig. 7. Isoelectric focusing pattern of commercial corn zeins (Freeman Co. and Colorcon, Co.), compared with that of single variety corn zeins extracted from single corn kernels (Wilson, 1984). Lanes 1-4: commercial corn zeins. Lanes 5-7: single variety corn zeins (Wilson, 1984).

Table 2.  Virial Coefficients of Protein Activity Obtained by Non-Linear Regression Analyses of NMR Relaxation Data[*]

| Parameter | Zein/$H_2O$, pH 11.5 Unheated | Zein/$H_2O$, pH 11.5 80°C for 20 Mins. | Zein/$H_2O$, pH 11.5 90°C for 20 Mins. | Soy Isolate, pH 11.5 | 80:20 Zein/Soy pH 11.5 |
|---|---|---|---|---|---|
| $n_H \cdot (R_{2B}-R_{2F})$ | 62.624±10.778 | 8.924±3.511 | 11.5495 | 24.890 | 66.555+3.785 |
| $B_0$,(ml/g) | 6.905±0.758 | 11.749+1.534 | 9.527 | 8.980 | 5.692±0.214 |
| $B_2$,$(ml/g)^2$ | -20.547+3.2 | -34.124±6.032 | -25.409 | -31.473 | -10.387±0.522 |
| $B_3$ | 32.699+6.406 | 52.766+11.686 | 35.985 | 54.160 | --- |
| $B_4$ | -12.362+3.048 | -20.140+5.485 | -12.490 | -18.800 | --- |
| RMS[+] | 3.515 | 5.315 | 5.662 | 4.300 | 0.912 |

[*] 10 MHz NMR Transverse Relaxation Rates, $R_2$, measured with a CPMG Pulse Sequence

[+] RMS = Root Mean Square

Further work would be desirable for individually separated corn zein fractions, so that the values of $B_o$ may be directly related to the net charges of the individual protein species.

## NMR Relaxation Studies of Wheat Storage Proteins in Acidic Solutions. NMR of Related Food Systems

Wheat storage proteins are of prime importance to the food industry because of their widespread use in large amounts in human foods (e.g. bread, cakes, etc.).  Somewhat similar in their average amino acid composition to corn zeins, the wheat storage proteins are also highly hydrophobic.  Amongst them, wheat gliadins are thought to be globular in shape and they are soluble in acidic aqueous solutions at pH <3.5.  Rapid precipitation of wheat gliadins occurs when the pH is raised above ~3.7, or if 5-10 mM NaCl is added to the wheat gliadins solution.  The molecular weights of wheat gliadins range from ~20kD to ~80kD, with a typical, "number-average" molecular weight of about 37kD.

Their hydration behavior and weak interactions with carbohydrates were recently investigated by [1]H NMR by employing a paramagnetic ion ($Mn^{+2}$) probe (Mora-Gutierrez and Baianu, 1990a). The slow correlation time of water "bound" to the paramagnetic ion probe on the surface of wheat gliadins in solutions at pH = 3.4 was determined to be ~18ns, in agreement with the expected tumbling rate of wheat gliadins of ~37kD. Furthermore, the NMR relaxation data (Fig. 8 and Table 3) were interpreted as providing evidence for a preferential hydration (Arakawa and Timasheff, 1982) of the wheat gliadins in the presence of carbohydrates. Other spin probes were also monitored in wheat gluten and wheat starch by Electron Spin Resonance (ESR), and a significant decrease in the spin label binding was found upon heating (Pearce et al., 1987).

The more complex, hydrated wheat gluten exhibited multiple-exponential transverse relaxation behavior at low field (Baghdadi et al., 1979), although at high-field, the 200 MHz [1]H NMR transverse and longitudinal relaxation curves of $H_2O$ in gluten were single exponential (Baianu et al., 1982); several, broad [1]H NMR peaks were, however, resolved in the FT NMR spectrum of hydrated wheat gluten in $D_2O$ (Fig. 14). Deuterium NMR relaxation curves at low field was reported, however, to be single exponential for wheat doughs made with $D_2O$ (Leung et al., 1983). Single-exponential [1]H and [17]O NMR relaxation data were also obtained for hydrated wheat flour (Richardson et al.,

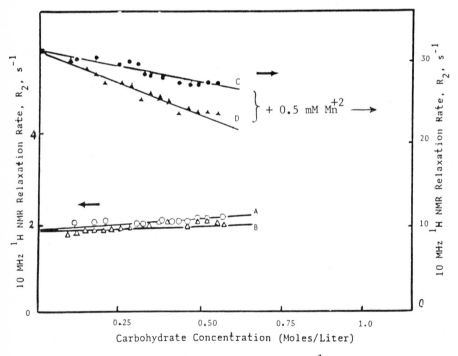

Fig. 8. Sucrose concentration dependence of 10 MHz [1]H NMR relaxation rates for 3mM wheat gliadins in an aqueous solution at pH 3.2, in the presence of 0.5 [C.D.] mM $Mn^{+2}$, paramagnetic ion probe (at fixed concentration). (A, B) control samples without addage $Mn^{+2}$.

Table 3. Calculated Values of the Relaxation Parameters for Water Bound to Wheat Gliadins at 25 ± 1°C and pH 3.2.

| $R_{ib}, s^{-1}$ [a] | $R_{ib} \cdot (Mn^{2+}), s^{-1}$ [b] | $\varepsilon_{ib}$ | $R_{ib}, (Mn^{2+})/R_{ib}$ [c] |
|---|---|---|---|
| $R_{ib} \cdot (H_2O) = 15.3$ | $R_{ib} \cdot (H_2O) = 77.3$ | $\varepsilon_{ib} \cdot (H_2O) = 10.5$ | 5.1 |
| $R_{ib} \cdot (HDO) = 11.3$ | $R_{ib} \cdot (HDO) = 53.5$ | $\varepsilon_{ib} \cdot (HDO) = 7.1$ | 4.7 |
| $R_{2b} \cdot (H_2O) = 38.5$ | $R_{2b} \cdot (H_2O) = 215.0$ | $\varepsilon_{2b} \cdot (H_2O) = 9.0$ | 5.6 |

[a] Calculated assuming a hydration number, $n_H = 0.34$ g $H_2O$/g total solids, or 750 moles $H_2O$/moles gliadin [MW (ave) ~ 40,000].

[b] With 0.5 mM $Mn^{2+}$ added to gliadin solutions of different concentrations.

[c] $R_{ib}, (Mn^{2+})$ was measured in the presence of 0.5 mM $Mn^{2+}$, whereas $R_{ib}$ was measured in the absence of added $Mn^{2+}$.

1985) and single-variety wheat grains with $D_2O$ (Fig. 9 of Baianu et al., 1982; also Baianu, 1983). [1]H NMR images (Fig. 10) of wheat grains with ~50 μm resolution and $T_1$ – contrast enhancement allowed the separation of the [1]H NMR signals from the germ oil in single wheat grains from that of the water in the grain (Baianu, 1983; Taylor et al., 1979). Further investigations of water NMR relaxation rates in these complex systems as a function of concentration, pH, temperature and magnetic field strength are necessary before drawing definitive conclusions. Because of the practical importance of these food materials an improved understanding of their hydration behaviors would have a significant economical impact.

## [1]H, [17]O and [23]Na NMR Studies of Muscle Proteins in Electrolyte Solutions

Because of their very large molecular weights and charged amino acid groups, muscle proteins such as myosin and, more generally, myofibrillar (MF) proteins present both a challenge to NMR studies of their hydration and a very interesting example of the effects of the protein charge-charge interactions on hydration. Recent theoretical calculations (Gilson and Honig, 1988) of charge-charge interactions in proteins indicate that for very large proteins that have charged amino acid groups one can expect significantly large potentials caused by charge-charge interactions to be acting even at long-range (> 20 Å). Such interactions may very well play important physiological roles in muscle contraction and are also technologically important because of their marked effect on the muscle protein functionality

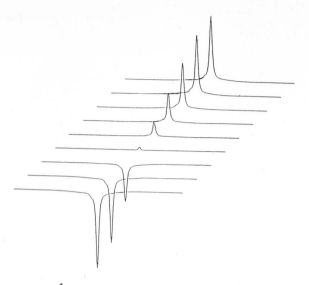

Fig. 9.   200 MHz $^1$H NMR longitudinal relaxation of hydrated wheat
(var. <u>M. Huntsman</u>) grains (40% $D_2O$ w/w); data obtained
by inversion recovery with an NT 200 (Nicolet Co.)
spectrometer.

in foods.  In this context, the effects of ions surrounding myosin or
myofibrillar proteins need be considered both from a physiological and a
technological standpoint.

The protein activity model discussed in Section 2A and recent calculations with a linearized Poisson-Boltzmann equation suggest that one
should carry out NMR relaxation studies of muscle proteins not only as a
function of protein concentration but also as a function of varying salt
concentration (for a fixed muscle protein concentration).  In the latter
case, neither Eq. (1) nor Eq. (2) apply, and a new equation that takes
into account the salt activity, ion binding, as well as the protein
activity, should be introduced:

$$R_{i,obs} - R_{i,F} = \frac{n_S}{1 + k_S \cdot c_S} \left( R_{i,B} - R_{i,F} \right)$$

$$+ k_T \cdot \left( R_{i,B} - R_{i,F} \right) \times \exp \left[ - \left( \frac{\partial m_3}{\partial m_2} \right)^2 \cdot c_S + B_1 c_S^{1/2} + B_2 \cdot c_S^2 \right], \qquad (3)$$

where $c_S$ is the ion concentration, $k_S$ is the ion binding constant,
$(\partial m_3 / \partial m_2)$ is the preferential binding coefficient for salt, $B_1$ and $B_2$ are
salt activity coefficients, and $k_T = c_p \cdot k_{exchange}$.  The above equation,
fits remarkably well all our $^1$H, $^{17}$O and $^{23}$Na NMR relaxation data both
for myosin and myofibrillar proteins (Figs. 11, 12, 13, respectively).

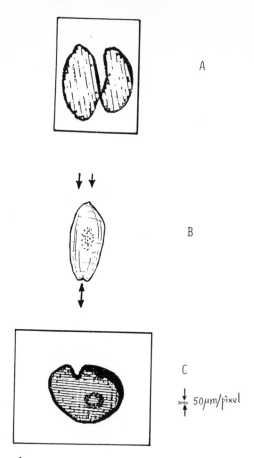

Fig. 10.  17 MHz $^1$H NMR image of fully hydrated, single variety wheat
grains [(moisture content ~40%; (data from Baianu, 1983;
Taylor et al., 1979)]. A. Two wheat grains side-by-side.
B. Schematics of a projection along the grain.
C. Cross-section of one wheat grain.

The hydration and ion binding parameters derived with this model by non-
linear regression analysis of the NMR relaxation data for myofibrillar
proteins are given in Table 4.  The analysis of the data not only gives
the hydration and bound ion numbers but also shows that water and sodium
ions are exchanged at the same rate.  Furthermore, good fits are obtained
only with Eq. (3) when it does <u>not</u> include any higher order virial
coefficients; this suggests that long-range, charge-charge (repulsive)
interactions dominate entirely the muscle protein activity effects on
hydration and ion binding.  Additional sorption equilibrium studies of MF
proteins were reported in a recent publication (Lioutas et al., 1988).
Ion specific effects are also expected, and experiments on ion specific
effects on myosin are in progress in our laboratories.

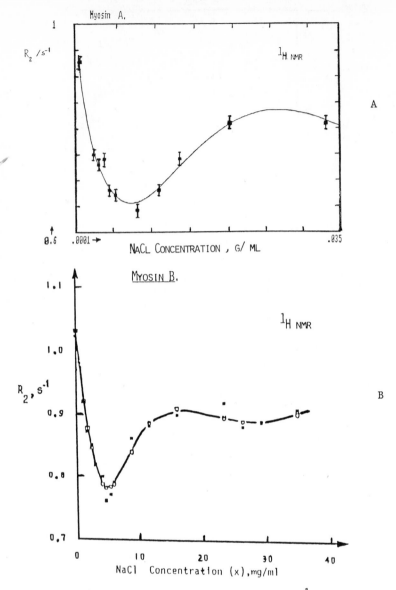

Fig. 11.  Salt concentration dependence of 10 MHz $^1$H NMR transverse
relaxation rates for purified myosin at pH 5.6 (fixed
conc., 4% w/w).  (Data from Mora-Gutierrez and Baianu,
1990b).  Other conditions as in Fig. 5.
A.  Myosin A
B.  Myosin B

HIGH-RESOLUTION NMR OF FOOD PROTEINS IN SOLUTION - A COMPARATIVE APPROACH
TO STRUCTURE-FUNCTIONALITY RELATIONSHIPS

The composition of protein containing food systems such as wheat
and corn glutens, wheat doughs, milk and soy isolate is extremely
complex, typically comprising 20 to 60, or more protein species.  Food
proteins are often insoluble in water forming aggregates that may have

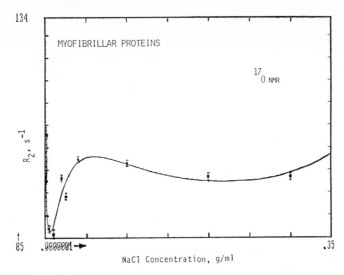

Fig. 12.  $^{17}$O NMR linewidth dependence on NaCl concentration for
myofibrillar proteins (fixed conc., 4% w/w) in aqueous
solutions at pD 7.4.  (Data from Lioutas et al., 1988;
linewidths were measured at half-height).

quite special, "functional" properties, such as the unique viscoelastic
properties of wheat gluten.  Furthermore, molecular weight distributions
of food proteins may range from 14,000 to several millions of daltons.

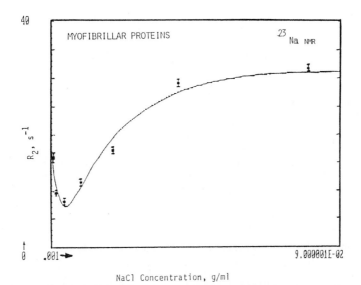

Fig. 13.  $^{23}$Na NMR linewidth dependence on NaCl concentration for
myofibrillar proteins (fixed conc., 4% w/w) in aqueous
solutions at pD 7.4.  (Data from Lioutas et al., 1988;
linewidths were measured at half-height).

Table 4.    Calculated Virial Coefficients of $^{23}$Na NMR Relaxation Data for Myofibrillar Protiens in $D_2O$ using Eq. (3) with a Simplex Algorithm for Nonlinear Regression Analysis.

A.   With the Activity Coefficients, $B_1$ and $B_2$, included:

| Parameter | R.M.S.* = 1.75 |
| --- | --- |
| $K_1$ | 840 ± 100 |
| $K_T$ | 73.8 ± 15.3 |
| $R_{2'bound'}$ s$^{-1}$ | 30.7 ± 3.3 |
| $(\partial m_3 / \partial m_2)^2$ | 0.022 ± 0.004 |
| $B_1$, ml/g | -2.2 ± 0.3 |
| $B_2$, ml/g | 4.0 ± 0.8 |

B.   Without including $B_1$ and $B_2$:

| Parameter | R.M.S.* = 1.75 |
| --- | --- |
| $K_1$ | 987 ± 250 |
| $K_T$ | 32.9 ± 1.6 |
| $R_{2'bound'}$ s$^{-1}$ | 32.3 ± 4.4 |
| $(\partial m_3 / \partial m_2)^2$ | 0.0122 ± 0.0022 |
| $B_1$, ml/g | 0 ± 0.0 |
| $B_2$, ml/g | 0 ± 0.0 |

This complexity of both physical and chemical characteristics of food proteins limits the resolution of the NMR spectra and makes difficult the analysis or assignment of such spectra. In the case of nuclei with a relatively narrow range of chemical shifts, such as protons, the NMR spectra are essentially limited to a few broad overlapping bands for a system such as a fully hydrated (~60% $D_2O$) wheat gluten (Fig. 14A), or to less than ten resolved [1]H NMR peaks for wheat gliadins in aqueous solution at pD 3.4 (Fig. 14B). By comparison, carbon-13 NMR spectra of these systems (Fig. 15) comprise a significantly larger number of resolved peaks than the [1]H NMR spectra, that can be assigned to distinct chemical groups. The assignments may not be, however, made to individual atomic sites or to a single amino acid, as it is the case with several single protein species of "low"

378

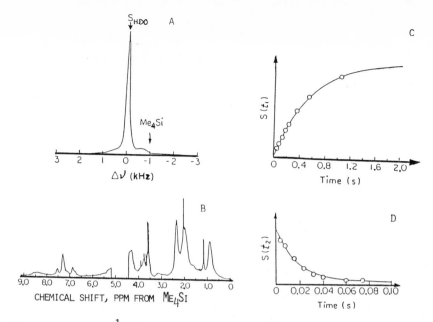

Fig. 14.  A.  200 MHz $^1$H NMR spectrum of wheat gluten (60% w/w hydrated with 40% $D_2O$ at 20°C; spectral width was 120 kHz (other details as in Fig. 5 of Baianu et al., 1982).

B.  200 MHz $^1$H NMR spectrum of wheat gliadins (5mM) in $D_2O$ at pD 3.4, (NT 200 NMR spectrometer, Nicolet Tech. Co).

C.  $T_1$ measurement on hydrated wheat gluten by inversion recovery of the water (HDO) peak.

D.  $T_2$ measurement on hydrated wheat gluten with the CPMG sequence.

molecular weight (~14,000) in solution (Wüthrich, 1976). In spite of such limitations, the resolution obtained in carbon-13 NMR spectra of food proteins is often greater than the resolution obtainable by other techniques (e.g. FT IR, Raman spectroscopy, etc.); the resolution is often sufficient to provide specific answers to questions concerning structure-functionality relationships. Furthermore, resolved $^{13}$C NMR spectra can also be obtained from food protein powders or "semi-solids" (gels, hydrated micelles, etc.) without the need to solubilize the proteins (which may be insoluble in aqueous solutions); in the latter case, solid state techniques are normally required to obtain high-resolution spectra by employing cross-polarization (CP) and magic-angle spinning (MAS), (Pines et al., 1973). Examples of such applications of the CP-MAS technique to recording $^{13}$C NMR spectra of gluten powders (Baianu and Förster, 1980), wheat and corn starch are shown in Figs. 15A to 15C, respectively; clearly, the $^{13}$C NMR resonances of wheat gluten proteins (Fig. 15A) are readily resolved from those of starch (Figs. 15B and 15C). The resolution of the $^{13}$C NMR peaks of food proteins (as well as other materials, Fig. 18) is much lower, however, in the CP/MAS spectra of the solids than it is in solution (Figs. 16A and 16B or 18A, respectively, compared with Fig. 18B).

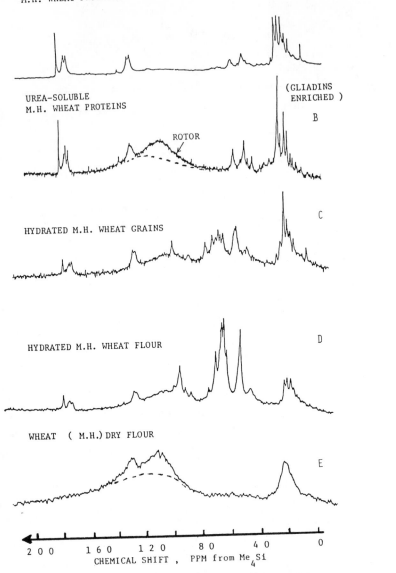

Fig. 15.  Proton broadband decoupled, 50 MHz solid-state Carbon-13
NMR spectra of hydrated wheat storage proteins and related
food systems (wheat glutens and wheat doughs).

A.  Wheat gliadins (60% w/w) mixed with $D_2O$;

B.  Wheat glutenins (60% w/w) mixed with $D_2O$;

C.  Wheat gluten (60% w/w) mixed with $D_2O$;

D.  Wheat dough (60% w/w) mixed with $D_2O$;

E.  "Dry" wheat flour (~7% moisture).

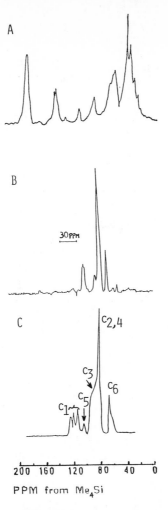

Fig. 16.   75 MHz CP/MAS, $^{13}C$ NMR spectra of wheat gluten proteins
and related food systems.  (Data in Figs. 16A and 16B are
from Baianu and Forster, 1980).  A.  Wheat gluten; B.
Wheat starch; C. Corn starch, (data from Baianu, 1985).
Data was obtained with Bruker, CXP 200 (Fig. 16C), and CXP
300 (Figs. 16A, 16B), NMR spectrometers, with magic-angle
sample spinning rates >3.5 kHz (5 and 10 mm diam.
Andrew-type rotors).

Figures 17A to 17G present a comparison of several groups of
practically important food and feed proteins in aqueous solutions of fully
hydrated mixtures:  wheat gliadins (A), wheat glutenins (B), soy glycinins
(C), soy conglycinins (D) and corn zeins (E).  The $^{13}C$ NMR spectra of amino
acids from protein hydrolyzates were also included for comparison in
Fig. 17F for wheat gliadin acid hydrolyzates and in Fig. 17G for soy
globulin hydrolyzates.  The amino acid peaks are very sharp in comparison
with the corresponding protein peaks and serve to simplify the assignments
of the $^{13}C$ resonances in the food protein spectra (Baianu et al., 1982;

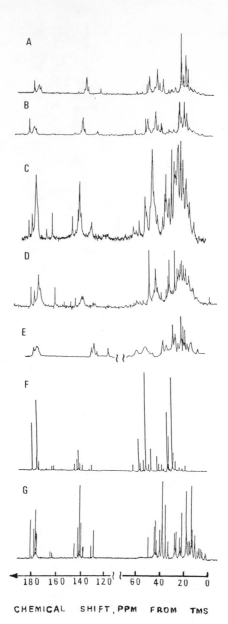

Fig. 17.  Comparison of proton broadband, or WALTZ–decoupled,
high–resolution (50 MHz) $^{13}$C NMR spectra of food proteins
in aqueous solutions (40% $D_2O$/60% $H_2O$ for $^2H$ lock).

A. Wheat gliadins; B. Wheat glutenins; C. Soy glycinins;
D. Soy conglycinins; E. Corn zeins; F. Amino acid hydrolyzates of wheat gliadins; G. Amino acid hydrolyzates of soy
globulins.

Baianu, 1989; Wüthrich, 1976; Kakalis and Baianu, 1985, 1989).  Additional
supporting information regarding the amino acid composition of these food
protein samples was obtained for such hydrolyzates by employing ion–exchange
chromatography (Augustine and Baianu, 1984; Kakalis and Baianu, 1985).

382

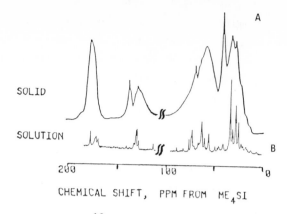

SOLID

SOLUTION

200                    100                    0

CHEMICAL SHIFT, PPM FROM ME$_4$SI

Fig. 18.   Comparison of a [13]C CP/MAS NMR spectrum of a wheat gliadin
powder sample (A) with the broadband proton decoupled [13]C
NMR spectrum of wheat gliadins in solution at pD 3.4, (B).
(Data from Baianu and Forster, 1980).

Molecular weight distributions of the food proteins were determined by gel
filtration chromatography (Schofield and Baianu, 1982; Baianu, 1989; Kakalis
and Baianu, 1985, 1989), and subunit compositions were determined by sodium
dodecyl sulphate-polyacrylamide gel electrophoresis (Baianu, 1989; and
references cited therein).   In the case of corn zeins the separation of the
intact, charged polypeptides was carried out by isoelectric focusing
(Myers-Betts and Baianu, 1990).

The [13]C NMR spectra in Figs. 17A to 17E show numerous and significant
differences between these important groups of food proteins, related to
their amino acid compositions (Figs. 17E and 17G and the ion-exchange
chromatography discussed above), molecular weight distributions, and to a
certain degree to the protein internal motions or conformation.   For
example, soy globulins that have substantially greater (50 kD to 1,200 kD)
molecular weights than wheat gliadins (27 kD to 52 kD in our preparation),
or corn zeins (22 kD to 44 kD), exhibit a number of resolved resonances that
are narrower than those of either wheat gliadins or corn zeins.   On the
basis of molecular wieght distributions alone one would expect the opposite;
furthermore, without the fast internal motions ($\tau_c$ < 1ns) in the soy
globulins such observations would not be possible.   [13]C NMR $T_1$ measurements
for the resolved, sharp peaks yielded results consistent with such fast,
internal motions (Kakalis and Baianu, 1989).   [15]N NMR spectra of wheat
gliadins in solution (Baianu et al., 1982) exhibited negative NOEs corresponding to correlation times of the order of 1 ns or less, much shorter
than the expected, overall correlation time of about 18 ns (or longer) for
wheat gliadins in aqueous solution at pD 3.4.   As shown previously (Baianu
et al., 1982), heating wheat gliadins in solution to 60°C for 30 min causes
irreversible changes to appear in the [13]C and [15]N NMR spectra (recorded at

383

room temperature after heating at 60°C) in comparison with that of unheated samples. Heating wheat gluten to 60°C in its hydrated state is known to cause substantial loss of its elastic properties, or "functionality", and therefore, the $^{13}$C NMR observations of irreversible changes of wheat gliadins with heating at 60°C may be relevant to functionality and to the effects of heat processing on food proteins. Marked heating effects can also be observed by NMR in other food materials, such as starch (Baianu, 1985), especially in relation to micro-rheology, gelatinization and retrogradation. High-resolution NMR observations are, therefore, important for an improved understanding and control of the functionality of major food components such as proteins and starch.

NEW DEVELOPMENTS OF NMR TECHNIQUES RELEVANT TO AGRICULTURAL AND FOOD CHEMISTRY APPLICATIONS

Substantial improvements in resolution and sensitivity are possible in NMR by two-dimensional techniques, as described in detail by several authors in this volume. For small proteins such as lysozyme ($M_w$ = 14,400 daltons), 2D-COSY $^1$H NMR, especially in the phase-sensitive mode with a double-quantum filter (Fig. 19), provides a wealth of structural, connectivity and dynamic information (Bax, 1984; Hull, 1987).

For larger food proteins, however, experiments such as indirectly detected ($^{13}$C or $^{15}$N) and chemical shift (e.g., $^1$H...$^{13}$C), correlated (CSC) spectra (Fig. 20) can provide greater sensitivity, increased resolution and improved spectral assignments. CSC (2D) NMR spectra and indirect $^{13}$C/$^{15}$N detection for purified food protein fractions would greatly facilitate progress with spectral assignments and composition analysis of food proteins. Selective multiple-quantum excitation NMR observations (Bax, 1984) could be employed to simplify dramatically the spectra of complex food materials such as B-starch and wheat gluten. Increasing the magnetic field is another well-known means for improving the NMR sensitivity and resolution; Fig. 21 presents a relevant example of the dramatic increase in resolution obtained for food proteins at high magnetic fields.

Pulsed gradient NMR techniques for nuclei other than the proton ($^2$H, $^{17}$O, $^{23}$Na) provide powerful tools for studying the restricted diffusion of water and ions in food systems, or the distribution of water in cereal grains, seeds and plant roots, leaves or stems.

The rapid development and application of these "new" NMR techniques to Food Chemistry and Agriculture hold promise for greater understanding of food material properties, structural characterization, controlled functionality of food components, quality control of foods, genetic engineering and nondestructive testing.

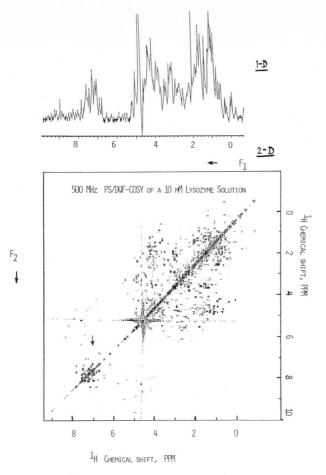

Fig. 19.   500 MHz $^1$H, 2D NMR spectrum of an 8 mM lysozyme solution in $D_2O$ recorded with a phase-sensitive, double-quantum filter, COSY pulse sequence. (Arrow points to the aromatic region of the 2D spectrum.)


CONCLUDING REMARKS

Unlike many analytical techniques that are able to determine just one material property, NMR techniques are <u>unique</u> in their ability to monitor <u>several</u> essential material properties related to both <u>structure and dynamics</u>.

Hydration of proteins and food systems is one such dynamic process that is being investigated in great detail by NMR relaxation techniques. During the last five years it has become increasingly important to consider the charge-charge interactions and protein activity in the interpretation of NMR data on hydration.   $^{17}O$ and $^2H$ NMR relaxation measurements provide more reliable means for probing the hydration of

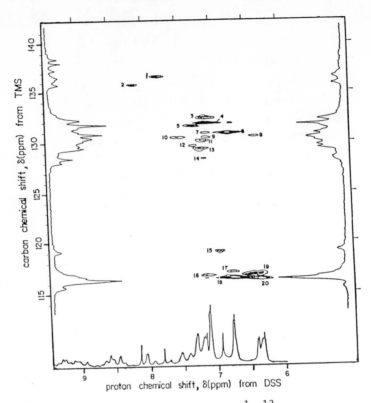

Fig. 20. Chemical shift correlated, 2D ($^1$H, $^{13}$C) NMR spectrum of a small protein (from Markley, 1987).

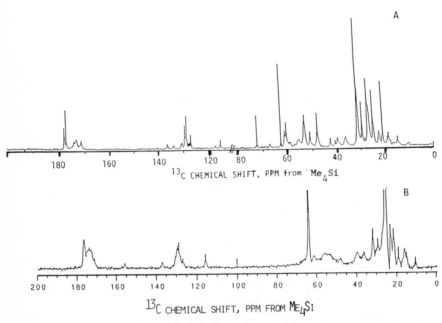

Fig. 21. 125 MHz $^{13}$C NMR spectra of cereal proteins recorded with broadband proton decoupling at 500 MHz, using a GE 500 NMR spectrometer. A. Wheat gliadins; B. Corn zeins.

foods than $^1$H NMR relaxation.  Determination of food protein activities would also allow a functionality index scale to be established, which would have technological applications in the food industry.

High-resolution NMR techniques, and especially $^{13}$C NMR, are now being employed to investigate structure-functionality relationships in food proteins, the effects of heat processing on food protein structure, activity and functionality, internal motions in food proteins, and other food related processes such as gelation, starch gelatinization/ retrogradation, food protein-carbohydrate interactions, and so on.

The development of 2D NMR and NMR Imaging techniques opens new dimensions in Agricultural and Food Chemistry research, with the promise of providing many of the answers to long-standing problems such as structure-functionality relationships in foods and the quantitative non-destructive testing of functional properties in foods.

REFERENCES

Arakawa, T., and Timasheff, S.N., 1982, Biochemistry, 21:6543-6548.
Augustine, M.E., and Baianu, I.C., 1984, Analysis of amino acid composition of cereal and soy proteins by high-field Carbon-13 NMR and ion-exchange chromatography., Proc. Fed. Amer. Soc. Exp. Biol., 43(3):672.
Augustine, M.E., and Baianu, I.C., 1986, High-resolution Carbon-13 Nuclear Magnetic Resonance Studies of Maize Proteins., J. Cereal Sci., 4:371-378.
Augustine, M.E., and Baianu, I.C., 1987, Basic studies of corn proteins for improved solubility and future utilization:  A physicochemical approach., J. Food Sci., 52(3):649-652.
Baghdadi, S., Derbyshire, W., and Baianu, I.C., 1979, $^1$H NMR relaxation studies of hydrated wheat gluten and wheat doughs.  (Unpublished results).
Baianu, I.C., 1981, Carbon-13 and proton NMR studies of wheat proteins., J. Sci. Food & Agric., 32:309-313.
Baianu, I.C., 1983, High-field Nuclear Magnetic Resonance studies of cereal proteins and related systems.  186th Natl. Meet., Amer. Chem. Soc. Symp. NMR Appl. Agric. Food Chem., Washington, DC.
Baianu, I.C., 1985, High-field NMR applications in the study of starch and chemically modified starch.  Invited presentation at the Natl. Conf. Starch Sci. and Technol., Orlando, FL.
Baianu, I.C., 1989, High-resolution NMR studies of food proteins.  Ch. 7 in:  "NMR in Agriculture", pgs. 167-218; P. Pfeffer and V. Gerasimowicz, .Eds., C.R.C. Press, Inc., Boca Raton.
Baianu, I.C. and Förster, H., 1980, Cross-polarization, high-field carbon-13 nuclear magnetic resonance techniques for studying physicochemical properties of wheat grain, flour, starch, gluten and wheat protein powders., J. Appl. Biociemistry, 2:347-354.
Baianu, I.C., Johnson, L.F., and Waddell, D.K., 1982, High-resolution, proton, carbon-13 and nitrogen-15 NMR studies of wheat proteins at high magnetic fields: Spectral changes with concentration and heating treatments of Flinor wheat gliadins in solution:  Comparison with gluten spectra., J. Sci. Food & Agric., 33:373-383.
Baianu, I.C., Lioutas, T., and Steinberg, M.P., 1985, Biophys. J., 47:30a.
Baianu, I., Bechtel, P., Lioutas, T., and Steinberg, M.P., 1986., Biophys. J., 49:328a.

Bax, A., 1984, "Two-dimensional Nuclear Magnetic Resonance in liquids."
  Delft Univ. Press., Delft.
Bodenhausen, G., 1981, Multiple-Quantum NMR., In Progr. NMR Spectrosc.,
  14:137-173.
Breitmeier, E., and Bauer, G., 1984, "$^{13}$C NMR Spectroscopy. A Working
  Manual with Exercises." Vol. 3. Harwood Acad. Publ., Chur, London,
  Paris, New York.
Bryant, R.G., 1978., NMR relaxation studies of solute-solvent
  interactions., Ann. Rev. Phys. Chem., 29:167-188.
Callaghan, P.T. et al., 1979, Biophys. J., 28:133-141.
Callaghan, P.T., 1988., "Dynamic NMR Imaging Techniques." Invited
  presentation, University of Illinois at Urbana.
Derbyshire, W., 1982, in: "Water - A Comprehensive Treatise." Franks,
  F. Ed., Vol. 8, pp. 68-102, Plenum, New York.
Fullerton, G.D. et al., 1986, An evaluation of the protein hydration by
  an NMR titration method., Biochim. Biophys. Acta, 869:230-246.
Gilson, M.K., and Honig, B.H., 1988, Energetics of charge-charge
  interactions in proteins, in: "Proteins: Structure, Function and
  Genetics." Vol. 3:32-52, Alan R. Liss, Inc., New York.
Gratton, E. et al., 1986, Biochem. Soc. Trans., 14:835-838.
Halle, B. et al., 1981, Protein hydration from water oxygen-17 magnetic
  relaxation., J. Amer. Chem. Soc., 103(3):500-508.
Hull, W.E., 1987, Ch. 2: Experimental aspects of two-dimensional NMR.,
  pp. 232-267 in: "Two-dimensional NMR Spectroscopy. Applications
  for chemists and biochemists." Croasmun, W.R. and Carlson, R.M.K.,
  Eds., VCH Publ. Inc., New York.
Kakalis, L., and Baianu, I.C., 1985, FASEB Proc., 44(2):532.
Kakalis, L., and Baianu, I.C., 1988, Oxygen-17 and deuterium NMR
  relaxation studies of lysozyme in solution. Field dispersion,
  concentration and pH/pD dependence of relaxation rates., Arch.
  Biochem. Biophys., 267:829-841
Kakalis, L., and Baianu, I.C., 1989, High-resolution carbon-13 NMR of soy
  globulins in solution, J. Agric. Food Chem., 37:1222-1227.
Kricheldorf, H.R. and Muller, D., 1984, Secondary Structures of peptides.
  Characterization of proteins by means of $^{13}$C CP/MAS spectroscopy.,
  Colloid & Polymer Science, 262:856-861.
Kumosinski, T., and Pessen, H., 1982, Arch. Biochem. Biophys.,
  218:286-302.
Kumosinski, T., and Pessen, H., 1985, In Methods Enzymology, Vol. 117.
  ,Academic Press, Orlando, FL.
Lelievre, J., and Callaghan, P.T., 1988, Amer. Chem. Soc. Meet., AGFC
  Div., Orono, Maine (abstr. only).
Leung, H.K., Magnuson, J.A., and Bruinsma, B.L., 1983, J. Food Sci.,
  48:95-98.
Levy, G.C., 1979, "Nitrogen-15 NMR Spectroscopy.", J. Wiley and Sons,
  New York.
Lioutas, T., Baianu, I.C., and Steinberg, M.P., 1986, Oxygen-17 and
  deuterium nuclear magnetic resonance studies of lysozyme hydration.,
  Arch. Biochem. Biophys., 247:68-75.
Lioutas, T., Baianu, I.C., and Steinberg, M.P., 1987, Sorption
  equilibrium and hydration studies of lysozyme: Water activity and
  360-MHz proton NMR measurements., J. Agric. Food Chem., 35:133-138.
Lioutas, T., Baianu, I.C., Bechtel, P., and Steinberg, M.P., 1988,
  Oxygen-17 and sodium-23 Nuclear Magnetic Resonance Studies of
  myofibrillar protein interactions with water and electrolytes in
  relation to sorption isotherms., J. Agric. Food Chem., 36:437-444.
Maciel, G. et al., 1981, J. Agric. & Food Chem., 29:135-139.
Markley, J.L., 1987, in: "Protein Engineering," Alan R. Lisa, Inc., pp.
  15-33.
Mora-Gutierrez, A., and Baianu, I.C., 1990, $^{1}$H NMR studies of
  preferential hydration of wheat gliadins in solutions with
  carbohydrates., subm., J. Food Microstructure.
Mora-Gutierrez, A., and Baianu, I.C., 1990b, [unpublished results].

388

Myers-Betts, P., and Baianu, I.C., 1987, $^{1}$H NMR relaxation studies of corn zeins in alkaline solutions: The effects of heat treatments on protein-protein and protein-solvent interactions. Proc. Biotechnol. Conf. Res. Dir. Biomol., Argonne Natl. Lab., pp. 174-180.

Myers-Betts, P., and Baianu, I.C., 1990, Determination of corn zeins in alkaline solutions from protein activity $^{1}$H nuclear spin relaxation data as a function of concentration and heat treatments, J. Agric. Food Chem., 477-483, in press.

Pearce, L.F. et al., 1987, An electron spin resonance study of stearic acid interactions in model wheat starch and gluten systems., J. Food Microstructure, 6:121-126.

Picullel, L., and Halle, B., 1986, Water spin relaxation in colloidal systems., J. Chem. Soc. Faraday Trans., 82:401-414.

Pines, A., Gibby, M.G., and Waugh, J.S., 1973, Proton-enhanced NMR of dilute spins in solids., J. Chem. Phys., 59:569-576.

Richardson, S.J., Baianu, I.C., and Steinberg, M.P., 1985, Relationship between $^{1}$H/$^{2}$H NMR and rheological characteristics of wheat flour suspensions., J. Food Sci., 50:1148-1153.

Richardson, S.J., Baianu, I.C., and Steinberg, M.P., 1986, Mobility of water in wheat flour suspensions as studied by proton and oxygen-17 NMR., J. Agric. Food Chem., 34:17-24.

Schaefer, J., Stejskal, E.O. and McKay, R.A., 1979, Biochem. Biophys. Res. Commun., 88:274-280.

Schofield, J.D. and Baianu, I.C., 1982, Solid-state, cross-polarization magic-angle spinning carbon-13 nuclear magnetic resonance and biochemical characterization of wheat proteins., Cereal Chem., 59(4):240-245.

Taylor, D., LeGrys, G.A., and Baianu, I.C., 1979. NMR imaging and $T_{1}$-relaxation studies of hydrated wheat grains. (Unpublished results).

Wilson, C.M., 1984, Cereal Chem., 61:198-204.

Wüthrich, K., 1976, "NMR in Biological Research: Peptides and Proteins." American Elsevier Publ. Co., Inc., New York.

Zimmerman, J.R., and Brittin, W.E., 1957., J. Phys. Chem., 61:1328-1333.

# $^1$H AND $^2$H NMR STUDIES OF WATER IN WORK-FREE WHEAT FLOUR DOUGHS

D. André d'Avignon, Chi-Cheng Hung, Mark T. L. Pagel,
Bradley Hart, G. Larry Bretthorst and Joseph J. H. Ackerman

Department of Chemistry
Washington University
St. Louis, MO   63130

## INTRODUCTION

In the baking industry, the type of wheat flour employed dictates to a
large extent the nature of the final baked product.  Breads, for example,
are prepared exclusively from hard wheat flour, while biscuits, cookies, and
cakes are generally derived from soft wheat flours.  A striking difference
between hard and soft flour, from a baker's perspective, is their difference
in water absorption [1].  Hard wheat flour can generally accommodate its own
weight in water, while a soft flour at comparable moisture levels forms a
soupy mixture and lacks the appropriate physical properties.  In this paper
we examine, by $^1$H and $^2$H NMR methods, the hydration of soft and hard wheat
flour doughs in an effort to better understand differences between flour
substrates in their interaction with water.

NMR relaxation methods have been employed extensively in efforts to
quantitate the interaction of water with macromolecules [2,3].  Not
surprisingly, there are differences of opinion regarding the interpretation
of NMR results relating to the interaction of water with complex macromolecules and substrates.  Multicomponent NMR relaxation behavior has been
observed for water associated with many materials, including muscle
tissue, [4,5] food products [6-8] and amorphous catalyst substrates [9,10].
Models have been proposed which allow interpretation of complex relaxation
decay in terms of discrete water states (bound, weakly bound, free, etc.)
and exchange processes.  Koenig and coworkers, through NMR dispersion
measurements, postulated water need not necessarily be "bound" (i.e.
hindered mobility) to macromolecules nor in exchange to account for its
relaxation properties [11,12].  Instead they postulated that the relaxation
behavior of water in close proximity to large molecules is governed by a

*NMR Applications in Biopolymers,* Edited by
J. W. Finley *et al.,* Plenum Press, New York, 1990

hydrodynamic effect as well as cross relaxation which can significantly shorten the relaxation time and, thus, mask the true correlation time of water. In this model, water senses (is coupled to) the slower motions (longer correlation times) of nearby macromolecules giving rise to a wide dispersion of water motional behavior. Bryant and coworkers have also proposed cross-relaxation as an efficient relaxation mechanism for water closely associated with protein [13,14]. They predict that this protein-associated water has a rotational correlation time on the order of pico-seconds, suggesting only a slight loss of motion when in the "bound state". From relaxation studies on small proteins, Fullerton has proposed four specific types of water "states" (i.e., domains of mobility). His predictions, based on proteins of known structure and numbers of specific "adsorption sites", are in remarkably close agreement with experimental NMR data as well as measurements from other physical methods [15,16]. Lillford, in contrast to those who subscribe to physically discrete water states as defined by NMR relaxation brought forth the concept of sample heterogeneity to account for multiexponential proton relaxation behavior [17]. In this model, complex relaxation behavior results when the concentration of "bound sites" is not constant over the volume through which a water molecule diffuses during its relaxation time.

In our studies of water hydration of work-free doughs we base our determination of water compartmentalization solely on multiexponential analysis of spin-spin relaxation behavior. We recognize that this analysis simplifies a very complex system, but it has the advantage of allowing us to label water domains consistent with hydration states in a general sense according to motional freedom. The water states, as defined by multi-exponential relaxation decay, can then be monitored as a function of flour type, moisture content, hydration time, and mixing.

The studies reported here are concerned primarily with the character-ization of work-free flour-water doughs. By work-free we mean precautions are taken to eliminate the effects of mechanical work (mixing) input on the malleable (semi-solid) dough matrix which may itself have secondary effects on the water hydration.

MATERIALS AND METHODS

Work-Free Dough Sample Preparation

For the preparation of work-free wheat flour doughs at fixed water concentrations [45% and 35%, g water/(g water + g dry flour)] ice was finely powdered in a freezer mill (model 6700, Spex Industries, Inc., Edison, NJ) at liquid nitrogen temperature for a period of five minutes and filtered to particles of less than 100 microns. Wheat flour (obtained from Nabisco Brands, Inc., East Hanover, NJ) was precooled and hand mixed with the ice powder at -40°C in a cold box (Lehrer Microtome Cryostat model LC-2000, Refrigeration for Sciences, Inc., Island Park, NY). While still at -40°C, component amounts of the ice/flour mixture were adjusted to produce a sample containing 35% or 45% total moisture based on dry weight (i.e., analysis of moisture content by drying at 100°C for 48 hours). The ice/flour mixture was then further milled at liquid nitrogen temperature for two additional minutes to ensure a homogeneous distribution of ice and flour. Separate 5mm o.d. NMR glass sample tubes were then loaded with the ice/flour mixture (100-150 mg each) and stored at -40°C. One hour prior to NMR studies, the ice/flour mixture was thawed at room temperature. Samples prepared in this manner are referred to as "work-free" doughs since no physical mixing of liquid state water and flour takes place.

For preparation of the work-free wheat flour doughs at 35% or 45% $D_2O$ content ($^2H$ = D) samples of hard wheat flour (ca. 14% moisture) were first placed in a vacuum oven at room temperature (ca. 25°C) for a period of two days. This resulted in a weight loss of nearly 13%, i.e., only 1% of the original $H_2O$ remained. The flour was rehydrated in a desiccator over $D_2O$ (99.8%-Merck Isotopes, Rahway, NJ) for a period of 24 hours and the sample evacuation/drying procedure and subsequent rehydration with $D_2O$ repeated. (This removal of hydration moisture from "as is" hard flour was not employed in the $H_2O$ hydration protocol, vide supra.) The final moisture content of the dough sample after the second rehydration was adjusted to 14% $D_2O$ total moisture (based on dry weight). Further preparation of the ice/flour mixture at 35% or 45% $D_2O$ ice content was carried out as described above except powdered $D_2O$ ice was used instead of $H_2O$.

In preparation of work-free ice-flour mixtures at 35% and 45% ($H_2O$ or $D_2O$) for time dependent hydration studies the procedure is identical with the following exceptions. Prior to NMR analysis, NMR tubes were removed from the cold box at -40°C to an ice-salt bath maintained at -3°C and held for 10 minutes. The samples were then transferred to an ice water bath at 0°C for one minute ($H_2O$) and at 4°C for one minute ($D_2O$), and finally to the NMR probe maintained at the desired temperature. The sample is held at the probe temperature for 6 minutes prior to the start of data collection.

For preparation of work-free ice-flour mixtures at variable $D_2O$ content (8%-55%), a slightly different procedure was employed. For doughs below 26% moisture the $H_2O$ depleted flour was hydrated in a desiccator over $D_2O$ for varying time periods to achieve the desired moisture content. For $D_2O$ levels above 26%, powdered ice ($D_2O$) was added to 26% $D_2O$-hydrated flour as described above. The work-free dough samples were equilibrated at room temperature for a period of at least one hour prior to NMR measurements. For all samples the % moisture was obtained gravimetrically by heating in an oven at 100°C for a period of two days.

## NMR Measurements

All measurements were performed at 7.05 tesla on a XL–300 NMR spectrometer (Varian Associates, Palo Alto, CA), $^1$H resonance frequency = 300 MHz, $^2$H resonance frequency = 45 MHz, equipped with a high power rf amplifier (model 2002A, Henry Radio, Los Angeles, CA) and a variable temperature accessory. The 90° pulse length was maintained at 13 μsec for both $^1$H and $^2$H NMR studies. The sweep width and number of complex time domain points collected were 40 KHz, 832 points for $^1$H and 10 KHz, 1024 points for $^2$H. The Carr–Purcell–Meiboom–Gill (CPMG) sequence $[90_x°-(\tau-180_y°-2\tau-180_y°-\tau)_n-$ half-echo-acquire] was employed for all spin-spin ($T_2$) measurements [18]. The inversion recovery method (180°-τ-90°-acquire) was used for the spin-lattice ($T_1$) determinations.

For all $T_2$ measurements the sequence-repetition-period (TR) was 3 x $T_1(^1H)$ and 5 x $T_1(^2H)$. At least four free induction decay (FID) transients or half-echoes were collected at each evolution time for both $T_1$ and $T_2$ measurements.

The measurements on fixed concentration doughs (either 35% or 45% moisture) during hydration were carried out at 3° or 30°C ($^1$H) and 7° and 30°C ($^2$H). For the freezing curve experiments, the dough was allowed to equilibrate at each temperature for a period of 30 minutes prior to data collection. For experiments monitoring time dependent hydration, $T_2$ assays were repeated every 6 and 4 minutes for $^1$H and $^2$H, respectively. For $^1$H measurements, fifty even half-echoes were collected (from 0.26 to 114.6 msec) for dough samples prepared at 45% moisture, while 30 even half-echoes where sampled (0.26-60.6 msec) on 35% moisture doughs. In all $T_2$ measurements (except when noted otherwise) the interpulse spacing (TE) or 2τ time was maintained at 100 μsec. For respective $^2$H measurements sixty even-echoes were collected on both 35% and 45% moisture doughs with 2τ time equal to 100 μsec.

The studies varying the 2τ value with a fixed water concentration of work-free doughs (45% moisture) at equilibrium were carried out at 30°C. Approximately 60 even half-echoes were collected at net evolution times ranging from 0.2 to 114 msec in cases where 100 μsec interpulse spacings were employed. In cases where longer 2τ values were used, fewer spin-echoes were sampled. The NMR collection conditions were customized for the $^2$H studies of work-free doughs ranging from 8-55% moisture. For samples at low moisture levels only 20 echoes were collected. Intermediate hydration levels required 40 echoes while 60 echoes were employed for the high moisture content samples.

## Data Analysis

The time-domain NMR signals (FID's or the latter half of each echo) were Fourier transformed (with line broadening of 200 Hz and 80 Hz for $^1$H and $^2$H, respectively), yielding an array of standard absorption mode frequency spectra collected at different net evolution times. The $^1$H NMR spectra were dominated by the intense water resonance except at long evolution times when, with the great diminution of the water resonance, there was some evidence of signals arising from fats, protein or carbohydrate (deuterium spectra showed no contamination from these other signals). The water resonance amplitude was quantified by digital integration. Magnetization relaxation for water in the dough systems was, thus, represented by a plot of frequency domain signal integral (I) as a function of net evolution time (t). In order to establish the number of distinct water states defined by $T_2$ decay, the data were fit to a sum of decaying exponentials each with a different decay time constant, Eq. (1),

$$I(t) = A + \sum_{i=1}^{m} B_i \, \exp\left(\frac{t}{T_{2i}}\right), \quad m = 1, \ 2 \ \text{or} \ 3. \tag{1}$$

The curve fitting was carried out by use of either RS/1 software (BBN Software, Northbrook, IL) which employs a non-linear least squares algorithm or by use of software developed in-house (vide infra). All magnetization decay profiles were described well by the sum of one, two or three (depending on TE's and hydration level) exponential terms. Two criteria were used to evaluate for, and discriminate between, one, two or three exponential-decay component behavior. The first method, employing a least-squares "goodness of fit" criterion, minimized the variance between the experimental data and the calculated decay curves for each term (i) of the exponential. If the additional exponential term failed to improve the goodness of fit based on a given level of variance criterion, then the additional exponent was designated insignificant. The second criteria employed Bayesian probability theory [19]. This method essentially calculates the model probability of one, two and three exponent fits to the signal decay data. In all cases probability theory greatly favored either one, two or three relaxation components such that assignment of the most probable model was unambiguous. Both probability theory and goodness of fit criteria yielded similar results. One other method was used to analyze the NMR data for multicomponent $T_2$ relaxation behavior on selected data sets. This method involved fitting the magnitude of the first complex time-domain data point in the (half) spin-echo collected at each net evolution time to the multicomponent exponential decay expression. The magnitude of the first time-domain data point should, in principle, represent the total transverse magnetization of the sample, i.e., the equivalent of integrating the entire frequency spectrum.

## RESULTS AND DISCUSSION

### Observation of Three Water States

Figures 1 and 2 show [1]H and [2]H $T_2$ decays (semilog plots) obtained from work-free hard flour dough samples after one hour equilibration using the CPMG method with 100 $\mu$sec interpulse spacings and digital integration of the frequency domain absorption spectrum. These transverse relaxation profiles both show strong evidence for three component exponential decay behavior for $H_2O$ or $D_2O$ hydrated samples. Previous reports on mixed flour dough revealed two component magnetization decay [7,8]. This observation defines three water domains in work-free dough that are characterized by different average correlation times (mobilities) and that are in limited communication (i.e., slow exchange) on the time scale presented by this experiment (i.e., 100 $\mu$sec). The solid, dashed, and dot-dashed lines in Figs. 1 and 2 represent, respectively, hydration states defined by a relatively long $T_2$ (23 msec – $H_2O$, 13 msec – $D_2O$), an intermediate $T_2$ (6 msec – $H_2O$, 5 msec – $D_2O$) and a short $T_2$ (0.4 msec – $H_2O$, $D_2O$). Work-free soft wheat flour doughs show similar water compartmentalization with slight differences in $T_2$ times. The fraction of total water represented by each domain (Table 1) is readily derived from the $B_i$ terms in Eq. (1) and is identical (within experimental error) for both [1]H and [2]H CPMG analysis with 100 $\mu$sec $2\tau$ times. We assign (label) the three water states in order of increasing $T_2$ to highly perturbed or immobilized water, intermediate perturbed water and relatively "free" or mobile water. This labeling reflects the normal association of a short $T_2$ with molecular immobility (long correlation time) while a relatively long $T_2$ is associated with greater mobility (short correlation time). This labeling scheme is consistent with prior assignment of tightly bound, intermediate bound, and free water. As water is believed to be physically bound to a macromolecule for only a very short time the terminology of "bound water" is probably not appropriate in these systems.

Table 1 includes the results of relaxation analysis carried out on sample preparations at 45% moisture where the magnitude of the first complex time domain point of the half-echo represents total transverse magnetization. This is essentially equivalent to the method commonly used for obtaining $T_2$ relaxation profiles, i.e., when only a time-domain signal is utilized in the absence of Fourier transformation. Again, as with the frequency-domain resonance area analysis, three water states are defined by three distinct $T_2$'s for both [1]H and [2]H measurements. This confirming observation of three exponential decay components is not unexpected as the first complex data point of the half-echo is equal to the summed amplitude (total integrated frequency-domain absorption intensity) of all NMR signals present (i.e., water). However, the frequency domain peak area determination provided greater signal-to-noise than did determination of the

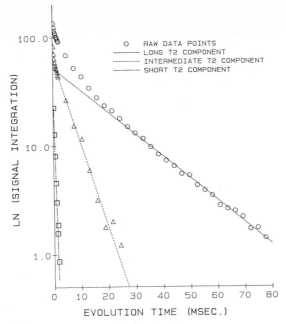

Fig. 1. Semilogarithmic plot of proton transverse magnetization decay generated via a CPMG pulse sequence with TE=100 μsec; the sample was work-free hard flour dough of 45% moisture ($H_2O$)

content prepared as described in the main text. The open circles represent the raw data (integrated frequency domain resonance area vs. evolution time) and the solid line through the open circles the best fit to the long $T_2$ component. The

open triangles represent the raw data after subtraction of the long $T_2$ component and the dashed line through the open

triangles the best fit to the intermediate $T_2$ component. The

open squares represent the raw data after subtraction of both the long and intermediate $T_2$ components. The dot-dashed line

through the open squares is the best fit to the short $T_2$ component.

magnitude of the first point in the half-echo and this led to greater precision (less scatter) in the relaxation decay profile. Parameter estimates taken from integrated resonance area decay curves are, thus, likely to be more accurate.

It is unrealistic to assign a physically unique state of water to each of the three observed $T_2$ components as it seems reasonable that a complex macromolecular system such as dough would have water present in numerous motionally distinct environments. Water environments are perhaps best characterized in terms of a distribution of adsorption energies (energy-wells) corresponding to a distribution of relaxation times (mobilities) [10]. That only three exponential relaxation decay components are resolved

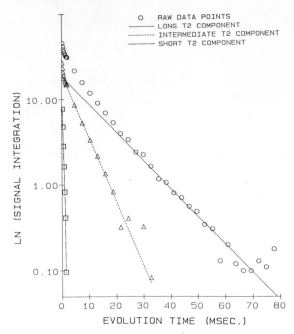

Fig. 2.  Semilogarithmic plot of deuteron transverse magnetization
decay generated via a CPMG pulse sequence with TE=100 μsec;
the sample was work-free hard flour dough at 45% moisture
($D_2O$) content prepared as described in the main text.  Other
aspects of Fig. 2 are as described in the caption to Fig. 1.

further suggests many water states possess similar average motional
character and/or that rapid chemical/physical exchange is present (rapid
with respect to $2\tau$) and serves to average the expected multitude of
hydration motional environments or perturbed water environments into three

Table 1.  Occupancy and $T_2$ times for the three water components observed
in 45% moisture work-free hard flour dough by both [1]H and [2]H

NMR measurements.  The results from both frequency domain and
time-domain first point magnitude methods are presented.  The
water concentration is expressed in g water/g dry solids (DS).

| Analysis Method | | Long $T_2$ Component Fraction (g $H_2O$/g DS) | Long $T_2$ (ms) | Medium $T_2$ Component Fraction (g $H_2O$/g DS) | Medium $T_2$ (ms) | Short $T_2$ Component Fraction (g $H_2O$/g DS) | Short $T_2$ (ms) | N' (trials) |
|---|---|---|---|---|---|---|---|---|
| Integration of frequency spectrum | [1]H | 0.28 | 20.6 | 0.31 | 6.0 | 0.23 | 0.4 | 5 |
| | [2]H | 0.31 | 12.7 | 0.29 | 5.6 | 0.22 | 0.4 | 5 |
| First complex point of FID | [1]H | 0.25 | 20.6 | 0.32 | 5.7 | 0.25 | 0.4 | 2 |
| | [2]H | 0.30 | 12.0 | 0.21 | 3.5 | 0.31 | 0.4 | 2 |

$T_2$-distinguishable states (vide infra). Another way to state this is that water samples many different environments rapidly compared to its relaxation rate and that what is observed is the average relaxation rate of all environments [10]. In this configuration, the highly perturbed water component represents a distribution of molecules with restricted motion (corresponding to deep wells on a potential energy surface) which are exchanged with one another at a rate in excess of the $1/T_2$ between each individual environment [20]. The water domain of intermediate mobility may represent a similar distribution of environments in which water is less affected by protein and carbohydrate macromolecules from the flour. The water compartment represented by an intermediate $T_2$ value may also cor-respond to entrapped water that is contained within starch granules and thus may have intermediate order. The compartment with the longest $T_2$ is "freezable" and is, thus, consistent with a free water phase (Fig. 3, Table 2). Lillford and coworkers propose a model which incorporates a distribution of probabilities of water exchanges as an explanation for multicompartment relaxation behavior [17]. In these models, multi-exponential relaxation decay is associated with sample heterogeneity. Alternatively, cross relaxation processes and not water mobility could account in part for the observed three exponential relaxation decay of water in work-free doughs [21,22]. The fact that [1]H and [2]H data are similar, i.e., both show 3 components with roughly the same fractional compart-mentalization, argues against this explanation as [2]H should be affected little by cross relaxation. The $T_2$ relaxation experiment "operationally" defines water as constrained to only three states of distinct "average" mobility which are apparently not in rapid chemical/physical exchange or such relaxation discrimination would not be possible.

The interpretation of NMR data regarding water mobility in complex macromolecular systems is largely model dependent.[2,3] The fast exchange model would attribute the observed $T_2$ times to weighted average of $T_2$ values for water states in rapid exchange [9,10]. Therefore, the $T_2$ measured for "free water" in dough represents the mole fraction weighted average $T_2$ from bulk water and some fraction of perturbed water (that is in rapid exchange with the bulk water) and is thus, considerably shorter than that of bulk water. The $T_2$ temperature dependence behavior for the long $T_2$ component, vide infra, suggests this component is in intermediate exchange. Since water adsorption by macromolecules is largely reversible it seems reasonable to postulate that all NMR-derived relaxation rates for water in dough likely result from average motional states modulated in part by chemical exchange with bulk water. This may be true even for the shortest $T_2$ observed, although it seems more reasonable to postulate that the highly immobilized fraction represents an averaging effect of "bound" water sites of varying $T_2$'s

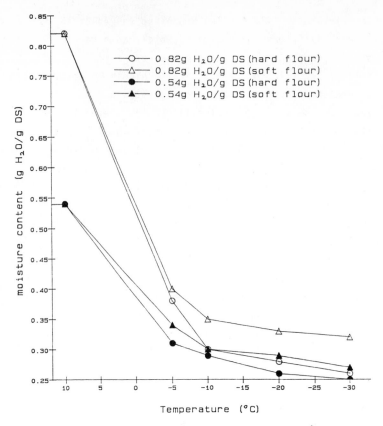

Fig. 3. Variable temperature curves as determined by [1]H transverse
relaxation measurements of work-free hard and soft wheat flour
doughs. The samples are prepared at 35% and 45% moisture
corresponding to 0.54 and 0.82 g $H_2O$/g dry solid,

respectively. NMR measurements were carried out by the CPMG
method employing TE=100 μsec. The sample was equilibrated at
each temperature for 30 minutes prior to data collection. The
open and filled circles correspond to NMR observable water for
the 45% and 35% moisture hard flour dough samples,
respectively. The open and filled triangles correspond to 45%
and 35% moisture soft flour dough samples, respectively.

in exchange with each other. Alternative models would predict water correlation times to be orders of magnitude shorter than predicted by an
exchange model due to cross relaxation or to a hydrodynamic effect [22,23].

In "as is" hard wheat flour (ca. 14% moisture) a single water $T_2$ of 0.7
msec is observed. This broad (~500 Hz $\Delta \nu_{1/2}$) water resonance is not
observed in flour dried under vacuum to ca. 1% $H_2O$. One expects water to be
largely associated with protein or carbohydrate in "as is" flour and the
short $T_2$ value observed is consistent with this, i.e., the $T_2$ is consistent
with greatest immobility. It is somewhat curious, however, that an even
shorter $T_2$ of 0.4 msec is observed for the highly perturbed water component
when the flour has been hydrated to work-free dough at 45% moisture. One

400

Table 2. Variable temperature studies of water components in work-free hard and soft flour doughs as measured by [1]H relaxation methods. The dough samples are equilibrated at each temperature for 30 minutes prior to relaxation measurements. The moisture content, given on a dry solids basis, is established by extrapolating the water components (1, 2 or 3) to zero evolution time.

45% moisture (0.82 g $H_2O$/g DS)

| Temp. (°C) | Hard Flour Component fractions (g $H_2O$/g DS) | $T_2$ (ms) | Soft Flour Component fractions (g $H_2O$/g DS) | $T_2$ (ms) |
|---|---|---|---|---|
| 30 | 0.29 | 21.4 | 0.35 | 22.0 |
|  | 0.34 | 7.0 | 0.27 | 5.9 |
|  | 0.19 | 0.43 | 0.20 | 0.38 |
| 10 | 0.26 | 22.1 | 0.27 | 24.8 |
|  | 0.33 | 6.8 | 0.33 | 6.5 |
|  | 0.23 | 0.40 | 0.22 | 0.39 |
| -5 | 0.26 | 2.5 | 0.27 | 2.3 |
|  | 0.12 | 0.40 | 0.13 | 0.31 |
| -10 | 0.19 | 1.6 | 0.24 | 1.6 |
|  | 0.11 | 0.36 | 0.11 | 0.34 |
| -20 | 0.17 | 1.0 | 0.20 | 1.1 |
|  | 0.11 | 0.3 | 0.13 | 0.33 |
| -30 | 0.26 | 0.8 | 0.19 | 0.7 |
|  |  |  | 0.13 | 0.23 |

35% moisture (0.54 g $H_2O$/g DS)

| Temp. (°C) | Hard Flour Component fractions (g $H_2O$/g DS) | $T_2$ (ms) | Soft Flour Component fractions (g $H_2O$/g DS) | $T_2$ (ms) |
|---|---|---|---|---|
| 30 | 0.10 | 15.6 | 0.13 | 13.9 |
|  | 0.30 | 5.3 | 0.26 | 4.1 |
|  | 0.14 | 0.36 | 0.15 | 0.33 |
| 10 | 0.09 | 17.4 | 0.13 | 14.5 |
|  | 0.30 | 5.3 | 0.27 | 3.9 |
|  | 0.15 | 0.38 | 0.14 | 0.37 |
| -5 | 0.19 | 1.74 | 0.21 | 1.53 |
|  | 0.12 | 0.41 | 0.13 | 0.48 |
| -10 | 0.18 | 1.56 | 0.30 | 1.19 |
|  | 0.11 | 0.42 |  |  |
| -20 | 0.15 | 0.93 | 0.29 | 0.75 |
|  | 0.11 | 0.28 |  |  |
| -30 | 0.15 | 0.79 | 0.27 | 0.48 |
|  | 0.10 | 0.25 |  |  |

would anticipate observing a somewhat longer $T_2$ than that found in the natural flour for this water component in dough, due to predicted exchange with less immobilized water. That this is not seen may simply reflect the substantial experimental inaccuracy in determining this short component, i.e., 0.4 ms and 0.7 ms $T_2$'s may be equivalent within the error of the relaxation analysis. However, there also may be structural changes in flour gluten or starch at 45% moisture content which account for this apparent discrepancy. Furthermore, the assumption of a direct $T_2$-mobility correlation is based on isotropic water motion.  This may be an over-simplification in the dough system.  Effects of cross relaxation which are reported for hydrated proteins also may contribute to the $^1$H $T_2$ differences observed for the immobile water component in "as is" flour and 45% moisture doughs [12,23].

The agreement between $^1$H and $^2$H NMR measured water component fractions is excellent (Tables 1 and 2), although the primary relaxation mechanism for each nuclide is different, i.e., through space nuclear dipole-dipole inter-action for $^1$H and through nuclear quadruple interaction with local electric field gradients for $^2$H.  For the $^1$H data both cross relaxation between water and other nearby protons and signal "contamination" due to contributions from mobile protons in protein, carbohydrate, and fats may occur in principle.  (Rigid protons in proteins and carbohydrates should exhibit $T_2$'s <200 μsec, and negligible signal contribution is expected from these species at longer evolution times.)  However, based on the $^2$H data for which such considerations are expected to be slight, these effects do not appear to alter the observed fractional compositions of water states as defined by $^1$H NMR.  Differences in $^1$H and $^2$H $T_2$'s (Table 1) can be rationalized by noting that the more efficient quadrupolar relaxation pathway is available only for the deuteron.

Freezing Studies

Table 2 lists results (%$H_2O$, $T_2$ times) of $^1$H NMR freezing studies carried out on work-free hard and soft flour doughs.  Above 0°C, three motionally distinct water compartments are readily separated in both flour types.  Upon cooling to -5°C, however, the long $T_2$ compartment disappears, i.e., at -5°C, the transverse relaxation decay is made up of two rather than three exponential terms.  Presumably the fraction of water which gave rise to the long $T_2$ component becomes NMR invisible, under our collection conditions, because at -5°C its mobility is greatly restricted (solid ice).  Curiously, the amount of water signal lost in going to -5°C is greater than the long $T_2$ component above 0°C suggesting some fraction of the highly and intermediately perturbed water is freezable as well.  A reasonable explana-tion for this observation is that the free water fraction is in exchange with water in the fractions characterized by shorter $T_2$ times.  As the free water freezes at -5°C, any water diffusing from a hydration site (on protein or carbohydrate) also becomes freezable, but no freely diffusible free water is then available to rehydrate the vacant hydration site.

Cooling the doughs to temperatures below -5°C causes further loss of NMR observable water.  The water loss occurs predominately from the inter-mediate $T_2$ compartment.  In some cases (soft flour 35% moisture at -10° and below and hard flour 45% moisture at -30°C) only one water component is

402

observed by NMR. In these cases the $T_2$ time is commonly intermediate
between the two shorter $T_2$ fractions characterized at higher temperatures.
We expect this results from a contribution of intermediate water component
being present but non-resolvable by our transverse relaxation analysis.
This is a common problem encountered in deconvolution of multiexponential
decays, i.e., when one component is present to only a small extent or when
decay time constants are similar.

The NMR-observable signal decreases further as the temperature is
lowered to -30°C. This observation is typical for freezing curves of many
food systems studied by NMR methods [24]. This NMR observation is
consistent with a loss of motion in the amorphous non-freezable water
component. For our freezing studies we held the sample at each temperature
for 30 minutes prior to analysis by NMR. We find the dough sample to be at
a "pseudo-equilibrium" under these conditions in that holding them for a
period of several hours results in no apparent change in the observable
water. It is possible that all of the non-freezable water we observe at,
say, -10°C would freeze if the sample was held below its freezing point for
an infinitely long period.

Figure 3 shows profiles of water (% of total sample) as a function of
temperature for work-free doughs. Note the freezing curves show clear
differences between hard and soft flour hydration, especially at lower
temperatures. Specifically, the amount of NMR observable signal is greater
for soft than hard flours for the same total water content. This result
suggests water in soft flour dough has greater mobility than that in
hard-flour. This may be the result of the greater protein content, commonly
associated with hard flours (the hard wheat flour has ~12% protein compared
to ~6% for soft wheat flour). Flour proteins (glutens) are known to have
high water adsorption capacity and may account in part for the differences
in non-freezable water fraction. Differential thermal analysis studies of
hard and soft flour doughs yield contradicting evidence for this, however,
as the high energy binding states (associated with NMR invisibility) are
felt to be associated with starch rather than protein [25].

Effect of Long Echo Time (TE)

In pulsed NMR, $T_2$ measurements are carried out by sampling refocused
transverse magnetization (spin-echoes). Refocusing methods eliminate signal
decay caused by static magnetic field inhomogeneities within the sample and
associated deleterious effects from molecular diffusion [26,27]. Additionally, chemical/physical exchange effects which limit the average lifetime of
a spin configuration are reduced via spin echo methods carried out at short
TE times [28]. As defined here, the $2\tau$ time (TE) or, alternatively, the
interpulse spacing, is the time between successive echos. Pulse sequences

which take advantage of refocused magnetization such as the CPMG method allow one to approach the natural $T_2$ decay which, in the absence of the effects mentioned above, reflects the mobility of a nucleus (molecule). Choice of the $2\tau$ employed in a multipulse experiment is especially important in cases where the molecule of interest is in exchange with one or more motionally unique environment(s). For example, consider water which is in exchange between a bound configuration (motional environment A) and free bulk water (motional environment B) with equivalent chemical shifts. If $2\tau$ is chosen to be significantly shorter than the lifetime and $T_2$ of water in each environment, the observed transverse magnetization will decay biexponentially. Analysis of the decay profile [see Eq. (1)] yields characteristic relaxation times ($T_{2A}$, $T_{2B}$) and the fraction of water in each environment (via $B_A$, $B_B$). If instead $2\tau$ is chosen significantly longer than the lifetime of water in each environment, the observed transverse magnetization decays with a profile defined by a single exponential time constant which represents the mole fraction weighted average of both water environments A and B (Eq. (2)).

$$\frac{1}{T_{2, observed}} = \frac{X_A}{T_{2A}} + \frac{X_B}{T_{2B}}. \tag{2}$$

More generally, for n motionally distinct environments for water all in rapid exchange relative to TE (all site lifetimes short relative to TE), the observed single exponential time constant is given by Eq. (3),

$$\frac{1}{T_{2, observed}} = \sum_{i=1}^{n} \frac{X_i}{T_{2i}} \tag{3}$$

where $X_i$ and $T_{2i}$ represent, respectively, the mole fraction and spin-spin relaxation time of spins in site i.

Table 3 and Fig. 4 show results of $T_2$ measurements on dough as a function of $2\tau$. Varying $2\tau$ causes a dramatic effect on (a) the number of water states defined by the relaxation experiment, (b) the apparent fractional composition of water states and (c) the $T_2$ associated with each state. These results suggest one must exercise caution in assigning water state compartmentization by NMR methods, especially when making comparisons to results from other physical methods, as the results obtain depend critically on the $2\tau$ chosen.

The loss of the tightly bound water component with increasing $2\tau$ is readily explained. In considering the CPMG sequence, the shortest $2\tau$ used (100 μsec) makes data acquisition possible at even-echo evolution-times of 200, 400, 600, ..., μsec and, thus, provides sufficient though certainly not optimal sampling of the short, immobilized, component (e.g., $3 \times T_2 = 1.2$ msec for 95% decay during which 6 evolution times are sampled at 2, 4, 6, 8,

Table 3. Water compartmentalization in 45% moisture work-free hard flour dough measured as a function of the echo time (2τ) in the CPMG sequence. The standard deviation for parameter estimates is given in parentheses for cases where multiple samples were evaluated at the same echo time, 2τ.

Proton (H₂O) Relaxation

| TE (2τ) | Long T$_2$ Component Fraction (g H$_2$O/g DS) | Long T$_2$ (ms) | Medium T$_2$ Component Fraction (g H$_2$O/g DS) | Medium T$_2$ (ms) | Short T$_2$ Component Fraction (g H$_2$O/g DS) | Short T$_2$ (ms) | N* (trials) |
|---|---|---|---|---|---|---|---|
| 100 μsec | 0.28(0.01) | 20.6(1.6) | 0.31(0.02) | 6.0(0.9) | 0.23(0.01) | 0.41(0.10) | 5 |
| 300 μsec | 0.34 | 20.9 | 0.48 | 5.0 | | | 1 |
| 600 μsec | 0.52 | 14.9 | 0.30 | 6.5 | | | 1 |
| 900 μsec | 0.57(0.04) | 11.1(1.1) | 0.25(0.04) | 3.1(0.6) | | | 5 |
| 1200 μsec | 0.52 | 18.9 | 0.30 | 5.6 | | | 1 |
| 1600 μsec | 0.32 | 17.1 | 0.50 | 5.8 | | | 1 |
| 2 msec | 0.20(0.05) | 16.7(0.5) | 0.62(0.05) | 6.0(0.02) | | | 5 |
| 4 msec | | 8.3(0.3) | | | | | 5 |

Deuteron (D₂O) Relaxation

| TE (2τ) | Long T$_2$ Component Fraction (g H$_2$O/g DS) | Long T$_2$ (ms) | Medium T$_2$ Component Fraction (g H$_2$O/g DS) | Medium T$_2$ (ms) | Short T$_2$ Component Fraction (g H$_2$O/g DS) | Short T$_2$ (ms) | N* (trials) |
|---|---|---|---|---|---|---|---|
| 100 μsec | 0.31(0.03) | 12.7(0.7) | 0.29(0.3) | 5.0(0.8) | 0.22(0.01) | 0.40(0.06) | 5 |
| 900 μsec | 0.56(0.06) | 9.2(0.6) | 0.26(0.06) | 2.0(1.2) | | | 5 |
| 2 msec | 0.25(0.06) | 14.2(1.6) | 0.57(0.06) | 6.2(0.5 | | | 5 |
| 4 msec | | 9.6(0.5) | | | | | 5 |

*Standard deviations for parameter estimates are given in parentheses for cases where multiple 0.1 cm samples were evaluated using the same echo time, TE.

10 and 12 x 2τ). The next longest TE used (300 μsec) allows data sampling at even-echo evolution-times of only 600, 900, 1200, ... μsec and, thus, largely misses sampling the rapidly decaying magnetization from the immobilized water (only 20% of the magnetization remains at the first even echo). Prior NMR reports of water in mixed dough systems found only two components likely due to choice of a long interpulse spacing value [7,8]. In characterizing water, one must choose a 2τ value which is not too short, as the echos will contain substantial contributions from protein magnetization. In using short 2τ times of less than ca. 300 μsec, there will also be some degree of spin-locking on the water magnetization. At an interpulse spacing of 100 μsec, spin locking effects should be significant. This will result in apparent T$_2$ times greater than the natural T$_2$'s for the water compartments in flour doughs but should not affect our ability to distinguish water components.

The derived fractional compositions of water states show an unusual trend with increasing 2τ value (Table 3, Fig. 4). The changes observed in going from 100 to 300 μsec TE are expected, as there is a difference in the number of components observed (3 → 2, vide supra). The apparent increase of free water with increasing TE and flip-flop in water domain occupancy fraction above 1.2 msec TE is less easily explained. This trend is observed for both [1]H and [2]H studies, suggesting that contributions from cross relaxation are unlikely. Possible explanations for observed (relaxation

405

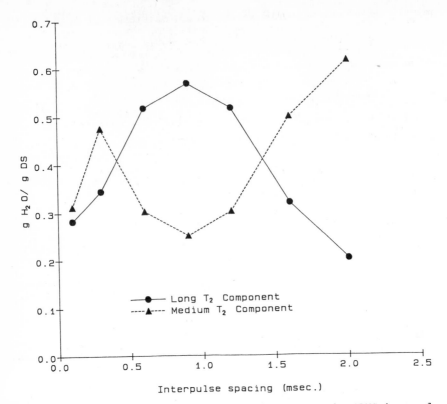

Fig. 4.   Dependence of water component occupancy on the CPMG interpulse spacing value. The transverse magnetization decay in hard flour dough, prepared at 45% moisture, is measured and analysed as the $2\tau$ time is increased. The filled circles and triangles represent the water occupancy [dry solids (DS) basis] vs. $2\tau$ as defined by the NMR method. Only the two (out of three observed) water components are represented at the $2\tau=100$ μsec data point. The longer $2\tau$'s chosen reveal only two components except at 4 msec when single exponential decay is observed.

measurement defined) increased occupancy of the free water fraction include decreased effects of spin-locking [29] and increased effects of water exchange [28] with increasing interpulse spacing. The corresponding decrease of $T_2$ times over the same regime ($2\tau = 100$ to $900$ μsec) is also consistent with water exchange rates being on the same time scale as the $2\tau$, i.e., water in intermediate exchange on the transverse magnetization refocusing time scale [5]. The apparent flip-flop of water state occupancy fractions at a $2\tau$ above 1.2 msec may reflect both exchange effects and inadequate sampling of the intermediate decay component. At a 4 msec inter-pulse spacing, only a single exponential decay, consistent with only one relaxation defined water state, is observed with $T_2$ intermediate between that found for free and intermediate perturbed water. This suggests these two water domains are in rapid exchange on a time scale of $2\tau$.

Hydration Studies

NMR relaxation studies of the hydration dynamics of work-free doughs (35% and 45% moisture) were carried out at both 30° and 3°C (7°C for $^2$H studies). Figure 5 shows representative time courses of the NMR defined water compartments. Table 4 includes water occupancy and $T_2$ times for 50 and 64 minute post-thawing $^1$H and $^2$H studies, respectively (hydration "equilibrium" was considered to be complete about one hour post-thawing). The reproducible feature in these studies is a loss of water from the free fraction (longest $T_2$ component) and a buildup of the medium $T_2$ water fraction, as shown in Fig. 5. The water associated with the shorter $T_2$ compartment also increases slightly during the same time course. The $^2$H data yielded more consistent results in this experiment than the $^1$H studies (less scatter in repetitive trials) but

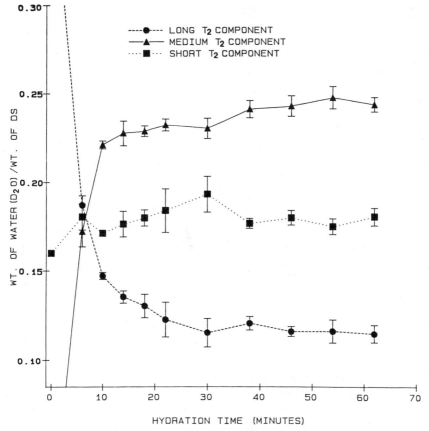

Fig. 5. Hydration profile for 35% $D_2O$ work-free hard flour dough at 7°C. Wheat flour-water ($D_2O$) mixtures are prepared at −40°C and allowed to thaw above the freezing point. The transverse magnetization decay is measured during the equilibration period and water occupancy for the three components is determined. The triangles, squares and circles represent water occupancy for the medium, short and long $T_2$ component, respectively. The vertical bar on each symbol represents one standard deviation based on five separate trials.

Table 4. Equilibrium (one hour post-thaw) water compartment analysis [dry solids (DS) basis] based on $^1$H and $^2$H transverse relaxation decay for work-free soft and hard flour doughs. The $^1$H and $^2$H data has a relative standard deviation of 15% and 7%, respectively, based on five trials for each measurement.

| Nuclide | Flour Type | Temperature of Hydration (°C) | Long $T_2$ * Component Fraction $\frac{\text{g H}_2\text{O(D}_2\text{O)}}{\text{g DS}}$ | Long $T_2$ (ms) | Medium $T_2$ Component Fraction $\frac{\text{g H}_2\text{O(D}_2\text{O)}}{\text{g DS}}$ | Medium $T_2$ (ms) | Short $T_2$ Component Fraction $\frac{\text{g H}_2\text{O(D}_2\text{O)}}{\text{g DS}}$ | Short $T_2$ (ms) |
|---|---|---|---|---|---|---|---|---|
| | | | | | 45% moisture (0.82 g Water/g DS) | | | |
| $^1$H | SOFT | 3° | 0.34 | 30.2 | 0.28 | 5.4 | 0.20 | 0.34 |
| $^2$H | SOFT | 7° | 0.33 | 23.0 | 0.24 | 3.4 | 0.25 | 0.44 |
| $^1$H | SOFT | 30° | 0.27 | 27.0 | 0.26 | 5.6 | 0.30 | 0.37 |
| $^2$H | SOFT | 30° | 0.31 | 16.7 | 0.27 | 3.7 | 0.24 | 0.37 |
| $^1$H | HARD | 3° | 0.26 | 22.1 | 0.27 | 5.8 | 0.29 | 0.39 |
| $^2$H | HARD | 7° | 0.27 | 24.7 | 0.30 | 4.4 | 0.25 | 0.51 |
| $^1$H | HARD | 30° | 0.28 | 20.5 | 0.25 | 6.3 | 0.30 | 0.40 |
| $^2$H | HARD | 30° | 0.28 | 16.0 | 0.29 | 4.4 | 0.23 | 0.45 |
| | | | | | 35% moisture (0.54 g Water/g DS) | | | |
| $^1$H | SOFT | 3° | 0.15 | 15.1 | 0.22 | 3.8 | 0.17 | 0.38 |
| $^2$H | SOFT | 7° | 0.10 | 14.5 | 0.25 | 2.5 | 0.20 | 0.51 |
| $^1$H | SOFT | 30° | 0.12 | 13.5 | 0.24 | 4.2 | 0.17 | 0.32 |
| $^2$H | SOFT | 30° | 0.17 | 11.1 | 0.20 | 2.8 | 0.17 | 0.38 |
| $^1$H | HARD | 3° | 0.16 | 12.9 | 0.20 | 4.0 | 0.18 | 0.37 |
| $^2$H | HARD | 7° | 0.11 | 15.0 | 0.24 | 2.8 | 0.18 | 0.48 |
| $^1$H | HARD | 30° | 0.08 | 16.6 | 0.26 | 5.0 | 0.19 | 0.36 |
| $^2$H | HARD | 30° | 0.14 | 11.5 | 0.25 | 3.2 | 0.14 | 0.39 |

* $^1$H water fractions have a relative standard deviation of about 15%.
$^2$H water fractions have a relative standard deviation of about 7%.

both methods show similar trends. The most notable feature is the very rapid uptake of free water (from the melted ice) by the dough. The time scale of our $T_2$ measurement is too long to accurately quantitate the three water compartments in the region where hydration is most rapid (less than 5 minutes after thawing). Because of this we are unable to make any distinction between the rate of hard and soft flour hydration. Based on the more gradual changes occurring after the first five minutes post-thawing, we can conclude that hydration in the absence of work appears to be complete after 40 minutes. Additionally, we have found no evidence for continued changes in the NMR-defined hydration compartments at times up to 3 hours post-thawing. The "equilibrium" water distribution shown in Table 4 reveals a substantial difference in occupancy of the free water compartment between doughs prepared at 35% and 45% moisture. This difference presumably reflects substantial saturation of the two shorter $T_2$ fractions at 45% moisture and, thus, accounts for the larger reservoir of mobile water. At the 35% moisture level both hard and soft flour doughs show less occupancy of the immobile water fractions by about 23% ($H_2$O/DS) compared to 45% moisture samples. The water component identified by the shortest $T_2$ also shows an increase in population

408

during hydration. If this component correlates with the short $T_2$ environment in "as is" flour, then it also reflects considerable hydration in the first 5 minutes post-thawing. For example, for 45% moisture doughs, this fraction reaches a value of ~50% greater, and for 35% moisture, a value of ~10% greater than the starting flour.

The NMR water relaxation measurements reveal only slight differences in hydration characteristics as a function of temperature or flour type. This observation is consistent with previous reports which failed to link NMR observables with baking properties [7,8]. In comparing samples prepared at different moisture levels, an increase in occupancy for all water components is observed at higher water content as expected. However, the increase in free water is striking (~100%), compared to domains characterized by shorter $T_2$ times. This observation is consistent with flour having a finite capacity for water absorption. The long $T_2$ component serves essentially as an overflow reservoir for water which cannot be accommodated elsewhere. At 45% moisture, the soft flour doughs on average have 10% more water in the free and less in the medium and tightly "bound" compartments than the hard flour doughs. This trend is consistent with the greater water absorption found in hard wheat flours. Additionally, the free water $T_2$ shows a reproducible decrease with decreasing moisture content (the medium $T_2$ shows this to some extent also) and increasing temperature. Both of these observations are consistent with the exchange models [5,9]. From Eq. (2) the observed $T_2$ can be seen to increase with an increase in long $T_2$ component (bulk water) if exchange is involved.

## Water ($D_2O$) Titration Experiments

Hard and soft work-free flours were hydrated with $D_2O$ over a range of 8-55% moisture (based on $D_2O$) and characterized by $^2H$ NMR relaxation methods. Figure 6 shows the $T_1$ dependence, while Figures 7 and 8 depict the $T_2$ relaxation results. The $T_1$ profile appears identical for both flour types while the $T_2$ analysis reveals a slight increase in the occupancy of the long $T_2$ water compartment for the soft compared to hard flour doughs. The results from the titration experiment show that water occupies the NMR defined hydration states preferentially, i.e., the bound compartment fills before the medium compartment, etc.

The spin-lattice relaxation vs. moisture concentration curve (Fig. 6) shows a break around 0.2 g $D_2O$/g dry solid (DS). The $T_1$ profile can be approximated by two linear regions from 0.1-0.2 and from 0.2-1.2 g $D_2O$/g DS. The $D_2O$ flour concentration of 0.2 g/DS is very close to the value at which the intermediate $T_2$ component begins to show occupancy (Figs. 7 and 8) and perhaps signifies a marked change in average water mobility. Below 0.2 g/DS moisture the NMR decay indicates that only the highest energy (lowest

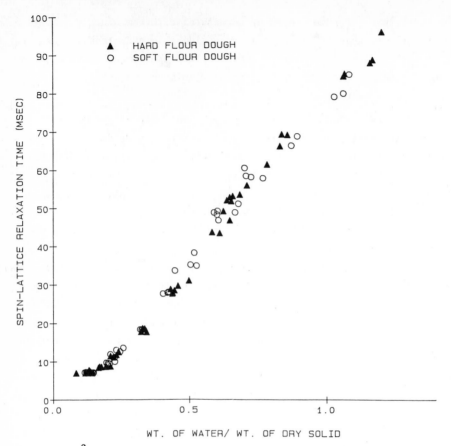

SPIN-LATTICE RELAXATION TIME (MSEC)

WT. OF WATER/ WT. OF DRY SOLID

Fig. 6.    The $^2$H dependence of hard and soft wheat flour doughs as a
function of water content.  Variable moisture content deuterium
exchanged wheat flour is hydrated for a period of at least one
hour prior to $^2$H NMR $T_1$ measurements (inversion recovery method)
at 30°C.  The triangles (▲) and circles (O) correspond to hard
and soft flour preparations, respectively.

mobility) sites are occupied.  These are defined by a $T_2$ time of 0.2-0.4 msec.
In fact, even at the highest moisture samples studied, the tightly bound
component is characterized by the same $T_2$ time.  In contrast, the $T_2$ values
for the medium and long $T_2$ components increase with increasing moisture
content (Fig. 9).  It seems reasonable to correlate the suggested break in the
$T_1$ profile to the onset of hydration-site occupancy by water with much greater
mobility than that of the shortest $T_2$ or highly immobilized compartment.  The
increase in $T_1$ and $T_2$ times shown for both the medium and "free" water
components with increased moisture is largely consistent with the predictions
of exchange with bulk water, i.e., as the mole fraction of bulk water is
increased (long $T_2$) the observed $T_2$'s become greater.  That the shortest $T_2$
compartment does not show this trend may indicate that it is in limited
exchange with bulk water.

410

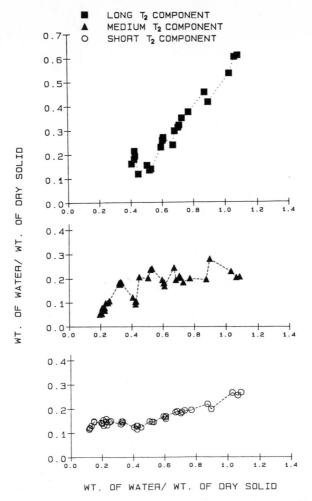

Fig. 7.  $D_2O$ titration of hydration sites in work-free hard wheat flour
dough.  The circles (O), triangles (▲), and squares (■)
represent moisture content for the respective short, medium
and long $T_2$ water compartments measured by $^2H$ NMR at 30°C.

The occupancy of water compartments as determined by transverse relaxation decay shows an initial rapid increase followed by a decrease and a
final increase (Figs. 7, 8).  The initial rapid occupancy is reasonable, as
one expects an abundance of vacant sites on the dry or partially hydrated
flour substrate.  The decrease is likely an artifact of our non-linear least
squares analysis.  It becomes difficult to obtain clean separation of multi-
exponential decays when either one component is in great excess or the decay
constants are similar.  At 0.2 g $D_2O$/g DS, for example, the short $T_2$ is
about 0.3 msec compared to the estimated medium $T_2$ fraction of about 0.8
msec.  As more hydration occurs (below the onset of the "free" water
compartment), the medium $T_2$ increases to ~2.5 msec which enables a cleaner

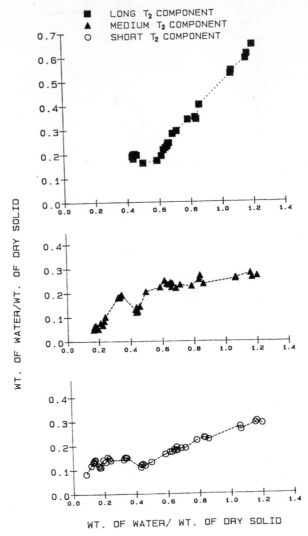

Fig. 8. $D_2O$ titration of hydration sites in work-free soft wheat flour dough. The circles (O), triangles (▲), and squares (■) represent moisture content for the respective short, medium and long $T_2$ water compartments measured by $^2H$ NMR at 30°C.

separation between decaying components. At hydration levels of 0.4 g $D_2O$/g DS, when the "free" water becomes observable, a similar discrimination problem is seen affecting both the tightly and medium bound sites.

Over the moisture range we investigated, we did not see clear evidence for saturation of any water compartment. The increase in occupancy of the two shorter $T_2$ compartments clearly slowed from its initial rapid growth, but on average was increasing slightly even at the higher concentration. The "free" water component increases linearly at a rapid rate above 0.5 g $D_2O$/g DS, suggesting few remaining hydration sites characterized by short $T_2$ time.

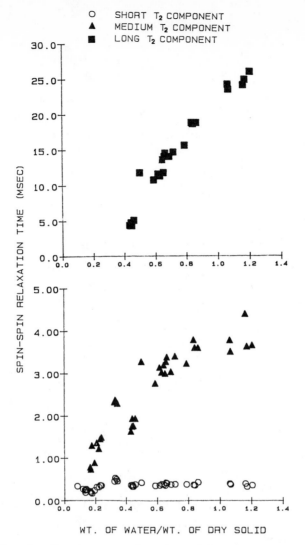

○     SHORT T₂ COMPONENT
▲     MEDIUM T₂ COMPONENT
■     LONG T₂ COMPONENT

Fig. 9.  $T_2$ time as a function of hydration level.

This $T_2$ study showed little difference between flour types, except slightly greater water occupancy of the "free" water compartment for soft flour doughs.  This same trend was noted earlier in studies of doughs prepared at 35% and 45% moisture content.

CONCLUSIONS

The NMR studies reported here show that work-free flour is characterized by three water compartments, instead of by two as was reported previously on mixed doughs.  Additionally, the water compartmentalization is found to depend strongly, and in an unexplained fashion, on the interpulse spacing ($2\tau$) employed.

The NMR studies do show slight differences between hard and soft flour types. These differences include: a) a lower amount of NMR-visible non-freezable water in hard flour doughs; b) slightly longer $T_2$ times for the two longer $T_2$ compartments in soft flour; and c) a slightly greater occupancy of the free water component in soft flour dough, especially at high moisture content. The NMR observations may simply reflect the different protein and/or carbohydrate contents of the flours and their relative ability to retain water. It is not clear, however, that product quality, from a baking perspective, can be linked to protein or carbohydrate content of a flour type.

## REFERENCES

1. A. H. Bloksma, 1972, Cereal Sci. Today, 17:380.
2. S. J. Richardson, I. C. Baianu and M. P. Steinberg, 1986, J. Agric. Food Chem., 34:17.
3. J. L. Finney, J. M. Goodfellow, and P. L. Poole, 1982, in "Structural Molecular Biology"; D.B. Davies, W. Saenger, S.S. Danyluk, eds.; Plenum Press, New York.
4. P. S. Belton, R. R. Jackson, and K. J. Packer, 1972, Biochem. Biophys. Acta., 286:16.
5. H. A. Resing, A. N. Garroway, and K. R. Foster, 1976, in "Magnetic Resonance in Colloid and Interface Science", H.A. Resing and C.G. Wade, eds.; ACS, Washington, DC.
6. H. K. Leung, and M. P. Steinberg, 1979, J. Food Sci., 44:1212.
7. H. K. Leung, J. A. Magnuson, and B. L. Bruinsma, 1979, J. Food Sci., 44:1408.
8. H. K. Leung, J. A. Magnuson, and B. L. Bruinsma, 1983, J. Food Sci., 48:45.
9. J. R. Zimmerman, and W. E. Brittan, 1957, J. Phys. Chem., 61:1328.
10. H. A. Resing, 1968, Adv. Mol. Relaxation Processes, 3:199.
11. S. H. Koenig, K. Hallenga, and M. Shporer, 1975, Proc. Nat. Acad. Sci., 72:2667.
12. S. H. Koenig, 1980, in "Water in Polymers", S. P. Rowland, ed., Amer. Chem. Soc., Washington, DC.
13. R. G. Bryant, 1978, Ann. Rev. Phys. Chem., 29:167.
14. R. G. Bryant, 1980, Biophys. J., 32:80.
15. G. D. Fullerton, V. A. Ord, and I. L. Cameron, 1986, Biochem. Biophys. Acta., 869:230.
16. K. B. Lange, G. D. Fullerton, and I. L. Cameron, 1988, Seventh Annual Society of Magnetic Resonance in Medicine, San Francisco.
17. P. J. Lillford, A. H. Clark, and D. V. Jones, 1980, in "Water in Polymers", S.P. Rowland, ed., American Chemical Society, Washington, DC.
18. S. Meiboom, and P. Gill, 1958, Rev. Sci. Instr., 29:688.
19. G. L. Bretthorst, 1988, in Lecture Notes and Statistics, Vol. 48, Springer-Verlag, New York, NY.
20. J. S. Leigh, Jr., 1971, J. Magn. Res., 4:308.
21. H. T. Edzes, and E. T. Samalski, 1977, Nature (London), 265:521.
22. S. H. Koenig, R. G. Bryant, K. Halenga, and G. S. Jacob, 1978, Biochemistry, 17:4348.
23. R. G. Bryant, and W. M. Shirley, 1980, in "Water in Polymers", S.P. Rowland, ed., American Chemical Society, Washington, DC.
24. H. K. Leung, and M. P. Steinberg, 1979, J. Food Sci., 44:1213.
25. N. Bushuk, and V. K. Mehrotra, 1977, Cereal. Chem., 54:311.
26. E. L. Hahn, 1950, Phys. Rev., 80:580.
27. H. Y. Carr, and E. M. Purcell, 1954, Phys. Rev., 94:630.
28. Z. Luz, and S. Meiboom, 1963, J. Chem. Phys., 39:366.
29. G. E. Santyr, R. M. Honkelman, and M. J Bronskill, 1988, J. Magn. Reson., 79:28.

# CHARACTERIZATION OF WATER IN FOODS BY NMR

Shelly J. Richardson Schmidt

Division of Foods and Nutrition
University of Illinois, Urbana

## INTRODUCTION

Water is probably the most important component of a food system
because it influences so many process variables and product characteristics. For example, water is a key component in determining the amount
of energy necessary for many unit operations, such as freezing, dehydration and freeze-drying; water strongly influences chemical changes, such
as protein denaturation (Nakano and Yasui, 1976), Maillard browning
(Labuza and Saltmarch, 1981) and enzyme activity (Drapron, 1985); water
is the determining factor in rheological behavior (Urbanski, 1981); water
is extensively involved in chemical, physical, nutritional (Kirk, 1981)
and microbial changes during storage; finally, water is involved in the
kinesthetic attributes of the food during consumption (Katz and Labuza,
1981). This extensive involvement of water in food processing and
stability has made it an essential focus of study from many directions
and for many years.

It has been recognized for some time that the actual water content
of a food is an imprecise indicator of stability (van den Berg and Bruin,
1981; Franks, 1982); rather it is the "nature," "state" or "availability"
of the water that determines its ability to participate in deteriorative
and other processes. What is needed is a precise measure of this
"availability."

Currently, the most widely used measure of this "availability" of
water in foods is water activity ($a_w$), which is related to the vapor
pressure or "volatility" of the water. However, despite this popularity
and the concept of $a_w$ as a precise indicator of biological viability and
quality of food products, several problems still exist. For example, the
substantiation of $a_w$ from thermodynamic principles is based on assumptions

*NMR Applications in Biopolymers,* Edited by
J. W. Finley *et al.,* Plenum Press, New York, 1990

that the system to be measured is at constant temperature and pressure and is in thermodynamic equilibrium (van den Berg and Bruin, 1981). However, food systems especially are in violation of the last assumption of thermodynamic equilibrium. Three of the major sources of this violation are: 1) the complexity of food systems (multicomponent and/or multiphase), 2) instability due to delayed crystallization (Makower and Dye, 1956; Karl, 1975; Saltmarch and Labuza, 1980; Flink, 1983) and 3) hysteresis (Kapsalis, 1981). Consequently, a true $a_w$, which by definition is an equilibrium value, does not exist in the system. Thus, what is often measured as $a_w$ is only a pseudo-water activity value.

Another area where the concept of $a_w$ leaves much to be desired is in relation to microbial viability. Usually the minimum $a_w$ for growth of microorganisms is reported as a range of values because the precise value depends on several additional parameters, such as, how the medium was prepared, i.e., adsorption or desorption (Kapsalis, 1981; van den Berg and Bruin, 1981; Franks, 1982), environmental conditions, i.e., temperature, pH and Eh (Leistner and Rodel, 1976; Troller, 1985), and chemical agents utilized to adjust the $a_w$ (Gould and Measures, 1977; Christian, 1981). Thus, a more precise measure of the "availability" of the water in a food is still much needed.

One of the most successful techniques to probe this "availability" aspect of water, as well as its quantity, in biological systems is Nuclear Magnetic Resonance (NMR) spectroscopy (Fuller and Brey, 1968; Eisenstadt and Fabry, 1978; Richardson et al., 1986). The extensive usefulness of NMR for studying water is attributed to two major factors. First of all, NMR is a rapid, sensitive, direct and, most importantly, a noninvasive, nondestructive instrumental technique. The only other noninvasive method for characterizing water in foods is dielectric relaxation. All other methods, such as vacuum oven and differential scanning calorimetry are destructive of the sample. This invasion of the sample causes serious problems with both the reliability and interpretation of the data (Richardson and Steinberg, 1987).

Secondly, steady advancements in NMR technology have expanded the avenues and techniques available to study water. NMR techniques for the study of water now range from determining the quantity of water present using proton ($^1H$) Pulsed NMR (Pande, 1975) to determining the detailed structure and dynamic characteristics of water using deuterium ($^2H$) and oxygen-17 ($^{17}O$) Pulsed Fourier Transform NMR (Walmsley and Shporer, 1978; Halle and Wennerstrom, 1981; Richardson et al., 1987a), $^1H$ and $^2H$ Pulsed Field Gradient NMR (Callaghan et al., 1983a; Callaghan and Lelievre, 1986) and NMR Imaging (Pykett et al., 1982; Rothwell et al., 1984; Rothwell, 1985).

Two basic types of NMR spectroscopy are available: wide-line (also called continuous wave) and pulsed. Essentially only the latter is used today because it is more versatile and powerful (Campbell and Dwek, 1984). Therefore, only pulsed NMR is considered here.

The noninvasive nature of NMR coupled with its ability to both quantitate and characterize water makes it an ideal tool for the investigation of water in foods on both theoretical and applied bases. The objectives of this review are to explain the theory of NMR techniques relevant to water and to present studies illustrating various applications of NMR to the investigation of water in foods and food constituents by pulsed NMR. Several other reviews discussing a wide variety of NMR uses in studying water in biological systems are available (Steinberg and Leung, 1975; Richards and Franks, 1977; Byrant, 1978; Mathur-De-Vré, 1979; Weisser, 1980; Nagashima and Suzuki, 1981; von Meerwall, 1983, 1985; Horman, 1984; Richardson and Steinberg, 1987; Saenger, 1987; Stilbs, 1987). The present review is limited to foods.

NMR OF WATER

Nuclear Magnetic Resonance (NMR) spectroscopy is based on the measurement of resonant, radio-frequency energy absorption by nonzero nuclear spins in the presence of an externally applied constant magnetic field ($B_o$). Only a charged atomic nucleus possesses a nonzero spin value (I); only such a nucleus has a magnetic moment ($\mu_n$) and can interact with an externally applied magnetic field. In the absence of such a magnetic field (no $B_o$), these nuclei are randomly oriented. However, when placed in a magnetic field, these nuclei become aligned either with the field or opposed to the field. These two orientations have different energies; the nucleus aligned with the field is at a lower energy than the one in the opposed position. The total number of aligned and opposed energy level positions is equal to (2I + 1). The population of nuclei at each energy level is dictated by the Boltzmann distribution. Water has four nuclei which possess a nonzero spin: proton ($^1H$; I = 1/2), deuterium ($^2H$ or D; I = 1), tritium ($^3H$; I = 1/2) and oxygen-17 ($^{17}O$; I = 5/2). For example, $^1H$ with I = 1/2 has a total of two energy levels; one energy level corresponding to nuclei aligned with the field (-1/2) and the other corresponding to nuclei opposed to the field (+1/2).

In addition to the interaction of nonzero spin nuclei with the magnetic field, nuclei with I > 1/2 ($^2H$ and $^{17}O$ for water) also possess an electric quadrupole moment (Q) which allows these nuclei to interact with the electric fields produced by neighboring electrons and nuclei (Campbell and Dwek, 1984). This quadrupole interaction is often very strong and therefore dominates the NMR behavior of both $^2H$ and $^{17}O$

nuclei. Several references discuss in detail NMR theory and techniques
(Abragam, 1961; Laszlo, 1983; Campbell and Dwek, 1984; Kemp, 1986;
Derome, 1987).

There are three basic variations of pulsed NMR: Pulsed NMR, with or
without Fourier Transform, Pulsed-Field Gradient NMR and Imaging NMR.
All three variations have been applied to study the behavior of water in
biological systems; each of these is discussed below. A brief outline of
the theory of each variation is presented followed by a detailed
discussion of its application to water.

PULSED NMR

Theory

In pulsed NMR the sample is placed in an applied magnetic field, $B_o$.
As discussed above, this causes the nuclei with nonzero spins to align
with or opposed to the applied $B_o$. Conventionally, the direction of $B_o$
is defined as the z axis of a set of Cartesian coordinates. The $B_o$ field
should be as homogeneous as possible so as to apply a uniform magnetization to each nucleus in the sample. To improve the effective $B_o$ field,
the tube which contains the sample is often spun about its vertical or z
axis (Campbell and Dwek, 1984). Next, the nuclei in the sample are then
irradiated with a radio frequency pulse ($B_1$) generated by a rotating
magnetic field at right angles to $B_o$. The frequency of this incoming $B_1$
is set to correspond to the resonant or Larmor frequency ($\nu$) of the
nucleus to be observed. We now have two magnetizations at right angles,
exerting a force on each other. The net result is that the spins of all
nuclei being probed become aligned in a single direction. The final
aligned direction is dependent on the time, length and strength of $B_1$.
For example, a 90° or $\pi/2$ pulse rotates the sample magnetization by 90°
from $B_o$ and a 180° or $\pi$ pulse rotates the sample magnetization by 180°
from $B_o$. The $B_1$ pulse is thus defined in terms of the angle of magnetic
rotation from $B_o$. After the pulse has been applied and shut-off, the
spin system returns to its equilibrium position (i.e., prepulse
condition) in the $B_o$ field through dephasing (i.e., losing its alignment
generated by the $B_1$ pulse) via two relaxation processes; longitudinal or
spin-lattice and transverse or spin-spin relaxation. Longitudinal
relaxation, $T_1$, is defined as the time constant that characterizes the
rate at which the z vector component of magnetization returns its
equilibrium value. Transverse relaxation, $T_2$, is defined as the time
constant that characterizes the rate of decay of the magnetization in the
x-y plane. Several authors present extensive discussions of both

relaxation processes (Berliner and Reuben, 1980; Martin et al., 1980; Kemp, 1986; Derome, 1987).

After the $T_1$ relaxation process has occurred, that is, after the equilibrium magnetization has been reestablished, the process may be repeated a number of times, so as to improve the signal-to-noise ratio. The induced voltage generated by the spin dephasing relaxation process is monitored in the x-y plane by a resonant radio frequency coil. The result is a plot of voltage intensity or magnetization amplitude against time called the free induction decay (FID). The FID is a complex super-imposition of the relaxation behavior from all relaxing nuclei. Since it is a measure of the NMR signal with time it is often said to be the time domain plot. However, since there is a reciprocal relation between time and frequency, the time domain FID signal can be converted to the frequency domain. This is done using the mathematical method known as Fourier Transformation (FT). The result is a plot of amplitude against frequency, which is the usual manner of reporting NMR data. This conversion from the time domain to the frequency domain for a spin echo pulse sequence is illustrated in Fig. 1.

When applying pulsed NMR to the study of water in foods, two questions need to be explored: 1) Which nucleus should be probed? and 2) What useful information can be extracted from the NMR spectra and how can it be interpreted?

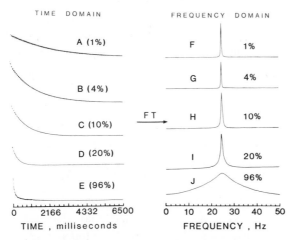

Fig. 1.  Fourier Transformation (FT) of time domain signals (A-E) to frequency domain spectra (F-J) for [1]H NMR (360 MHz) of lysozyme at concentrations ranging between 1 and 96%. (Adapted from Lioutas et al., 1987.)

<u>Nucleus To Be Probed</u>. As discussed above there are four possible water nuclei which may be probed: $^1H$, $^2H$, $^3H$ and $^{17}O$. Table 1 lists the important NMR values for these nuclei. The decision as to which nucleus should be probed is dependent on a variety of factors, scientific as well as economic. The majority of studies which were done 15 to 20 years ago used low field $^1H$ NMR. However, since that time, problems with the interpretation of $^1H$ NMR relaxation data from macromolecular systems at both low and high fields (small and large $B_o$) were found. The major concern is the contribution of different relaxation mechanisms to the line width of the water peak in the $^1H$ NMR Fourier transform spectra of such complex systems. This concern has been studied by a number of investigators: Kalk and Berendsen (1976), Edzes and Samulski (1978), Koenig et al. (1978) and Peemoeller et al. (1984). Two relaxation mechanisms have been identified as significant contributors to the line width of the water peak: cross-relaxation and proton exchange.

Cross-relaxation is caused by dipole-dipole interactions between the protons of the water and the protons of the macromolecule, resulting in a <u>transfer or exchange of magnetization</u>. On the other hand, proton exchange is the <u>physical exchange</u> of protons between distinct states of water, i.e., "bound" and "free" water. This is also called chemical exchange. These two processes are schematically illustrated in Figs. 2a and 2b.

Overall the affect of both processes is the same: a distortion of the true relaxation value. In cross-relaxation the relaxation rates of all the coupled protons tend to become equal (Berendsen, 1975); this results in a decrease in the apparent relaxation time. This shorter relaxation time causes the water to appear to be bound to a greater

Table 1. Important NMR Values for the Four Water Nuclei

| Isotope | Spin (I) | Natural Abundance (%) | Frequency at $B_o$ = 2.35 Telsa (MHz) | Sensitivity Relative to Proton |
|---|---|---|---|---|
| $^1H$ | 1/2 | 99.98 | 100.000 | 1.00 |
| $^2H$ | 1 | $1.5 \times 10^{-2}$ | 15.351 | $9.65 \times 10^{-3}$ |
| $^3H$ | 1/2 | 0 | 106.663 | 1.21 |
| $^{17}O$ | 5/2 | $3.7 \times 10^{-2}$ | 13.557 | $2.91 \times 10^{-2}$ |

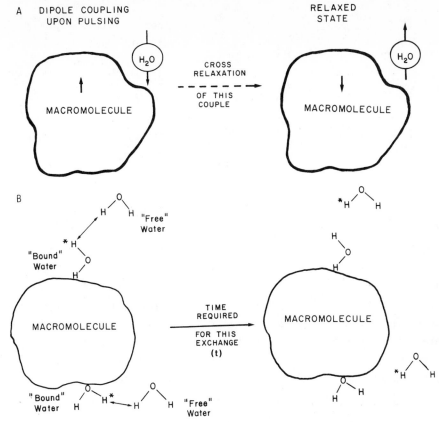

A DIPOLE COUPLING UPON PULSING

RELAXED STATE

CROSS RELAXATION

OF THIS COUPLE

$H_2O$

MACROMOLECULE

$H_2O$

MACROMOLECULE

B

"Free" Water

"Bound" Water

MACROMOLECULE

TIME REQUIRED

FOR THIS EXCHANGE (t)

MACROMOLECULE

"Bound" Water

"Free" Water

Fig. 2a. Schematic representation of cross-relaxation. Solid arrows represent the proton magnetic dipole moments from the macromolecule and the water components. Coupled moments cross-flip as the water proton transfers its magnetization energy to the macromolecular proton. (From Richardson and Steinberg, 1987.)

2b. Schematic representation of chemical exchange between water molecules. There is a physical exchange between a free water proton (H) and a proton of water bound to the macromolecule (*H). Time t is the time required for this physical exchange to occur. (From Richardson and Steinberg, 1987.)

degree than it actually is in the system being measured. In chemical exchange, several situations may be encountered. For example, if we have only two populations of water, "bound" and "free," with different relaxation times but their protons physically exchange at a rate <u>greater than</u> that of the NMR frequency (i.e., the time required for exchange is very short), then the observed relaxation rate is a weighted average over both water populations. The resultant spectrum has only one peak. However, if the protons physically exchange at a rate <u>less than</u> that of the NMR frequency (i.e., the time required for exchange is long), then the observed relaxation rate for each water population can be separated and the resultant spectrum has two peaks. Thus, the magnitude of the effect of chemical exchange is dependent on several interrelated factors, three of which are: 1) the rate of exchange between water populations, 2) the number of nuclei in each population and 3) the intrinsic relaxation rates of nuclei in each population (Berliner and Reuben, 1980). Chemical exchange can also occur between water states and available prototrophic residues on the macro-molecule (Halle et al., 1981). Since a large quantity of water is usually present, the exchange between water states usually dominates chemical exchange.

These concerns with $^1$H NMR along with the steady advances in NMR technology have engendered an interest in probing the other three water nuclei.

The application of both $^2$H and $^{17}$O nuclei for NMR water investigations have been most fruitful (Halle and Wennerstom, 1981; Halle et al., 1981; Laszlo, 1983; Lioutas et al., 1986; Richardson et al., 1986). To date $^3$H has not received much attention but remains a possible nucleus for explora-tion of radioactive effects of tritiated water. Information regarding $^3$H NMR is given by Mantsch et al. (1977), Brevard and Kintzinger (1978) and Mathur-De-Vré et al. (1982).

Since both $^2$H and $^{17}$O nuclei are quadrupolar, their relaxation pro-cesses are dominated by electric field gradient interactions (Kemp, 1986) and are not affected by the problem of cross-relaxation experienced by the $^1$H nuclei (Halle et al., 1981). However, $^2$H nuclei still experience chemical exchange.

From this brief discussion, it is seen that the $^{17}$O nucleus has important advantages over the other nuclei. These advantages are further discussed by Halle et al. (1981). One concern with $^{17}$O NMR is that the $^{17}$O relaxation is influenced by proton exchange broadening which is the result of spin-spin coupling between $^1$H and $^{17}$O nuclei. Because of the large quadrupole coupling constant, this spin-spin coupling of oxygen to protons is seldom resolved into separate $^{17}$O peaks, but manifest itself as extensive broadening of the existing $^{17}$O peak(s). This broadening has been shown to affect only the transverse ($T_2$) relaxation process (Glasel, 1972) and its magnitude is dependent on pH (Meiboom, 1961; Rabideau and Hecht, 1967; Kintzinger and Marsmann, 1981; Halle and Karlstrom, 1983; Richardson, 1989). Figure 3 illustrates the effect of pH on $T_2^{-1}$ (or $R_2$) of water. At neutral pH the proton exchange is slow and $T_2^{-1}$ is approximately three times larger than at either acidic (pH < 5.5) or basic (pH > 8.5) conditions. In using $^{17}$O NMR to study water mobility, it is important to decrease or eliminate these spin-spin coupling

422

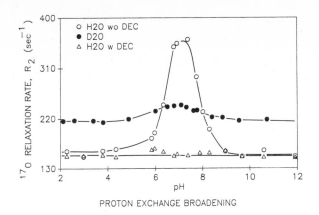

Fig. 3.  Dependence of $^{17}O$ transverse relaxation rate ($R_2$) on pH in $H_2O$ without (wo) $^1H$ decoupling ($\circ$), in $D_2O$ ($\bullet$) and in $H_2O$ with (w) $^1H$ decoupling ($\blacktriangle$).  (From Richardson, 1989.)

effects, so as to accurately determine $T_2$ relaxation times.  As can be seen in Fig. 3, the coupling effect is decreased by the use of deuterium oxide ($D_2O$) as a solvent (Halle et al., 1981; Lioutas, 1984; Richardson et al., 1986) and is eliminated by the use of proton decoupling with the instrument (Earl and Niederberger, 1977; Richardson, 1989).

The importance of eliminating proton exchange broadening in order to accurately measure $T_2$ values obtained by standard linewidth at half-height measurements is illustrated in Fig. 4.  The $^{17}O$ NMR transverse relaxation rate [$T_2^{-1}$ or $R_2$ ($sec^{-1}$)] with and without $^1H$ decoupling with increasing instant starch concentration (Dura-Jel) in water (pH = 5.9 to 6.3) are compared.  As can be seen, there is a dramatic difference between the observed $^{17}O$ $R_2$ with $^1H$ decoupling and that without $^1H$ decoupling.  Without decoupling, for all starch concentrations, $R_2$ was less than that for water alone.  This would mean that the water alone showed less mobility than the water associated with the instant starch.  However, with $^1H$ decoupling, the $R_2$ increased with increasing concentration, showing that the mobility of the water actually decreased with increasing starch concentration.  The $^1H$ decoupling eliminated the proton exchange broadening, allowing the $R_2$ values to be accurately determined and in turn correctly interpreted.

Principles of $^{17}O$ NMR are discussed in detail elsewhere (Rodger et al., 1978; Kintzinger and Marsmann, 1981; Kintzinger, 1983; Kemp, 1986).

Extraction and Interpretation of NMR Data.  Four parameters are usually extracted from pulsed NMR spectra:  1) chemical shift ($\delta$), 2) spin-spin coupling (J), 3) area or intensity of the signal and 4) the two relaxation times ($T_1$ and $T_2$).  For studying water, the two relaxation times are used most often.  Both $T_1$ and $T_2$ can be determined by multipulse experiment

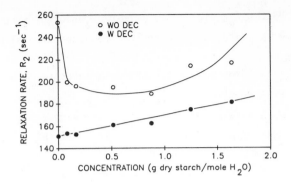

Fig. 4.   Dependence of $^{17}O$ NMR transverse relaxation rate ($R_2$) in
$H_2O$ on instant starch concentration (Dura–Jel), without
(wo) $^1H$ decoupling ($\circ$) and with (w) $^1H$ decoupling ($\bullet$).

methods.   Some of the more common methods include:  Inversion–recovery
Fourier transform and Progressive–saturation Fourier transform for $T_1$;
Carr–Purcell, Carr–Purcell–Meiboom–Gill (CPMG) and Spin–echo for $T_2$.   These
and other methods are reviewed by Martin et al. (1980).  Besides these
multipulse methods, $T_2$ is often estimated by the linewidth at half–height
($\Delta\nu_{1/2}$; in Hz) times $\pi$ and is designated as $T_2^*$ (Derome, 1987).  However,
$T_2^*$ is often reported as $T_2$.  The experimental procedures should state
clearly how $T_2$ was obtained, by multipulse or linewidth methods.  The
linewidth at half–height can be obtained from the NMR spectrum of a single
pulse or a multipulse experiment.  $T_2^*$ is the sum of the pure $T_2$ plus all of
the inhomogeneities in the sample and magnet.  Therefore, how close $T_2^*$ is
to $T_2$ depends on the magnitude of these inhomogeneities (Kemp, 1986).

T_1 and $T_2$ relaxation times can be used individually or can be used
together to calculate the rotational correlation time, $\tau_c$ (Halle et al.,
1981; Kumosinski and Pessen, 1982).  There is no general solution to the
relation between $T_1$, $T_2$ and $\tau_c$.  Specific mathematical relationships must
be developed for each mechanism of spin relaxation and for each nucleus
probed (Kuntz and Kauzmann, 1974).

For model or simple systems, the detailed quantum mechanics have
been elucidated.  However, as the complexity of the system under investigation increases so does the complexity of the relation between $T_1$, $T_2$
and $\tau_c$.  Thus, when studying the behavior of water in complex systems,
such as food, it is often necessary to introduce some simplifying
assumptions along with basic interpretive modeling.  Finney et al. (1982)
caution that one consequence of having to use models to interpret NMR
data from complex systems is that the results obtained are very dependent
on the model used to interpret the data.  Two different models used to

analyze the same data could yield very different results! Thus, great
care must be taken when drawing conclusions from experimental data.
Since a recent review (Richardson and Steinberg, 1987) includes a
comprehensive discussion of models for NMR data interpretation, only a
brief discussion will be given here.

The two-(or more) state fast exchange model has been the major model
used to interpret NMR data (Finney et al., 1982). If the exchange times
between the states (i.e., "bound" and "free") is faster than the NMR time
scale, then the relaxation behavior yields a single relaxation time:

$$T_{obs}^{-1} = \sum_i P_i T_i^{-1} \tag{1}$$

where $T_{obs}^{-1}$ is the observed relaxation time, $T_i^{-1}$ is the relaxation time
of the $i_{th}$ state, and $P_i$ is the probability that the nucleus is found in
that state (Cooke and Kuntz, 1974). If we take the simplest case of two
possible states of water, let us say "bound" and "free" water, then
Eq. (1) has been shown to simplify to (Derbyshire, 1982; Richardson and
Steinberg, 1987):

$$T_{obs}^{-1} = P_B\left(T_B^{-1} - T_F^{-1}\right) + T_F^{-1} \tag{2}$$

where $T_B^{-1}$ is the relaxation rate of the bound water, $P_B$ is the probability of water being bound, $T_F^{-1}$ the relaxation rate of the free water, and
$P_F$ is the probability of water being free (Note: $P_B + P_F = 1$). Thus,
the two-state model with fast exchange, predicts a linear relationship
between the observed relaxation rate ($T_{obs}^{-1}$) and the probability of the
water being in the bound state ($P_B$), where $P_B$ can be related to concentration. Estimated values of $P_B$ are often made using sorption isotherm
data and related isotherm equations (Lioutas et al., 1986; Richardson et
al., 1986). It has been found that the two-state model with fast
exchange often holds for dilute solutions (Derbyshire, 1982); however at
higher concentrations deviations from linearity have been observed
(Woodhouse, 1974; Halle et al., 1981; Richardson et al., 1986). Three
mechanisms to account for this departure from linearity have been
proposed: 1) a change in hydration number (Derbyshire, 1982), 2) a
change in relaxation rate of the bound water ($T_B^{-1}$) (Derbyshire, 1982)
and 3) charge repulsion or charge fluctuation (Kumosinski and Pessen,
1982).

The major application of the third mechanism is the use of activities in place of concentrations when dealing with systems which strongly
deviate from ideality. Kumosinski and Pessen (1982) applied their theory
to β-lactoglobulin A and found that plotting the relaxation rate against
the activity of β-lactoglobulin A resulted in a linear relationship

whereas plotting against the concentration of β-lactoglobulin A was curvilinear (i.e., deviated from linearity). The Kumosinski Model has been applied to a variety of other systems such as lysozyme with and without salt (Baianu, 1989), wheat flour suspensions (Richardson et al., 1986) and milk proteins (Farrell et al., 1987).

## Applications

Pulsed NMR has been applied to water dynamics and structure in a variety of systems from pure water, to biological systems, to food components and complex food systems (Glasel, 1972; Hansen, 1976; Bryant, 1978; Mathur-De Vre, 1979; Leung et al., 1979, 1983; Finney et al., 1982; Richardson et al., 1987b). The scope of this research ranges from very basic to applied. Since the majority of the basic as well as some of the applied research has been recently reviewed (Richardson and Steinberg, 1987), only the applied research will be covered here.

Protein - Muscle. Pulsed NMR relaxation times have been used to study both isolated muscle protein (Nakano and Yasui, 1979; Yasui et al., 1979) and intact muscles (Lillford et al., 1980a, 1980b; Currie et al., 1981; Suzuki, 1981). Nakano and Yasui (1976) observed a dramatic difference in myosin denaturation during dehydration depending on the presence or absence of sucrose. Without sucrose, the extent of myosin denaturation increased with decreasing water activity, until the multilayer region was reached, and then it decreased with further decreasing water activity down to the monolayer region. With addition of sucrose, denaturation was completely inhibited. In order to further investigate these findings, Nakano and Yasui (1979) measured the $^1$H NMR $T_1$ and $T_2$ relaxation times of water in myosin suspensions, with and without 0.1 M sucrose, during dehydration. However, despite the dramatic difference observed in their previous study (Nakano and Yasui, 1976), the mobilities, as well as the populations of water protons calculated from the $T_1$ and $T_2$ measurements, showed little differences between the presence and the absence of sucrose over the entire moisture range studied (91 to 6% moisture); the only exception was that $T_1$ in the presence of sucrose demonstrated smaller values in the multilayer region (10 to 40% moisture) than those without sucrose. The authors did not draw any definitive conclusions, but stated that further studies were needed in order to correlate the protein denaturation to the state of the water associated with the protein molecules.

Yasui et al. (1976) compared the $^1$H $T_2$ relaxation times of myosin solutions (10 mg/ml) before and after gelation at two temperatures (50 and 60°C) and at two pH (6.0 and 7.0). All of the $T_2$ relaxation plots were nonexponential, indicating multiphase behavior, i.e., more than one species of water, and the exchange rate between the different species was slow compared to their relaxation times. The water specie with the longest relaxation time, i.e., the most mobile fraction was examined. It was found that the mobility of this water specie was more restricted after gelation than before gelation and that the degree of this restriction was greater in the gel formed at pH 6.0 than in that formed at pH 7.0. This general decrease in mobility after gelation was supported by morpho- logical observations from scanning electron micrographs which showed that gelation resulted in an extensive network structure. These structures could trap the bulk water, rendering it less mobile.

Changes in water structure in postmortem rat and beef muscle was done by Currie et al. (1981) using [1]H NMR $T_1$ relaxation values. Figure 5 is typical of the results obtained for $T_1$ values as a function of time postmortem for the rat gastrocnemius muscle. However, the numerical changes in $T_1$ varied from rat to rat: in the five rats studied the initial rise in $T_1$ from one hour postmortem to the peak of rigor varied from 5 msec to 31 msec and the negative slope of the $T_1$ values following the peak at rigor produced different regression coefficients, ranging from –0.36 msec/hr. to –3.69 msec/hr. A significant relationship (correlation coefficient –0.95) between the initial rise in $T_1$ and the negative slope (post rigor) for each rat was found. If the initial rise in $T_1$ was large the negative slope was minimal (i.e., little change in $T_1$ with time past rigor). This relationship

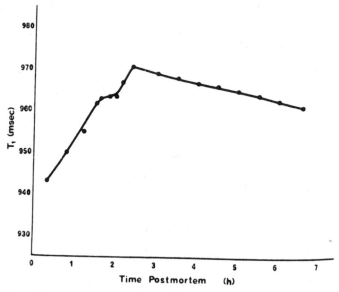

Fig. 5. Plot of $T_1$ (msec) vs time postmortem (hours) for rat gastrocnemius. (From Currie et al., 1981.)

led the authors to suggest that once the factors contributing to the highly structured state of water in the living organism have lost their influence, then the restructuring of the available water following the extremes during rigor becomes minimal, but was highly influenced by prerigor events. It was also reported that the greater the initial rise in $T_1$ and the lower the negative slope post rigor, the more rapid the drop in pH after slaughter. The beef muscle tested exhibited similar behavior to the rat muscle, with the initial rise in $T_1$ varying from 8.5 msec to 70 msec and the regression coefficients (post rigor) varying from -0.22 msec/hr. to -6.0 msec/hr.

Protein - Vegetable Proteins. Two $^1$H NMR methods were used to probe the state of water in soy protein concentrate (Hansen, 1976). The first method measured $T_2$ relaxation time of water in hydrated protein samples; $T_2^{-1}$ values were plotted against solids content (g solids/g water). The resultant plot consisted of two linear segments; the intersection was evaluated as the bound water value (0.26 g water/g solids). In the second method, a bound water value was obtained by measuring the water $^1$H NMR signal amplitude as a function of decreasing temperature. Here, the NMR bound water value corresponded to the amount of water which did not freeze at temperatures below 0°C, where the amount of water was determined from the amplitude of the water NMR $^1$H signal. The value for the soy concentrate was 0.26 g water/g solids at -50°C. These NMR results were compared to both the BET (Brunauer et al., 1938) and Bradley (1936) isotherm equations for characterization of water.

Soya protein fibers (composed of protein, water, oil and salt) were studied by $^1$H NMR (Lillford et al., 1980a). Nonexponential relaxation was detected for both $T_1$ and $T_2$ relaxation times; they developed a theory to explain these nonexponential processes.

Water in wheat flour doughs and breads has been investigated using $^1$H, $^2$H and $^{17}$O NMR (Leung et al. 1979, 1983; Richardson et al., 1986). Leung et al. (1979) determined the $^1$H NMR $T_2$ relaxation time of water protons in wheat flour doughs made with both hard and soft wheat flour. They reported nonexponential decay of the $^1$H signal for each dough tested and resolved the relaxation curves into two components; a long and a short component. The long component represented the more mobile water fraction with a $T_2$ of about 60 nsec, whereas the short component represented the less mobile water fraction with a $T_2$ of about 20 nsec. This short component accounted for about 0.62 g water/g solid. They observed only a small variation in the magnitude of these two components throughout the moisture range of 0.64-0.93 g water/g solid and concluded that the mobility of each water fraction, as measured by $^1$H NMR, did not change appreciably with water content. These observed $T_2$ values appeared to be independent of both flour strength (hard vs. soft) and mixing time.

In a subsequent study, Leung et al. (1983) used $^2$H NMR to study water in both doughs and breads. For the doughs, $^2$H NMR $T_1$ and $T_2$ values were measured as a function of both wheat flour type (hard vs. soft) and moisture content. Compared to their $^1$H study where two water fractions were resolved, only one water fraction was able to be resolved; they attributed this to the different relaxation mechanisms for $^1$H and $^2$H. Hard and soft wheat flour doughs showed similar $^2$H $T_1$ and $T_2$ values with increasing moisture content. For the bread, $T_1$ and $T_2$ values were measured as a function of storage time. The results showed a decrease in both $T_1$ and $T_2$ with time, indicating an overall decrease in water mobility and an increase in water binding during the staling process.

Richardson et al. (1985) tested the hypothesis that the rheological properties of a wheat flour–water systems are related to the mobility of its water components as measured by $^{17}$O NMR $T_2$ relaxation time. They reported an inverse relationship between the consistency coefficient of the power law equation and water mobility as measured by $T_2^{-1}$. Three distinct regions were observed, each exhibiting a linear relationship with water mobility decreasing as apparent viscosity increased. No overall relation between the $T_2^{-1}$ and apparent viscosity was observed.

Richardson et al. (1986), employing both $^1$H and $^{17}$O high–field NMR $T_2$ measurements, studied water mobility in wheat flour suspensions and doughs (30–95% moisture). The isotropic two–state model with fast exchange was used to interpret these data by means of the Derbyshire (1982) and Kumosinski (Kumosinski and Pessen, 1982) models. As expected, the two nuclei yielded different absolute relaxation rates due to their different relaxation mechanisms, but showed the same trends in their dependence of $T_2^{-1}$ on increasing flour concentration in both water and deuterium oxide. From the $^{17}$O NMR data, a correlation time of 16.7 psec was calculated for the water "bound" by the wheat flour. This value is in good agreement with the $^{17}$O NMR correlation times calculated for lysozyme (Lioutas et al., 1986) and other proteins (Halle et al., 1981).

Protein – Milk. Brosio et al. (1983) used low resolution (20 MHz) $^1$H NMR to study the hydration mechanism of powdered milk samples. The spin–echo decay curve ($T_2$) was found to be nonexponential. This nonexponentiality was accounted for by the presence of three proton populations: $T_{2S}$ (solid components), $T_{2B}$ (bound water) and $T_{2F}$ (free water). Based on this three component analysis, both the relaxation times and the relative abundance of each population were calculated for milk samples at concentrations from 0.10–1.80 ml water/g solids. The $T_{2S}$ values at all concentrations were almost the same, ranging from 0.031–0.043 msec. However, both $T_{2B}$ and $T_{2F}$ values varied with increasing concentration.

Lelievre and Creamer (1978) used low field (20 MHz) [1]H NMR $T_1$ and $T_2$ relaxation times to study the formation and syneresis of renneted milk gels. Simple exponential decays of nonequilibrium magnetization were found for all samples, except those gels that had undergone syneresis. The isotropic two-state (bound and free) model with fast exchange was used to interpret the exponential relaxation results. When the gels exhibited visible syneresis of whey, two relaxation times were observed. One relaxation time was associated with the protons in the gel while the other was due to the protons in the liquid expelled from the gel phase. Measurements of the NMR signal attenuation on freezing to determine the bound water showed 0.3 g water bound to each gram of dry casein micelles. There was no change in this bound water value after syneresis.

Samuelsson and Hueg (1973) developed a method for measuring the rate of solution of dried milk using proton NMR. This method was based on the curve-linear relation between the [1]H NMR signal and the percent dry matter (e.g., milk powder) in solution.

Protein - Egg.  The [1]H NMR $T_1$ and $T_2$ relaxation times and self-diffusion coefficient of water in hen egg white and yolk were investigated by James and Gillen (1972). All $T_1$ and $T_2$ values exhibited single exponential behavior. The average $T_1$ and $T_2$ values for egg yolk (0.067 sec and 0.024 sec, respectively) and egg white (1.185 sec and 0.44 sec, respectively) were significantly less than that measured for distilled water (2.83 sec for $T_1$; $T_2$ was not measured but was assumed to be the same as $T_1$).

Goldsmith and Toledo (1985) monitored the gelation of egg albumin with [1]H NMR $T_1$ relaxation times. $T_1$ measurements were taken as a function of heating time and temperature for a 10% egg albumin dispersion. For each temperature (60-90°C), $T_1$ declined rapidly at first (first 5 mins of heating), then leveled off asymptotically. The asymptotic value of $T_1$, as well as the heating time required to reach it, decreased as the heating temperature increased. The effect of heating time and temperature on gel strength was also measured and gel strength was related to $T_1$; they reported a high negative correlation between these parameters.

Protein - Other.  Other protein containing systems studied included casein (Leung et al., 1976; Lang and Steinberg, 1983) collagen (Migchelsen and Berendsen, 1973; Hoeve, 1980; Renou et al., 1983; Grigera and Bienkiewicz, 1984; Maquet et al., 1984) and protein-based sols and gels (Lambelet et al., 1988).

Carbohydrates - Simple Sugars.  The water associated with simple
sugar carbohydrates has been studied by $^1$H, $^2$H and $^{17}$O NMR.  Harvey and
Symons (1976, 1978) used $^1$H NMR (100 MHz) to investigate the hydration of
several monosaccharides; glucose, mannose, galactose, ribose, sorbose and
fructose.  Based on chemical shifts as a function of temperature and the
effect of added dimethyl sulfoxide, they concluded that each hydroxyl
group of the monosaccharide bonds an average of two water molecules.
Additional proton studies of simple sugars include proton exchange and
hydration studies of glucose, mannose and galactose (Bociek and Franks,
1979), and low-field (10 MHz) $T_1$ and $T_2$ relaxation times of sucrose,
fructose and corn syrup (Mora-Gutierrez, 1987).

Tait et al. (1972c) used $^{17}$O NMR to study $T_2$ relaxation times of
water associated with glucose, ribose and mannose as a function of
temperature.  Water-sugar systems were adjusted with p-hydroxybenzoic
acid to pH 3.5 to inhibit line broadening due to $^{17}$O-$^1$H spin-spin
coupling.  The $^{17}$O NMR $T_2$ relaxation times, as measured from the
linewidths, increased in monosaccharide solutions compared to pure water
and were greater for hexose solutions than pentose and ribose solutions.
The $^{17}$O NMR $T_2^{-1}$ of glucose with increasing concentration as a function
of temperature was also measured.  $T_2^{-1}$ increased with increasing
concentration in a nonlinear manner.  This nonlinear behavior, however,
became more linear with increasing temperature from 10°C to 80°C.

Suggett et al. (1976), studying molecular motion and interactions in
several mono- and di-saccharide solutions, employed $^{17}$O NMR relaxation
for the solvent and $^1$H, $^2$H, $^{13}$C and $^{17}$O relaxation for the solute.  These
NMR data were used to test alternative models for the resolution of
complimentary dielectric relaxation rates (Suggett, 1976).

Richardson et al. (1987c) determined the mobility of water in
sucrose solutions ranging in concentration from 5-80% sucrose by $^2$H and
$^{17}$O NMR $T_2$ relaxation times.  Figure 6 shows the dependence of the $^{17}$O
$R_2$ ($R_2 = T_2^{-1}$) in deuterium oxide on increasing sucrose concentration.
Four regions of water mobility were observed.  Region I, from 5-40%
sucrose, exhibited a linear behavior of $R_2$ with concentration and was
described by the isotropic two-state model with fast exchange.  Region
II, from 40-60% sucrose showed nonlinear behavior.  This deviation from
linearity was hypothesized to be due to the formation of intermolecular
hydrogen bonds between water and sucrose, hydrogen bond bridging of water
between sucrose molecules, and sucrose-sucrose hydrogen bonding.  Region
III, from 60-80% sucrose, showed a slight rise in $R_2$ between 60 and 67%
sucrose; thereafter it shows a horizontal line that represents a mixture

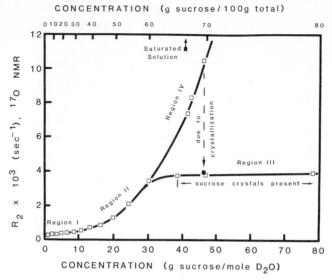

Fig. 6.   Effect of dissolved and crystalline sucrose content in $D_2O$ on the [17]$O$ NMR transverse relaxation rate ($R_2$).   (From Richardson et al., 1987c.)

of sucrose crystals in a saturated sucrose solution.  Region IV was a smooth continuation of Region II and represents samples in the supersaturated condition.  This continued increase in the $R_2$ response for the supersaturated condition was attributed to the increased formation of the hydrogen bonding structures begun in Region II.  The most supersaturated solution (last point in Region IV) was allowed to crystallize and the [17]$O$  NMR $R_2$ remeasured. The $R_2$ value shown by the closed box at the same sucrose concentration fell on the line for Region III, a saturated solution with crystals.  This confirmed that the mobility of water in a supersaturated-sucrose solution differs markedly from that in the corresponding saturated solution with crystals.

Lai and Richardson (1989) investigated the crystallization of lactose over time and water activity in skim milk powder by [2]$H$ NMR.  Figure 7 shows a plot of moisture gain or lost (g water/100 g solid) and [2]$H$ NMR $R_2$ for dried skim milk powder at 0.54 water activity against time.  Both the moisture content and [2]$H$ NMR $R_2$ showed a significant change with time.  The decrease in $R_2$ corresponded to an increase in the mobility of water molecules in the system.  The authors ascribed these changes in the mobility of the water to the transition in the lactose from the amorphous to crystalline state.  As the system absorbed water from the saturated salt solution ($a_w$ = 0.54) an increase in water mobility was observed (decrease in $R_2$).  This water then could act as a solvent for reorientation of the

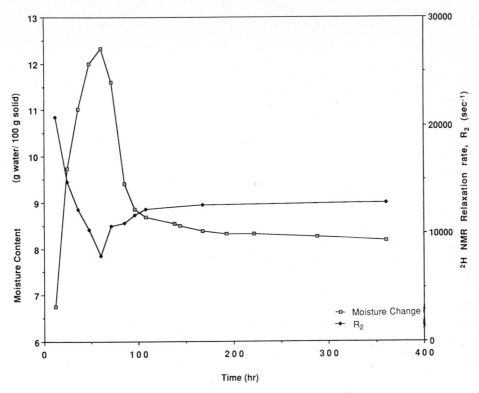

Fig. 7. Moisture change (g water/100 g solid) and $^2$H NMR transverse
relaxation rate ($R_2$, sec$^{-1}$) plotted against time for skim
milk powder at water activity 0.54 at 20°C. (From Lai and
Richardson, 1989.)

lactose molecules, providing for their transition from the amorphous to
crystalline form. As the reorientation of the lactose molecules proceeded,
a portion of the initially absorbed water was released and reabsorbed by the
saturated salt solution (increase in $R_2$ after 2.5 days). As equilibrium of
the system was obtained, the $R_2$ leveled off to a constant value of 12,800
sec$^{-1}$.

    Carbohydrates - Starches. NMR has been extensively applied to several
aspects of starch hydration (Tait et al., 1972a; Lechert and Hennig, 1976;
Lechert et al., 1980; Lechert, 1981; Richardson and Steinberg, 1987).
Studies have been done on native starch-water systems (Tait et al., 1972b;
Hennig and Lechert, 1974; Leung et al., 1976; Hennig, 1977; Henning and
Lechert, 1977; Lechert, 1981; Schwier and Lechert, 1982; Lang and Steinberg,
1983; Richardson et al., 1987a, 1987d), starch gelatinization (Jaska, 1971;
Lelievre and Mitchell, 1975; Nakazawa et al., 1980; Lechert, 1981; Callaghan
et al., 1983b) and starch retrogradation (Nakazawa et al., 1983, Richardson,
1988a). Leung et al. (1976) investigating corn starch found that $^1$H NMR $T_1$

values showed single exponential behavior, while $T_2$ values showed single exponential decay for samples containing less than 45% water (dry basis) and multiphase behavior at higher moisture contents. These relaxation rates were plotted against water activity; $T_1$ was reported to be independent of $a_w$, while $T_2$ increased exponentially with $a_w$. Richardson et al. (1987a), investigating the mobility of water in corn starch powders using both $^2H$ and $^{17}O$ NMR, found $R_2$ for both nuclei to be linear against water activity. In case of $^{17}O$ NMR data, this linear behavior extended from $a_w$ 0.97 to 0.80, below which the $^{17}O$ NMR measurement could not be made because of bandwidth limitations on the spectrometer used. In case of the $^2H$ NMR data, there were two linear regions: Region A, $a_w$ 0.99 to 0.23, and Region B, $a_w$ 0.23 to 0.11. This linear relation in Region A was hypothesized to be due to the hydrophobic nature of the starch granules, which results in diffusion limited water motion. The intersection between Regions A and B was hypothesized to represent the end of the monolayer.

Callaghan et al. (1983b) measured $^1H$ NMR $T_1$ and $T_2$ relaxation rates as well as $^{13}C$ NMR spectra and water diffusion coefficients of gelatinized wheat starch pastes. These pastes were prepared by heating starch-water suspensions at 95°C for 1 hour. Both $T_1$ and $T_2$ relaxations were characterized by single time constants. These results were interpreted by the isotropic two-state model with fast exchange. Pastes which differed markedly with respect to their rheological properties showed no significant variations in their $T_1$ or $T_2$ values.

The retrogradation behavior of gelatinized glutinous and non-glutinous rice starch and nonglutinous rice at 3°C was studied by $^1H$ NMR $T_1$ and $T_2$ relaxation times by Nakazawa et al. (1983). Changes in $T_1$ and $T_2$, the correlation time, and the fraction of bound water during the retrogradation process were small for the nonglutinous rice starch and rice. However, a significant change in water behavior was observed during retrogradation of the glutinous rice starch. The most dramatic changes occurred around 1 g water/g dry matter. For example, the fraction of bound water increased from 0.1 to 0.23 and the correlation time from $5 \times 10^{-8}$ to $12 \times 10^{-8}$ sec at three days after gelatinization.

Richardson (1988a) examined the molecular mobilities of three instant starch gels using high-field $^1H$ decoupled $^{17}O$ and $^{13}C$ NMR as affected by concentration and storage conditions. The mobility of water was studied by $^{17}O$ NMR and the mobility of the carbohydrate polymer by $^{13}C$ NMR. The $^1H$ decoupled $^{17}O$ NMR $R_2$ was found to be linear with increasing instant starch concentration from a thin suspension (0.5% starch, wet basis) to a very thick paste (9.0% starch, wet basis) for two of the instant starches: the pregelatinized instant starch (Dura-Jel) and the crosslinked and substituted alcohol modified instant starch

(Mira-Thik). The third instant starch, which was alcohol modified (Mira-Gel), exhibited a linear relationship between the $^{17}O$ NMR $R_2$ and concentration up to approximately 5% starch, after which a departure from linearity was observed. The 5% starch samples of all three starches were subjected to three storage conditions: 1) room (23 $\pm$ 2°C) and 2) low (4 $\pm$ 1°C) temperature storage for 6 and 12 days and 3) a six-cycle freeze-thaw treatment. Every starch under each storage condition, except one, showed an increase in the $^{17}O$ NMR $R_2$ over the $R_2$ of the samples measured fresh, despite visually observable differences in the degree of retrogradation. No specific trends in the $^{17}O$ NMR $R_2$ were observed for either time or temperature of storage.

The overall attributes of the $^{13}C$ NMR spectra for both Dura-Jel and Mira-Gel starches after both room- and low-temperature storage remained similar to that of the fresh spectrum (Fig. 8), despite the fact that the Dura-Jel starch did not exhibit any syneresis, while the Mira-Gel showed extensive syneresis. The authors concluded that there remained a sufficient number of water molecules surrounding (i.e., hydrating) the starch chains in the retrograded state to retain the same mobility as in the soluble gel-state. Similar findings for amylodextrin were reported by Jane (1985). It was found that the $^{13}C$ NMR spectrum of retrograded amylodextrin was nearly identical to that of amylodextrin in solution. Jane (1985) attributed this similarity in spectra to little conformational change about the glycosidic bond between the two states, as well as sufficient hydration of the amylodextrins in both solution and retrograded conditions.

On the other hand, the $^{13}C$ NMR spectra for the freeze-thaw samples (Dura-Jel, Mira-Jel and Mira-Thik) showed two basic patterns (Fig. 9). For the starch that did not exhibit freeze-thaw stability (Mira-Gel), a large change in the spectrum between fresh and freeze-thawed conditions was observed. However, for the two starches which did exhibit freeze-thaw stability (Dura-Jel and Mira-Thik), no changes in the $^{13}C$ NMR spectra were observed. The observed changes in the nonfreeze-thaw stable starch were attributed to polymer conformational and chemical changes during freezing, giving rise to the observed spectral changes in all of the carbon environments except carbon 6 (located at 63.05 ppm).

Carbohydrates - Other Complex. Other complex carbohydrates studied by NMR include glycogen (Brittain and Geddes, 1978), cellulose (Hsi et al., 1979), agar (Woessner et al., 1970; Labuza and Busk, 1979), agarose (Child and Pryce, 1972; Derbyshire and Duff, 1973; Ablett et al., 1976; Lillford et al., 1980a) and carrageenan (Ablett et al., 1976; Labuza and Busk, 1979).

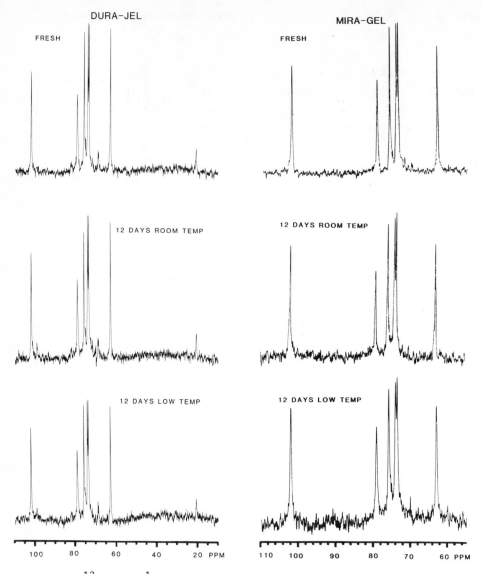

Fig. 8. $^{13}$C fully $^{1}$H decoupled NMR spectra of Dura–Jel and Mira–Gel corn starches at 5% concentration fresh, and after storage at room temperature (23°C) and low temperature (4°C) for 12 days. (From Richardson, 1988a.)

Carbohydrates - Simple and Complex Mixtures. Lang and Steinberg (1983) employed low field $^{1}$H NMR $T_1$ and $T_2$ relaxation times to characterize the water bound by polymers (e.g., starches and proteins) called polymer water and solutes (e.g., sugars and salts) called solute water and their mixtures. Under the instrumental parameters used, their results indicated a distinct difference in the NMR signal response for these different water states. When they plotted the first minus second

436

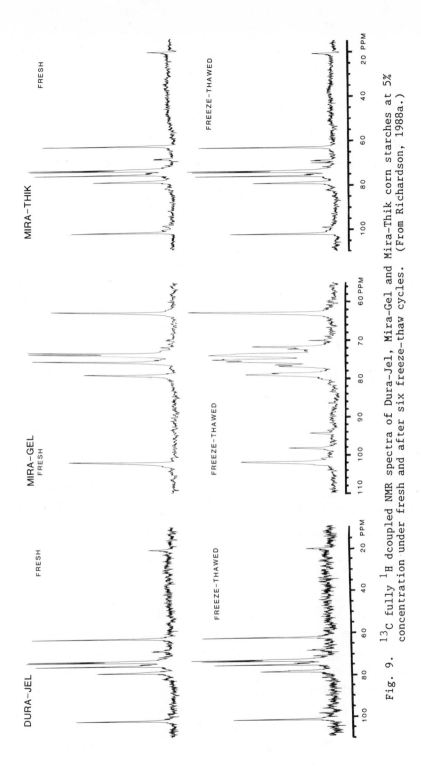

Fig. 9.  $^{13}C$ fully $^1H$ dcoupled NMR spectra of Dura-Jel, Mira-Gel and Mira-Thik corn starches at 5% concentration under fresh and after six freeze-thaw cycles.  (From Richardson, 1988a.)

437

pulse NMR signals with increasing variable time delay (VTD) for samples of starch, sucrose and a 90:10 mixture of starch: sucrose (all at $a_w$ = 0.91), the starch gave a constant signal, the sucrose showed a rapidly decreasing signal and the starch: sucrose mixture gave a signal between the two individual components, close to that for the starch.

Richardson et al. (1987b), expanding the work of Lang and Steinberg (1983), studied the mobility of water in polymer-solute (starch and sucrose) systems with $^{17}O$ and $^{2}H$ high-field (250 MHz) NMR. The linewidth and shape of the $^{17}O$ NMR peak was shown to monitor the mobility of the water associated with both the starch and the sucrose.

Freezing and Thawing Behavior. The use of pulsed NMR to study the behavior of water during freezing and thawing has been relatively limited (Kuntz, 1971; Lynch and Webster, 1979; Katayama and Fujiwara, 1980; Nagashima and Suzuki, 1981, 1984, 1985; Suzuki and Nagashima, 1982; Cameron et al., 1985). A few studies have been done with food components (Kuntz, 1971; Nagashima and Suzuki, 1981, 1984, 1985; Suzuki and Nagashima, 1982).

Nagashima and Suzuki (1981, 1984) reported a method using low frequency (20 MHz) pulsed $^{1}H$ NMR to characterize bound water. This method consisted of measuring the free induction decay (FID) amplitude with decreasing (or increasing) temperature and relating this to the unfrozen water (UFW) content of the sample (in g water/g solids) at each temperature. They have improved this method to include computer analysis of the FID and measurement of not only the quantity of UFW, but its mobility (i.e., $T_2$ relaxation times) (Nagashima and Suzuki, 1985). The FID curve was resolved into two components characterized by long and short relaxation times, $T_{2L}$ (liquid-state protons) and $T_{2S}$ (solid-state protons). The population of liquid-state protons was converted to a UFW content. The plotted output was presented as a UFW content on a logarithmic scale as a function of temperature; the plot was referred to as a freezing curve when the measurements were taken as the temperature was decreasing and as a thawing curve when the temperature was increasing. These authors have presented an extensive number of such freezing and thawing curves for a variety of food components and systems (e.g., amino acids, beef, starch, sugar-alcohols and fruits) and, where appropriate, discussed their relation to each other (e.g., amino acids) as well as their relation to other properties of the system, such as hysteresis (e.g., egg yolk), rate of cooling (e.g., D-mannitol), gelatinization, retrogradation (e.g., waxy cornstarch) and further processing (e.g., freeze-dried coffee). Such a freezing curve for beef extract is shown in Fig. 10. Beef extract did not freeze (Fig. 10 curve A); decreasing UFW content at low temperatures (about -50°C) was

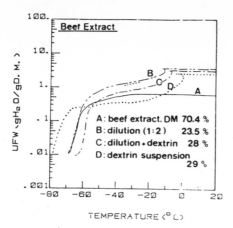

Fig. 10.  Relation of unfreezable water (UFW) content, as determined by [1]H NMR response, to freezing of:  A) beef extract, B) beef extract, diluted with water (1:2), C) beef extract, diluted with water (1:2) and a dextrin suspension, and D) beef extract with a dextrin suspension. (From Nagashima and Suzuki, 1985.)

hypothesized to be due to vitrification (i.e., glassy solidification of the entire system).  An aqueous dilution results in partial freezing of water but increases the UFW content (Fig. 10 curve B).  With the addition of a dextrin suspension along with the dilution, the UFW content decreased more than expected (Fig. 10 curve C).  However, addition of a dextrin suspension alone yielded the fastest decrease in UFW (until about -58°C) (Fig. 10 curve D).  Addition of a dextrin suspension to beef extract produced a porous, soluble and full-flavored freeze-dried product.

Emulsions.  Pulsed NMR spectroscopy has been applied to investigate a variety of emulsion properties, such as the measurement of the oil and water content (Hester and Quine, 1977; Brosio et al., 1981; Brosio et al., 1982; Shih, 1983), the hydrophilic and lipophilic balance (HLB) (Ben-Et and Tatarsky, 1972), the stability of the emulsion (Trumbetas et al., 1976, 1977, 1978) and the state of the water in the emulsion (Hansen, 1974; Arai and Watanabe, 1985).  Only the first and the last of these references directly investigated the water component of the emulsion; therefore, only these are discussed below.

Brosio et al. (1982) presented a low-resolution (20 MHz) [1]H NMR method for the determination of oil and water content in oil and water (O/W) emulsions.  The method was based on an analysis of the $T_1$ magnetization decay curve.  Due to the different relaxation times of oil and water the $T_1$ curve consisted of two distinct components, $T_{1F}$ (decay component of the oil) and $T_{1S}$ (decay component of the water).  From these

$T_1$ values the relative hydrogen abundance of the two components was determined and, in turn, the percentage of each component was calculated.

Hansen (1974) investigated the state of water in water-in-oil micro-emulsions made with $D_2O$ by $^2H$ NMR $T_1$ and $T_2$ relaxation times. All relaxation curves yielded simple exponential functions, i.e., single $T_1$ and $T_2$ values. Hansen plotted both $T_1$ and $T_2$ against increasing water droplet radius, $r_w$ ($r_w$ increased as the molar ratio of water to surfactant increased). Both $T_1$ and $T_2$ increased with increasing droplet size, indicating an increased average reorientation rate of water molecules with increasing droplet size. These data were further interpreted according to the isotropic two-state model with fast exchange. The water was found to exist in two rapidly exchanging states: 1) about a 1-A thick monolayer of water molecules of low mobility associated with the surfactant polar groups at the aqueous interface and 2) bulk water with mobility similar to that of free water.

Arai and Watanabe (1985), as part of their investigation of a new surfactant (an enzymatically modified gelatin), measured the pulsed NMR $T_1$ and $T_2$ values of several corn oil/surfactant emulsions. From these values they calculated the average rotational correlation time ($\tau_c$) for the bound water. The emulsion made with the modified gelatin had the slowest $\tau_c$ (1.01 x $10^{-8}$ sec) compared to the other surfactants (e.g., Tween-80 and sodium soy proteinate) which had $\tau_c$ ranging from 1.9 to 5.3 x $10^{-9}$ sec. Arai and Watanabe (1985) also used the freezing and thawing curve method developed by Nagashima and Suzuki (1981) to measure the freeze-thaw behavior of their surfactant.

Moisture Determination. The theory and application of pulsed NMR for moisture determination has been reviewed by Pande (1975), Rollwitz (1985) Richardson and Steinberg (1987), and Richardson (1988b).


PULSED-FIELD GRADIENT NMR

The motion of molecules, apart from convection-like phenomena (for example, a heat or density gradient inherent or imposed on the system), is usually partitioned into rotational motion (i.e., internal molecular motion) and translational motion (i.e., diffusional motion). Both of these molecular motions are reflected in the characteristic NMR relaxation times, $T_1$ and $T_2$, respectively. Under the most favorable conditions (usually over a wide range of temperatures), it is possible to partition the observed relaxation times and obtain a measure of each molecular motion (von Meerwall, 1983). Partitioning or, more appropriately, disentanglement of these relaxation times is very difficult and becomes infinitely complex when a multicomponent solution is analyzed.

However, it is possible to use pulsed-NMR to directly measure the translational motion of molecules. This motion can be characterized by a single parameter, D, the self-diffusion coefficient.

## Theory

There are several NMR techniques currently available to measure the translational motion or self-diffusion of molecules, such as the nuclear spin-echo (SE) method and the pulsed-field gradient spin-echo (PGSE) method. Several current reviews discuss these techniques in detail (von Meerwall, 1983, 1985; Stokes, 1984; Tyrrell and Harris, 1984; Blum, 1986; Stilbs, 1987). Only the PGSE method will be discussed here.

The NMR technique for measuring self-diffusion coefficients has several advantages: 1) No special labeling and thus no special sample preparation or handling is required, 2) It is possible to measure the diffusion coefficient over several orders of magnitude, 3) It is possible, using Fourier transform techniques, to simultaneously determine as many diffusion coefficients as there are resolvable resonances, and 4) Measurements can usually be done in less than one hour. These and other advantages are discussed by Blum (1986).

The fundamental principle underlying the PGSE method is that nuclear magnetic resonance can label a nuclear magnetic moment (a spin) with respect to its position in the sample space, (i.e., the NMR tube) via its characteistic Larmor or precessional frequency. In regular pulsed NMR, as discussed previously, the magnetic field, $B_o$, is uniform over the entire sample. In this case, ignoring complicating effects (i.e., chemical shifts and coupling constants), all like-nuclei will resonate at the same frequency. In case of PGSE, in addition to this uniform $B_o$, another, smaller magnetic field, G (T/m), is applied to the sample (Note: $B_o \gg G$). However, instead of being uniformly applied to the sample, G is applied as a linear magnetic-field gradient. The result is a linear variation in the magnetic field ($B_G$) (Fig. 11). In turn, the precessional frequencies of the nuclei become a function of their position, x, within the magnetic field and consequently within the sample. Thus, the spins will experience different net magnetic fields when G is applied (i.e., $B_G$ ranging between $B_o \pm G$) depending on their position in the sample. For example, in Fig. 11, spin A at position -x will experience a net magnetic field that is weaker (i.e., $B_o - G$) than spin B at position 0 (i.e., $B_o$) or spin C at position +x, which experiences the strongest net magnetic field (i.e., $B_o + G$). The relaxation behavior of these spins as they are affected by the variation of these net magnetic fields with time, is measured in the PGSE experiment.

441

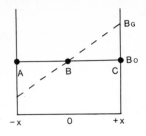

SAMPLE IN NMR MAGNET

Fig. 11. Net magnetic field strength experience by a sample within the NMR magnet, as a function of position, ± x, in the presence ($B_G$) and absence ($B_o$) of the field gradient. The sample has a radius 2x and the average magnetic field is $B_o$. Spin A at position -x will experience a net magnetic field (nmf) that is weaker (nmf = $B_o$ - G) than spin B at position 0 (nmf = $B_o$) or spin C at position +x, which experiences the strongest (nmf = $B_o$ + G).

The basic PGSE experiment consists of a two radio-frequency pulse Hahn-echo experiment (90°-Δ-180°) with two identical magnetic field gradient pulses of magnitude G, duration δ and separation Δ. The field gradient pulses are applied during the dephasing and rephasing segments of the spin-echo cycle. The review by Blum (1986) gives an excellent description of the experimental details.

The amplitude of the resultant NMR resonance is given by (Blum, 1986):

$$A = A_o \exp (-\gamma^2 G^2 \beta D) \tag{3}$$

where $A_o$ is a constant which incorporates the signal intensity for $\Delta = 0$, the J-modulation effects and effect of the transverse relaxation time [$\exp (-2\Delta/T_2)$], $\gamma$ is the magnetogyric ratio (rad/sT), G the strength of the linear magnetic-field gradient (T/m), $\beta$ is equal to $\delta^2$ ($\Delta-\delta/3$) and D is the self-diffusion coefficient. To measure D, a series of spectra are taken for a fixed time Δ and varying δ (and thus varying β) with known γ and G. The log of the resultant amplitude, A, of these spectra is plotted against β. The slope of this plot is $-\gamma^2 G^2 D$. Thus, D can be obtained knowing γ and G.

The majority of pulse-field gradient NMR experiments have been done probing the $^1H$ nucleus because $^1H$ offers the greatest sensitivity and has a high gyromagnetic ratio. However, other nuclei have also been probed, i.e., $^{13}C$, $^2H$, $^{19}F$ and $^7Li$ (Callaghan et al., 1983a; Callaghan, 1984).

Applications

Pulsed-field gradient NMR (PGNMR) has been applied to a variety of biological systems from water diffusion in cells (Chang et al., 1973) to water diffusion in cheese (Callaghan and Jolley, 1983). The studies reviewed here will be limited to water diffusion in food components and systems.

Proteins. James and Gillen (1972) determined D, $T_1$ and $T_2$ of water in egg yolk, egg white and egg albumin solutions. The observed D values relative to the D value of pure water ($D_o$) were 0.25 in egg yolk, 0.80 in egg white and 0.88 in a 10% egg albumin solution. The self-diffusion constants were analyzed in terms of a two-state model of "immobilized" and ordinary water. The conclusion was drawn from comparing the D values with the relaxation times that a large part of the decreased mobility of was due to hydration of the biopolymers.

Nystrom et al. (1981) also measured D, $T_1$ and $T_2$ for small molecules, including water, dioxane and t-butanol in the gel systems cellulose/$H_2O$ and polyacrylamide/$H_2O$. The temperature dependence of both D and $T_1$ for all penetrant molecules were represented by an Arrhenius type relationship, with or without the presence of the polymer. In the presence of the polymer, the linear relationship was parallel to the relation without the polymer but shifted toward small values. The cellulose-water system yielded a $D/D_o$ value of 0.72; the other penetrant molecules showed very similar $D/D_o$ values. These authors, by use of the Mackie and Mears equation, which relates $D/D_o$ to the polymer volume fraction, hypothesized that the observed reduction in D for the small molecules (including water) in the presence of the polymer gels was due mainly to the obstruction of the diffusant by the polymer. This is called the obstruction effect.

PGSE measurements of D for water and dioxane in solutions of and in thermally-induced gels of bovine serum albumin (BSA) as a function of concentration (5-20% w/w) were made by Brown and Stilbs (1982). $D/D_o$ for water in both solution and gel of BSA was found to decrease linearly with volume fraction BSA. This linear relation suggests that the dominant mechanism for the decrease in D is the obstruction effect. These results are similar to those reported by Nystrom et al. (1981).

Carbohydrates. Several PGNMR studies have been carried out on carbohydrate-water systems (Basler and Lechert, 1974; Lechert et al., 1980; Callaghan et al., 1983b; Callaghan, 1984; Callaghan and Lelievre, 1985, 1986).

443

Basler and Lechert (1974) determined the diffusion of water in corn starch gels as a function of concentration (50-95% water) and temperature (1 to 47°C). They found that the water molecules showed uniform and unrestricted diffusion with the same activation energy as bulk water. The boundaries of the swollen starch grains were not a barrier to diffusion.

Callaghan et al. (1983b) measured D as a function of wheat starch at two concentrations. No significant differences were observed for D despite differences in their rheological properties. In this and subsequent studies (Callaghan, 1984; Callaghan and Lelievre, 1985, 1986) Callaghan and coworkers demonstrated the use of solvent and starch self-diffusion coefficients as effective probes for the determination of motion, size and shape of the starch. In a suspension of macromolecules, one can independently monitor the diffusion of solvent or macromolecule. When the macromolecule signal is measured, a deuterated solvent is required. Of particular interest to this review is solvent diffusion, especially when the solvent is water.

The reduction in D for solvent molecules in a macromolecule may be attributed to two main factors: 1) The obstruction effect and 2) the solvation effect. In the obstruction effect the solvent diffusion is reduced due to the diversion of the solvent molecules around the macro-molecules, whereas in the solvation effect the reduction in D is due to a direct interaction between the solvent and the macromolecule. Both effects have been incorporated in a model by Wang (1954) in which the solvent self-diffusion coefficient, $D^{solv}$, is related to a macromolecular shape factor.

Callaghan and Lelievre (1985) applied this model to wheat starch amylopectin-water and wheat starch amylopectin-DMSO systems. Figure 12 is a plot of the normalized solvent self-diffusion, $D^{solv}/D_0^{solv}$ ($D_0^{solv}$ is the self-diffusion coefficient of pure water), versus amylopectin weight fraction, $w$. The slope of these lines was related by the Wang model to the difference in the shape factor of amylopectin in DMSO versus water. In the DMSO the amylopectin molecules were highly planar. In contrast, in water the amylopectin was an aggregate with a more spherical shape and had a volume some 400 times larger than a single molecule. Other macromolecular-water systems studied included wheat starch pastes (Callaghan et al., 1983b) poly-D-glucose, dextrin and glycogen (Callaghan and Lelievre, 1986).

Other Systems. Callaghan and co-workers have used PGNMR to investigate diffusion of water in the endosperm tissue of wheat grains as a function of water content (Callaghan et al., 1979) and diffusion of fat

444

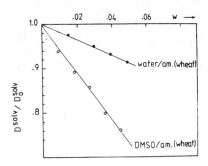

Fig. 12.  Dependence of solvent diffusion on amylopectin concentration, w (g/g), for wheat starch amylopectin in DMSO and in water.  (From Callaghan and Lelievre, 1985.)

and water in cheddar and swiss cheese (Callaghan and Jolley, 1983).  In the wheat endosperm study, a model for capillary confined water diffusion was used to fit the data and yield a unique D value at each water content.  D varied from $1.8 \times 10^{-10}$ $m^2$/s for the lowest moisture content to $1.2 \times 10^{-9}$ $m^2$/s at the highest moisture content.  In the cheese study, no significant difference in diffusion coefficient was observed between the two cheeses, cheddar and swiss.  The measured D values were about one-sixth of the value in bulk water at 30°C.  They reported strong evidence that water diffusion was confined to the protein surface.

## Potential Applications

The applications of pulsed-field gradient NMR (PGNMR) to the diffusion of water in food components and systems has been rather limited compared to the use of regular pulsed NMR.  However, there are several potential applications where the PGNMR technique could be useful.  For example, most of the chemical, microbiological and physical reactions which take place in foods depend on the availability of water and its solvent abilities.  Thus, it seems reasonable that there would be a direct relationship between the translational mobility of the water (i.e., rate of water movement) and, e.g., a reaction rate.  Lechert et al. (1980) suggested that self-diffusion measurements, such as the ones they determined for potato-starch, may be important for a number of reactions dependent on water diffusion (i.e., the Maillard Browning reaction).

This proposed relationship between the translational mobility of water and the availability of the water to participate in chemical reactions microbiological growth or physical changes is a hypothesis which must be tested.

Nuclear Magnetic Resonance Imaging (NMRI) is the newest of the pulsed NMR techniques. Lauterbur (1973) was the first to demonstrate the image construction potential of NMR. He showed that, by applying controlled magnetic field gradients, the spatial distribution of hydrogen-containing fluids could be mapped with NMR. Since that time NMRI has evolved at a very rapid pace, especially for medical applications (Mansfield and Maudsley, 1977; Mansfield and Morris, 1982; Budinger and Lauterbur, 1984; Morris, 1986).

## Theory

NMRI produces planar or three dimensional images of an object by mapping the spin density and/or spin relaxation times, $T_1$ and $T_2$, from selected regions. Chemical shifts and fluid flow velocity can also be imaged (Mansfield and Morris, 1982). Most of the NMRI work has been done using $^1$H because of its high sensitivity to detection and ubiquitous nature. However, the other nonzero spin nuclei of water could also be imaged.

A number of methods have been developed to produce NMR images. Brunner and Ernst (1979) classified the various imaging methods according to the spatial element that was detected in each step of the imaging sequence: 1) sequential point, 2) sequential line, 3) sequential plane and 4) simultaneous volume. Lauterbur (1973) used a sequential plane method called 2-D projection reconstruction. In this method a linear magnetic field gradient is applied to the sample similar to that discussed above in the section on Pulsed-Gradient NMR. This gradient causes the resonant frequencies of the nuclear spins to take on a spatial dependence. The overall effect of the gradient is to product a projection of spin density perpendicular to the direction of the gradient. The 2-D NMR image in the projection reconstruction method is produced by combining projections from different gradient angles. To obtain these different angles, the gradients are rotated electronically. Detailed NMRI theory is given by Bottomly (1982), Mansfield and Morris (1982), Andrew (1983) and Rothwell (1985).

The majority of NMRI has been applied to macroscopic objects with dimensions in excess of 1 cm. However, the use of NMRI in the submillimeter regime (coined microscopic NMRI) is currently being investigated (Aguayo et al., 1986; Hall et al., 1986; Eccles and Callaghan, 1986; Eccles et al., 1988). Also the use of NMRI for measurement of flow is being investigated in both medical and nonmedical applications (Redpath et al., 1984; Kose et al., 1985; O'Donnell, 1985; Cho et al., 1986; Ridgway and Smith, 1986; Callaghan et al., 1988).

## Applications

By far the majority of NMRI applications have been done in the medical field. Nonmedical areas of science are just beginning to receive attention (Rothwell et al., 1984; Rothwell and Gentempo, 1985; Eccles and Callaghan, 1986; Perez et al., 1989). Applications of NMRI have been demonstrated in petrophysics (Rothwell and Vinegar, 1985; Vinegar, 1986), polymer absorption chemistry (Rothwell et al., 1984) and agricultural products (Perez et al., 1989; Eccles et al., 1988).

Rothwell et al. (1984) investigated the potential use of NMRI to non-invasively determine the distribution of fluids, such as water, in polymeric materials (glass-fiber-reinforced epoxy-resin composites). Only the protons from the water were observed and mapped since protons belonging to the rigid polymer matrix had very short $T_2$ relaxation times and were not observed. They concluded that NMRI was a powerful technique due to its noninvasive nature and its ability to map fluid distribution in a solid matrix. They suggested using NMRI to study absorption and diffusion processes in plastic polymers. This suggestion should be equally applicable to food polymers.

Perez et al. (1989) used NMRI spectroscopy techniques to obtain a MRI of water density in a plane through a whole apple (Fig. 13) and to non-invasively measure transient moisture profiles during a one-dimensional drying experiment of a slice of apple (Fig. 14). A Fourier imaging sequence was used to obtain the transient moisture profiles of the apple slice at one-hour intervals during 8 hours of drying. The apple slices were dried directly in the magnet (2 tesla) with a specially constructed wind tunnel drier. The sample holder was a capillary tube filled with water. The tube held the apple slice sample in place as it shrank during drying, and the water in the tube provided a reference to which the rest of the sample could be contrasted. The profile represents the changes in moisture content with time for a line or rod (0.1 mm x 2 mm) which extends from the top to the bottom of the sample and intersects one of the capillary holder tubes. Spatial resolution within the rod was 0.67 mm. The void at the center of the graph was due to the presence of the capillary tube. While the two tall peaks surrounding the void were the result of an increase in the aqueous solution in the intercellular space caused by disruption of the cellular structure by insertion of the capillary tube. The moisture profile on either side of these peaks showed a more uniform decrease in moisture than the characteristic steep parabolic moisture profiles predicted by theory. The characteristic parabolic moisture profile was manifested after 8 hours of drying (bottom curve in Fig. 14). Drying curves (moisture content versus time) were also obtained from the images and compared to drying curves

447

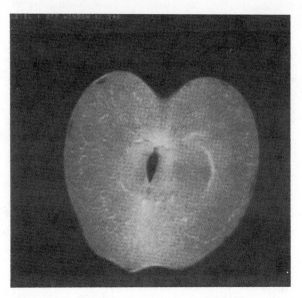

Fig. 13.   Magnetic resonance image of water density in a plane
through a whole apple.  (From Perez et al., 1989.)

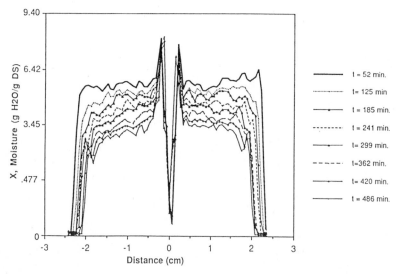

Fig. 14.   Transient moisture profile as a function of time and
position within the apple rod section.  Top half of the
rod is shown with negative distance values.  (From Perez
et al., 1989.)

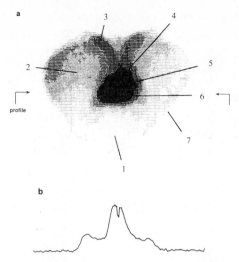

Fig. 15.   (a) Image of transverse section of a wheat grain indicating the
           location of its major structural features.  The endosperm
           (regions 1, 2 and 3) is the organ in which starch is deposited.
           The outer layer of the endosperm (the aleurone layer), the seed
           coat (testa) and the wall of the fruit (pericarp) are not
           resolved and together correspond to region 7.  Longitudinal
           transport of water and solutions of nutrients is provided by
           xylem and phloem vessels in vascular bundles scattered through
           region 4.  Nutrients travel from the vascular bundles into the
           endosperm via the chalaza and nucellar projection (region 5)
           and through a liquid-filled cavity occupying part of region 6.
           (b) Intesity profile taken along the line indicated in (a).
           (From Eccles et al., 1988.)

Table 2.   Localized Water Self-Diffusion Normal to Transverse
           Section of Wheat Kernel (From Eccles et al., 1988).

| Region[*] | Identification | $m^2 s^{-1}$ $D$ |
|---|---|---|
| 1 | Doral endosperm | 5 (3) x $10^{-10}$ |
| 2 | Check endosperm | 7.0 (5) x $10^{-10}$ |
| 3 | Ventral endosperm | 7 (1) x $10^{-10}$ |
| 4 | Vascular bundle + chalaza | 10.1 (5) x $10^{-10}$ |
| 5 | Nucellar projection | 5 (2) x $10^{-10}$ |
| 6 | Endosperm cavity | 10.6 (8) x $10^{-10}$ |
| 7 | Aleurone layer + testa + pericarp | 9 (3) x $10^{-10}$** |

[*]  Numbers correspond to Fig. 15.

**  These data are subject to errors associated with defining a consistent
    region close to the image boundary.

obtained by periodically weighing the sample using identical drying conditions. The trends shown by the two drying curves were very similar.

Eccles et al. (1988), incorporating a pulsed field gradient spin echo sequence in a NMRI experiment, measured the self-diffusion coefficient of water as a function of position in wheat grain. Images were obtained at 60 MHz at 28°C on 1.3 mm thick transverse sections of a wheat grain and were recorded as 16 bit 256 x 256 pixel arrays. Figure 15a shows an image obtained at zero pulsed gradient along with an intensity profile along the designated line in the image (Fig. 15b). The differences in water intensities in the endosperm tissue and the vascular bundle area are apparent. Other features of the grain are identified in the legend in Fig. 15.

The localized self-diffusion experiment was performed by varying the pulsed gradient. The water self-diffusion coefficients obtained are summarized in Table 2. The results indicated that the water in the wheat grain in all regions was significantly less mobile than free water ($2.3 \times 10^{-9}$ $m^2s^{-1}$ at 28°C) and were consistent with bulk averages obtained in previous work (Callaghan et al., 1979). Water motion appeared to be most severely hindered in the endosperm, whereas water was less hindered in regions 4, 5 and 6, the route taken by nutrients moving into the endosperm. The authors also suggested that with the use of appropriate selective excitation it should be possible to measure the localized transport of the selected molecular species in any desired direction.

To date there have been few NMRI applications specific to water in food components or food systems. However, as with many other fields, there are a number of important potential applications. For example, NMRI would be an excellent technique with which to monitor the behavior of water in food throughout growth, processing and storage as well as measure internal structure and moisture content of the food system. A specific application has been proposed to use NMRI to monitor the movement of water during drying of grains and pasta (Litchfield, 1987).

REFERENCES

Ablett, S., Lillford, P. J., Baghdadi, S. M. A., and Derbyshire, W., 1976, NMR relaxation in polysaccharide gels and films, in: "Magnetic Resonance in Colloid and Interface Science," Resing, H. A., and Wade, C. G. (Eds.), ACS Symposium Series 34, American Chemical Society, Washington, DC.
Abragam, A., 1961, The Principles of Nuclear Magnetism, Clarendon Press, Oxford.
Agvayo, J. B., Blackband, S. J., Schoeniger, J., Mattingly, M. A., and Hintermann, M., 1986, Nuclear magnetic resonance imaging of a single cell, Nature, 322:190.
Andrew, E. R., 1983, NMR imaging, Acc. Chem. Res., 16:114.

Arai, S., and Watanabe, M., 1985, An enzymatically modified protein as a new surfactant and its function to interact with water and oil in an emulsion system, in: "Properties of Water in Foods," Simatos, D., and Multon, J. L., (Eds.), Martinus Nijhoff Publishers, Boston, MA.

Basler, V. W., and Lechert, H., 1974, Diffusion of water in starch gels, Starch, 26(2):39.

Baianu, I. C., 1989, Personal communication, University of Illinois, Champaign - Urbana, IL.

Ben-Et, G., and Tatarsky, D., 1972, Application of NMR for the determination of HLB values of nonionic surfactants, JAOCS, 49:499.

Berendsen, H. J. C., 1975, Specific interactions of water with biopolymers, in: "Water - A Comprehensive Treatise, Vol. 5. Water in Disperse Systems," Franks, F. (Ed.), Plenum Press, New York.

Berliner, L. J., and Reuben, J., (Eds.), 1980, Biological Magnetic Resonance, Plenum Press, New York.

Blum, F. D., 1986, Pulsed-gradient spin-echo nuclear magnetic resonance spectroscopy, Spectroscopy, 1(5):32.

Bociek, S., and Franks, F., 1979, Proton exchange in aqueous solutions of glucose, Faraday Trans. I., 2:262.

Bottomly, P. A., 1982, NMR imaging techniques and applications: A review, Rev. Sci. Instrum., 53:1319.

Bradley, R. S., 1936, Polymolecular adsorbed films, I. The adsorption of argon on salt crystals at low temperatures and the determination of surface fields, J. Chem. Soc., 1467.

Brevard, C., and Kintzinger, J. P., 1978, Deuterium and tritium, in: "NMR and the Periodic Table," Harris, R. K., and Mann, B. E., (Eds.), Academic Press, New York.

Brittain, T., and Geddes, R., 1978, Water binding by glycogen molecules, Biochem. Biophys. Acta., 543:258.

Brosio, E., Altobelli, G., Yu, S. Y., and DiNola, A., 1983, A pulsed low resolution NMR study of water binding to powdered milk, J. Fd. Technol., 18:219.

Brosio, E., Conti, F., DiNola, A., Scalzo, M., and Zulli, E., 1982, Oil and water determination in emulsions by pulsed low-resolution NMR, JAOCS, 59(1):59.

Brosio, E., Conti, F., DiNola, A., Scorano, O., and Balestrieri, F., 1981, Simultaneous determination of oil and water content in olive husk by pulsed low resolution nuclear magnetic resonance, J. Fd. Technol., 16:629.

Brown, W., and Stilbs, P., 1982, Self-diffusion measurements on bovine serum albumin solutions and gels using a pulsed-gradient spin-echo NMR technique, Chemica Scripta, 19:161.

Brunauer, S., Emmett, P. H., and Teller, E., 1938, Adsorption of gases in multi-molecular layers, J. Am. Chem. Soc., 60:309.

Brunner, P., and Ernst, R. R., 1979, Sensitivity and performance time in NMR imaging, J. Magn. Reson., 33:83.

Bryant, R. G., 1978, NMR relaxation studies of solute-solvent interactions, Ann. Rev. Phys. Chem., 29:167.

Budinger, T. F., and Lauterbur, P. C., 1984, Nuclear magnetic resonance technology for medical studies, Science, 226:288.

Callaghan, P. T., 1984, Pulsed field gradient nuclear magnetic resonance as a probe of liquid state molecular organization, Aust. J. Phys., 37:359.

Callaghan, P. T., and Jolley, K. W., 1983, Diffusion of fat and water in cheese as studied by pulsed field gradient nuclear magnetic resonance, J. Colloid and Interface Sci., 93(2):521.

Callaghan, P. T., Jolley, K. W., and Lelievre, J., 1979, Diffusion of water in the endosperm tissue of wheat grains as studied by pulsed field gradient nuclear magnetic resonance, Biophys. J., 28:133.

Callaghan, P. T., Le Gros, M. A., and Pinder, D. N., 1983a, The measurement of diffusion using deuterium pulsed field gradient nuclear magnetic resonance, J. Chem. Phys., 79(12):6372.

Callaghan, P. T., Jolley, K. W., Lelievre, J., and Wong, R. B. K., 1983b,

Nuclear magnetic resonance studies of wheat starch pastes, J. Colloid and Interface Sci., 92(2):332.

Callaghan, P. T., and Lelievre, J., 1985, The size and shape of amylopectin: A study using pulsed-field gradient nuclear magnetic resonance, Biopolymers, 24:441.

Callaghan, P. T., and Lelievre, J., 1986, The influence of polymer size and shape on self-diffusion of polysaccharides and solvents, Analytica Chimica Acta., 189:145.

Callaghan, P. T., Eccles, C. D., and Xia, Y., 1988, NMR microscopy of dynamic displacements: k-space and q-space imaging, J. Phys. E: Sci. Instrum., 21:820.

Cameron, I. L., Hunter, K. E., Ord, V. A., and Fullerton, G. D., 1985, Relationships between ice crystal size, water content and proton NMR relaxation times in cells, Physiological Chem. and Physics and Medical NMR, 17:371.

Campbell, I. D., and Dwek, R. A., 1984, Biological Spectroscopy, Chapt. 6, The Benjamin/Cummings Publishing Co., Inc., Menlo Park, CA.

Chang, D. C., Roscharch, H. E., Nichols, B. L., and Hazlewood, C. F., 1973, Implications of diffusion coefficient measurements for the structure of cellular water, Annals N.Y. Academy of Sciences, 204:434.

Child, T. F., and Pryce, N. G., 1972, Steady-state and pulse NMR studies of gelatin in aqueous agarose, Biopolymers, 11:409.

Cho, Z. H., Oh, C. H., Kim, Y. S., Mun, C. W., Nalcioglu, O., Lee, S. J., and Chung, M. K., 1986, A new nuclear magnetic resonance imaging technique for unambiguous unidirectional measurement of flow velocity, J. Appl. Phys., 60:1256.

Christian, J. H. B., 1981, Specific solute effects on microbial/water relations, in: "Water Activity: Influences on Food Quality," Rockland, L. B., and Stewart, G. F., (Eds.), Academic Press, New York.

Cooke, R., and Kuntz, I. D., 1974, The properties of water in biological systems, Ann Rev. Biophys. Bioeng., Mullins, L. J., (Ed.), 9035:95.

Currie, R. W., Jordan, R., and Wolfe, F. H., 1981, Changes in water structure in postmortem muscle, as determined by NMR $T_1$ values, J. Food Sci., 46:822.

Derbyshire, W., and Duff, I. D., 1973, NMR of agarose gels, Chem. Soc. Faraday Discuss., 57:243.

Derbyshire, W., 1982, The dynamics of water in heterogeneous systems with emphasis on subzero temperatures, in: "Water – A Comprehensive Treatise. Vol. 7. Water and Aqueous Solutions at Subzero Temperatures," Franks, F., (Ed.), Plenum Press, New York.

Derome, A. E., 1987, Modern NMR Techniques for Chemistry Research, Pergamon Press, New York.

Drapron, R., 1985, Enzyme activity as a function of water activity, in: "Properties of Water in Foods," Simatos, D., and Multon, J. L., (Eds.), Martinus Nijhoff Publishers, Boston, MA.

Earl, W. L., and Niederberger, W., 1977, Proton decoupling in $^{17}$O nuclear magnetic resonance, J. Magn. Reson., 27:351.

Eccles, C. D., and Callaghan, P. T., 1986, High-resolution imaging: The NMR microscope, J. Magn. Reson., 68:393.

Eccles, C. D., Callaghan, P. T., and Jenner, C. F., 1988, Measurement of the self-diffusion coefficient of water as a function of position in wheat grain using nuclear magnetic resonance imaging, Biophys. J., 53:77.

Edzes, H., and Samulski, E., 1978, The measurement of cross-relaxation effects in proton NMR spin lattice relaxation of water in biological systems: Hydrated collagen and muscle, J. Magn. Reson., 31:207.

Eisenstadt, M., and Fabry, M. E., 1978, NMR relaxation of the hemoglobin-water proton spin system in red blood cells, J. Magn. Reson., 29:591.

Farrell, H. M., Pessen, H., and Kumosinski, T. F., 1987, Water interactions with varying molecular states of milk proteins: $^2$H NMR relaxation studies, 82nd Annual Meeting of the American Dairy Association, Paper #D116, Columbia, Missouri, June 21-24.

Finney, J. L., Goodfellow, J. M., and Poole, P. L., 1982, The structure and dynamics of water in globular proteins, in: "Structural Molecular Biology - Methods and Applications," Davies, D. B., Saenger, W., and Danyluk, S. S., (Eds.), Plenum Press, New York.

Flink, J. M., 1983, Structure and structure transitions in dried carbohydrate materials, in: "Physical Properties of Foods," Peleg, M., and Bagley, E.B., (Eds.), AVI Publishing Co., Inc., Westport, Connecticut.

Franks, F., 1982, Water activity as a measure of biological viability and quality control, Cereal Foods World, 27(9):403.

Fuller, M. E., and Brey, W. S., 1968, Nuclear magnetic resonance study of water sorbed on serum albumin, J. Biol. Chem., 243(2):1968.

Glasel, J. A., 1972, Nuclear magnetic resonance studies on water and ice, in: "Water - A Comprehensive Treatise Vol. 1," Franks, F., (Ed.), Plenum Press, New York.

Goldsmith, S. M., and Toledo, R. T., 1985, Studies on egg albumin gelation using nuclear magnetic resonance, J. Food Sci., 50:59.

Gould, G. W., and Measures, J. C., 1977, Water relations in single cells, Philos. Trans. R. Soc. London Ser., B278:151.

Grigera, J. R., and Bienkiewicz, K. J., 1984, Hydration of collagen. Support for the exchange model, Studia Biophysica, 103(3):195.

Hall, L. D., Luck, S., and Rajanayagam, V., 1986, Construction of a high-resolution NMR probe for imaging with submillimeter spactial resolution, J. Magn. Reson., 66:349.

Halle, B., Andersson, T., Forsen, S., and Lindman, B., 1981, Protein hydration from oxygen-17 magnetic relaxation, J. Am. Chem. Soc., 103:500.

Halle, B., and Karlstrom, G., 1983, Prototropic charge migration in water, Part 1., J. Chem. Soc. Faraday Trans., 279:1031.

Halle, B., and Wennerstrom, H., 1981, Interpretation of magnetic resonance data from water nuclei in heterogeneous systems, J. Chem. Phys., 75(4):1928.

Hansen, J. R., 1974, High-resolution and pulsed nuclear magnetic resonance studies of microemulsions, J. Phys. Chem., 78(3):256.

Hansen, J. R., 1976, Hydration of soybean protein, J. Agric. Food Chem., 24(6):1136.

Harvey, J. M., and Symons, M. C. R., 1976, Proton magnetic resonance study of the hydration of glucose, Nature, 261:435.

Harvey, J. M., and Symons, M. C. R., 1978, The hydration of monosaccharides - An NMR study, J. Solution Chem., 7(8):571.

Hennig, V. H. J., 1977, NMR - Investigations of the role of water for the structure of native starch granules, Starch, 29:1.

Hennig, V. H. J., and Lechert, H., 1974, Measurements of the magnetic relaxation times of the protons in native starches with different water contents, Starch, 26(7):232.

Hennig, H. J., and Lechert, H., 1977, DMR study of $D_2O$ in native starches of different origins and amylose of type B, J. Colloid Interface Sci., 62(2):199.

Hester, R. E., and Quine, D. E. C., 1977, Quantitative analysis of food products by pulsed NMR, Rapid determination of oil and water in flour and feedstuffs, J. Sci. Fd. Agric., 28:624.

Hoeve, C. A. J., 1980, The structure of water in polymers, in: "Water in Polymers," Rowland, S. P., (Ed.), ACS Symposium Series 127, American Chemical Society, Washington, DC.

Horman, I., 1984, NMR Spectroscopy, in: "Analysis of Foods and Beverages: Modern Techniques," Charalambous, G., (Ed.), Academic Press, New York.

Hsi, E., Vogt, G. F., and Bryant, R. G., 1979, Nuclear magnetic resonance study of water adsorbed on cellulose, J. Colloid and Interface Sci., 70(2):338.

James, T. L., and Gillen, K. T., 1972, Nuclear magnetic resonance relaxation time and self-diffusion constant of water in hen egg white and yolk, Biochem. Biophys. Acta., 286:10.

Jane, J. L., 1985, $\alpha$-amylose action and $^{13}$C NMR studies on amylose-V complexes and retrograded amylose, Ph.D. Thesis, Iowa State University, Ames, IA.

Jaska, E., 1971, Starch gelatinization as detected by proton magnetic resonance, Cereal Chem., 48:437.

Kalk, A., and Berendsen, H. J. C., 1976, Proton magnetic relaxation and spin diffusion in proteins, J. Magn. Reson., 24:343.

Kapsalis, J. G., 1981, Moisture sorption hysteresis, in: "Water Activity: Influences on Food Quality," Rockland, L. B., and Stewart, G. F., (Eds.), Academic Press, New York.

Karel, M., 1975, Physico-chemical modification of the state of water in foods - A speculative survey, in: "Water Relations of Foods," Duckworth, R. B., (Ed.), Academic Press, New York.

Katayama, S., and Fujiwara, S., 1980, NMR study of the freezing/thawing mechanism of water in polyacrylamide gel, J. Phys. Chem., 84:2320.

Katz, E. E., and Labuza, T. P., 1981, Effect of water activity on the sensory chrispress and mechanical deformation of snack food products, J. Food Sci., 46:403.

Kemp, W., 1986, NMR in Chemistry: A Multinuclear Introduction, MacMillian Education Ltd., London.

Kintzinger, J. P., 1983, Oxygen-17 NMR, in: "NMR of Newly Accessible Nuclei, Vol. 2," Laszlo, P., (Ed.), Academic Press, New York.

Kintzinger, J. P., and Marsmann, H., 1981, Oxygen-17 and Silicon-29, Springer-Verlag, New York.

Kirk, J. R., 1981, Influence of water activity on stability of vitamins in dehydated foods, in: "Water Activity: Influences on Food Quality," Rockland, L. B., and Stewart, G. E., (Eds.), Academic Press, New York.

Koenig, S. H., Bryant, R. G., Hallenga, K., and Jacobs, G. S., 1978, Magnetic cross-relaxation among protons in protein solutions, Biochemistry, 17:4348.

Kose, K., Satoh, K., Inouye, T., and Yasuoka, H., 1985, NMR flow imaging, J. Phys. Soc. Japan, 54:81.

Kumosinski, T. F., and Pessen, H., 1982, A deuteron and proton magnetic resonance relaxation study of $\beta$-lactoglobulin A association: Some approaches to Scatchard hydration of globular proteins, Arch. Biochem. Biophys., 218(1):286.

Kuntz, I. D., 1971, Hydration of macromolecules, IV, Polypeptide conformation in frozen solutions, J. Am. Chem. Soc., 93(2):516.

Kuntz, I. D., and Kauzmann, W., 1974, Hydration of protein and polypeptides, Adv. Protein Chem., 28:239.

Labuza, T. P., and Busk, G. C., 1979, An analysis of the water binding in gels, J. Food Sci., 44:1379.

Labuza, T. P., and Saltmarch, M., 1981, The nonenzymatic browning reactions as affected by water in foods, in: "Water Activity: Influences on Food Quality," Rockland, L. B., and Stewart, G. E., (Eds.), Academic Press, New York.

Lai, H. M., and Richardson, S. J., 1989, Lactose crystallization in skim milk powder observed by hydrodynamic equilibria, scanning electron microscopy and $^{2}$H nuclear magnetic resonance, Accepted, J. Food Sci.

Lambelet, P., Berrocal, R., Desarzens, C., Froehlicher, I., and Ducret, F., 1988, Pulsed low-resolution NMR investigations of protein sols and gels, J. Food Sci., 53(3):943.

Lang, K. W., and Steinberg, M. P., 1983, Characterization of polymer and solute bound water by pulsed NMR, J. Food Sci., 48:517.

Laszlo, P., (Ed.), 1983, NMR of Newly Accessible Nuclei, Vol. 1, Academic Press, New York.

Lauterbur, P. C., 1973, Image formation by induced local interactions: Examples employing nuclear magnetic resonance, Nature, 242:190.

454

Lechert, H. T., 1981, Water binding on starch: NMR studies on native and gelatinized starch, in: "Water Activity: Influences on Food Quality," Rockland, L. B., and Stewart, G. F., (Eds.), Academic Press, New York.

Lechert, H., and Hennig, H. J., 1976, NMR investigations on the behavior of water in starches, in: "Magnetic Resonance in Colloid and Interface Science," Resing, H. A., and Wade, C. G., (Eds.), ACS Symposium Series 34, American Chemical Society, Washington, DC.

Lechert, H., Maiwald, W., Kothe, R., and Basler, W. D., 1980, NMR – Study of water in some starches and vegetables, J. Food Processing and Preservation, 3:275.

Leistner, L., and Rodel, W., 1976, The stability of intermediate moisture foods with respect to microorganisms, in: "Intermediate Moisture Foods," Davies, R., Birch, G. G., and Parker, K. J., (Eds.), Applied Science Publishers, Ltd., London.

Lelievre, J., and Creamer, L. K., 1978, An NMR study of the formation and syneresis of renneted milk gels, Milchwissenschaft, 33(2):73.

Lelievre, J., and Mitchell, J., 1975, A pulsed NMR study of some aspects of starch gelatinization, Starch, 27(4):113.

Leung, H. K., Magnuson, J. A., and Bruinsma, B. L., 1979, Pulsed nuclear magnetic resonance study of water mobility in flour doughs, J. Food Sci., 44:1408.

Leung, H. K., Magnuson, J. A., and Bruinsma, B. L., 1983, Water binding of wheat flour doughs and breads as studied by deuteron relaxation, J. Food Sci., 48:95.

Leung, H. K., Steinberg, M. P., Wei, L. S., and Nelson, A. I., 1976, Water binding of macromolecules determined by pulsed NMR, J. Food Sci., 41:297.

Lillford, P. J., Clark, A. H., and Jones, D. V., 1980a, Distribution of water in heterogeneous food and model systems, in: "Water in Polymers," Comstock, M. J., (Ed.), ACS Symposium Series 127, American Chemical Society, Washington, D.C.

Lillford, P. J., Jones, D. V., and Rodger, G. W., 1980b, Water in fish, in: "Advances in Fish Science and Technology," Connell, J. J., (Ed.), Fishing News Books Ltd., England.

Lioutas, T. S., 1984, Interaction among protein, electrolytes and water determined by nuclear magnetic resonance and hydrodynamic equilibria., Ph.D. Thesis, University of Illinois, Urbana.

Lioutas, T. S., Baianu, I. C., and Steinberg, M. P., 1986, Oxygen-17 and deuterium nuclear magnetic resonance studies of lysozyme hydration, Arch. Biochem. Biophys., 247(1):136.

Lioutas, T. S., Baianu, I. C., and Steinberg, M. P., 1987, Sorption equilibrium and hydration studies of lysozyme: Water activity and 360-MHz proton NMR measurements, J. Agric. Food Chem., 35:133.

Litchfield, B., 1987, Personal communication, University of Illinois, Urbana.

Lynch, L. J., and Webster, D. S., 1979, An investigation of the freezing of water associated with wool keratin by NMR methods, J. Colloid and Interface Sci., 69(2):238.

Makower, B., and Dye, W. B., 1956, Equilibrium moisture content and crystallization of amorphous sucrose and glucose, J. Agric. Food Chem., 4:72.

Mansfield, P., and Maudsley, A. A., 1977, Medical imaging by NMR, J. Br. Radiol., 50:188.

Mansfield, P., and Morris, P. C., 1982, NMR imaging in biomedicine, in: "Advances in Magnetic Resonance, Suppl. 2," Waugh, J. S., (Ed.), Academic Press, New York.

Mantsch, H. H., Saito, H., and Smith, I. C. P., 1977, Deuterium magnetic resonance, applications in chemistry, physics and biology, Adv. NMR Spectra, 11(4):211.

Maquet, J., Thevenau, H., Djabourov, M., and Papon, P., 1984, [1]H NMR study of gelatin gels, Int. J. Biol. Macromol., 6:162.

Martin, M. L., Delpuech, J. J., and Martin, G. J., 1980, Practical NMR Spectroscopy Chap. 7, Heyden and Son Ltd., London.

Mathur-De Vré, R., 1979, The NMR studies of water in biological systems, Prog. Biophys. Molec. Biol., 35:103.

Mathur-De Vré, R., Grimee-Declerck, and Lejeune, P., 1982, An NMR study of isotope distribution and the state of water in the hydration layer of DNA, in: "Biophysics of Water," Franks, F., and Mathias, S. F., (Eds.), John Wiley & Sons, Ltd., New York.

Meiboom, S., 1961, Nuclear magnetic resonance study of the proton transfer in water, J. Chem. Phys., 34(2):375.

Migchelsen, C., and Berendsen, H. J. C., 1973, Proton exchange and molecular orientation of water in hydrated collagen fibers, An NMR study of $H_2O$ and $D_2O$, J. Chem. Phys., 59(1):296.

Mora-Gutierrez, A., and Baianu, I. C., 1987, Physical and chemical studies of structure-functionality relationships of carbohydrate and carbohydrate wheat protein mixtures, Masters Thesis, University of Illinois, Urbana.

Morris, P. G., 1986, Nuclear Magnetic Resonance: Imaging in Medicine and Biology, Clarendon Press, Oxford.

Nagashima, N., and Suzuki, E., 1981, Pulsed NMR and state of water in foods, in: "Water Activity: Influence on Food Quality," Rockland, L. B., and Stewart, G. F., (Eds.), Academic Press, New York.

Nagashima, N., and Suzuki, E., 1984, Studies of hydration by broad-line pulsed NMR, Appl. Spectroscopy Rev., 20(1):1.

Nagashima, N., and Suzuki, E., 1985, Computed instrumental analysis of the behavior of water in foods during freezing and thawing, in: "Properties of Water in Foods," Simatos, D., and Multon, J. L., (Eds.), Martinus Nijhoff Publishers, Boston, MA.

Nakano, H., and Yasui, T., 1976, Denaturation of myosin-ATPase as a function of water activity, Agric. Biol. Chem., 40(1):107.

Nakano, H., and Yasui, T., 1979, Pulsed nuclear magnetic resonance studies of water in myosin suspension during dehydration, Agric. Biol. Chem., 43:89.

Nakazawa, F., Takahashi, J., Noguchi, S., and Kato, M., 1980, Water binding in gelatinized nonglutinous and glutinous rice starch determined by pulsed NMR, J. Home Econ. of Japan, 31(8):541.

Nakazawa, F., Takahashi, J., Noguchi, S., and Takada, M., 1983, Pulsed NMR study of water behavior in retrogradation process of rice and rice starch, J. Home Econ. of Japan, 34(9):566.

Nystrom, B., Moseley, M. E., Brown, W., and Roots, J., 1981, Molecular motion of small molecules in cellulose gels studied by NMR, J. Applied Polymer Sci., 26:3385.

O'Donnell, M., 1985, NMR blood flow imaging using multiecho phase constrast sequences, Med. Phys., 12:59.

Pande, A., 1975, Handbook of Moisture Determination and Control: Principles, Techniques and Applications, Vol. 2, Chap. 7, Marcel Dekker, Inc., New York.

Peemoeller, H., Kydon, D. W., Sharp, A. R., and Schreiner, L. J., 1984, Cross relaxation at the lysozyme-water interface: An NMR line-shape-relaxation correlation study, Can. J. Phys., 62:1002.

Perez, E., Kavten, R., and McCarthy, M. J., 1989, Noninvasive measurement of moisture profiles during the drying of an apple, in: "Drying '89," Mujumdar, A. S., (Ed.), Hemisphere Publishing Co., New York.

Pykett, I. L., Newhouse, J. H., Buonanno, F. S., Brady, T. J., Goldman, M. R., Kistler, J. P., and Pohost, G. M., 1982, Principles of nuclear magnetic resonance imaging, Radiology, 143:157.

Rabideau, S. W., and Hecht, H. G., 1967, Oxygen-17 NMR linewidths as influenced by proton exchange in water, J. Chem. Phys., 47(2):544.

Redpath, T. W., Norris, D. G., Jones, R. A., and Hutchison, J. M. S., 1984, A new method of NMR flow imaging, Phys. Med. Biol., 29:891.

Renou, J. P., Alizon, J., Dohri, M., and Robert, H., 1983, Study of the water-collagen system by NMR cross-relaxation experiments, J. Biochem. Biophys. Methods, 7:91.

Richards, R. E., and Franks, F., (Eds.), 1977, A discussion on water structure and transport in biology, Phil. Trans. R. Soc. Lond. B., 278:1.

Richardson, S. J., 1988a, Molecular mobilities of instant starch gels determined by oxygen-17 and carbon-13 nuclear magnetic resonance as affected by concentration and storage conditions, J. Food Sci., 53(4):1175.

Richardson, S. J., 1988b, Determination of moisture by pulsed nuclear magnetic resonance, NMR Short Course at the American Oil Chemists Society Annual Meeting, May 3-6, Phoenix, AZ.

Richardson, S. J., 1989, Contribution of proton exchange to the oxygen-17 nuclear magnetic resonance transverse relaxation rate in water and starch-water systems, Cereal Chem., 66(3):244.

Richardson, S. J., Baianu, I. C., and Steinberg, M. P., 1985, Relation between oxygen-17 NMR and rheological characteristics of wheat flour suspensions, J. Food Sci., 50:1148.

Richardson, S. J., Baianu, I. C., and Steinberg, M. P., 1986, Mobility of water in wheat flour suspensions as studied by proton and oxygen-17 nuclear magnetic resonance, J. Agric. Food Chem., 34(1):17.

Richardson, S. J., Baianu, I. C., and Steinberg, M. P., 1987a, Mobility of water in starch powders by nuclear magnetic resonance, Starch, 39(6):198.

Richardson, S. J., Baianu, I. C., and Steinberg, M. P., 1987b, Mobility of water in starch-sucrose systems determined by deuterium and oxygen-17 nuclear magnetic resonance, Starch, 39(9):302.

Richardson, S. J., Baianu, I. C., and Steinberg, M. P., 1987c, Mobility of water in sucrose solutions determined by deuterium and oxygen-17 nuclear magnetic resonance measurements, J. Food Sci., 52(3):806.

Richardson, S. J., Baianu, I. C., and Steinberg, M. P., 1987d, Mobility of water in corn starch suspensions determined by nuclear magnetic resonance, Starch, 39(3):79.

Richardson, S. J., and Steinberg, M. P., 1987, Applications of nuclear magnetic resonance, in: "Water Activity: Theory and Applications to Foods," Rockland, L. B., and Beuchat, L. R., (Eds.), Marcel Dekker, Inc., New York.

Ridgway, J. P., and Smith, M. A., 1986, A technique for velocity imaging using magnetic resonance imaging, Br. J. Radiol., 59:603.

Rodger, C., Sheppard, N., McFarlane, C., and McFarlane, W., 1978, Group VI - Oxygen, sulfur, selenium and tellurium, in: "NMR and the Periodic Table," Harris, R. K., and Mann, B. E., (Eds.), Academic Press, New York.

Rollwitz, W., 1985, Using radiofrequency spectroscopy in agricultural applications, Agric. Engr., May :12.

Rothwell, W. P., 1985, Nuclear magnetic resonance imaging, Applied Optics, 24(23):3958.

Rothwell, W. P., and Gentempo, P. P., 1984, Concepts in nonmedical applications of NMR imaging, Paper MD1-1, Topical Meeting on Industrial Applications of Computed Tomography and NMR Imaging, Optical Society of America, Aug. 13 and 14, Hecla Island, Manitoba, Canada.

Rothwell, W. P., and Gentempo, P. P., 1985, Nonmedical applications of NMR imaging, Bruker Reports, 1:46.

Rothwell, W. P., Holecek, D. R., and Kershaw, J. A., 1984, NMR Imaging: Study of fluid absorption by polymer composites, J. Polymer Sci., Polymer Letters Edition, 22:241.

Rothwell, W. P., and Vinegar, H. J., 1985, Petrophysical applications of NMR imaging, Applied Optics, 24(23):3969.

Saenger, W., 1987, Structure and dynamics of water surrounding biomolecules, Ann. Rev. Biophys. Biophys. Chem., 16:93.

Saltmarch, M., and Labuza, T. P., 1980, Influence of relative humidity on

the physicochemical state of lactose in spray-dried sweet whey
powders, J. Food Sci., 45:1231.

Samuelsson, E. G., and Hueg, B., 1973, Nuclear magnetic resonance (NMR)
as a method for measuring the rate of solution of dried milk,
Milchwissenschaft, 28(6):329.

Schwier, V. I., and Lechert, H., 1982, X-ray and nuclear magnetic resonance investigations on some structure problems of starch, Starch,
34(1):11.

Shih, J. M., 1983, Determination of the oil and water content of rice by
pulsed NMR, IBM Instruments, Inc., Danbury, CT.

Steinberg, M. P., and Leung, H., 1975, Some applications of wide-line and
pulsed NMR investigations of water in foods, in: "Water Relations
of Foods," Duckworth, R. B., (Ed.), Academic Press, New York.

Stilbs, P., 1987, Fourier transform pulsed-gradient spin-echo studies of
molecular diffusion, Progress in NMR Spectroscopy, 19:1.

Stokes, H. T., 1984, Study of diffusion in solids by pulsed nuclear
magnetic resonance, in: "Nontraditional Methods in Diffusion,"
Murch, G. E., Birnbaum, H. K., and Cost, J. R., (Eds.), The
Metallurgical Society of AIME, New York.

Suggett, A., 1976, Molecular motion and interactions in aqueous
carbohydrate solutions, III, A combined nuclear magnetic and
dielectric-relaxation strategy, J. Solution Chem., 5(1):33.

Suggett, A., Ablett, S., and Lillford, P. J., 1976, Molecular motion and
interactions in aqueous carbohydrate solutions, II, Nuclear-
magnetic-relaxation studies, J. Solution Chem., 5(1):17.

Suzuki, E., and Nagashima, N., 1982, Freezing-thawing hysteresis
phenomena of biological systems by the new method of proton magnetic
resonance, Bull. Chem. Soc. Jpn., 55(9):2730.

Suzuki, T., 1981, State of water in sea food, in: "Water Activity:
Influences on Food Quality," Rockland, L. B., and Stewart, G. F.,
(Eds.), Academic Press, New York.

Tait, M. J., Ablett, S., and Franks, F., 1972a, An NMR investigation of
water in carbohydrate systems, in: "Water Structure at the
Water-Polymer Interface," Jellinck, H. H. G., (Ed.), Plenum Press,
New York.

Tait, M. J., Ablett, S., and Wood, F. W., 1972b, The binding of water on
starch, an NMR investigation, J. Colloid and Interface Sci.,
41(3):594.

Tait, M. J., Suggett, A., Franks, F., Ablett, S., and Quickenden, P. A.,
1972c, Hydration of monosaccharides: A study by dielectric and
nuclear magnetic relaxation, J. Solution Chem., 1(2):131.

Troller, J. A., 1985, Effect of $a_w$ and pH on growth and survival of

Staphylococcus aureus, in: "Properties of Water in Foods," Simatos,
D., and Multon, J. L., (Eds.), Martinus Nijhoff Publishers, Boston,
MA.

Trumbetas, J., Fioriti, J. A., and Sims, R. J., 1976, Application of
pulsed NMR to fatty emulsions, JAOCS, 53:722.

Trumbetas, J., Fioriti, J. A., and Sims, R. J., 1977, Nuclear magnetic
resonance (NMR), JAOCS, 54:433.

Trumbetas, J., Fioriti, J. A., and Sims, R. J., 1978, Use of pulsed
nuclear magnetic resonance to predict emulsion stability, JAOCS,
55:248.

Tyrrell, H. J. V., and Harris, K. R., 1984, Diffusion in Liquids: A
Theoretical and Experimental Study, Chapter 5, Butterworth and Co.
Publishers, Ltd., Boston, MA.

Urbanski, G. E., 1981, Rheological properties of soybean and soybean-
solute systems, Ph.D. Thesis, University of Illinois, Urbana.

Van den Berg, C., and Bruin, S., 1981, Water activity and its estimation
in food systems: Theoretical aspects, in: "Water Activity:
Influences on Food Quality," Rockland, L. B., and Stewart, G. F.,
(Ed.), Academic Press, New York.

Vinegar, H. J., 1986, X-ray CT and NMR imaging of rocks, J. Petroleum Technology, March :257.

von Meerwall, E. D., 1983, Self-diffusion in polymer systems, measured with field-gradient spin-echo NMR methods, Advances in Polymer Sci., 54:1.

von Meerwall, E. D., 1985, Pulsed and steady field gradient NMR diffusion measurements in polymers, Rubber Chem. and Technology, 58:527.

Walmsley, R. H., and Shporer, M., 1978, Surface-induced NMR line splittings and augmented relaxation rates in water, J. Chem. Phys., 68(6):2584.

Wang, J. H., 1954, Theory of the self-diffusion of water in protein solutions, A new method for studying the hydration and shape of protein molecules, J. Am. Chem. Soc., 76:4755.

Weisser, H., 1980, NMR-Techniques in studying bound water in foods, in: "Food Process Engineering Vol. 1; Food Processing Systems," Linko, P., Malkki, Y., Olkku, J., and Larinkari, J., (Eds.), Applied Science Pub. Ltd., London.

Woessner, D. E., Snowden, B. S., and Chiu, Y. C., 1970, Pulsed NMR study of the temperature hysteresis in the agar-water system, J. Colloid and Interface Sci., 34(2):283.

Woodhouse, D. R., 1974, NMR in systems of biological significance, Ph.D. Thesis, University of Nottingham.

Yasui, T., Ishioroshi, M., Nakano, H., and Samejima, K., 1979, Changes in shear modulus, ultrastructure and spin-spin relaxation times of water associated with heat-induced gelation of myosin, J. Food Sci., 44:1201.

# USE OF $^{13}$C AND $^{17}$O NMR TO STUDY WHEAT STARCH-WATER-SUGAR INTERACTIONS WITH INCREASING TEMPERATURES

D. Sobczynska, C. Setser(a), H. Lim(a), L. Hansen(a) and
J. Paukstelis(b)

(a) Department of Foods and Nutrition
Kansas State University
(b) Department of Chemistry
Kansas State University
Manhattan, KS  66506

## INTRODUCTION

Nuclear magnetic resonance (NMR) is a very sensitive probe of molecular motion in the fluid state.  NMR spectroscopic techniques reported in this overview have been used to study starch-water-sugar systems at the molecular level.  The increased onset temperature of starch gelatinization in the presence of sugars is well known, but not so well understood.  Several mechanisms and variations of those mechanisms have been suggested including 1) a competition between the sugars and starch for the available water and thus changes in the free water volume (Derby et al., 1975; Hoseney et al., 1977; Slade and Levine, 1988b), 2) an inhibition of starch swelling by the sugars (D'Appolonia, 1972; Bean and Yamazaki, 1978; Savage and Osman, 1978; Wooton and Bamunuarachchi, 1980; Lelievre, 1984), which might be related to the competition for water, and 3) a penetration of the starch granule by the sugars and interactions leading to a stabilization of the granule that requires more energy to disrupt (Spies and Hoseney, 1982).  Studies of the starch-sugar-water systems have used the amylograph for macro-level measurements of viscosity changes (Lund, 1984), microscopic techniques to observe the loss of birefringence and increased swelling (Bean and Osman, 1959; Watson, 1977; Bean and Yamazaki, 1978; Bean et al., 1978), and differential scanning calorimetry (DSC) to measure melting temperatures and enthalpies (Wooton and Bamunuarachchi, 1980; Spies and Hoseney, 1980).

Wide-line nuclear magnetic resonance measures the energy absorbed by particular types of nuclei from a radio frequency field when resonance is achieved.  The early NMR techniques were used to obtain information on the mobility of the water protons in various systems, and thus, the binding of water.  The mobility of water in a system can be inferred from measurements of longitudinal and transverse relaxation times, $T_1$ and $T_2$, respectively.

*NMR Applications in Biopolymers*, Edited by
J. W. Finley *et al.*, Plenum Press, New York, 1990

However, the instrument is set so that only the sharp signal due to resonance of protons in the liquid state is measured. As a result, neither the hydrogen atoms of the solid in a sample, high molecular weight solutes, nor those of the most firmly bound water molecules, which behave as part of the solid, contribute to the signal (Duckworth, 1981).

Currently, $^{17}O$ NMR is considered more useful than $^{1}H$ NMR for understanding the state of water in heterogenous systems. Transverse relaxation rates ($R_2 = 1/T_2$) from $^{17}O$ directly and invasively monitor the molecular motions of the water molecule and are used to study the mobility states of water in food systems (Richardson et al., 1987c). The advantages of $^{17}O$ over proton and deuteron magnetic relaxation techniques include a large relaxation effect because of the strong quadrupolar interaction and no disturbance by cross-relaxation between the bound water protons and protons of the solute. In addition, $^{17}O$ relaxation of water is not influenced by proton exchange with non-water nuclei except for a narrow pH range around neutrality. However, $^{17}O$ relaxation is affected by exchange of entire water molecules. The broadening of the $^{17}O$ signals around neutrality because of the proton exchange can be eliminated by proton decoupling (Halle et al., 1981; Richardson et al., 1986, 1987a, 1987b).

Water in heterogenous systems like the starch-water-sugar system usually is distributed between several environments. Each region is characterized by different intrinsic relaxation rates and resonance frequencies. In the case of a starch-water-sugar system, three different states of water are noted: 1) water associated with starch molecules (polymer water), 2) water associated with sugar (solute water) and 3) bulk water, which also is referred to as free water.

Previous NMR studies on starch systems involved heating the system to a defined temperature and investigating the associated changes from the heating (Lechert, 1981; Richardson et al., 1986). New NMR techniques had to be developed to study dynamic changes in the starch-water-sugar systems as the heating occurred. The temperature range studied corresponded to that used in thermal processing of a food product and the experiments were designed to elucidate the mechanism(s) operative in gelatinization processes of starch in the presence of sugars.

METHODS

The method developed by Hansen et al. (1987) was to prepare starch (Aytex P, prime wheat starch, Ogilvie Mills) and sugar dispersions substituting 0.15% (w/w) xanthan gum (Keltrol T, Kelco) solution for water as the solvent to prevent starch precipitation. Checks with DSC and NMR indicated that the low concentration of xanthan gum used in the dispersion had no effect on the starch gelatinization onset temperature or on chemical shifts compared to a starch-distilled water system. In the studies reported, sugars used were either sucrose, glucose or fructose [reagent grade, Mallinckrodt, Inc. (for $^{13}C$ studies) or Fisher (for $^{17}O$ water studies)] at concentrations of 0.5, 1.0, 1.5 or 2.0 M (14, 28, 43 and 57%, respectively). Dispersions contained either 30% starch and 70% sugar solution or 10% starch and 90% sugar solution (w/w) unless indicated as otherwise.

## $^{13}C$ NMR Measurements

Sufficient carbon atoms ($^{13}C$) of both sucrose and starch were needed for strong NMR signals; this was the basis for the starch-sugar ratio used in all investigations. In addition, the dispersions had to remain sufficiently fluid to be injected into a 5-mm (id) NMR sample tube. Gelatinization of starch, as evidenced by use of DSC techniques, was apparent during the temperature range used at the different molar sugar concentrations studied. The starch-sugar dispersions were prepared and allowed to equilibrate at room temperature for approximately 8 hrs. before measurement in the NMR spectrometer.

NMR measurements were made following the procedure of Hansen et al. (1987). All NMR chemical shifts were determined relative to the external $Me_2SO$ standard of 50% $D_2O$:50% $Me_2SO$. In order to obtain a stronger lock signal and better resolution with the solubilized starch experiments, the external reference used was 75% $D_2O$:25% $Me_2SO$. The standard was not affected by the solutions, so the same standard was used for every sample. Chemical shifts using $^{13}C$ NMR only determined changes in interactions between the sugar and starch as a function of temperature, and did not determine initial interactions. Samples were held 5 min. at each temperature before data acquisition was started to ensure temperature equilibrium and to obtain reproducible results.

At each temperature, differences were noted between the chemical shifts of the carbon atoms of the sugar in solution alone and the chemical shifts of the carbon atoms of the sugar in the starch-sugar dispersions. The assumption was made that an interaction occurring between the sugar and the starch would alter the environment near the sugar carbon atoms. Thus, the chemical shift of the carbon atoms of the starch-sugar dispersions would differ from that of the carbon atoms in the sugar solution alone. A plot of the differences in chemical shifts as a function of temperature should produce a line with approximately zero, or constant, slope, if no inter-actions had occurred. Deviations in slope for a particular carbon would indicate possible interaction sites; the larger the deviation, the stronger the interaction. Deviations of less than one Hz (measurements were made to $\pm$ 0.1 Hz) were not thought to be significant and were considered within the experimental error of the measurements.

Chemical shifts of carbon atoms of the sucrose molecule had been assigned previously (Bock and Lemieux, 1982; Hull, 1982). To clarify whether a carbon atom was associated with the fructose or glucose moiety of the sucrose molecule, in all chemical shift discussions, carbon atoms have been labelled F1 through F6 and G1 through G6. That is, F1 represents the C1 carbon of the fructose moiety and G1 represents the C1 carbon of the glucose moiety.

## $^{17}O$ Measurements

A low natural abundance of oxygen's one magnetically active isotope ($^{17}O$, 0.037%) necessitated the use of 0.5% enriched $^{17}O$ (Isotec Inc.) water in the studied systems unless otherwise noted. Enrichment enabled detection of oxygen mobility of only water and not starch or sucrose in the starch–water–sucrose system.

Data for $^{17}O$ enriched water solutions were acquired with the following conditions: a spectrometer frequency of 54.234 MHz, a sweep width of 5000 Hz, a 90° pulse width of 27.0 μs, 4°K data points, broad band $^{1}H$ decoupling and a 205 msec acquisition time with no delay. Under those conditions adequate signal-to-noise was achieved in 64 pulses. Samples were held 5 min. at each 2°K temperature interval before data acquisition. Line-shapes were evaluated in terms of least squares fits of the peaks to a Lorentzian equation after application of a 10 Hz line broadening. Line-widths ($\Delta V_{obs}$) were half-height values for the water peak at each temperature. Transverse relaxation rates $R_2$ ($sec^{-1}$) were calculated according to Richardson, et al. (1987a).

$$R_2 \left(s^{-1}\right) = \pi \, \Delta V_{obs} = 1/T_2$$

$R_2$ was determined for each spectrum using the fit of the peak to a Lorentzian lineshape, which was confirmed by direct measurements from the plot.

## Calculations of Free, Solute and Polymer Water for Theoretical Model

A model was constructed to relate the NMR measurements to what occurred to the water in a starch–sucrose system at varying concentrations of each. Correlation times for the water molecules in wheat flour were calculated to be 16.7 picoseconds (ps) for a quadrupole coupling constant of I=6.67 MHz, or 45.8 ps for I=4.03 MHz (Richardson et al., 1986). Correlation times for the sucrose–water systems were calculated to range from 100 ps to 9.5 ps depending on the hydration number assumed (Richardson et al., 1987c). Correlation time for free water at 300°K is 2.40 ps (Halle and Wennerström, 1981). In these states, the water molecules exchange a few orders of magnitude faster than the intrinsic relaxation rates and inverse resonance frequency. Inverse resonance frequency was equal to 18.4 nanoseconds (ns); thus, it is slow enough to fulfill the extreme narrowing condition. In extreme narrowing situations, the observed relaxation rates can be written as population weighted averages:

$$R_{stsu} = \left[\left(P_{sol}\right)\left(R_{sol}\right)\right] + \left[\left(P_{pol}\right)\left(R_{pol}\right)\right] + \left[\left(P_{f_1}\right)\left(R_f\right)\right]$$

where R=relaxation, st=starch, su=sucrose, sol=solute, pol=polymer, f=pure or free water, $f_1$=free water with starch and sucrose in the system, and P=portion.

The intrinsic relaxation rate for the bulk water in a heterogenous system is, by definition, equal to the measured rate for a sample containing only the aqueous medium. The intrinsic relaxation rates for solute water were determined from the rates measured for a saturated solution of a solute, e. g., sucrose. The intrinsic relaxation rate for polymer water is equal to the rate measured for a saturated solution of a polymer such as starch.

To calculate the portion of theoretical (th) free water, $P_f^{th}$, in a starch-water-sucrose system, the following assumptions had to be made:

1) The interaction in a starch-water-sucrose system was negligible between starch and sucrose in the temperature range studied, and

2) The amount of water associated with starch, $P_{pol}$, and that associated with sucrose, $P_{sol}$, were the same in a starch-water-sucrose system as they were for starch-water and sucrose-water systems individually at the same temperature and concentrations.

The portion of free water was calculated from the equation,

$$P_f^{th} = 1 - \left(P_{pol}\right) - \left(P_{sol}\right) . \tag{1}$$

This value was compared to the value calculated from observed experimental (obs) data using the same assumptions given above.

$$R_{stsu} = \left(P_{sol} R_{sol}\right) + \left(P_{pol} R_{pol}\right) + \left(P_f^{obs} R_f\right) \tag{2}$$

$$P_{f_1}^{obs} = \frac{R_{stsu} - \left(P_{sol} R_{sol}\right) - \left(P_{pol} R_{pol}\right)}{R_f} \tag{2a}$$

In both cases, the portion of the polymer water, $P_{pol}$, and the portion of the solute water, $P_{sol}$, were calculated from the same data used for the separate systems of starch-water and sucrose-water, respectively.

$$R_{st} = \left(P_{pol} R_{pol}\right) + \left(P_{f_2} R_f\right), \tag{3}$$

where $f_2$ = free water in the starch system, and

$$R_{su} = \left(P_{sol} R_{sol}\right) + \left(P_{f_3} R_f\right), \tag{4}$$

where $f_3$ = free water in the sucrose system.

465

RESULTS

## Sucrose - Starch Interactions

The effects of different molar concentrations of sucrose on chemical shifts in starch-sucrose dispersions using $^{13}$C NMR are illustrated in Figs. 1 through 4. In general, as sucrose concentration increased, the temperature increased at which major chemical shift differences occurred. Interestingly, the major deviations in slope occurred prior to the onset temperature of starch gelatinization. Onset temperatures of starch, measured with the DSC, as altered by 0.5 M, 1.0 M, 1.5 M, and 2.0 M sucrose solutions, were 61.0 ± 0.1, 67.8 ± 0.3, 77.5 ± 0.7, and 88.4 ± 0.2°C, respectively. After the onset of starch gelatinization, relatively few deviations in slope were noted, which was especially true for the glucose carbons.

A comparison of the systems with the different molar concentrations of sucrose indicated that G1, G5, G6, F1, and possibly F3 and F6 had significant chemical shift differences at all concentrations investigated. At all concentrations, chemical shift differences of G1, G5, and F6 decreased, whereas the chemical shift differences of G6, F1, and F3 increased as a function of increasing temperature. The differences in chemical shifts of these carbon atoms were large enough (>1 Hz) to be considered possible interaction sites with starch. Results of solubilized starch investigations indicated that the starch granule was necessary for interactions or changes in interactions as a function of temperature to occur (Hansen et al., 1989).

Research findings of Bock and Lemieux (1982) supported the idea that chemical shift differences were a result of starch-sucrose interactions and not conformational changes in the sucrose molecule. They stated that sucrose in solution is a rigid molecule, stabilized by a hydrogen bond between OH-1$^f$ and O-2$^g$, which exists in the crystal lattice, with only slight flexing of the furanoid ring and different degrees of freedom of rotation around the hydroxymethyl to carbon bonds. Thus, differences in chemical shifts of sucrose between the sucrose solution and starch-sucrose dispersions would not be the result of conformational changes in the sucrose molecules, but likely of an interaction between the sucrose and starch.

## Glucose - Starch Interactions

In Figs. 5 and 6, chemical shift differences for glucose-starch systems are given for 0.5 and 1.5 M glucose concentrations, respectively. Again the large number of changes prior to onset temperature of gelatinization are obvious. Major chemical shifts occurred at 1.0 and 2.0 M concentrations in a similar manner, and all indicated that C1a, C4 and C6 were involved. Most chemical shifts occurred prior to onset temperature and then leveled off with the exception of the C1a; however, this carbon was the hardest to resolve. Chemical shift differences with glucose systems were numerically

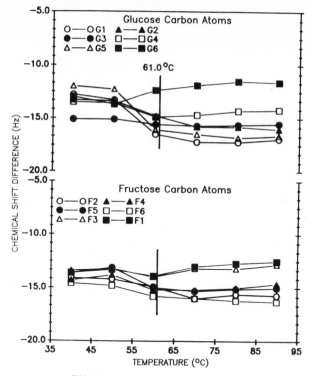

70%–0.5M SUCROSE SOL.:30% STARCH

Fig. 1. Chemical shift differences between sucrose carbon atoms in a
70% 0.5 M sucrose solution with 30% starch (w/w) dispersion
and in a 0.5 M sucrose solution as a function of increasing
temperature. Spectra and chemical shift data were acquired
using the following NMR instrument settings: 512 pulses;
pulse width 6.0 sec (60°); 0.49 sec delay between pulses; line
broadened 5.0 hertz. Chemical shifts are all upfield relative
to reference sucrose. The upper portion of the figure shows
the glucose carbon atoms and the bottom portion illustrates
the fructose carbon atoms of the sucrose molecule. Vertical
line represents gelatinization onset temperature of the system
as measured by DSC; onset temperature of a 70% distilled
water: 30% starch system is 55.2°C.

larger than those with the sucrose systems, yet the chemical shift differences of the C6 were notably similar in both systems. The slope of the
line of the differences in chemical shifts for C6 of glucose changed in a
positive direction, but once onset temperature of starch gelatinization was
reached, the C6 glucose line became horizontal while the G6 sucrose line
continued to increase.

Fructose – Starch Interactions

Fructose-starch interactions were more difficult to interpret than
either sucrose-starch or glucose-starch systems because fructose exists in

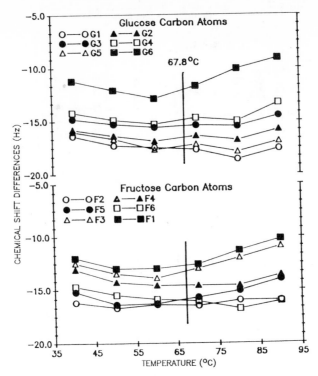

Fig. 2. Chemical shift differences between sucrose carbon atoms in a 70% 1.0 M sucrose solution with 30% starch (w/w) dispersion and in a 1.0 M sucrose solution as a function of increasing temperature. All other conditions are the same as for Fig. 1.

four configurations: alpha-D-fructofuranose, β-D-fructofuranose, alpha-D-fructopyranose, and β-D-fructopyranose. Strong enough signals were obtainable only at 2.0 M concentration for the β-D-fructofuranose (Fig. 7) and the β-D-fructopyranose (Fig. 8). The significant chemical shift differences were noted between 60° and 70°C and were particularly notable for

C1 in both the β-D furanose and pyranose configurations.

## Water Mobility Studies

Although an interaction between the sugars and starch was suggested by the $^{13}C$ studies, a three-way interaction among the sugar, starch, and water was not eliminated. Interestingly, as the molar concentration of sucrose solutions increased, the differences or changes in slope decreased. The strongest interactions and most carbon atoms (i.e., G1, G2, G3, etc.) were involved at the lowest sucrose concentration investigated. Hansen et al. (1987) suggested that perhaps only a limited number of interaction sites on the starch granules were available and, at higher sucrose concentrations,

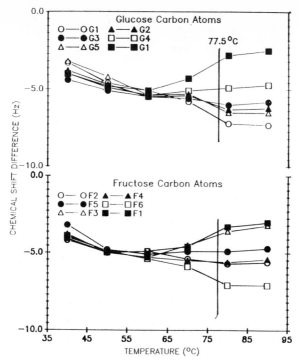

Fig. 3.  Chemical shift differences between sucrose carbon atoms in a
70% 1.5 M sucrose solution with 30% starch (w/w) dispersion
and in a 1.5 M sucrose solution as a function of increasing
temperature.  All other conditions are the same as for Fig. 1.


more sucrose atoms existed than sites for interaction.  Thus, only a portion
of the sucrose interacted with the starch, and chemical shift changes
appeared to be fewer.

Our results confirm the earlier, related research by Richardson et al.
(1987c) using [17]O-NMR techniques, who found that as sucrose concentration
increased, relaxation rates of water increased, indicating decreased water
mobility.  They attributed the decrease in water mobility to: 1) formation
of intermolecular hydrogen bonds between sucrose and water; 2) formation of
water bridges between sucrose molecules; and 3) extensive sucrose to sucrose
hydrogen bonding.  If extensive hydrogen bonding was occurring between
sucrose and water or sucrose and sucrose, then less sucrose likely would
bind or interact with starch.

A calculated percentage of starch mobile enough to be measured by [13]C
NMR was determined to be 82% of the starch in the 0.5 M sucrose-starch
dispersion but about 62% for all other concentrations of sucrose based on C2
intensities (Hansen et al., 1989).  Callaghan and co-workers (1983) reported

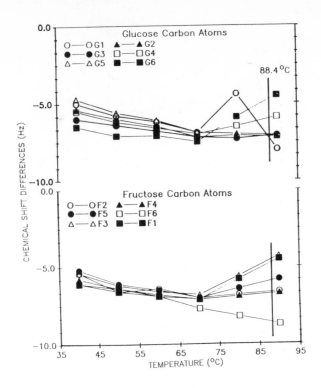

70%-2.0M SUCROSE SOL.:30% STARCH

Fig. 4. Chemical shift differences between sucrose carbon atoms in a
70% 2.0 M sucrose solution with 30% starch (w/w) dispersion
and in a 2.0 M sucrose solution as a function of increasing
temperature. All other conditions are the same as for Fig. 1.

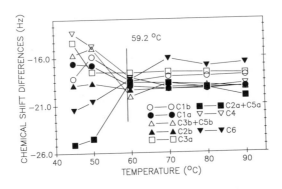

70% 0.5M GLUCOSE:30% STARCH

Fig. 5. Chemical shift differences between glucose carbon atoms in a
70% 0.5 M glucose solution with 30% starch (w/w) dispersion
and in a 0.5 M glucose solution as a function of increasing
temperature. All other conditions are the same as for Fig. 1.

470

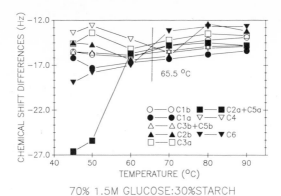

Fig. 6.  Chemical shift differences between glucose carbon atoms in a
         70% 1.5 M glucose solution with 30% starch (w/w) dispersion
         and in a 1.5 M glucose solution as a function of increasing
         temperature.  All other conditions are the same as for Fig. 1.

that 60% of the starch in a 10% starch-water paste was mobile enough to give
NMR signals.  A 0.5 M sucrose solution would have more water than solutions of
greater molar concentrations, and possibly this additional water increased the
mobility of starch, enabling it to contribute more to peaks.  Also, at lower
molar concentrations, more soluble starch would be leached from the granule.
Solubles leached from the granule are more mobile than intact starch and presumably contribute more to the signals.  Whether or not the amount of starch
contributing to the signals affected the chemical shifts of the sucrose has
not been determined.

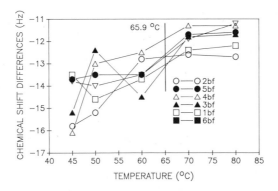

70% 2.0M FRUCTOSE SOLUTION:30% STARCH
Beta—Furanose Carbon Atoms

Fig. 7.  Chemical shift differences between β-D-fructofuranose carbon
         atoms in a 70% 2.0 M fructose solution with 30% starch (w/w)
         dispersion and in a 2.0 M fructose solution as a function of
         increasing temperature.  All other conditions are the same as
         for Fig. 1.

Figure 9 depicts results of $^{17}$O transverse relaxation rate, or $R_2$, of
the 30% starch-water-sugar system with increasing temperatures and sucrose
concentrations from 308°-348°K (35°-75°C) taken at 2°K intervals. When the
system contained only starch, $R_2$ increased in the 323°-327°K (50°-54°C)
range. When water is bonded tightly to the substrate, it is highly
immobilized, reducing the $T_2$, thus increasing the corresponding $R_2$ (Leung,
et al., 1979), which was noted in the starch shortly after 50°C. In the
systems reported in this work, the addition of sucrose to the starch system
resulted in an increased $R_2$, indicating less water mobility with increasing
sucrose concentration. The $R_2$ maximum also occurred at higher temperatures
as concentration increased, which corresponds both to the progressively
higher onset temperatures with increased concentrations of sucrose noted in

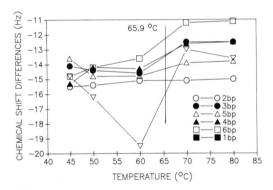

Fig. 8.  Chemical shift differences between β-D-fructopyranose carbon
atoms in a 70% 2.0 M fructose solution with 30% starch (w/w)
dispersion and in a 2.0 M fructose solution as a function of
increasing temperature.  All other conditions are the same as
for Fig. 1.

the DSC measurements and to the interactions of sucrose carbon atoms noted
in the $^{13}$C studies. If the concentration of sugar in the system is
increased, $R_2$ might be expected to increase proportionally. The observed
curves (Fig. 9) suggest that the amount $R_2$ increased depends upon the
starch-sucrose concentration in the starch-water-sucrose system.

The calculated portions of free water from the $^{17}$O measurements, $P_f^{obs}$,
[Eq. (1)] with changing temperatures are shown for 30% starch with 1.5, 1.0
and 0.5 M sucrose and for 10% starch with 1.5 and 1.0 and 0.5 M sucrose in
Fig. 10. The measured water mobility curves from the $^{17}$O data deviated from
the theoretical curves shown in Fig. 11. The amount of free water

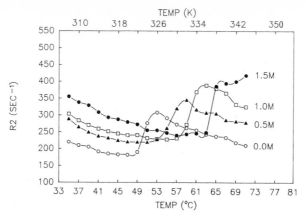

Fig. 9.  Transverse relaxation rate of the 30% starch-water-sucrose
system as a function of increasing temperature with increasing
sucrose concentrations:   0 - no sucrose; ▲ - 0.5 M sucrose;
□ - 1.0 M sucrose; and ● - 1.5 M sucrose.  All measurements
were made with enriched $^{17}$O water.

calculated from the observed data did not change linearly, but the changes
in area varied and were striking.  The amount of free water, $P_f^{th}$, for all
systems is similar, and indicates less free water would be present as the
amount of dispersed substrates increases:   10% starch: 0.5 M sucrose > 10%
starch: 1.0 M sucrose > 10% starch: 1.5 M sucrose > 30% starch: 0.5 M
sucrose > 30% starch: 1.0 M sucrose > 30% starch: 1.5 M sucrose.

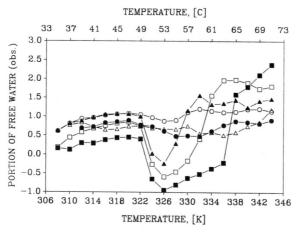

Fig. 10.  Calculated portion of free water based on observations of
sucrose starch-water system for 0 - 10% starch, 0.5 M sucrose;
● - 10% starch, 1.0 M sucrose; Δ - 10% starch, 1.5 M sucrose;
▲ - 30% starch, 0.5 M sucrose; □ - 30% starch, 1.0 M sucrose;
and ■ - 30% starch, 1.5 M sucrose as a function of increasing
temperature.

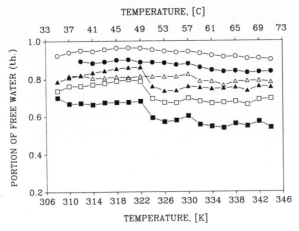

Fig. 11. Theoretical calculated portion of free water from starch-water
and sucrose-water systems for O – 10% starch, 0.5 M sucrose;
● – 10% starch, 1.0 M sucrose; Δ – 10% starch, 1.5 M sucrose;
▲ – 30% starch, 0.5 M sucrose; □ – 30% starch, 1.0 M sucrose;
and ■ – 30% starch, 1.5 M sucrose as a function of
increasing temperature.

Furthermore, in most systems, $P_f^{th}$ decreases slowly with increasing
temperature from 318°– 344°K (41°–67°C). The $P_f^{obs}$ curves for these systems
calculated using the polymer water and solute water data obtained from
starch-water and sucrose-water observations [Fig. 12, based on Eqs. (3) and
(4)] and from data obtained for starch-water-sucrose systems (Eq. 2a) give
different results. Only trends shown for the model systems in these curves
(Figs. 10–12) should be noted and discussed in comparing the observed curve,
calculated from $^{17}O$ measurements, to the theoretical curve.

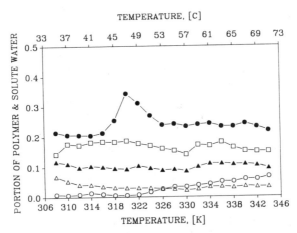

Fig. 12. Calculated portion of polymer water for O – 10% starch-
water and ● – 30% starch-water systems and solute water
for Δ – 0.5 M sucrose-water, ▲ – 1.0 M sucrose-water and
□ – 1.5 M sucrose-water systems as a function of
increasing temperature.

For all systems, increasing temperatures between 308°-322°K (35°-49°C) resulted in slightly increasing amounts of free water (Fig. 10). The amount of free water in the solution is similar when the concentrations of sucrose are the same even with different starch concentrations. In this temperature range, sucrose is the more influential of the two dispersed phases on water mobility. This was noted in the model described as the decrease of free water with increasing sucrose concentrations, and the decrease is nearly linear in this temperature range with the increase in sucrose concentration.

After the initial increase of free water, a decrease follows, which is sharpest in systems containing 30% starch. The minimum level is at 326°K (53°C). Apparently the starch and sucrose are competing for the available water at the high levels of starch and not enough is actually available as indicated in the model by negative amounts. Shiotsubo and Takahashi (1984) found insufficient water for complete starch gelatinization if the starch concentration was greater than 30%.

When the ratio of starch and sucrose was similar as with 30% starch: 1.5 M sucrose and 10% starch: 0.5 M sucrose, the decrease in the amount of water was similar to the subsequent increase. These observations suggest that in all starch-water-sucrose systems, the decrease of free water at 322°-326°K (49-53°C) is associated with the presence of starch in systems. Secondly, differences were observed because of sucrose at all temperatures. The temperature at which the highest level of free water was noted apparently is related to the concentration of both sucrose and starch (Fig. 10). Another interesting observation can be made regarding melting; the onset temperatures as determined by DSC appears to occur a few degrees after the water reaches its maximum mobility. For example, onset temperature of the 0.5 M sucrose: 30% starch system was 61.0°C and maximum water mobility was at 59°C. For the 1.0 M sucrose: 30% starch system, onset temperature was determined to be 67.8°C and maximum water mobility was about 63°-63°C. Further work at varied starch: sucrose ratios and at higher temperatures is needed to evaluate this observation more extensively.

According to the work of Bean and Yamazaki (1978), starch granules swell at about 331°K (58°C), just beyond the minimum free water temperature range. Apparently, according to this work, the concentration of sucrose influences the amount of mobile water present when the process starts and the temperature at which the process is completed. For a given sucrose concentration, the concentration of starch determined the amount of highly mobile water after the starch granule swelling was completed. After the free water reached a maximum at approximately the gelatinization onset temperature, as determined by DSC measurements (Hansen et al., 1987), and in agreement with Jaska (1971), the amount of free water in the system again slowly decreased. The slightly decreased level of free water after onset temperature of gelatinization could be related to an immobilization of the water molecules in a developing gel network, which was observed visually, particularly in the 30% starch system, at the completion of the experiment.

A comparison (Fig. 13) of the $^{17}O$ enriched 30% starch: 1.5 M sucrose system with the same system prepared with natural abundance oxygen, which measured both sucrose and water oxygen atoms, indicated that sucrose mobility changed at about 330°K (57°C), whereas water mobility changed after 334°K (61°C). In the natural abundance $^{17}O$ water system, the relaxation rates of both water and sucrose molecules were observed; whereas, in the enriched $^{17}O$ system only the relaxation rates of water molecules are seen. If sucrose mobility did not change in the temperature range shown in Fig. 13, one would observe higher relaxation rates for the natural abundance systems throughout the range of temperatures studied rather than the lower rate that was observed after 334°K (61°C). In observations made on starch-water systems without any added sucrose, the high relaxation rates of the natural abundance system did parallel the rates for the enriched system over the entire range of temperatures (305°-347°K). Thus, an early change in starch mobility was not indicated. The temperature when sucrose mobility starts to decrease indicated on Fig. 13 corresponds to the temperature range when starch-sucrose interactions were occurring (Fig. 3) based on the $^{13}C$ data. One could hypothesize that the sucrose of the solution interacts with starch granules just prior to starch-water interactions.

DISCUSSION

The water uptake by starch granules is controlled by the chemical potential of water. In general, for any chemical reaction, as the

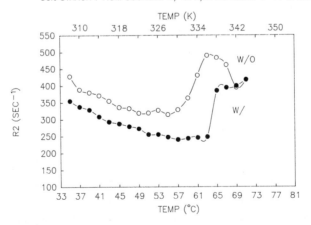

30% STARCH : 1.5M SUCROSE W/ & W/O ENRICHED O–17 WATER

Fig. 13. Transverse relaxation rate of the 30% starch, 1.5 M sucrose-water systems with natural water (O) and with 0.5% O-17 enriched water (●) as a function of increasing temperature.

difference between chemical potentials gets larger, the reactivity will increase. In the reported measurements, the heating rate was 2°K/5min. According to Shiotsubo and Takahashi (1984) at a heating rate slower than 0.5°K/min, gelatinization is approaching equilibrium. Thus, these measurements are assumed to have been made after an equilibrium state had been reached, and therefore, after the chemical potential of water outside and inside the granules was essentially equal.

The $^{17}$O relaxation rate data demonstrate effects of sucrose on gelatinization temperature. Data and calculated curves from the $^{13}$C and $^{17}$O NMR studies seem to indicate the following possible modes of action. Native starch initially has a moisture content of 6-10% w/w. The first layers are the bonded capillary water and the outer layers are considered mobile. When the starch granules are heated, the capillary water chemical potential increases as the average free energy increases. The mobile water inside the granule contains the dissolved amylose and salts. Thus, the solute concentration inside the granules is high and the activity of water is low compared to the activity of the water with low solute concentrations outside the starch granules. At low temperatures, limited hydration of starch granules occurs. With increased temperatures and no or low concentrations of sucrose, swelling continues because of the large difference in the free water concentration gradient. This corroborates the work of Lelievre (1984) that low concentrations of sugars alter the swelling of the starch granules.

In systems with higher concentrations of sucrose, one possibility is that a lowered chemical potential of water in the sucrose solutions precludes the same extent of hydration (swelling) of the granule that occurs in the absence of sucrose. The water mobility data and calculations given in Figs. 9 and 10 add support to the hypotheses of Derby et al. (1975), Hoseney et al. (1977) and Slade and Levine (1988b) that changes in free water volume and a competition between the sucrose and starch for the available water occur. If the starch concentration and sucrose concentration are both high, not enough water is present at low temperatures to equilibrate the system as noted in the large decrease in free water at 326°K in the 30% starch: 1.5 M sucrose system (Fig. 10). If the concentration of sucrose is high relative to the concentration of starch (10% starch: 1.5 M sucrose), more sucrose is present than binding sites, apparently, and fewer chemical shifts were noted in the $^{13}$C studies.

Another possibility is that sucrose directly interacts with the glucose residues of the starch to increase the rigidity, and, thereby, increases the temperature needed to "mobilize" the starch chains. Both the $^{17}$O enriched/natural abundance data (Fig. 13) and the $^{13}$C data (Figs. 1-4) lend support to the hypothesis that sucrose interacts with the starch approximately 10°-20°C prior to the onset temperature of gelatinization creating a need for more energy to break the starch-sucrose-water bonds. Possibly this interaction could be with the amylose chains dissolved in the granular capillary water, which, in turn, gives more rigidity to the granule. With both sugars and amylose in the granule, little oscillation can occur and additional energy is needed to break hydrogen bonds. The delay in onset temperature of gelatinization (DSC studies), and delays in the decrease of sucrose and water mobility (Fig. 9) as sucrose concentration of the system increases are noted in this regard. The sucrose also could compete directly with the water for the hydration sites of the glucose residues. Each of these possibilities implies an interaction of sucrose with the starch, that is, an equilibrium process wherein sucrose is distributed between the free and bound state. That would change its $^{17}O$ relaxation rate compared to the sucrose solution. As temperature is increased, the water might become more mobile and the segmental mobility of chains would increase, which would cause more hydrogen bonds to be broken until onset temperature is reached. Likely, the maximum free water point was observed in the system at the point when the swollen granules rupture. Once the bonds are broken (onset temperature of gelatinization), the water would again be mobile until finally it would be immobilized in a gel network with cooling.

CONCLUSIONS

Our results suggest that sugar-starch interactions might occur during the heating of starch. Major changes in chemical shifts occurred in the $^{13}C$ data 10°-20°C prior to the temperature range of starch gelatinization measured by the DSC. As molar concentrations of sucrose increased, the number of carbon atoms involved (i.e., G1, G2, G3, etc.) and the strength of the interactions appeared to decrease. According to the $^{17}O$ enriched/ natural abundance data, sucrose mobility also changed prior to onset temperature, which corresponds to the changes suggested by the $^{13}C$ data for sucrose-starch interactions. The data further indicate a competition of starch and sucrose for water as the concentration of sucrose and the resultant starch: sucrose ratio influenced the temperature at which the water mobility was at its maximum.

ACKNOWLEDGMENTS

The authors are appreciative of the helpful comments and suggestions given by Dr. Delbert Mueller, Department of Biochemistry, Kansas State University. We gratefully acknowledge the financial support of Nabisco Brands Inc., the Kansas Agricultural Experiment Station, and the United States Department of Agriculture.

# REFERENCES

Bean, M. M. and Osman, E. M., 1959, Behavior of starch during food preparation. II. Effects of different sugars on the viscosity and gel
strength of starch pastes. Food Res., 24:665.

Bean, M. M. and Yamazaki, W. T., 1978, Wheat starch gelatinization in
sugar solutions. I. Sucrose; light microscopy and viscosity
effects. Cereal Chem., 55:936.

Bean, M. M., Yamazaki, W. T. and Donelson, D. H., 1978, Wheat starch
gelatinization in sugar solutions. II. Fructose, glucose, and
sucrose cake performance. Cereal Chem., 55:945.

Bock, K. and Lemieux, R., 1982, The conformational properties of sucrose
in aqueous solution: Intramolecular hydrogen-bonding. Carbohydr.
Res., 100:63.

Callaghan, P. T., Jolley, K. W., Lelievre, J. and Wong, R. B. K., 1983,
Nuclear magnetic resonance studies of wheat starch pastes. J.
Colloid and Interface Sci., 92:332.

D'Appolonia, B. L., 1972, Effect of bread ingredients on starch gelatinization properties as measured by the amylograph. Cereal Chem.,
49:532.

Derby, R. I., Miller, B. S., Miller, B. F. and Trimbp, H. B., 1975,
Visual observations of wheat-starch gelatinization in limited water
systems. Cereal Chem., 52:702.

Duckworth, R. B., 1981, in: "Water activity influences on food quality,"
(Rockland, L. B., Stewart, G. F., ed.) p. 295-317. Academic Press.,
Inc.

Halle, B., Anderson, T., Forsen, S. and Lindman, B., 1981, Protein
hydration from water oxygen-17 magnetic relaxation. J. Am. Chem.
Soc., 103:500.

Halle, B. and Wennerström, H., 1981, Interpretation of magnetic resonance
data from water nuclei in heterogenous systems. J. Chem. Phys.,
75:1928.

Hansen, L. M., Paukstelis, J. V. and Setser, C. S., 1987, [13]C nuclear
magnetic resonance spectroscopic methods for investigating sucrose-
starch interactions with increasing temperature. Cereal Chem.,
64:449.

Hansen, L. M., Setser, C. S., and Paukstelis, J. V., 1989, Investigations
of sugar-starch interactions using [13]C-NMR. I. Sucrose. Cereal
Chem., 66:411.

Hoseney, R. C., Atwell, W. A. and Lineback, D. R., 1977, Scanning
electron microscopy of starch isolated from baked products. Cereal
Foods World., 22:56.

Hull, W. E., 1982, Two-Dimensional NMR. Bruker Analytische Messtechack.,
Karlsruhe, West Germany.

Lechert, H. T., 1981, in: "Water Activity Influences on Food quality,"
L. B. Rockland and G. F. Stuart, (Ed.), pp. 223-245. Academic
Press, Inc., New York.

Lelievre, J., 1984, Effects of sugars on the swelling of crosslinked
potato starch. J. Colloid and Interface Sci., 101:225.

Leung, H. K., Magnuson, J. A. and Bruinsma, B. L., 1979, Pulsed nuclear
magnetic resonance study of water mobility in flour doughs. J.
Food Sci., 44:1408.

Lund, D., 1984, Influence of time, temperature, moisture, ingredients,
and processing conditions on starch gelatinization. CRC Crit. Rev.
in Food Sci. and Nutr., 20:249.

Richardson, S. J., Baianu, I. C. and Steinberg, M. P., 1986, Mobility of
water in wheat flour suspensions as studied by proton and oxygen-17
nuclear magnetic resonance. J. Agr. and Food Chem., 34:17.

Richardson, S. J., Baianu, I. C. and Steinberg, M. P., 1987a, Mobility of
water in starch-sucrose systems determined by deuterium and
oxygen-17 NMR. Starke, 39:302.

Richardson, S. J., Steinberg, M. P., De Vor, R. E. and Sutherland J. W., 1987b, Characterization of the oxygen-17 nuclear magnetic resonance water mobility response surface. J. Food Sci., 52:189.

Richardson, S. J., Baianu, I. C. and Steinberg, M. P., 1987c, Mobility of water in sucrose solutions determined by deuterium and oxygen-17 nuclear magnetic resonance measurements. J. Food Sci., 52:806.

Savage, H. L. and Osman, E. M., 1978, Effects of certain sugars and sugar alcohols on the swelling of corn starch granules. Cereal Chem., 55:447.

Shiotsubo, T. and Takahashi, K., 1984, Differential thermal analysis of potato starch gelatinization. Biol. Chem., 48:9.

Slade, L. and Levine, H., 1988a, Non-equilibrium melting of native granular starch: Part I. Temperature location of the glass transition associated with gelatinization of A-type cereal starchs. Carbohy. Polymers, 8:183.

Slade, L. and Levine, H., 1988b, Recent advances in starch retrogradation, in: "Recent Developments in Industrial Polysaccharides," S. S. Stivala, V. Crescenzi, and I.C.M. Dea (Ed.), Gordon and Breach Science, New York. In press.

Spies, R. D. and Hoseney, R. C., 1982, Effect of sugars on starch gelatinization. Cereal Chem., 59:128.

Sterling, C., 1964, Starch-primulin fluorescence. Protoplasma, 59:180.

Wooton, M. and Bamunuarachchi, A., 1980, Application of differential scanning calorimetry to starch gelatinization. III. Effect of sucrose and sodium chloride. Starke 32:126.

APPLICATION OF THE LOW RESOLUTION PULSED NMR "MINISPEC" TO ANALYTICAL
PROBLEMS IN THE FOOD AND AGRICULTURE INDUSTRIES

Philip J. Barker(a)* and Henry J. Stronks(b)

(a) The Broken Hill Proprietary Co. Ltd., BHP Melbourne
    Research Labs, 245 Wellington Road, Mulgrave, Victoria,
    3170, AUSTRALIA
(b) Bruker Spectrospin (Canada) Ltd., 555 Steeles Avenue, East,
    Milton, Ontario, L9T 1Y6, CANADA

INTRODUCTION

   While the principles of NMR are increasingly well-known and today form a
normal part of any undergraduate chemistry course, it is still true to say
that the average chemist will associate NMR with a technique of considerable
sophistication for resolving problems of structural analysis.  Unfortunately,
low resolution NMR and its applications receive little attention, but the
industrial interest in this area is considerable and growing rapidly.

   The development of analytical methodology in the food and agricultural
industries is currently directed towards automation and instrumental
analytical techniques.  The goals are achieved by placing less reliance upon
wet methods which are solvent based (i.e., extraction) or sometimes
potentially hazardous (e.g., Karl Fischer titration); less reliance on
skilled laboratory personnel; emphasis on rapid and accurate analysis with
high sample throughput leading to a statistical approach to quality control.

   There are many types of analyses in areas of the food and agricultural
industries that are seemingly amenable to low resolution study.  These
applications concern problems where direct structural analysis is of
secondary importance compared to determination of the total amount of some
important quantity in a sample, the size of which is a statistically
significant representation of the bulk material.  For example in the
oilseeds industry, at the processing stage there is more interest in
exactly how much extractable oil is contained in a given batch of seeds,
and not in the structure or quantification of different individual

---

* Formerly at:  Bruker Analytische Messtechnik, Am Silberstreifen, D-7512
  Rheinstetten, West Germany

*NMR Applications in Biopolymers,* Edited by
J. W. Finley *et al.,* Plenum Press, New York, 1990

components. If such an analysis can be performed as rapidly and accurately as possible, the time required to finalize financial transactions with the growers is greatly reduced.

The potential of low-resolution NMR has been recognized for many years; the first reviews on low resolution NMR in the confectionary industry appeared as early as 1957 (Conway et al., 1957). In the beginning, most work was directed at moisture measurement in various foodstuffs, but then interest developed in the area of total fat determination and other applications. As in the high resolution area, continuous wave instruments were the first low resolution NMR analyzers to be commercially available, but since the appearance of the first commercial pulsed low resolution NMR instruments in 1971, there has been a steady increase in the percentage of pulsed NMR users for the reasons of speed, ease of sample preparation and flexibility for adaptation to a wide range of applications.

There is actually no NMR textbook which describes low resolution NMR theory and applications adequately, however a recent review on high resolution NMR applications in the food industry (Horman, 1984) made the telling observation that over 50% of papers published in the food area were on low resolution NMR results and methods. The degree of current interest in low resolution applications was amply demonstrated in May 1988 when the American Oil Chemist Society ran a short course entitled "Low Resolution Pulsed NMR in the Food Industry."

PRINCIPLES OF LOW RESOLUTION NMR

Low resolution NMR spectrometers function by measuring the quantity of hydrogen atoms in a sample of interest. These protons are most commonly associated with a fat (oil) component or a moisture component in a processed or raw foodstuff, and these components may be in solid or liquid phases.

The time evolution of the NMR signal after a single 90° pulse contains most of the important information for analytical purposes, and is characterized by the following properties:

1) The initial amplitude of the signal is proportional to the total number of hydrogen nuclei in the sample.

2) The signals due to nuclei in different chemical or physical phases decay at different rates, which means that for a sample with protons in more than one phase, the observed signal is a superposition of more than one component.

3) Each component decays with its characteristic time constant, the relaxation time.

Most applications, and the principles behind them, derive directly from consideration of these three well known basic principles.

In many cases a spin-echo sequence is required to affect the application, and in more sophisticated research and development areas relaxation time measurements are appropriate.

TO SUMMARIZE

In pulsed low resolution NMR applications, the focus is upon applications where the quantity or quantities of interest may be obtained by manipulation of a signal amplitude (or amplitudes) derived from a free induction decay (an FID), a spin-echo, or a complete relaxation curve without the necessity for a frequency domain spectrum or Fourier Transformation.

DIFFERENT CLASSES OF APPLICATION

Although the number and variety of applications for pulsed low resolution NMR is increasing rapidly, most may be categorized as one of three types:

1) Ratio applications: weight independent measurements with no stringent probehead requirements (constant active volume) such as the Solid Fat Content (SFC) of edible fats for margarines.

2) Absolute applications: where the absolute amount of some substance (e.g., oil in seeds) is required in a statistically representative sample. Here an "absolute" probehead is required, capable of providing a pulse of a high degree of homogeneity over the whole sample.

3) Relaxation time applications: where the relaxation time or the initial amplitudes derived from the relaxation data are measured and correlated with some property of the sample of interest. Again there are no stringent probehead requirements unless weight normalized measurements are required.

The first two of these are very attractive methods because as will be shown below, evaluation is most often dependent on a simple amplitude measurement and requires no sophisticated mathematical manipulations. The time scale of the experiment is also exceedingly rapid; for example the oil in seeds measurement requires only 20 seconds to give a result derived from the average of 9 transients.

INSTRUMENTAL CONSIDERATIONS

A typical instrument for pulsed low resolution applications is shown in Fig. 1, below. This figure depicts a Bruker "MINISPEC" PC 120, with a configuration similar to that used in the chocolate industry for determination of total fat content of finished chocolate.

Fig. 1.  PC 100 Series Minispec Process Analyzer.

The two boxes house the electronics of the instrument, with controls for operator input, on the right and the magnet box, containing the (permanent) magnet with probehead assembly on the left.  The units are separated to enable the magnet to be installed in hostile environments up to 30 meters from the electronics.

The electronics box is a modular design containing a power supply and two radio frequency (r.f.) parts (the modulator and power amplifier) as self contained units, and three digital cards (the ADC, interface, and micro-processor boards).  The modular design philosophy makes for easy maintenance, and in the event of failure of one of the r.f. or digital components the defective component may be simply exchanged.

The electronic box also contains two interface ports:  one for uni-directional or bidirectional interfacing with an external personal computer; the other for an electronic balance to achieve automatic weight transfer to the on-board RAM.  A checkout cable is provided so that the analog signal may be directly observed on an oscilloscope.

The magnet box contains the permanent magnet, which is fitted with a set of coils for field adjustment, the probehead, the temperature control electronics, and the receiver module.  The temperature inside the box is maintained at 40°C in order to achieve field stability and also to maintain a constant operating temperature for the receiver electronics.  Three potentiometers on the rear of the magnet box adjust the field and the analog offset in diode and phase sensitive detection (PSD) modes.

The keyboard for operator input is on the front of the instrument. All instrument functions and numerical inputs are controlled with these keys. 90° and 180° pulse widths are adjusted with two potentiometers.  All key-board functions and pulse widths may be locked with internal dip switches so that unskilled operators may use the instrument without changing the measurement characteristics.

All internal functions are controlled by the microprocessor board. The microprocessor is of the 8085 type and the MINISPEC master program is housed on the board in a series of ROM chips. Application EPROMs housed behind the keyboard are interchangeable by the user and program the microprocessor for the experiment to be performed. The application EPROM contains the pre-determined pulse sequence which is programmed in a series of steps containing the pulses to be set, the sample pulses for amplitude measurement, a loop set-up, and loop counter function. The durations between the steps may be set by the user, but each EPROM has default values preset which are optimum if a dedicated application is used. The evaluation routines are also programmed by the application EPROM, and when the experiment is optimized and set-up correctly, sample insertion starts the measurement and the result is displayed after a few seconds.

Thus a pulse sequence for solid fat determination (SFC) consists of a 90° pulse, and two sample pulses after 11 and 70 microseconds, respectively. The time evolution is represented in Fig. 2. The calculation routine is then performed as described in the following section on applications.

Pulse sequences may have up to 14 steps, enabling many different sequences to be performed. In normal applications only four or five steps are needed, but in some modified CPMG sequences all the available steps and the loop counter are required.

All results may be automatically transferred, at any data transmission baud rate, to an external personal computer via the serial I/O which in the standard version is one way. This one way interface may also be used to transfer the signal amplitudes measured by any of the relaxation time applications EPROMs to a personal computer for multiexponential decomposition (the on-board calculation routines only give monoexponential fits).

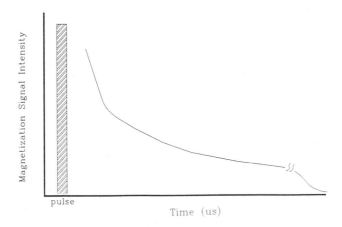

Fig. 2.   Time Evolution of NMR Response after a Single 90° Pulse.

Alternatively, with a modified microprocessor board and a freely programmable EPROM, the I/O port may be initialized as a two-way port for control of all functions by the external personal computer.

FIELD AND FREQUENCY CONSIDERATIONS

The MINISPEC is normally available as a 20 MHz instrument (0.47 Tesla) and with this instrument a sample diameter of up to 18 mm is possible. For samples where an 18 mm sample diameter does not give a satisfactory statistical representation (i.e., poor sample homogeneity) then 25 mm and 40 mm probeheads are available. However, at 0.47 Tesla the larger air gaps would lead to difficulties in achieving the same field stability and homogeneity and solving these problems would then lead to unacceptable cost increases. Therefore for 25 mm and 40 mm sample sizes the field strength is reduced to 0.23 Tesla, the frequency to 10 MHz, and the concomitant reduction of Signal/Noise ratio is offset by the larger sample volumes.

The magnetic field homogeneity is of necessity comparatively bad in order to reduce costs, but as no frequency modulated FID is required, this is unimportant. The homogeneity is, however, strictly controlled and must be between critical limits. In the SFC application, a ratio is made between the signal amplitudes measured at 11 and 70 microseconds after the 90° pulse. The calculation assumes that there is no appreciable decay in the signal due to field inhomogeneity between 11 and 70 microseconds. Therefore a correction factor is automatically calculated from the ratio between the two signal amplitudes of a mineral oil sample and all 70 microsecond measurements are corrected by this factor. In principle, the ratio should be as close to 1.0 as possible and if greater than 1.025 the field homogeneity is unacceptable. If the homogeneity deteriorates with time, a self test routine detects this and the instrument ceases to function. Simple adjustments to the field can then be made by the user until satisfactory field strength and homogeneity is attained.

The SFC measurement thus defines a certain minimum requirement for field homogeneity, after which homogeneity is unimportant because field homogeneity independent methods may be employed. Many applications are based upon a single spin echo method, for example, and real $T_2$ data is obtained by the standard Hahn echo or CPMG measurements.

Unlike high resolution studies, the observed "FID" from a liquid sample is thus a measure of $T_2^*$ and not the real $T_2$. In fact the overall field homogeneity is measured by the time taken for a 4 Volt NMR signal from a mineral oil sample to reach half height, and characteristically this value is about 1 millisecond. The real $T_2$ of the same sample is of the order of 150 milliseconds at 20 MHz.

PROBEHEADS

As was implied in section "Different Classes of Application" above, different probehead requirements are necessary in different types of applications.

The fundamental requirement of probehead construction for SFC determinations is for a relatively short dead time for the system. The sample size for SFC is standard at 10 mm diameter, and the dead time for the system must be less than 10 microseconds in order to be able to measure the signal at 11 microseconds. The actual measurement is weight independent, and performed by filling the tubes to a depth to ensure a constant active r.f. volume and therefore a short (5-10 mm) coil is used, which gives a profile similar to that in Fig. 3 (solid) when a 5 mm oil sample is passed through the coil. With this type of probehead, the 90° pulse width is of the order of 1-2 microseconds and the total dead time around 8-9 microseconds.

When an "absolute" application is required, the probehead characteristics are much more critical. The r.f. pulse homogeneity must be constant over the whole of the sample volume, and sufficient to ensure that slightly different filling heights give the same signal response per gram of sample. In addition, the whole of the sample must be within this critical pulse volume. In order to achieve this a much longer, tightly wound r.f. coil is employed, which gives a profile such as that in Fig. 3 (absolute). The active volume resulting from this type of probehead is about 30 mm deep, which for a 40 mm diameter sample tube allows around 40 mL of sample to be measured. The effect of the longer r.f. coil however is to increase the dead time of the system to anywhere between 15 microseconds (13 mm diameter) and 25 microseconds (40 mm diameter), and thus applications requiring amplitude measurements in the solid part of the decay are not feasible.

"solid-liquid" coil

"absolute" coil

Fig. 3.   Probehead Characteristics.

Choice of absolute probehead size is dependent on the type of sample to be measured and particularly on the statistical significance of the sampling size. An excellent example of this effect can be seen in Table 1, where the results from a series of measurements on whole rape seeds are shown. Both sets of results were measured at 10 MHz, but in 25 mm and 40 mm diameter probeheads, the measuring volumes contain around 8 g and 24 g, respectively. The difference between measuring the same seeds in these different sample sizes is reflected in the standard deviation for the measurement which shows that three times more measurements are needed for the 8 g samples to give the same arithmetic mean as for the 24 g samples; in addition the standard deviation is twice as large.

APPLICATIONS

It would be difficult in a short article such as this to describe all the possible applications of the low resolution instrument. The focus of this section will be how the difference in relaxation times in a sample which contains protons in two different phases enables the development of simple routine applications. The three examples described in detail range from the simplest application, the measurement of SFC, through a routine "absolute" application, the measurement of oil and moisture in seeds, to a more sophisticated measurement involving biexponential decomposition of a $T_2$ relaxation curve where two liquid phases are present, the measurement of moisture in marzipan. The common factor in all these applications concerns an evaluation of signals from two measured phases having relaxation times which are sufficiently different to allow evaluation of either or both of them.

SOLID FAT CONTENT OF EDIBLE FATS

Historically, low resolution pulsed NMR was developed specifically for the problem of rapid measurement of the solids content of edible fats. Most naturally occurring fats and oils are rather complex mixtures of different triglycerides and rather than having a discrete melting point, they actually melt over a considerable temperature range. Thus, when the solids content of a fat is measured at different temperatures, the melting characteristics of the fat may be evaluated. This is a critical measurement in most areas of the edible fats industry. The margarine industry, for example, requires different melting characteristics for different types of margarines (hard margarines, soft spreadable margarines, etc.). In the baking industry, the shortening ability of fats also requires very critical melting properties. In the chocolate industry, the quality control of cocoa butter, the development of cocoa butter replacers (CBR's) and cocoa butter substitutes (CBS's) is only possible with strict control of the melting characteristics.

Table 1. Effect of sample size on statistical significance: evaluation of 40 consecutive results upon different rape seed samples from the same batch.

MINISPEC PC 110/100/25 RTa

| ID | WEIGHT (g) | % OIL (1) | 24 G AVERAGE (2) |
|----|-----------|-----------|------------------|
| 1  | 8.123 | 40.72 | 40.58 |
| 2  | 8.135 | 40.74 | |
| 3  | 8.265 | 40.30 | |
| 4  | 8.009 | 40.71 | |
| 5  | 8.132 | 40.43 | 40.56 |
| 6  | 8.242 | 40.53 | |
| 7  | 8.074 | 40.76 | |
| 8  | 8.216 | 40.69 | 40.54 |
| 9  | 8.235 | 40.17 | |
| 10 | 8.365 | 40.30 | |
| 11 | 8.323 | 40.53 | 40.49 |
| 12 | 8.213 | 40.64 | |
| 13 | 8.160 | 40.51 | |
| 14 | 8.320 | 40.39 | 40.43 |
| 15 | 8.262 | 40.40 | |
| 16 | 8.147 | 40.74 | |
| 17 | 8.236 | 40.41 | 40.56 |
| 18 | 8.124 | 40.52 | |
| 19 | 8.166 | 40.62 | |
| 20 | 8.113 | 40.62 | 40.62 |
| 21 | 8.367 | 40.62 | |
| 22 | 8.274 | 40.45 | |
| 23 | 8.253 | 40.48 | 40.50 |
| 24 | 8.392 | 40.58 | |
| 25 | 8.222 | 40.61 | |
| 26 | 8.143 | 40.92 | 40.56 |
| 27 | 8.260 | 40.70 | |
| 28 | 8.258 | 40.40 | |
| 29 | 8.120 | 40.73 | 40.61 |
| 30 | 8.354 | 40.69 | |
| 31 | 8.192 | 40.75 | |
| 32 | 8.350 | 40.31 | 40.55 |
| 33 | 8.404 | 40.59 | |
| 34 | 8.362 | 40.41 | |
| 35 | 8.291 | 40.37 | 40.43 |
| 36 | 8.185 | 40.50 | |
| 37 | 8.291 | 40.45 | |
| 38 | 8.157 | 40.70 | 40.58 |
| 39 | 8.281 | 40.60 | |

mean of 8 gram samples = 40.55    $\Delta = 0.75$    sigma = 0.16
mean of 24 gram samples = 40.55    $\Delta = 0.31$    sigma = 0.08

The previously used method was dilatometry, which through study of the expansion of melted fats gave an indirect measurement of the solids content of a fat, the Solid Fat Index (SFI). The method was time consuming, requiring skilled personnel and in particular ranges could give results up to 15% from the real percentage of solids. Continuous wave NMR could only give an indirect number derived from integration of the liquid NMR signal.

By using pulsed NMR it was possible to develop a direct method using a short dead time, whereby from a measurement in the solid part of the NMR signal ($T_2$ of the solid fat is around 10-15 microseconds at 20 MHz) and a measurement in the liquid part of the decay (liquid fat having a $T_2$ approaching 100 milliseconds or more), a direct estimate of the real solids content could be made. The measurement principle is explained in Fig. 4, which represents a typical time evolution of the signal after a single 90° pulse. The amplitude of the signal at 11 microseconds (Amp 1) is obviously proportional to the total number of protons in both liquid and solid phases, while the signal of 70 microseconds (Amp 2) is proportional to the total number of protons in only the liquid phase. Therefore the ratio of these signals must in some way be proportional to the real solid fat content of the sample. In fact, a further factor is added to compensate for the dead time of the instrument as depicted in equation (i).

$$\text{Percentage Solids} = \frac{F \ (AMP \ 1 - AMP \ 2)}{F \ (AMP \ 1 - AMP \ 2) + AMP \ 2} \times 100 \qquad \text{(i)}$$

Calculation of this so called F-factor and the calibration of the instrument are carried out simultaneously using a set of standard oil in polymer samples of known solid/liquid ratio. Linearity of the instrument is checked over a range from 0% to 90% solids.

In theory the proton densities of different fat compositions will be slightly different and a different F-factor could be used, but in practice the single F-factor calculated from the standard samples is sufficient for quality control of nearly all fats.

The major advantage in the method is in the speed of measurement, which for a single sample is 6 seconds; no weighing or measurement of totally liquid fat samples is required. In this application, the actual measurement is almost secondary to the sample tempering procedures. Tempering is

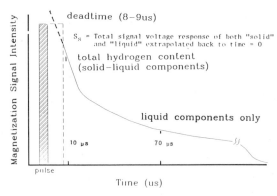

Fig. 4.   Calculation of the Solid Fat Content.

necessary in order to ensure a uniform crystallization of fats which is reproducible between different laboratories and types of fat. Of necessity this is a rather time consuming procedure, and cumbersome in a quality control situation. Tempering methods have been standardized by several standards organizations and may be of a series type, where a single fat sample is tempered at each temperature to be measured finishing with a measurement (by dilatometry or NMR at 80°C), or a parallel type where the fat is divided into tubes placed at each temperature. The MINISPEC method has enabled a parallel tempering method to be adopted which saves several hours compared with the series method, and thus greatly increases the sample turnover in quality control situations.

ANALYSIS OF OIL-CONTAINING SEEDS

With the increasing demand for unsaturated fats and oils, the oil bearing crops industry has expanded dramatically over recent years. In the edible oils sector the major crops have become soya, sunflower and rape, and there are several potential areas for low resolution pulsed NMR applications. One obvious application is in the area of crop improvement where, in general, seeds with a high oil content will yield crops with similarly high oil content. Probably the most important area though is in the seed transaction areas, where the seed millers pay growers on oil content; therefore the analysis must be reliable, quick and accurate. In addition the miller controls his residues after pressing or extraction, not only to check the efficiency of his extractors, but also to control the meals which are subsequently used in animal feed.

In oilseeds, the oil is present in an essentially totally liquid state, with a $T_2$ similar to the $T_2$ of the extracted oil (for sunflower seeds at 10 MHz, between 110 msec and 120 msec). Recalling the situation for the SFC measurement, if only the carbohydrates and proteins in the seeds contribute to the solid phase one might think that either the signal at 70 microseconds is proportional to the liquid phase and therefore to the oil content of the seeds or that the inverse of the SFC should be proportional to the oil content. Unfortunately the situation is not that simple because moisture is always present to some degree, and the samples would need to be dried to completion for both assumptions to be true.

Drying the seeds is not easy, and Fig. 5 shows the FID's of peanut kernels dried at different temperatures and the relationships between different drying methods. Drying is not at all desirable in these measurements for two reasons: firstly because on complete drying the seeds become nonviable (for crop improvement); secondly because complete drying requires four to eight hours at 105°C and this is an unacceptable delay for the seed miller handling thousands of tons of seeds per day.

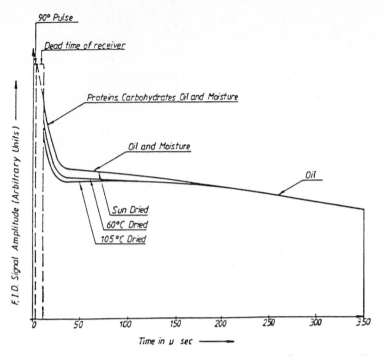

Fig. 5.   Effect of Different Drying Methods on FID from Peanut Kernels.


The situation may be solved assuming that seeds for shipment are to be measured. Seeds for shipment necessarily contain some "bound" water but over a certain limit shipment is not possible as any "free" water leads to bio-degradation (rotting, etc.). This limit varies between seed types from around 10% for rape seeds to 13% for sunflower. The actual physical state of the moisture in the seeds is not solid, but a food chemist would probably call it "bound" water, with a $T_2$ somewhere between 300 and 1500 microseconds. Thus returning to the theme of this section on applications, a simple method may be evaluated to measure the oil based solely on the fact that the relaxation times of the moisture and oil are significantly different.

A $T_2$ relaxation curve measured with a Hahn Echo sequence is shown in Fig. 6; this was recorded from a sample of rape seeds, over the time scale 0-20 milliseconds with logarithmically incremented tau values. Decomposition of the curve leads to (for this sample) a $T_2$ for moisture of 400 microseconds and for the oil of 90 milliseconds. This means that at some point on the curve greater than 2 milliseconds, there will be no contribution to the amplitude of a single spin echo from the moisture phase. This fact forms the basis of the oil (and moisture) measurement in seeds.

Calibration of the instrument is achieved using either extracted oil samples or seed samples of known oil content. An applications chip is used with a preset single spin echo pulse sequence and the default value of tau

492

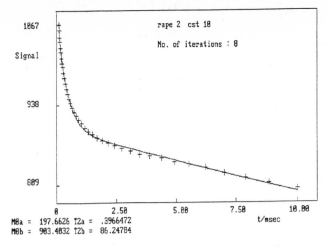

MØa = 197.6626  T2a = .3966472
MØb = 903.4032  T2b = 86.24784

Fig. 6.   $T_2$ Determination of Rape Seeds.

is 3500 microseconds.  This gives a single spin echo at 7 milliseconds, which in most cases is well into the region of the $T_2$ curve where only the oil contributes.  Instrumental requirements are determined by the nature of the application.  Normally a 40 mm diameter absolute probe is used for seed analysis to ensure a statistical representation of the sample.

In the calibration mode of operation, the echo amplitude is measured and after three calibration samples (covering the range of oil values normally encountered), and every subsequent calibration sample, a linear regression is made to calculate slope, intercept and linear correlation coefficient.  (A calibration using three samples of seeds of known oil content is shown in Fig. 7).  The weight of each calibration sample is entered followed by the known percentage of oil.  A correlation is made between signal per gram of sample versus given percentage of oil.  In measure mode only the weight of an unknown sample is entered, following which the instrument measures the echo amplitude, calculates signal per gram and then calculates the absolute percentage of oil in the unknown sample. Reproducibility of the instrument can be seen in Table 2, which shows the results of 10 successive measurements on a single sample.

Considering the actual sequence of events in the pulse sequence, it soon becomes obvious from Fig. 8 that the moisture can be measured simul- taneously.  At a time 70 microseconds after the initial 90° pulse, the signal is proportional to both the liquid phases, therefore the difference between this signal and the 7 millisecond echo signal must be in some way proportional to the moisture content.

For the oilseeds industry this method has had considerable consequences in that the simultaneous oil and moisture determination requires only 20

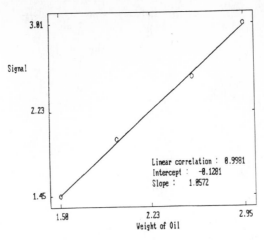

Fig. 7. Calibration Curve Determined Using Various Weights of Oil.

seconds measurement time, and only sample weighing as preparation. With a personal computer hook-up some plants are already achieving throughputs of 600 samples per day.

Acceptance of the pulsed NMR methods by standards authorities in Europe is well under way: in France and Germany, pulsed NMR is already well established in standard methods; in England, the FOSFA Institution is researching a new protocol for pulsed NMR; and, in Europe, in general the Community Bureau of Reference in Brussels has an advisory committee on NMR in oilseeds analysis.

Table 2. Reproducibility on 3 different samples of the same batch.

MINISPEC PC 110/125/40 RTa

| SAMPLE | 1 | 2 | 3 |
|---|---|---|---|
| WEIGHT | 11.769 | 10.748 | 10.804 |
| oil % #1 * | 40.811 | 40.971 | 40.921 |
| #2 | 40.881 | 40.914 | 41.014 |
| #3 | 40.825 | 40.935 | 40.963 |
| #4 | 40.869 | 40.856 | 40.980 |
| #5 | 40.801 | 40.886 | 40.965 |
| #6 | 40.780 | 41.000 | 40.995 |
| #7 | 40.906 | 40.986 | 40.973 |
| #8 | 40.825 | 40.871 | 40.893 |
| #9 | 40.753 | 40.765 | 40.881 |
| #10 | 40.848 | 40.884 | 40.938 |
| #11 | 40.832 | 40.884 | 40.986 |

* MINISPEC calibrated against three pure oil samples weighing:   4.152 g
    4.585 g
    5.187 g

Resultant calibration curve: slope         274.4
            intercept      .00842
            correlation    .99996

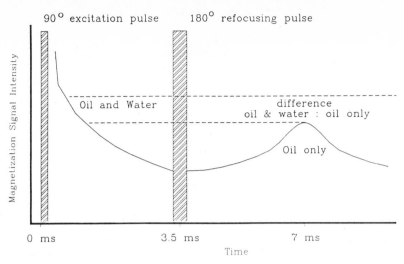

Fig. 8.   Spin Echo Sequence for Oil Bearing Seeds.

ANALYSIS OF MOISTURE IN MARZIPAN

The two application types discussed so far have involved methods based on ratios of $10^3$–$10^4$ (for SFC) and $10^2$–$10^3$ (for oil seeds) between the relaxation times of two components.  If the ratio is lower, more sophisticated measurements are required.  If, for example, the moisture content of seeds were 20%, free water would certainly be present and the relaxation time of the moisture phase as a whole would be considerably increased.  The position of a single echo, the amplitude of which is due to only one component, becomes more difficult to determine, and as the position from time zero becomes longer, diffusion effects become important and the apparent echo amplitude may not be proportional to the oil content.  A similar situation is encountered in marzipan.

Marzipan is a high cost, high profit delicacy in the confectionery industry with a large market in Western Europe.  Marzipan forms the basis for "petits fours" in France, Christollen in Germany, Christmas cake in England and is increasingly popular as a filling for candy bars.  Marzipan contains almond oil, moisture, sugar and a type of filling material made from flour, starch, etc.  The oil and moisture can be present in roughly equal amounts but the latter must be strictly controlled to be below 17%. Currently there is no rapid method for moisture determination and the standard drying oven method requires at least four hours.

Figure 9 shows the results of a least-squares biexponential decomposition of $T_2$ data obtained from a typical marzipan sample.  The data was obtained via a CPMG sequence (tau = 500 µsec) measuring the amplitude of every second echo in an 80 echo train.  The raw data in the form of the tau values and their corresponding signal intensities were transferred over the interface to a personal computer where the fit was made.

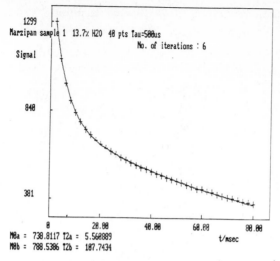

Fig. 9.  $T_2$ Determination of Typical Marzipan Sample.


The ratio between the $T_2$s for the component phases is around 20.  This measurement was performed using an "absolute" type probehead, and therefore the MO values should be comparable between samples.  Thus for calibration samples, plotting MO/gram for the moisture component (short $T_2$) against drying oven percentage should give a linear correlation.  The data for such an approach is plotted out in Fig. 10, which reveals an excellent linear correlation.

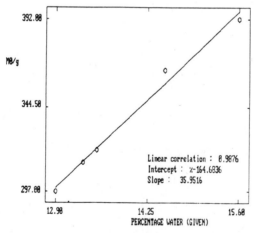

Fig. 10.  Calibration Curve for Determination of Moisture in Marzipan.

While the approach outlined in a stepwise fashion above is somewhat complex, a personal computer program is able to perform all the calculation routines. An operator needs only to insert the weight of the sample, and after the calculation routines are performed internally, receives only the moisture percentage as the result.

The time required for such a measurement depends on the speed of the computer but on a personal computer (AT compatible) it took about 40 seconds for the decomposition and around a minute in total from sample insertion to print out of result. This is actually a comparatively slow MINISPEC measurement but again is obviously significantly faster than the existing methods, and represents a considerable savings in time without loss of accuracy for the marzipan producer.

FUTURE DEVELOPMENTS

In the previous three sections some analyses with a common theme have been discussed. The principles involved run throughout the extensive array of MINISPEC applications. The principles outlined for the seeds measurements may be applied to many different areas. As an example the chocolate industry involves a number of processes whereby the fat content may be controlled; the principles and evaluation methods behind the measurement are the same but the tau value is normally only 200 or 500 microseconds. The situation can also be much simpler when only one phase is to be measured, and there are many applications where only a single sampling point on an FID or spin echo is required.

A ratio of $T_2$s of about 20:1 was reached (Fig. 9) and for the $T_2$ decomposition in general $T_2$ differences of between 6 and 10:1 may be resolved. There are areas however where the $T_2$s are even less different such as analysis of fat and moisture in meat. Some recent studies in meat research have indicated that 20 MHz FT-NMR is very promising. The first FT data on fresh meat has already been published (Renou et al., 1987) in which the fat and water signals could be adequately separated and integrated, leading to a simultaneous analysis of both moisture and fat in only a few minutes.

In the brewing industry recent studies indicate that analysis of alcohol in fermentation musts, in wines and spirits may be performed at 20 MHz with an essentially unmodified MINISPEC, without FT (Guillou and Tellier, in press). This method is based on comparison of a single spin echo amplitude at time $(2J)^{-1}$ calculated from high resolution measurements and comparison of this signal with the echo amplitude on a CPMG train at the same time on a pure alcohol sample. This methodology is based on a sound knowledge of high resolution principles, and perfectly illustrates how input from high resolution specialists may lead to low resolution solutions for a wide range of analytical problems.

CONCLUSIONS

In this contribution we have attempted to partially bridge the gap between the high resolution user. We have attempted to make it stimulating enough for the treatments which sometime deter the laboratory analyst from adopting NMR as a routine method.

The scope and potential for NMR is outstanding in major industry, and in the food manufacturing and processing areas particularly, and not only for low resolution NMR. High resolution spectroscopy is also beginning to play a significant role in routine analysis, and thus the adoption of automated deuterium NMR at 300 MHz for the routine authentication of wine is already at an advanced stage of development in the major wine producing countries of Europe.

ACKNOWLEDGMENTS

We would both like to thank Dr. John W. Finley for the invitation to attend and contribute to this meeting. We would like to thank our many friends throughout the food industries, without whose samples we would not have been able to work. We would like to thank the Bruker organization for their support including the applications team at Bruker, Canada. Finally one of us (P.J.B.) deeply regrets being unable to address the meeting, owing to his move to Australia (address noted).

REFERENCES

Conway, T. F., Cohee, R. F., and Smith, R. J., 1957, Manufacturing Confectioner, 37:27.
Guillou, M., and Tellier, C., Analytical Chem., in press.
Horman, L., 1984, in "Analysis of Foods and Beverages," S. Charalambous Ed., Academic Press Inc., 205-264.
Renou, J. P., Brioguet, A., Gattelier, Ph., and Kopp, J., 1987, Int. J. Food Science and Technology, 22:169.

AUTOMATIC USE OF SMALL NUCLEAR MAGNETIC RESONANCE SPECTROMETERS FOR
QUALITY CONTROL MEASUREMENTS

Robert M. Pearson[a] and John Q. Adams[b]

[a] 3590 Churchill Court, Pleasanton, California 94566
[b] 340 Glorietta Blvd., Orinda, California 94563

INTRODUCTION

The titles and abstracts of the papers presented at this meeting
describe the application of nuclear magnetic resonance to research of
importance in the agricultural/food processing industries.  For the most
part these papers have described the use of sophisticated, complex and
expensive spectrometers in these studies.  The complexity of these techniques makes it difficult for the average worker in the agricultural/food
processing industry to apply nuclear magnetic resonance without the
assistance of an expert trained in the field.  The cost of this type of NMR
equipment makes some sort of a central corporate facility imperative.  These
conditions tend to isolate the agricultural/food processing chemist from
the application of NMR techniques to his or her routine problems.

We would like to discuss an alternative approach to the use of magnetic
resonance to problems in the agricultural/food processing industry.  This is
the use of smaller, much less expensive and easier to run small magnetic
resonance spectrometers in this area.  We will, in particular, discuss the
use of these spectrometers to plant quality control problems.

ADVANTAGES OF NMR FOR QUALITY CONTROL MEASUREMENTS

Nuclear magnetic resonance is almost uniquely applicable to process
control problems in the agricultural field, having many advantages over
other better known process control devices.  Among these advantages include:

The Presentation of the Sample to the Device

The detector in a magnetic resonance spectrometer is just a coil of
wire which contains the sample.  This means that any sample which can pass
through the coil can be automatically sampled and presented to the instrument.

## Measurement of Unground Samples

Since practically all agricultural samples are nonconducting no skin depth problems are encountered. This means that the spectrometer "sees" the whole sample, not just the surface of the particles. Thus, samples do not have to be ground for the determination of all the moisture in the bulk of the sample. The moisture content of whole kernel grains can be determined as easily as in a ground material.

## Ease of Calibration

There are no extinction coefficients in magnetic resonance. All hydrogen atoms produce signal intensities of equal strength after dead times have been accounted for. Thus, the NMR signal from the moisture in the sample can be directly related to that from the oil without adjusting for the effects of complex extinction coefficients. We will discuss the effects of dead time on the signal for bound hydrogen in detail later in this paper.

## Quantitative Analysis without Weighing the Sample

The various kinds of hydrogen in most agricultural samples can be quantitatively determined without weighing the sample. This is very important to the application of magnetic resonance to process control and will be discussed in detail later in this paper.

## Quantitative Determination of Free and Bound Water

We will show how the concentration of both bound and free moisture can be determined in the same sample. By bound water we mean the amount of hydrogen bonded to carbon atoms, not the usual bound water discussed in this field. The concentration of "bound water" as defined in the agricultural field can probably also be determined with the use of NMR relaxation measurements.

## Simultaneous Determination of Both Fat and Moisture

We shall show how both fat and moisture can be quantitatively determined without weighing the sample.

## TIME DOMAIN MAGNETIC RESONANCE

Our efforts in the application of time domain NMR spectroscopy to quality control problems have been described (Pearson and Parker, 1984; Pearson et al., 1987). The first of these papers described the use of modified commercial spectrometers for quality control measurements in an aluminum oxide application, whereas the second paper gave examples of the use of these techniques in the agricultural industry.

We will define time domain NMR as the simple use of "on resonance" free induction decays to quantitatively measure the amounts of the different hydrogen populations in the samples being studied. In these studies the resonance condition described by the Larmor equation is exactly maintained. This produces NMR spectra whose maximum intensities are a direct measure of the concentration of that hydrogen population in the sample. Hydrogen populations with different spin-spin relaxation times will produce NMR signals with different decay rates and are easily identified in this spectra. Thus, it is a trivial matter to measure the relative concentrations of free and chemically bound water in most materials, including many of interest in the agricultural field.

Figure 1 is a time domain NMR spectrum of moist wheat. This spectrum consists of a fast decay due to the hydrogens bonded to carbon atoms in the sample followed by a slow decay from the exchangeable hydrogens in the sample including the free water. We mean by free water all the hydrogens which share an oxygen with another hydrogen. That is the total "water" content of the sample.

The spectrum shown in Fig. 1 was obtained by magnetizing the sample with a short and powerful pulse of RF energy then measuring the decay of this magnetization as a function of time. Both the probe and the receiver must be allowed to recover after the pulse before data can be taken. This delay time is known as the dead time of the spectrometer. At best it is about equal to ten times the Q of the coil in hertz. Thus, a 10MHz spectrometer will have at least a 10 usec. dead time.

The spectrum shown in Fig. 1 was obtained at 200 MHz on a spectrometer that had about a 1 usec. dead time. The amount of data lost in this one microsecond is very small so that Fig. 1 represents essentially the total NMR signal from the moist wheat. This sample was about 12% moisture so that one would expect the portion of the signal from the total water to be about 1/5 of the total intensity, where in fact it is about 1/3 of the total intensity. Thus the portion of the signal from the water in the sample is larger than it should be by a considerable amount. It is our opinion that the slow decay in these samples contains not only the signal from the hydrogen in the water molecules in the sample but also the signal from the hydrogens contained in the exchangeable hydroxyl groups in the starch. Thus the slow decay represents the signal from the total exchangeable hydrogens in this wheat sample.

We also know something about the rate of exchange between the hydrogens in the water and that in the hydroxyl groups. A chemically bound hydroxyl group would produce a fast decay in a hydrogen NMR spectrum. The fact that the hydrogens in the hydroxyl groups in the sample are contributing to the slow decay in this spectrum means that they are exchanging at a rate faster than the width of the line which would be produced in the frequency domain

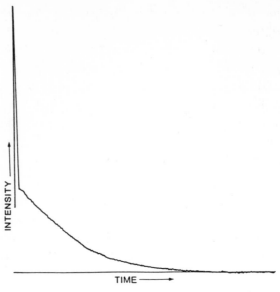

Fig. 1.  Free Induction Decay of Moist Wheat.

NMR spectrum of a unexchanged hydroxyl group.  That would be about ten kilohertz in this case.  So the spectrum shown in Fig. 1 is that of moist wheat in which the hydrogens in the hydroxyl groups of the starch are exchanging with the hydrogens in the water molecules faster than about 10 kilohertz.

The fast portion of the decay signal shown in Fig. 1 decays completely away in less than 50 microseconds.  This means that on a low field spectrometer, say at 10MHz, about half this fast decay will occur before the probe recovers and thus will be lost.  This prohibits the use of absolute intensities to measure the ratios of the various hydrogen populations in these samples without correcting for the spectrometer dead time.  One simple way to do this is to use calibration curves derived from known samples which have been analyzed by independent methods.

The fact that the dead time of the spectrometer is inversely proportional to its frequency points out one important difference between a high resolution NMR spectrometer designed to run samples in solution and a true solids spectrometer.  High resolution spectrometers often have probes with a Q of several hundred.  This gives long dead times and makes the measurement of the fast decay in a solid sample impossible.

Time domain nuclear magnetic resonance is really a "return to the basics".  It is limited to the use of free induction decays and the measurement of relaxation times to differentiate between hydrogen populations in the sample.  Since this work does not involve the use of chemical shift measurements, Fourier transforms of the data are not required.  This coupled with the reduced homogeneity and line width requirements of the magnet

greatly lowers the sophistication of the spectrometer, its cost and makes it
much easier to operate.

APPLICATIONS

Time domain nuclear magnetic resonance is capable of measuring three
fundamental properties of the hydrogen atoms in a sample. These are their
free induction decays, their spin-lattice relaxation times and their spin-
spin relaxation times. The measurement of chemical shifts and coupling
constants are really outside the ability of time domain NMR at least as we
shall use the term here. We will now describe how the measurement of free
induction decays and relaxation times allow the use of a small NMR spectrometer for plant quality control. The work we shall describe today is based
on our own experience in this field. This paper is not intended to be a
review of the work others have done.

Free Induction Decay Analysis

Our first application of time domain NMR for plant quality control
occurred in the aluminum industry. This work formed the basis for our
understanding of the nature of time domain NMR spectra of hydroxyl groups
and other hydrogen containing functional groups including the ones found in
agricultural materials. We would therefore like to discuss this work
in some detail. It has been published (Pearson and Parker, 1984).

Figure 2 shows the various types of hydrogen atoms found in an aluminum
oxide. It shows hydrogen atoms bonded to water molecules, hydrogen bonded
to surface hydroxyl groups and hydrogen atoms located in lattice hydroxyl
groups. Figure 3 shows the time domain NMR spectrum of such a sample. This
spectrum consists of a fast decay due to the chemically bound water in the
sample followed by a slower decay from hydrogens in the physically adsorbed
water in the sample. Each of these two types of hydrogen can be quantitatively measured from their intensities in the spectrum.

The compound character of the NMR spectrum of aluminum oxide is a
result of a slow exchange between the hydrogens in the water and the hydrogens of the hydroxyl groups. The hydrogens in these two populations are
exchanging at a rate slower than about ten kilohertz.

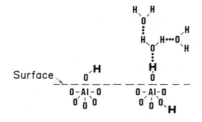

Fig. 2. Types of Hydrogen in Aluminum Oxide.

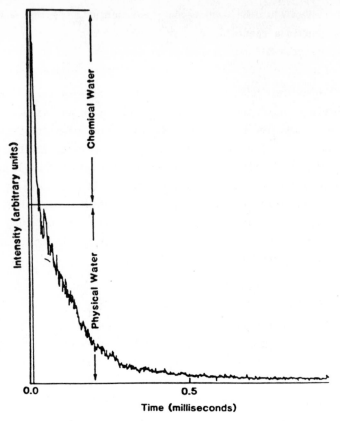

Fig. 3.   Time Domain NMR Spectrum of Moist Aluminum Oxide.


Since the spectrometer "sees" only the material contained in the coil
there are two ways to calibrate these spectra for quantitative results.
These consist of limiting the sample to a known weight all contained within
the coil, or measuring the bulk density of the sample and accounting for
changes in density in the calculation of the amount of hydrogen in the
sample according to Eq. (1).

$$\% \text{ HYDROGEN IN SAMPLE} = \%(H)std \times \frac{(I)sam}{(I)std} \times \frac{(B.D.)std}{(B.D.)sam} \tag{1}$$

where: (I)sam = Intensity of NMR spectrum from sample

(I)std = Intensity of NMR spectrum from standard

(B.D.)std = Bulk density of standard

(B.D.)sam = Bulk density of sample


In the aluminum oxide application the bulk density of the sample was
constant enough not to affect the results within experimental error, so

absolute intensities could be used without correcting for changes in bulk density. But this is a special case and is not true in general. The similarity between the NMR spectra of moist wheat and that of moist aluminum oxide indicates that methods developed in the aluminum industry could be applied to agricultural studies.

In agricultural applications we have no reason to assume that the samples would pack in the NMR tube with constant bulk density and the measurement of the density or weight of the sample is at least as difficult to do as the measurement of the NMR signal of the sample. Therefore it is very advantageous for us to develop NMR methods which will allow the quantitative determination of the hydrogen populations without density or weight measurements.

It has been shown (Pearson et al., 1987) that the intensity of the fast decay due to hydrogens in C-H bonds can be used as a measure of the amount of sample in the spectrometer coil. Thus, a plot of the ratio of the slow decay to the fast decay in these spectra vs. the moisture content of analyzed samples produces the calibration curve required for quantitative analysis of these samples. Such a curve is shown in Fig. 4. This curve was generated by adding moisture to the original wheat and analyzing all the samples by standard gravimetric methods. Note that the point corresponding to the original wheat falls on the same line as the points for the moisture added samples.

We now have a method which can produce quantitative results entirely from the NMR spectrum of the sample without the need to weigh the sample or measure its bulk density. It thus shows promise as a plant quality control method.

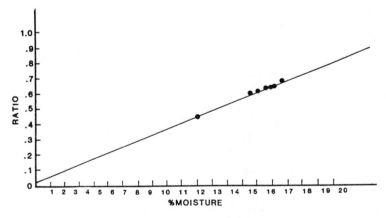

Fig. 4. Calibration Curve for Wheat.

## Spin-Spin Relaxation Time Methods

The above method developed for the determination of moisture in wheat did not take into account fat content. It appears that the oil content of wheat is low enough and consistent enough not to effect the accuracy of the moisture analysis. This is certainly not the case for the analysis of corn.

In solid samples the spin-spin relaxation time is generally much shorter than the spin-lattice relaxation time. Thus, it is $T_2$ which controls the decay rate of the free induction decay in most solid samples and in all the agricultural and food samples we have looked at.

In a physical mixture of oil and water the decay rate of each phase would depend on its viscosity. The more fluid water would decay slower than the more viscous oil. But remember we have shown that the hydrogens in the water are exchanging with the hydrogens of the hydroxyl groups of the starch. This gives the combined population of the two types of hydrogen a spin-spin relaxation time which is a weighted average of the two populations and causes it to decay at a rate faster than the oil in the sample. At least faster than the liquid portion of the oil. Thus, a simple Hahn Spin Echo (Hahn, 1950) can be used to separate the NMR signal from the oil and the water in corn samples. This has been known for some time; our efforts in this area were reported in Pearson et al. (1987).

The method described in Pearson et al. (1987) quantitatively determines both fat and moisture in corn samples without weighing the sample or controlling its density. This method can also used as a basis for plant quality control.

## Spin-Lattice Relaxation Methods

In theory quality control methods can also be based on the measurement of the spin-lattice relaxation times of the various hydrogen populations found in a solid sample. Our work in the field is just beginning, but I would like to report some preliminary results here.

Figure 5 shows $T_1$ data for a sample of vegetable oil. It shows the presence of two hydrogen populations, one with a spin-lattice relaxation time of 0.420 seconds and one with a $T_1$ of 0.129 seconds. They are in almost equal concentrations.

Figure 6 shows spin-lattice relaxation data for the hydrogen populations found in a common soda cracker. This data can be interpreted as consisting of three hydrogen populations with relaxation times of 0.462 secs., 0.1212 secs. and 0.0722 secs. Extrapolation to the zero time axis for each relaxation time gives relative concentrations of 6.72%, 42.9% and 50.3% for the slow, intermediate and fast decays, respectively.

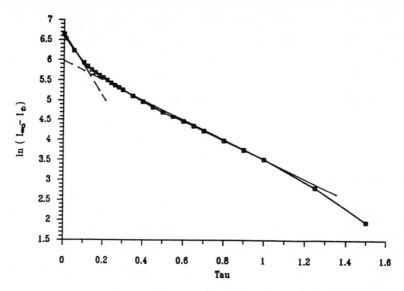

Fig. 5.  Vegetable Oil – Spin–Lattice Relaxation Data.

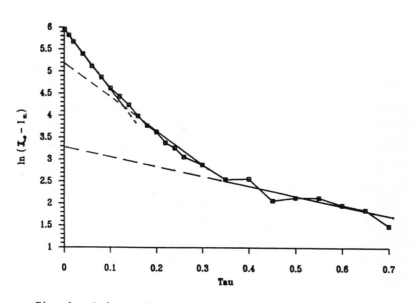

Fig. 6.  Soda Crackers – Spin–Lattice Relaxation Data.

According to Langes Handbook of Chemistry the nominal composition of a soda cracker is 4.0% water, 6.6% protein, 8.8% fat and 68.8% carbohydrate. By using average values of the hydrogen contents of each of these substances one can calculate that 69.8% of the hydrogen in the sample is associated with the carbohydrate, 15.6% with the fat, 7.41% with the protein and 7.10% with the water. We have the following information:

- The $T_1$ data can be divided into three hydrogen populations.

- In the wheat sample the hydroxyl hydrogens exchange with the water hydrogens fast enough to average the two signals from each population.

- The vegetable fat we measured showed two relaxation times.

- We can assume that the nonexchangeable hydrogens in the carbohydrates and the proteins will have similar relaxation times.

- The carbohydrate contains 10 hydrogen atoms per repeating group of which 4 are located in exchangeable hydroxyl groups. Thus we would expect one population to consist of:
  (69.8)(0.6)+(15.6)(0.5)+7.41 = 57.1%
  of the hydrogens in the sample.

One population to consist of:
(69.8)(0.4)+7.10 = 35.0%
of the hydrogens in the sample.

And one to consist of:
(15.6)(0.5) = 7.10%
of the hydrogens in the sample.

It thus appears that the fast decay is from the nonexchangeable hydrogens in the sample, the intermediate decay is from the exchangeable hydrogens in the sample and the slow decay is from the fat in the sample.

A similar analysis of a corn starch sample known to contain 11.31% moisture produced $T_1$ data with two decay rates, the slower of which was due to only water, so it does not appear that the hydroxyl hydrogens of the starch are exchanging with the water hydrogens fast enough to average the two populations. This is not necessarily inconsistent with the data from the cracker sample, but it does demonstrate that much more work needs to be done in this area before we can be sure of the interpretation of these data.

SUMMARY

We have demonstrated how the use of small NMR spectrometers can be of value in the agricultural/food processing industry. Some of the methods we have described in this paper have been applied to in-plant process control

(Pearson and Parker, 1984; Pearson et al., 1987). The advantages of NMR over other more established forms of plant control have been described.

In recent years several companies have sold small NMR spectrometers interfaced with a PC. As improved software and hardware become available for these computers their use in conjunction with small nuclear magnetic resonance spectrometers will enable this combination to make an important contribution to plant quality control in the food and agricultural process industries.

The future appears bright for these techniques.

REFERENCES

Hahn, E. L., 1950, Phys. Rev., 80:580.
Pearson, R. M., and Parker, T. L., 1984, "The Use of Small Nuclear
    Magnetic Resonance Spectrometers as On-Line Analyzers for Rotary
    Kiln Control," Light Metals, Pages 81-97.
Pearson, R. M., Ream, L. R., Adams, J., and Job, C., 1987, "The Use of
    Small Nuclear Magnetic Resonance Spectrometers for On-Line Process
    Control," Cereal Foods World, vol. 32, no. 11, p. 822.

hemicellulose, 287, 289, 290, 294, 295, 297,
heteronuclear correlation, 20
heteronuclei, 21
hexafluoroisopropanol, 118
hexagonal, 265
  lattice, 268
highly perturbed, 396
  water 400
HMBC, 22
HMQC, 21
HOHAHA, 18, 19, 22
homonuclear correlation, 18
homonuclear line narrowing, 256
hordein, 199
hormones, 117
hydration, 183, 192, 195, 259, 260, 361, 362, 363, 365, 367, 368, 370, 372, 373, 375, 392, 393, 394, 396, 398, 403, 407, 408, 410, 429, 433, 443, 464, 477, 478
hydrolysis, 91
hysteresis, 438

immobilized water, 396, 401
in vivo NMR, 255
INEPT, 22, 343
influence of heating, 220
insulin, 319, 321
internuclear distance, 135
intersaccharide bonds, 309
intestinal calcium-binding
  protein, 232
isomaltotriose, 49
isotope, 85
  shift, 86, 92
ISPA, 99, 101, 102, 104, 106, 115, 104, 137, 141, 142, 147, 148, 149, 151
  calculation, 97
Iterative Relaxation Matrix
  Analysis (IRMA), 147

Jump and Return, 65, 67, 68, 70

Karplus, 18, 96
Karplus-type, 12
katanosin B, 123
KINFIT, 89

α-LA, 231, 232, 233, 244, 245, 247, 248, 250, 251
β-lactoglobulin, 176, 425
lactose, 432, 433
lanthanide NMR shift reagents, 248
Laser Photo-CIDNP, 241
legumes, 288
ligand binding sites, 138
lignin, 288, 289, 290, 291, 292, 294, 296, 297, 299, 300

lipid, 304, 312
lipid bilayers, 264
liver, 317, 318, 319, 320, 324, 350
lond lengths, 135
low resolution NMR, 482
lumiflavin, 241
lysine, 341
Lysobactin, 95, 120, 121
lysozyme, 186, 244, 261, 263, 363, 364, 367, 369, 384, 426

β-maltose, 49
maltose, 156, 157
maltotetraose, 157
maltotriose, 160
D-mannose, 52, 54, 87, 210, 216, 431
margarines, 488
Marzipan, 495
membrane, 264, 305
membrane bilayer, 303, 306
methyl β-D-glucofuranoside, 9
methyl β-D-ribofuranoside, 9
methyl β-D-xylofuranoside, 9
methyl β-D-mannofuranoside, 9
methyl 2,3-di-O-methanesulfonyl-
  α-D-glucopyranoside, 30, 45
micelles, 183, 188, 189, 190, 191, 192, 193, 195
milk, 426, 429
milk powder, 432
MINISPEC, 483, 486, 497
mitochondria, 334
MLEV-17, 22
mobile water, 396
moisture measurement, 482
molecular fluctuation, 309
molecular-modelling, 13
monoglycosyl, 309
multilamellar structures, 306
muscle, 317, 427, 428
  proteins, 361, 373, 426
mutarotation, 85
myofibrillar, 373, 374, 375
myosin, 373, 374
  denaturation, 426

N-acetyl-D-galactosamine, 356
N-acetyl-D-glucosamine, 210, 216, 231, 356
N-lactalbumin, 186, 231
n-space, 55
NAD(H), 333
napharelin, 117, 119
nearest neighbor, 54, 55
neuropeptides, 117
neutron scattering, 185
Nicotiana tabacum, 334
NMR images, 373, 446
2D-NMR, 17, 135, 156, 157, 234
NMRI, 446, 447, 450

starch 155, 163, 166, 170, 288, 372, 379, 423, 433, 434, 435, 436, 438, 444, 463, 464, 465, 466, 467, 469, 473, 475, 476
state of water, 397, 425
steric interactions, 8
steric limitations, 135
stokes radii, 195
stokes radius, 192, 193, 194, 195
submicelles, 175, 176, 183, 185, 188, 190, 191, 193, 195
sucrose, 91, 426, 431, 438, 463, 464. 466, 469, 473, 475, 477
sugars, 297
surface-coil, 317, 324, 325, 350
syn-periplanar, 10
syneresis, 435

tendon, 273, 274, 275, 279, 282
timothy, 289, 292, 295, 299
TOCSY, 18
torsion angles, 159
triclinic, 265
tripalmitin, 264, 265
tristearin, 264
troponin C, 232, 233
tryptophan, 102
trypsin inhibitor, 143

UDP, 356
UDP-sugars, 350, 352, 354, 355, 359
UDPG, 333, 346
unfreezable water, 170, 172

urine, 47
UTP, 333, 354

vacuolar, 333
vacuolar pH, 336
vacuole, 330
vic-diol, 9
vitrification, 439
volatile fatty acids, 287

Waltz-16, 42, 318
WALTZ sequence, 202
water activity, 426, 434
water activity ($a_w$), 415

water diffusion, 362, 445
water mobility, 399, 422, 468, 469
water states, 396, 391, 399, 404
water suppression, 63, 65, 67, 78
wheat doughs, 362
wheat flour, 362, 426, 464
wheat flour doughs, 428
wheat grains, 362, 450
wheat starch, 362

x-ray, 135, 137, 156, 163, 167, 176, 264, 273, 367
x-ray diffraction, 166
xanthan gum, 462

yeast, 330, 342, 345
yeast cell-walls, 47

zeins, 361, 369, 370, 371, 381, 382, 383